The Finite-Difference Time-Domain Method for Electromagnetics with MATLAB® Simulations

The ACES Series on Computational Electromagnetics and Engineering (CEME)

Andrew F. Peterson, PhD – Series Editor

The volumes in this series encompass the development and application of numerical techniques to electrical systems, including the modeling of electromagnetic phenomena over all frequency ranges and closely-related techniques for acoustic and optical analysis. The scope includes the use of computation for engineering design and optimization, as well as the application of commercial modeling tools to practical problems. The series will include titles for undergraduate and graduate education, research monographs for reference, and practitioner guides and handbooks.

Titles in the Series

Elsherbeni and Demir – The Finite-Difference Time-Domain Method for Electromagnetics with MATLAB® Simulations, 2nd Edition (2015)

Elsherbeni, Nayeri, and Reddy – Antenna Analysis and Design Using FEKO Electromagnetic Simulation Software (2014)

Yu, Yang, and Li – VALU, AVX and GPU Acceleration Techniques for Parallel FDTD Methods (2013)

Warnick – Numerical Methods for Engineering: An Introduction Using MATLAB® and Computational Electromagnetics (2011)

The Finite-Difference Time-Domain Method for Electromagnetics with MATLAB® Simulations

ACES Series

2nd Edition

Atef Z. Elsherbeni
Colorado School of Mines

Veysel Demir
Northern Illinois University

SciTech Publishing
an imprint of the IET

scitechpub.com

SciTech PUBLISHING
an imprint of the IET

Published by SciTech Publishing, an imprint of the IET
www.scitechpub.com
www.theiet.org

First published 2009 (978-1-89112-171-5)
Reprinted 2013
Second edition 2015

ISBN 978-1-61353-175-4 (hardback)

Typeset in India by MPS Limited
Printed in the UK by CPI Antony Rowe Ltd

To my wife, Magda, my daughters, Dalia and Donia, and my son, Tamer.
To the memory of my parents.

Atef Z. Elsherbeni

To my parents, Abdurrahman and Aysan, my wife, Minmei, and
my daughter, Laureen, and my son, Ronnie.

Veysel Demir

Contents

List of figures

List of tables

Preface

The contents of this book have evolved gradually and carefully from many years of teaching graduate level courses in computational electromagnetics generally, and the Finite-Difference Time-Domain (FDTD) method specifically. The authors have further refined and developed the materials by teaching numerous short courses on the FDTD method at various educational institutions and at a growing number of international conferences. The theoretical, numerical, and programming experience of the second author, Veysel Demir, has been a crucial factor in bringing the book to successful completion.

Objective

The objective of the book is to introduce the powerful FDTD method to students and interested researchers and readers. An effective introduction is accomplished using a step-by-step process that builds competence and confidence in developing complete working codes for the design and analysis of various antennas and microwave devices. This book will serve graduate students, researchers, and those in industry and government who are using other electromagnetics tools and methods for the sake of performing independent numerical confirmation. No previous experience with finite-difference methods is assumed of readers to use this book.

Important topics

The main topics in the book include the following:

- finite-difference approximation of the differential form of Maxwell's equations
- geometry construction in discrete space, including the treatment of the normal and tangential electric and magnetic field components at the boundaries between different media
- outer-boundary conditions treatment
- appropriate selection of time and spatial increments
- selection of the proper source waveform for the intended application
- correct parameters for the time-to-frequency-domain transformation
- simulation of thin wires

- representation of lumped passive and some active elements
- total field/scattered field formulation
- definition and formulation of fields from near and far zone
- modeling of dispersive material
- analysis of periodic structures
- treatment of nonuniform grid
- use of graphical processing units for accelerating the simulations.

MATLAB® code organization

In the first edition of this book the chapters are presented in such a way that, by adding/developing a new part of the code, chapter by chapter, at the end a well-developed FDTD simulation package can be constructed. In the second edition, the first 11 chapters are either the same as the ones in the first edition or extended with additional topics, therefore one can progressively develop a package by studying the first 11 chapters.

Chapters 12–16 in the second edition are advanced topics in FDTD, and it is not feasible to follow the progressive code development approach as implementation of these advanced concepts, one over another, makes the code development cumbersome and deviate reader's attention from learning a concept to getting lost in a pile of complex code. As an alternative, the functional code that is developed until Chapter 11 is reorganized as a base code and it is being used as the start point for each of the advanced topics chapters.

Use of MATLAB®

The development of the working code is based on the MATLAB programming language due to its relative ease of use, widespread availability, familiarity to most electrical engineers, and its powerful capabilities for providing graphical outputs and visualization. The book illustrates how the key FDTD equations are derived, provides the final expressions to be programmed, and also includes sample MATLAB codes developed for these equations. None of the specialized MATLAB tool boxes is required to run the presented codes in this book.

The MATLAB M-files for all programming examples shown in the text are available from the publisher on request by emailing books@theiet.org.

Key strengths

The strengths of the book can be summarized in four points:

- First, the derivations of the FDTD equations are presented in an understandable, detailed, and complete manner, which makes it easy for beginners to grasp the FDTD concepts. Further to that point, regardless of the different treatments required for various objects, such as dielectrics, conductors, lumped elements, active devices, or thin wires, the FDTD updating equations are provided with a consistent and unified notation.

- Second, many three-dimensional figures are presented to accompany the derived equations. This helps readers visualize, follow, and link the components of the equations with the discrete FDTD spatial domain.
- Third, it is well known that readers usually face difficulties while developing numerical tools for electromagnetics applications even though they understand the theory. Therefore, we introduce in this book a top-down software design approach that links the theoretical concepts with program development and helps in constructing an FDTD simulation tool as the chapters proceed.
- Fourth, fully worked out practical examples showing how the MATLAB codes for each example is developed to promote both the understanding and the visualization of the example configuration, its FDTD parameters and setup procedure, and finally the frequency and/or time domain corresponding solutions.

At the end of each chapter, readers will find a set of exercises that will emphasize the key features presented in that chapter. Since most of these exercises require code developments and the orientation of geometries and sources can be chosen arbitrarily, a suggested configuration of each in almost all of these exercises is provided in a three-dimensional figure.

The authors hope that with this coverage of the FDTD method, along with the supplied MATLAB codes in a single book, readers will be able to learn the method, to develop their own FDTD simulation tool, and to start enjoying, with confidence, the simulation of a variety of electromagnetic problems.

Instructor resources

Upon adopting this book as the required text, instructors are entitled to obtain the solutions of the exercises located at the end of each chapter. Most of these solutions are in the form of MATLAB files, and can be obtained by contacting the publisher at books@theiet.org.

Errors and suggestions

The authors welcome any reader feedback related to suspected errors and to suggestions for improving the presentation of the topics provided in this book.

Acknowledgements

The authors would like to first thank God for giving us the endurance to initiate and complete this book. Many thanks go to our family members for their support, patience, and sacrifice during the preparation and completion of this book. The authors greatly acknowledge the co-authorship of Chapter 14 with Drs. Khaled Elmahgoub and Fan Yang in the second edition of the book, and the contribution of Dr. Matthew J. Inman to Chapter 12 in the first edition of the book. They would also like to thank Dr. Branko Kolundzija for providing a copy of WIPL-D software package to use for verifications for some of the presented examples.

The first author acknowledges the support and advice over many years of the late Charles E. Smith. Without his continuous encouragement and support this book would have not been finalized. He also thanks many of his colleagues, postdoctoral fellows and visiting scholars who worked with him in topics related to FDTD, the graduate students who participated in translating ideas into reality, and the undergraduate students who participated in research projects with this author, in particular, Allen Glisson, Fan Yang, Abdel-Fattah A. Elsohly, Joe LoVetri, Veysel Demir, Matthew J. Inman, Chun-Wen Paul Huang, Clayborne D. Taylor, Mohamed Al Sharkawy, Vicente Rodriguez-Pereyra, Jianbing (James) Chen, Adel M. Abdin, Bradford N. Baker, Asem Mokaddem, Liang Xu, Nithya L. Iyer, Shixiong Dai, Xuexun Hu, Cuthbert Martindale Allen, Khaled Elmahgoub, and Terry Gerald. Chapter 14 is co-authored by Drs. Khaled Elmahgoub and Fan Yang.

The authors thank Dudley Kay and Joanna Hughes of SciTech/IET Publishing for their encouragement, patience, and useful suggestions during the course of the development of the manuscript and production of this book.

CHAPTER 1

Introduction to FDTD

Computational electromagnetics (CEM) has evolved rapidly during the past decade to a point where now extremely accurate predictions can be given for a variety of electromagnetic problems, including the scattering cross-section of radar targets and the precise design of antennas and microwave devices. In general, commonly used CEM methods today can be classified into two categories. The first is based on differential equation (DE) methods, whereas the second is based on integral equation (IE) methods. Both IE and DE solution methods are based on the applications of Maxwell's equations and the appropriate boundary conditions associated with the problem to be solved. The IE methods in general provide approximations for IEs in terms of finite sums, whereas the DE methods provide approximations for DEs as finite differences.

In previous years, most numerical electromagnetic analysis has taken place in the frequency domain where time-harmonic behavior is assumed. Frequency domain was favored over time domain because a frequency-domain approach is more suitable for obtaining analytical solutions for canonical problems, which are used to verify the numerical results obtained as a first step before depending on a newly developed numerical method for generating data for real-world applications. Furthermore, the experimental hardware available for making measurements in past years was largely confined to the frequency-domain approach.

The recent development of faster and more powerful computational resources allowed for more advanced time-domain CEM models. More focus is directed toward DE time-domain approaches as they are easier to formulate and to adapt in computer simulation models without complex mathematics. They also provide more physical insight to the characteristics of the problems.

Therefore, an in-depth analysis and implementation of the commonly used time-domain DE approach, namely, the finite-difference time-domain (FDTD) method for CEM applications, is covered in this book, along with applications related to antenna designs, microwave filter designs, and radar cross-section analysis of three-dimensional targets.

The FDTD method has gained tremendous popularity in the past decade as a tool for solving Maxwell's equations. It is based on simple formulations that do not require complex asymptotic or Green's functions. Although it solves the problem in time, it can provide frequency-domain responses over a wide band using the Fourier transform. It can easily handle composite geometries consisting of different types of materials including dielectric, magnetic, frequency-dependent, nonlinear, and anisotropic materials. The FDTD technique is easy to implement using parallel computation algorithms. These features of the FDTD method have made it the most attractive technique of CEM for many microwave devices and antenna applications.

FDTD has been used to solve numerous types of problems arising while studying many applications, including the following:

- scattering, radar cross-section
- microwave circuits, waveguides, fiber optics
- antennas (radiation, impedance)
- propagation
- medical applications
- shielding, coupling, electromagnetic compatibility (EMC), electromagnetic pulse (EMP) protection
- nonlinear and other special materials
- geological applications
- inverse scattering
- plasma

1.1 The finite-difference time-domain method basic equations

The starting point for the construction of an FDTD algorithm is Maxwell's time-domain equations. The differential time-domain Maxwell's equations needed to specify the field behavior over time are

$$\nabla \times \vec{H} = \frac{\partial \vec{D}}{\partial t} + \vec{J}, \tag{1.1a}$$

$$\nabla \times \vec{E} = -\frac{\partial \vec{B}}{\partial t} - \vec{M}, \tag{1.1b}$$

$$\nabla \cdot \vec{D} = \rho_e, \tag{1.1c}$$

$$\nabla \cdot \vec{B} = \rho_m, \tag{1.1d}$$

where $\vec{E}$ is the electric field strength vector in volts per meter, $\vec{D}$ is the electric displacement vector in coulombs per square meter, $\vec{H}$ is the magnetic field strength vector in amperes per meter, $\vec{B}$ is the magnetic flux density vector in webers per square meter, $\vec{J}$ is the electric current density vector in amperes per square meter, $\vec{M}$ is the magnetic current density vector in volts per square meter, ρ_e is the electric charge density in coulombs per cubic meter, and ρ_m is the magnetic charge density in webers per cubic meter.

Constitutive relations are necessary to supplement Maxwell's equations and characterize the material media. Constitutive relations for linear, isotropic, and nondispersive materials can be written as

$$\vec{D} = \varepsilon \vec{E}, \tag{1.2a}$$

$$\vec{B} = \mu \vec{H}, \tag{1.2b}$$

where ε is the permittivity, and μ is the permeability of the material. In free space

$$\varepsilon = \varepsilon_0 \approx 8.854 \times 10^{-12} \text{ farad/meter},$$

$$\mu = \mu_0 = 4\pi \times 10^{-7} \text{ henry/meter}.$$

We only need to consider curl equations (1.1a) and (1.1b) while deriving FDTD equations because the divergence equations can be satisfied by the developed FDTD updating equations [1]. The electric current density $\vec{J}$ is the sum of the conduction current density $\vec{J}_c = \sigma^e \vec{E}$ and the impressed current density $\vec{J}_i$ as $\vec{J} = \vec{J}_c + \vec{J}_i$. Similarly, for the magnetic current density, $\vec{M} = \vec{M}_c + \vec{M}_i$, where $\vec{M}_c = \sigma^m \vec{H}$. Here σ^e is the electric conductivity in siemens per meter, and σ^m is the magnetic conductivity in ohms per meter. Upon decomposing the current densities in (1.1) to conduction and impressed components and by using the constitutive relations (1.2) we can rewrite Maxwell's curl equations as

$$\nabla \times \vec{H} = \varepsilon \frac{\partial \vec{E}}{\partial t} + \sigma^e \vec{E} + \vec{J}_i, \tag{1.3a}$$

$$\nabla \times \vec{E} = -\mu \frac{\partial \vec{H}}{\partial t} - \sigma^m \vec{H} - \vec{M}_i. \tag{1.3b}$$

This formulation treats only the electromagnetic fields $\vec{E}$ and $\vec{H}$ and not the fluxes $\vec{D}$ and $\vec{B}$. All four constitutive parameters ε, μ, σ^e, and σ^m are present so that any linear isotropic material can be specified. Treatment of electric and magnetic sources is included through the impressed currents. Although only the curl equations are used and the divergence equations are not part of the FDTD formalism, the divergence equations can be used as a test on the predicted field response, so that after forming $\vec{D} = \varepsilon \vec{E}$ and $\vec{B} = \mu \vec{H}$ from the predicted $\vec{E}$ and $\vec{H}$ fields, the resulting $\vec{D}$ and $\vec{B}$ must satisfy the divergence equations.

Equation (1.3) is composed of two vector equations, and each vector equation can be decomposed into three scalar equations for three-dimensional space. Therefore, Maxwell's curl equations can be represented with the following six scalar equations in a Cartesian coordinate system (x, y, z):

$$\frac{\partial E_x}{\partial t} = \frac{1}{\varepsilon_x}\left(\frac{\partial H_z}{\partial y} - \frac{\partial H_y}{\partial z} - \sigma_x^e E_x - J_{ix}\right), \tag{1.4a}$$

$$\frac{\partial E_y}{\partial t} = \frac{1}{\varepsilon_y}\left(\frac{\partial H_x}{\partial z} - \frac{\partial H_z}{\partial x} - \sigma_y^e E_y - J_{iy}\right), \tag{1.4b}$$

$$\frac{\partial E_z}{\partial t} = \frac{1}{\varepsilon_z}\left(\frac{\partial H_y}{\partial x} - \frac{\partial H_x}{\partial y} - \sigma_z^e E_z - J_{iz}\right), \tag{1.4c}$$

$$\frac{\partial H_x}{\partial t} = \frac{1}{\mu_x}\left(\frac{\partial E_y}{\partial z} - \frac{\partial E_z}{\partial y} - \sigma_x^m H_x - M_{ix}\right), \tag{1.4d}$$

$$\frac{\partial H_y}{\partial t} = \frac{1}{\mu_y}\left(\frac{\partial E_z}{\partial x} - \frac{\partial E_x}{\partial z} - \sigma_y^m H_y - M_{iy}\right), \tag{1.4e}$$

$$\frac{\partial H_z}{\partial t} = \frac{1}{\mu_z}\left(\frac{\partial E_x}{\partial y} - \frac{\partial E_y}{\partial x} - \sigma_z^m H_z - M_{iz}\right). \tag{1.4f}$$

The material parameters ε_x, ε_y, and ε_z are associated with electric field components E_x, E_y, and E_z through constitutive relations $D_x = \varepsilon_x E_x$, $D_y = \varepsilon_y E_y$, and $D_z = \varepsilon_z E_z$, respectively. Similarly, the material parameters μ_x, μ_y, and μ_z are associated with magnetic field components H_x, H_y, and H_z through constitutive relations $B_x = \mu_x H_x$, $B_y = \mu_y H_y$, and $B_z = \mu_z H_z$, respectively. Similar decompositions for other orthogonal coordinate systems are possible, but they are less attractive from the applications point of view.

The FDTD algorithm divides the problem geometry into a spatial grid where electric and magnetic field components are placed at certain discrete positions in space, and it solves Maxwell's equations in time at discrete time instances. This can be implemented by first approximating the time and space derivatives appearing in Maxwell's equations by finite differences and next by constructing a set of equations that calculate the values of fields at a future time instant from the values of fields at a past time instant, therefore constructing a time marching algorithm that simulates the progression of the fields in time [2].

1.2 Approximation of derivatives by finite differences

An arbitrary continuous function can be sampled at discrete points, and the discrete function becomes a good approximation of the continuous function if the sampling rate is sufficient relative to the function's variation. Sampling rate determines the accuracy of operations performed on the discrete function that approximates the operations on the continuous functions as well. However, another factor that determines the accuracy of an operation on a discrete function is the choice of the discrete operator. Most of the time it is possible to use more than one way of performing an operation on a discrete function. Here we will consider the derivative operation.

Consider the continuous function given in Figure 1.1(a–c), sampled at discrete points. The expression for the derivative of $f(x)$ at point x can be written as

$$f'(x) = \lim_{\Delta x \to 0} \frac{f(x + \Delta x) - f(x)}{\Delta x}. \tag{1.5}$$

However, since Δx is a nonzero fixed number, the derivative of $f(x)$ can be approximately taken as

$$f'(x) \approx \frac{f(x + \Delta x) - f(x)}{\Delta x}. \tag{1.6}$$

The derivative of $f(x)$ is the slope of the dashed line as illustrated in Figure 1.1(a). Equation (1.6) is called the *forward difference* formula since one forward point $f(x + \Delta x)$ is used to evaluate $f'(x)$ together with $f(x)$.

It is evident that another formula for an approximate $f'(x)$ can be obtained by using a backward point $f(x - \Delta x)$ rather than the forward point $f(x + \Delta x)$ as illustrated in Figure 1.1(b), which can be written as

$$f'(x) \approx \frac{f(x) - f(x - \Delta x)}{\Delta x}. \tag{1.7}$$

This equation is called the *backward difference* formula due to the use of the backward point $f(x - \Delta x)$.

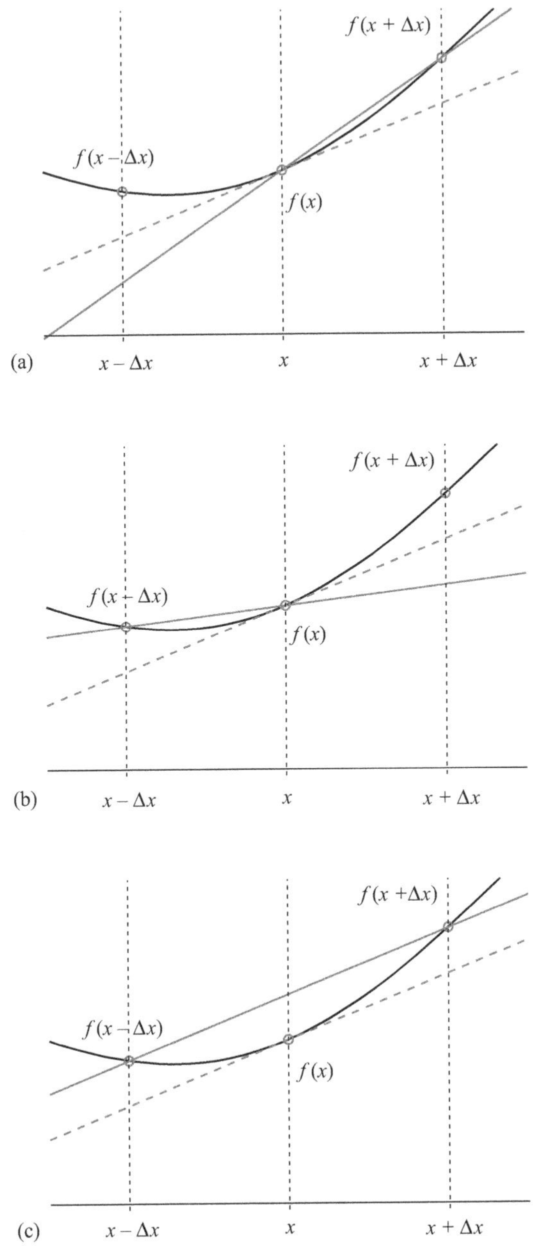

Figure 1.1 (a) Approximation of the derivative of $f(x)$ at x by finite differences: forward difference; (b) approximation of the derivative of $f(x)$ at x by finite differences: backward difference; (c) approximation of the derivative of $f(x)$ at x by finite differences: central difference.

The third way of obtaining a formula for an approximate $f'(x)$ is by averaging the forward difference and backward difference formulas, such that

$$f'(x) \approx \frac{f(x+\Delta x) - f(x-\Delta x)}{\Delta x}. \tag{1.8}$$

Equation (1.8) is called the *central difference* formula since both the forward and backward points around the center are used. The line representing the derivative of $f(x)$ calculated using the central difference formula is illustrated in Figure 1.1(c). It should be noted that the value of the function $f(x)$ at x is not used in central difference formula.

Examination of Figure 1.1 immediately reveals that the three different schemes yield different values for $f'(x)$, with an associated amount of error. The amount of error introduced by these difference formulas can be evaluated analytically by using the Taylor series approach. For instance, the Taylor series expansion of $f(x+\Delta x)$ can be written as

$$f(x+\Delta x) = f(x) + \Delta x f'(x) + \frac{(\Delta x)^2}{2} f''(x) + \frac{(\Delta x)^3}{6} f'''(x) + \frac{(\Delta x)^4}{24} f''''(x) + \cdots. \tag{1.9}$$

This equation gives an exact expression for $f(x+\Delta x)$ as a series in terms of Δx and derivatives of $f(x)$, if $f(x)$ satisfies certain conditions and infinite number of terms, theoretically, are being used. Equation (1.9) can be rearranged to express $f'(x)$ as

$$f'(x) = \frac{f(x+\Delta x) - f(x)}{\Delta x} - \frac{\Delta x}{2} f''(x) - \frac{(\Delta x)^2}{6} f'''(x) - \frac{(\Delta x)^3}{24} f''''(x) - \cdots. \tag{1.10}$$

Here it can be seen that the first term on the right-hand side of (1.10) is the same as the forward difference formula given by (1.6). The sum of the rest of the terms is the difference between the approximate derivative given by the forward difference formula and the exact derivative $f'(x)$, and hence is the amount of error introduced by the forward difference formula. Equation (1.10) can be rewritten as

$$f'(x) = \frac{f(x+\Delta x) - f(x)}{\Delta x} + O(\Delta x), \tag{1.11}$$

where $O(\Delta x)$ represents the error term. The most significant term in $O(\Delta x)$ is $\Delta x/2$, and the order of Δx in this most significant term is one. Therefore, the forward difference formula is *first-order accurate*. The interpretation of first-order accuracy is that the most significant term in the error introduced by a first-order accurate formula is proportional to the sampling period. For instance, if the sampling period is decreased by half, the error reduces by half.

A similar analysis can be performed for evaluation of the error of the backward formula starting with the Taylor series expansion of $f(x-\Delta x)$.

$$f(x-\Delta x) = f(x) - \Delta x f'(x) + \frac{(\Delta x)^2}{2} f''(x) - \frac{(\Delta x)^3}{6} f'''(x) + \frac{(\Delta x)^4}{24} f''''(x) + \cdots. \tag{1.12}$$

This equation can be rearranged to express $f'(x)$ as

$$f'(x) = \frac{f(x) - f(x - \Delta x)}{\Delta x} + \frac{\Delta x}{2} f''(x) - \frac{(\Delta x)^2}{6} f'''(x) + \frac{(\Delta x)^3}{24} f''''(x) - \cdots. \quad (1.13)$$

The first term on the right-hand side of (1.13) is the same as the backward difference formula and the sum of the rest of the terms represents the error introduced by the backward difference formula to the exact derivative of $f(x)$, such that

$$f'(x) = \frac{f(x) - f(x - \Delta x)}{\Delta x} + O(\Delta x). \quad (1.14)$$

The order of Δx in the most significant term of $O(\Delta x)$ is one; hence the backward difference formula is *first-order accurate*.

The difference between the Taylor series expansions of $f(x + \Delta x)$ and $f(x - \Delta x)$ can be expressed using (1.9) and (1.12) as

$$f(x + \Delta x) - f(x - \Delta x) = 2\Delta x f'(x) + \frac{2(\Delta x)^3}{6} f'''(x) + \cdots. \quad (1.15)$$

This equation can be rearranged to express $f'(x)$ as

$$f'(x) = \frac{f(x + \Delta x) - f(x - \Delta x)}{2\Delta x} - \frac{(\Delta x)^2}{6} f'''(x) + \cdots = \frac{f(x + \Delta x) - f(x - \Delta x)}{2\Delta x} + O\left((\Delta x)^2\right), \quad (1.16)$$

where the first term on right-hand side is the same as the central difference formula given in (1.8) and the order of Δx in the most significant term of the error $O((\Delta x)^2)$ is two; hence the central difference formula is *second-order accurate*. The interpretation of second-order accuracy is that the most significant term in the error introduced by a second-order accurate formula is proportional to the square of sampling period. For instance, if the sampling period is decreased by half, the error is reduced by a factor of four. Hence, a second-order accurate formula such as the central difference formula is more accurate than a first-order accurate formula.

For an example, consider a function $f(x) = \sin(x)e^{-0.3x}$ as displayed in Figure 1.2(a). The exact first-order derivative of this function is

$$f'(x) = \cos(x)e^{-0.3x} - 0.3\sin(x)e^{-0.3x}.$$

This function $f(x)$ is sampled with a sampling period $\Delta x = \pi/5$, and approximate derivatives are calculated for $f(x)$ using the forward difference, backward difference, and central difference formulas. The derivative of the function $f'(x)$ and its finite difference approximations are plotted in Figure 1.2(b). The errors introduced by the difference formulas, which are the differences between $f'(x)$ and its finite difference approximations, are plotted in Figure 1.2(c) for the sampling interval $\Delta x = \pi/5$. It is evident that the error introduced by the central difference formula is smaller than the errors introduced by the forward difference

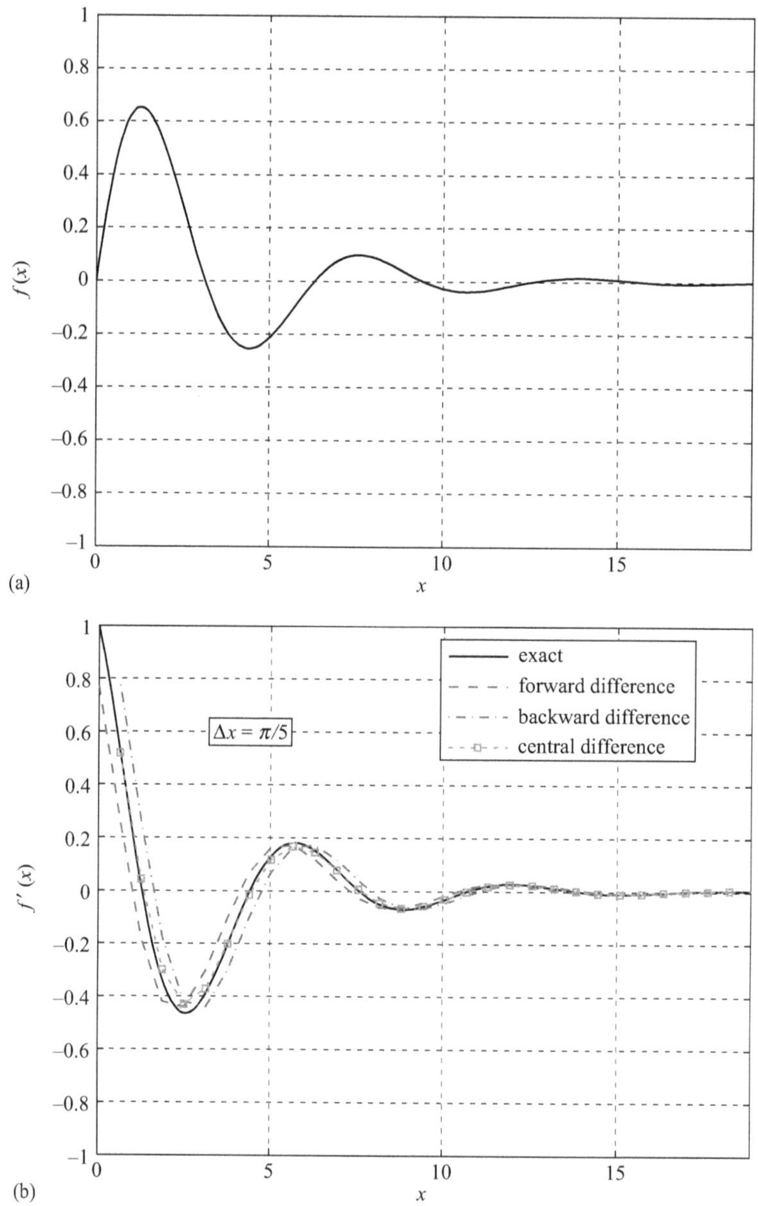

Figure 1.2 (a) $f(x), f'(x)$, and differences between $f'(x)$ and finite difference approximations of $f'(x)$ for $\Delta x = \pi/5$ and $\Delta x = \pi/10$: $f(x) = \sin(x)e^{-0.3x}$; (b) $f(x), f'(x)$, and differences between $f'(x)$ and finite difference approximations of $f'(x)$ for $\Delta x = \pi/5$ and $\Delta x = \pi/10$: $f'(x) = \cos(x)e^{-0.3x} - 0.3\sin(x)e^{-0.3x}$ and finite difference approximations of $f'(x)$ for $\Delta x = \pi/5$; (c) $f(x), f'(x)$, and differences between $f'(x)$ and finite difference approximations of $f'(x)$ for $\Delta x = \pi/5$ and $\Delta x = \pi/10$: error($f'(x)$) for $\Delta x = \pi/5$; (d) $f(x), f'(x)$, and differences between $f'(x)$ and finite difference approximations of $f'(x)$ for $\Delta x = \pi/5$ and $\Delta x = \pi/10$: error($f'(x)$) for $\Delta x = \pi/10$.

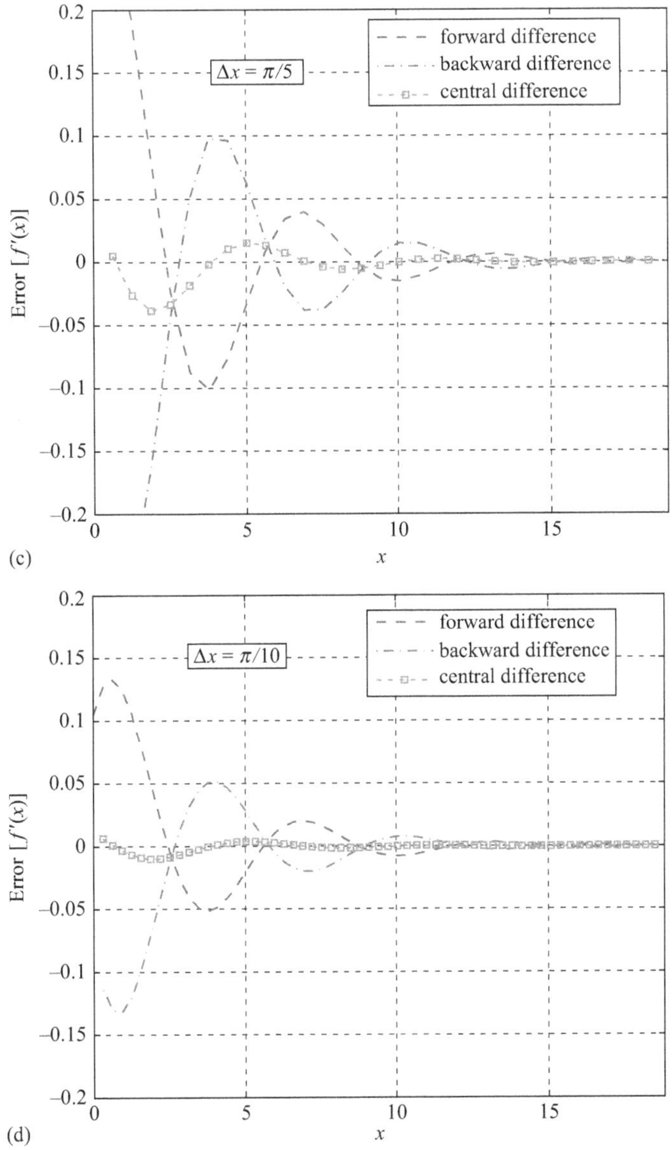

Figure 1.2 *(Continued)*

and backward difference formulas. Furthermore, the errors introduced by the difference formulas for the sampling period $\Delta x = \pi/10$ are plotted in Figure 1.2(d). It can be realized that as the sampling period is halved, the errors of the forward difference and backward difference formulas are halved as well, and the error of the central difference formula is reduced by a factor of four.

The MATLAB® code calculating $f(x)$ and its finite difference derivatives, and generating the plots in Figure 1.2 is shown in Listing 1.1.

Listing 1.1 MATLAB code generating Figure 1.2(a–d)

```
% create exact function and its derivative
N_exact = 301; % number of sample points for exact function
x_exact = linspace(0,6*pi,N_exact);
f_exact = sin(x_exact).*exp(-0.3*x_exact);
f_derivative_exact = cos(x_exact).*exp(-0.3*x_exact) ...
                     -0.3*sin(x_exact).*exp(-0.3*x_exact);

% plot exact function
figure(1);
plot(x_exact,f_exact,'k-','linewidth',1.5);
set(gca,'FontSize',12,'fontweight','demi');
axis([0 6*pi -1 1]); grid on;
xlabel('$x$','Interpreter','latex','FontSize',16);
ylabel('$f(x)$','Interpreter','latex','FontSize',16);

% create exact function for pi/5 sampling period
% and its finite difference derivatives
N_a = 31; % number of points for pi/5 sampling period
x_a = linspace(0,6*pi,N_a);
f_a = sin(x_a).*exp(-0.3*x_a);
f_derivative_a = cos(x_a).*exp(-0.3*x_a) ...
                 -0.3*sin(x_a).*exp(-0.3*x_a);

dx_a = pi/5;
f_derivative_forward_a  = zeros(1,N_a);
f_derivative_backward_a = zeros(1,N_a);
f_derivative_central_a  = zeros(1,N_a);
f_derivative_forward_a(1:N_a-1) = ...
        (f_a(2:N_a)-f_a(1:N_a-1))/dx_a;
f_derivative_backward_a(2:N_a) = ...
        (f_a(2:N_a)-f_a(1:N_a-1))/dx_a;
f_derivative_central_a(2:N_a-1) = ...
        (f_a(3:N_a)-f_a(1:N_a-2))/(2*dx_a);


% create exact function for pi/10 sampling period
% and its finite difference derivatives
N_b = 61; % number of points for pi/10 sampling period
x_b = linspace(0,6*pi,N_b);
f_b = sin(x_b).*exp(-0.3*x_b);
f_derivative_b = cos(x_b).*exp(-0.3*x_b) ...
                 -0.3*sin(x_b).*exp(-0.3*x_b);
dx_b = pi/10;
f_derivative_forward_b  = zeros(1,N_b);
f_derivative_backward_b = zeros(1,N_b);
f_derivative_central_b  = zeros(1,N_b);
f_derivative_forward_b(1:N_b-1) = ...
        (f_b(2:N_b)-f_b(1:N_b-1))/dx_b;
f_derivative_backward_b(2:N_b) = ...
        (f_b(2:N_b)-f_b(1:N_b-1))/dx_b;
f_derivative_central_b(2:N_b-1) = ...
        (f_b(3:N_b)-f_b(1:N_b-2))/(2*dx_b);

```

```
% plot exact derivative of the function and its finite difference
% derivatives using pi/5 sampling period
figure(2);
plot(x_exact, f_derivative_exact, 'k', ...
     x_a(1:N_a-1), f_derivative_forward_a(1:N_a-1), 'b--', ...
     x_a(2:N_a), f_derivative_backward_a(2:N_a), 'r-.', ...
     x_a(2:N_a-1), f_derivative_central_a(2:N_a-1), ':ms', ...
     'MarkerSize',4, 'linewidth',1.5);
set (gca,'FontSize',12,'fontweight','demi');
axis([0 6*pi -1 1]);
grid on;
legend('exact', 'forward␣difference',...
     'backward␣difference','central␣difference');
xlabel('$x$','Interpreter','latex','FontSize',16);
ylabel('$f''(x)$','Interpreter','latex','FontSize',16);
text(pi,0.6,'$\Delta␣x␣=␣\pi/5$','Interpreter', ...
     'latex','fontsize',16, 'BackgroundColor','w','EdgeColor','k');

% plot error for finite difference derivatives
% using pi/5 sampling period
error_forward_a  = f_derivative_a - f_derivative_forward_a;
error_backward_a = f_derivative_a - f_derivative_backward_a;
error_central_a  = f_derivative_a - f_derivative_central_a;

figure(3);
plot(x_a(1:N_a-1), error_forward_a(1:N_a-1), 'b--', ...
     x_a(2:N_a), error_backward_a(2:N_a), 'r-.', ...
     x_a(2:N_a-1), error_central_a(2:N_a-1), ':ms', ...
     'MarkerSize',4, 'linewidth',1.5);

set (gca,'FontSize',12,'fontweight','demi');
axis([0 6*pi -0.2 0.2]);
grid on;
legend('forward␣difference', 'backward␣difference', ...
     'central␣difference');
xlabel('$x$','Interpreter','latex','FontSize',16);
ylabel('error␣$[f''(x)]$','Interpreter','latex','FontSize',16);
text(pi,0.15,'$\Delta␣x␣=␣\pi/5$','Interpreter', ...
     'latex','fontsize',16, 'BackgroundColor','w','EdgeColor','k');

% plot error for finite difference derivatives
% using pi/10 sampling period
error_forward_b  = f_derivative_b - f_derivative_forward_b;
error_backward_b = f_derivative_b - f_derivative_backward_b;
error_central_b  = f_derivative_b - f_derivative_central_b;

figure(4);
plot(x_b(1:N_b-1), error_forward_b(1:N_b-1), 'b-- , ...
     x_b(2:N_b), error_backward_b(2:N_b), 'r-.', ...
     x_b(2:N_b-1), error_central_b(2:N_b-1), ':ms', ...
     'MarkerSize',4, 'linewidth',1.5);

set (gca,'FontSize',12,'fontweight','demi');
```

```
axis([0 6*pi -0.2 0.2]);
grid on;
legend('forward difference','backward difference', ...
    'central difference');
xlabel('$x$','Interpreter','latex','FontSize',16);
ylabel('error $[f''(x)]$','Interpreter','latex','FontSize',16);
text(pi,0.15,'$\Delta x = \pi/10$','Interpreter', ...
    'latex','fontsize',16,'BackgroundColor','w','EdgeColor','k');
```

Values of the function $f(x)$ at two neighboring points around x have been used to obtain a second-order accurate central difference formula to approximate $f'(x)$. It is possible to obtain formulas with higher orders of accuracy by including a larger number of neighboring points in the derivation of a formula for $f'(x)$. However, although there are FDTD formulations developed based on higher-order accurate formulas, the conventional FDTD is based on the second-order accurate central difference formula, which is found to be sufficiently accurate for most electromagnetics applications and simple in implementation and understanding.

It is possible to obtain finite-difference formulas for approximating higher-order derivatives as well. For instance, if we take the sum of the Taylor series expansions of $f(x+\Delta x)$ and $f(x-\Delta x)$ using (1.9) and (1.12), we obtain

$$f(x+\Delta x)+f(x-\Delta x) = 2f(x) + (\Delta x)^2 f''(x) + \frac{(\Delta x)^4}{12} f''''(x) + \cdots. \tag{1.17}$$

After rearranging the equation to have $f''(x)$ on the left-hand side, we get

$$\begin{aligned} f''(x) &= \frac{f(x+\Delta x)-2f(x)+f(x-\Delta x)}{(\Delta x)^2} - \frac{(\Delta x)^2}{12} f''''(x) + \cdots \\ &= \frac{f(x+\Delta x)-2f(x)+f(x-\Delta x)}{(\Delta x)^2} + O((\Delta x)^2). \end{aligned} \tag{1.18}$$

Using (1.18) we can obtain a central difference formula for the second-order derivative $f''(x)$ as

$$f''(x) \approx \frac{f(x+\Delta x)-2f(x)+f(x-\Delta x)}{(\Delta x)^2}, \tag{1.19}$$

which is second-order accurate due to $O(\Delta x)^2$.

Similarly, some other finite difference formulas can be obtained for the first- and second-order derivatives with different orders of accuracy based on different sampling points. A list of finite difference formulas is given for the first- and second-order derivatives in Table 1.1 as a reference (Figure 1.3).

Table 1.1 Finite difference formulas for the first- and second-order derivatives where the function f with the subscript i is an abbreviation of $f(x)$. Similar notation can be implemented for $f(x + \Delta x)$, $f(x + 2\Delta x)$, etc., as shown in Figure 1.3. FD, forward difference; BD, backward difference; CD, central difference.

Derivative $\partial f/\partial x$		Derivative $\partial^2 f/\partial x^2$	
Difference scheme	**Type error**	**Difference scheme**	**Type error**
$\frac{f_{i+1} - f_i}{\Delta x}$	FD $O(\Delta x)$	$\frac{f_{i+2} - 2f_{i+1} + f_i}{(\Delta x)^2}$	FD $O(\Delta x)$
$\frac{f_i - f_{i-1}}{\Delta x}$	BD $O(\Delta x)$	$\frac{f_{i+2} - 2f_{i-1} + f_{i-2}}{(\Delta x)^2}$	BD $O(\Delta x)$
$\frac{f_{i+1} - f_{i-1}}{2\Delta x}$	CD $O((\Delta x)^2)$	$\frac{f_{i+1} - 2f_i + f_{i-1}}{(\Delta x)^2}$	CD $O((\Delta x)^2)$
$\frac{-f_{i+2} + 4f_{i+1} - 3f_i}{2\Delta x}$	FD $O((\Delta x)^2)$	$\frac{-f_{i+2} + 16f_{i+1} - 30f_i + 16f_{i-1} - f_{i-2}}{12(\Delta x)^2}$	CD $O((\Delta x)^4)$
$\frac{3f_{i+1} - 4f_{i-1} + f_{i-2}}{2\Delta x}$	BD $O((\Delta x)^2)$		
$\frac{-f_{i+2} + 8f_{i+1} - 8f_{i-1} + f_{i-2}}{12\Delta x}$	CD $O((\Delta x)^4)$		

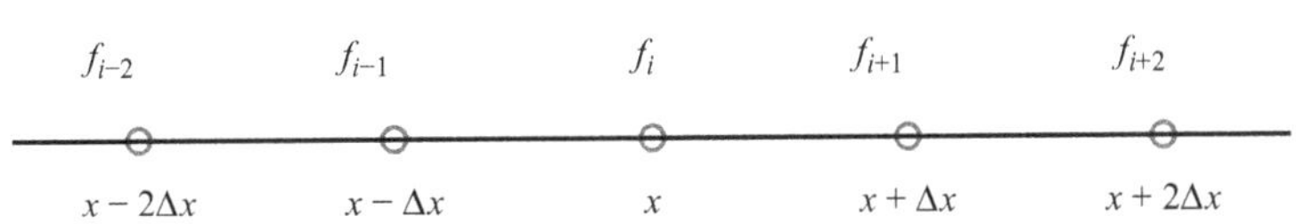

Figure 1.3 Sample points of $f(x)$.

1.3 FDTD updating equations for three-dimensional problems

In 1966, Yee originated a set of finite-difference equations for the time-dependent Maxwell's curl equations system [2]. These equations can be represented in discrete form, both in space and time, employing the second-order accurate central difference formula. As mentioned before, the electric and magnetic field components are sampled at discrete positions both in time and space. The FDTD technique divides the three-dimensional problem geometry into cells to form a grid. Figure 1.4 illustrates an FDTD grid composed of $(N_x \times N_y \times N_z)$ cells. A unit cell of this grid is called a Yee cell. Using rectangular Yee cells, a stepped or "staircase" approximation of the surface and internal geometry of the structure of interest is made with a space resolution set by the size of the unit cell.

The discrete spatial positions of the field components have a specific arrangement in the Yee cell, as demonstrated in Figure 1.5. The electric field vector components are placed at

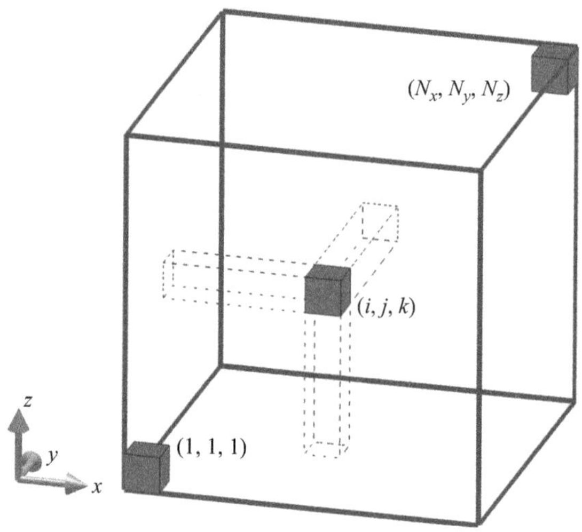

Figure 1.4 A three-dimensional FDTD computational space composed of $(N_x \times N_y \times N_z)$ Yee cells.

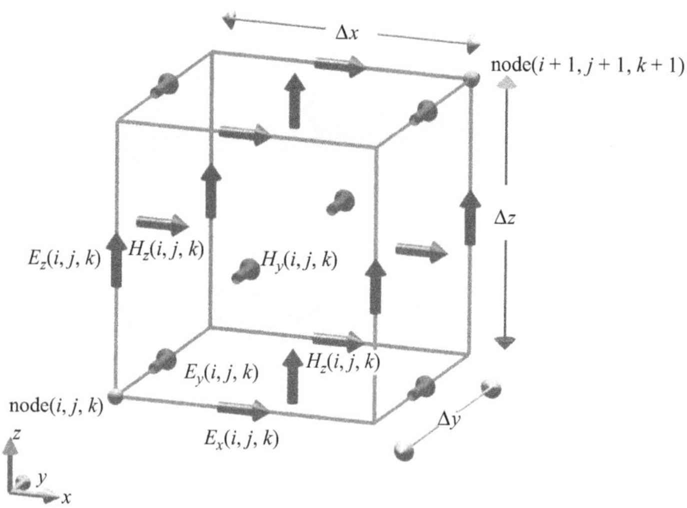

Figure 1.5 Arrangement of field components on a Yee cell indexed as (i, j, k).

the centers of the edges of the Yee cells and oriented parallel to the respective edges, and the magnetic field vector components are placed at the centers of the faces of the Yee cells and are oriented normal to the respective faces. This provides a simple picture of three-dimensional space being filled by an interlinked array of Faraday's law and Ampere's law contours. It can be easily noticed in Figure 1.5 that each magnetic field vector is surrounded by four electric field vectors that are curling around the magnetic field vector, thus simulating Faraday's law. Similarly, if the neighboring cells are also added to the picture, it would be apparent that each electric field vector is surrounded by four magnetic field vectors that are curling around the electric field vector, thus simulating Ampere's law.

Figure 1.5 shows the indices of the field components, which are indexed as (i, j, k), associated with a cell indexed as (i, j, k). For a computational domain composed of uniform Yee cells having dimension Δx in the x direction, Δy in the y direction, and Δz in the z direction, the actual positions of the field components with respect to an origin coinciding with the position of the node (1, 1, 1) can easily be calculated as

$$
\begin{aligned}
E_x(i,j,k) &\Rightarrow ((i-0.5)\Delta x,\ (j-1)\Delta y,\ (k-1)\Delta z),\\
E_y(i,j,k) &\Rightarrow ((i-1)\Delta x,\ (j-0.5)\Delta y,\ (k-1)\Delta z),\\
E_z(i,j,k) &\Rightarrow ((i-1)\Delta x,\ (j-1)\Delta y,\ (k-0.5)\Delta z),\\
H_x(i,j,k) &\Rightarrow ((i-1)\Delta x,\ (j-0.5)\Delta y,\ (k-0.5)\Delta z),\\
H_y(i,j,k) &\Rightarrow ((i-0.5)\Delta x,\ (j-1)\Delta y,\ (k-0.5)\Delta z),\\
H_z(i,j,k) &\Rightarrow ((i-0.5)\Delta x,\ (j-0.5)\Delta y,\ (k-1)\Delta z).
\end{aligned}
$$

The FDTD algorithm samples and calculates the fields at discrete time instants; however, the electric and magnetic field components are not sampled at the same time instants. For a time-sampling period Δt, the electric field components are sampled at time instants $0, \Delta t, 2\Delta t, \ldots, n\Delta t, \ldots$; however, the magnetic field components are sampled at time instants $\frac{1}{2}\Delta t, (1+\frac{1}{2})\Delta t, \ldots, (n+\frac{1}{2})\Delta t, \ldots$. Therefore, the electric field components are calculated at integer time steps, and magnetic field components are calculated at half-integer time steps, and they are offset from each other by $\Delta t/2$. The field components need to be referred not only by their spatial indices which indicate their positions in space, but also by their temporal indices, which indicate their time instants. Therefore, a superscript notation is adopted to indicate the time instant. For instance, the z component of an electric field vector positioned at $((i-1)\Delta x, (j-1)\Delta y, (k-0.5)\Delta z)$ and sampled at time instant $n\Delta t$ is referred to as $E_z^n(i,j,k)$. Similarly, the y component of a magnetic field vector positioned at $((i-0.5)\Delta x, (j-1)\Delta y, (k-0.5)\Delta z)$ and sampled at time instant $(n+\frac{1}{2})\Delta t$ is referred to as $H_y^{n+1/2}(i,j,k)$.

The material parameters (permittivity, permeability, electric, and magnetic conductivities) are distributed over the FDTD grid and are associated with field components; therefore, they are indexed the same as their respective field components. For instance, Figure 1.6 illustrates the indices for the permittivity and permeability parameters. The electric conductivity is distributed and indexed the same as the permittivity, and the magnetic conductivity is distributed and indexed the same as the permeability.

Having adopted an indexing scheme for the discrete samples of field components in both time and space, Maxwell's curl equations (1.4) that are given in scalar form can be expressed in terms of finite differences. For instance, consider again (1.4a):

$$\frac{\partial E_x}{\partial t} = \frac{1}{\varepsilon_x}\left(\frac{\partial H_z}{\partial y} - \frac{\partial H_y}{\partial z} - \sigma_x^e E_x - J_{ix}\right).$$

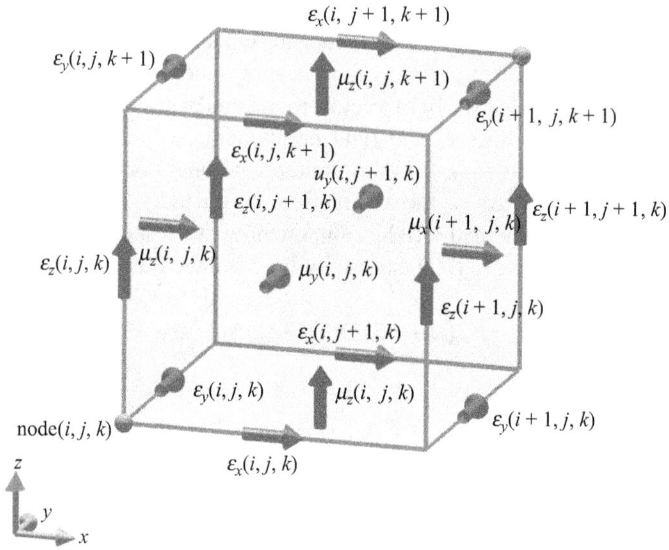

Figure 1.6 Material parameters indexed on a Yee cell.

The derivatives in this equation can be approximated by using the central difference formula with the position of $E_x(i, j, k)$ being the center point for the central difference formula in space and time instant $(n + \frac{1}{2})\Delta t$ as being the center point in time. Considering the field component positions given in Figure 1.7, we can write

$$\begin{aligned}\frac{E_x^{n+1}(i,j,k) - E_x^n(i,j,k)}{\Delta t} &= \frac{1}{\varepsilon_x(i,j,k)} \frac{H_z^{n+\frac{1}{2}}(i,j,k) - H_z^{n+\frac{1}{2}}(i,j-1,k)}{\Delta y} \\ &\quad - \frac{1}{\varepsilon_x(i,j,k)} \frac{H_y^{n+\frac{1}{2}}(i,j,k) - H_y^{n+\frac{1}{2}}(i,j,k-1)}{\Delta z} \\ &\quad - \frac{\sigma_x^e(i,j,k)}{\varepsilon_x(i,j,k)} E_x^{n+\frac{1}{2}}(i,j,k) - \frac{1}{\varepsilon_x(i,j,k)} J_{ix}^{n+\frac{1}{2}}(i,j,k). \end{aligned} \tag{1.20}$$

It has already been mentioned that the electric field components are defined at integer time steps; however, the right-hand side of (1.20) includes an electric field term at time instant $(n + \frac{1}{2})\Delta t$, that is, $E_x^{n+1/2}(i,j,k)$. This term can be written as the average of the terms at time instants $(n + 1)\Delta t$ and $n\Delta t$, such that

$$E_x^{n+\frac{1}{2}}(i,j,k) = \frac{E_x^{n+1}(i,j,k) + E_x^n(i,j,k)}{2}. \tag{1.21}$$

Using (1.21) in (1.20) and arranging the terms such that the future term $E_x^{n+1}(i,j,k)$ is kept on the left side of the equation and the rest of the terms are moved to the right-hand side of the equation, we can write

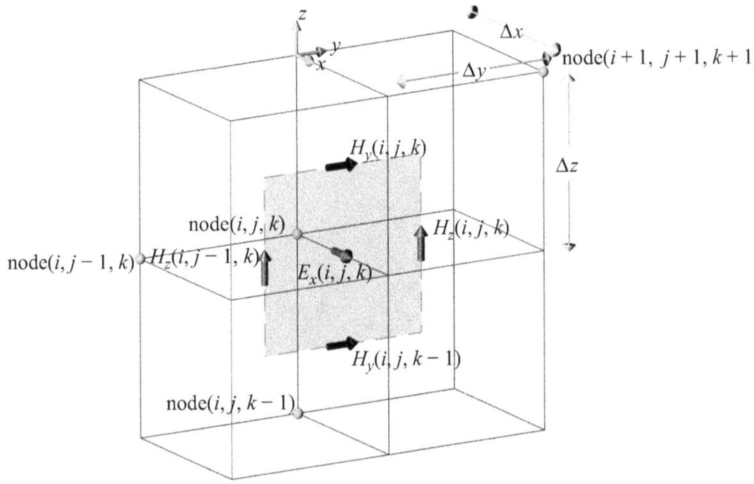

Figure 1.7 Field components around $E_x(i, j, k)$.

$$
\begin{aligned}
\frac{2\varepsilon_x(i,j,k) + \Delta t\sigma_x^e(i,j,k)}{2\varepsilon_x(i,j,k)} E_x^{n+1}(i,j,k) = {} & \frac{2\varepsilon_x(i,j,k) - \Delta t\sigma_x^e(i,j,k)}{2\varepsilon_x(i,j,k)} E_x^n(i,j,k) \\
& + \frac{\Delta t}{\varepsilon_x(i,j,k)\Delta y}\left(H_z^{n+\frac{1}{2}}(i,j,k) - H_z^{n+\frac{1}{2}}(i,j-1,k)\right) \\
& - \frac{\Delta t}{\varepsilon_x(i,j,k)\Delta z}\left(H_y^{n+\frac{1}{2}}(i,j,k) - H_y^{n+\frac{1}{2}}(i,j,k-1)\right) \\
& - \frac{\Delta t}{\varepsilon_x(i,j,k)} J_{ix}^{n+\frac{1}{2}}(i,j,k)
\end{aligned}
\tag{1.22}
$$

After some manipulations, we get

$$
\begin{aligned}
E_x^{n+1}(i,j,k) = {} & \frac{2\varepsilon_x(i,j,k) - \Delta t\sigma_x^e(i,j,k)}{2\varepsilon_x(i,j,k) + \Delta t\sigma_x^e(i,j,k)} E_x^n(i,j,k) \\
& + \frac{2\Delta t}{\left(2\varepsilon_x(i,j,k) + \Delta t\sigma_x^e(i,j,k)\right)\Delta y}\left(H_z^{n+\frac{1}{2}}(i,j,k) - H_z^{n+\frac{1}{2}}(i,j-1,k)\right) \\
& - \frac{2\Delta t}{\left(2\varepsilon_x(i,j,k) + \Delta t\sigma_x^e(i,j,k)\right)\Delta z}\left(H_y^{n+\frac{1}{2}}(i,j,k) - H_y^{n+\frac{1}{2}}(i,j,k-1)\right) \\
& - \frac{2\Delta t}{2\varepsilon_x(i,j,k) + \Delta t\sigma_x^e(i,j,k)} J_{ix}^{n+\frac{1}{2}}(i,j,k).
\end{aligned}
\tag{1.23}
$$

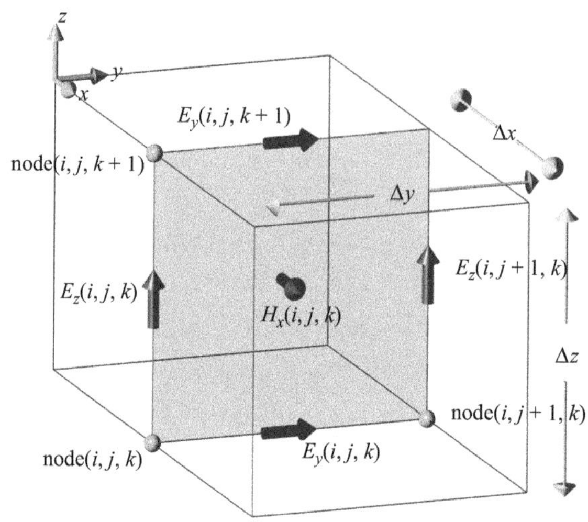

Figure 1.8 Field components around $H_x(i, j, k)$.

The form of (1.23) demonstrates how the future value of an electric field component can be calculated by using the past values of the electric field component, the magnetic field components, and the source components. This form of an equation is called an FDTD *updating equation*. Updating equations can easily be obtained for calculating $E_y^{n+1}(i,j,k)$ starting from (1.4b) and $E_z^{n+1}(i,j,k)$ starting from (1.4c) following the same methodology that has been used to obtain (1.23).

Similarly, updating equations can be obtained for magnetic field components following the same methodology. However, while applying the central difference formula to the time derivative of the magnetic field components, the central point in time shall be taken as $n\Delta t$. For instance, (1.4c), which is

$$\frac{\partial H_x}{\partial t} = \frac{1}{\mu_x}\left(\frac{\partial E_y}{\partial z} - \frac{\partial E_z}{\partial y} - \sigma_x^m H_x - M_{ix}\right),$$

can be approximated using finite differences based on the field positions (as shown in Figure 1.8) as

$$\begin{aligned}\frac{H_x^{n+\frac{1}{2}}(i,j,k) - H_x^{n-\frac{1}{2}}(i,j,k)}{\Delta t} &= \frac{1}{\mu_x(i,j,k)}\frac{E_y^n(i,j,k+1) - E_y^n(i,j,k)}{\Delta z} \\ &- \frac{1}{\mu_x(i,j,k)}\frac{E_z^n(i,j+1,k) - E_z^n(i,j,k)}{\Delta y} \\ &- \frac{\sigma_x^m(i,j,k)}{\mu_x(i,j,k)}H_x^n(i,j,k) - \frac{1}{\mu_x(i,j,k)}M_{ix}^n(i,j,k). \end{aligned} \tag{1.24}$$

After some manipulations, the future term $H_x^{n+\frac{1}{2}}(i,j,k)$ in (1.24) can be moved to the left-hand side and the other terms can be moved to the right-hand side such that

$$\begin{aligned}
H_x^{n+\frac{1}{2}}(i,j,k) = & \frac{2\mu_x(i,j,k) - \Delta t\sigma_x^m(i,j,k)}{2\mu_x(i,j,k) + \Delta t\sigma_x^m(i,j,k)} H_x^{n-\frac{1}{2}}(i,j,k) \\
& + \frac{2\Delta t}{(2\mu_x(i,j,k) + \Delta t\sigma_x^m(i,j,k))\Delta z}\left(E_y^n(i,j,k+1) - E_y^n(i,j,k)\right) \\
& - \frac{2\Delta t}{(2\mu_x(i,j,k) + \Delta t\sigma_x^m(i,j,k))\Delta y}\left(E_z^n(i,j+1,k) - E_z^n(i,j,k)\right) \\
& - \frac{2\Delta t}{2\mu_x(i,j,k) + \Delta t\sigma_x^m(i,j,k)} M_{ix}^n(i,j,k).
\end{aligned} \tag{1.25}$$

This equation is the updating equation for $H_x^{n+\frac{1}{2}}(i,j,k)$. Similarly, updating equations can easily be obtained for $H_y^{n+\frac{1}{2}}(i,j,k)$ starting from (1.4e) and $H_z^{n+\frac{1}{2}}(i,j,k)$ starting from (1.4f) following the same methodology used to obtain (1.25).

Finally, (1.4a)–(1.4f) can be expressed using finite differences and can be arranged to construct the following six FDTD updating equations for the six components of electromagnetic fields by introduction of respective coefficient terms:

$$\begin{aligned}
E_x^{n+1}(i,j,k) = & C_{exe}(i,j,k) \times E_x^n(i,j,k) \\
& + C_{exhz}(i,j,k) \times \left(H_z^{n+\frac{1}{2}}(i,j,k) - H_z^{n+\frac{1}{2}}(i,j-1,k)\right) \\
& + C_{exhy}(i,j,k) \times \left(H_y^{n+\frac{1}{2}}(i,j,k) - H_y^{n+\frac{1}{2}}(i,j,k-1)\right) \\
& + C_{exj}(i,j,k) \times J_{ix}^{n+\frac{1}{2}}(i,j,k),
\end{aligned} \tag{1.26}$$

where

$$\begin{aligned}
C_{exe}(i,j,k) &= \frac{2\varepsilon_x(i,j,k) - \Delta t\sigma_x^e(i,j,k)}{2\varepsilon_x(i,j,k) + \Delta t\sigma_x^e(i,j,k)}, \\
C_{exhz}(i,j,k) &= \frac{2\Delta t}{(2\varepsilon_x(i,j,k) + \Delta t\sigma_x^e(i,j,k))\Delta y}, \\
C_{exhy}(i,j,k) &= -\frac{2\Delta t}{(2\varepsilon_x(i,j,k) + \Delta t\sigma_x^e(i,j,k))\Delta z}, \\
C_{exj}(i,j,k) &= -\frac{2\Delta t}{2\varepsilon_x(i,j,k) + \Delta t\sigma_x^e(i,j,k)}.
\end{aligned} \tag{1.27}$$

$$\begin{aligned}
E_y^{n+1}(i,j,k) = & C_{eye}(i,j,k) \times E_y^n(i,j,k) \\
& + C_{eyhx}(i,j,k) \times \left(H_x^{n+\frac{1}{2}}(i,j,k) - H_x^{n+\frac{1}{2}}(i,j,k-1)\right) \\
& + C_{eyhz}(i,j,k) \times \left(H_z^{n+\frac{1}{2}}(i,j,k) - H_z^{n+\frac{1}{2}}(i-1,j,k)\right) \\
& + C_{eyj}(i,j,k) \times J_{iy}^{n+\frac{1}{2}}(i,j,k),
\end{aligned}$$

where

$$C_{eye}(i,j,k) = \frac{2\varepsilon_y(i,j,k) - \Delta t \sigma_y^e(i,j,k)}{2\varepsilon_y(i,j,k) + \Delta t \sigma_y^e(i,j,k)},$$

$$C_{eyhx}(i,j,k) = \frac{2\Delta t}{\Big(2\varepsilon_y(i,j,k) + \Delta t \sigma_y^e(i,j,k)\Big)\Delta z},$$

$$C_{eyhz}(i,j,k) = -\frac{2\Delta t}{\Big(2\varepsilon_y(i,j,k) + \Delta t \sigma_y^e(i,j,k)\Big)\Delta x},$$

$$C_{eyj}(i,j,k) = -\frac{2\Delta t}{2\varepsilon_y(i,j,k) + \Delta t \sigma_y^e(i,j,k)}. \tag{1.28}$$

$$\begin{aligned} E_z^{n+1}(i,j,k) = {} & C_{eze}(i,j,k) \times E_z^n(i,j,k) \\ & + C_{ezhy}(i,j,k) \times \Big(H_y^{n+\frac{1}{2}}(i,j,k) - H_y^{n+\frac{1}{2}}(i-1,j,k)\Big) \\ & + C_{ezhx}(i,j,k) \times \Big(H_x^{n+\frac{1}{2}}(i,j,k) - H_x^{n+\frac{1}{2}}(i,j-1,k)\Big) \\ & + C_{ezj}(i,j,k) \times J_{iz}^{n+\frac{1}{2}}(i,j,k), \end{aligned}$$

where

$$C_{eze}(i,j,k) = \frac{2\varepsilon_z(i,j,k) - \Delta t \sigma_z^e(i,j,k)}{2\varepsilon_z(i,j,k) + \Delta t \sigma_z^e(i,j,k)},$$

$$C_{ezhy}(i,j,k) = \frac{2\Delta t}{(2\varepsilon_z(i,j,k) + \Delta t \sigma_z^e(i,j,k))\Delta x},$$

$$C_{ezhx}(i,j,k) = -\frac{2\Delta t}{(2\varepsilon_z(i,j,k) + \Delta t \sigma_z^e(i,j,k))\Delta y},$$

$$C_{ezj}(i,j,k) = -\frac{2\Delta t}{2\varepsilon_z(i,j,k) + \Delta t \sigma_z^e(i,j,k)}. \tag{1.29}$$

$$\begin{aligned} H_x^{n+\frac{1}{2}}(i,j,k) = {} & C_{hxh}(i,j,k) \times H_x^{n-\frac{1}{2}}(i,j,k) \\ & + C_{hxey}(i,j,k) \times \Big(E_y^n(i,j,k+1) - E_y^n(i,j,k)\Big) \\ & + C_{hxez}(i,j,k) \times (E_z^n(i,j+1,k) - E_z^n(i,j,k)) \\ & + C_{hxm}(i,j,k) \times M_{ix}^n(i,j,k), \end{aligned}$$

where

$$
\begin{aligned}
C_{hxh}(i,j,k) &= \frac{2\mu_x(i,j,k) - \Delta t\sigma_x^m(i,j,k)}{2\mu_x(i,j,k) + \Delta t\sigma_x^m(i,j,k)},\\
C_{hxey}(i,j,k) &= \frac{2\Delta t}{\left(2\mu_x(i,j,k) + \Delta t\sigma_x^m(i,j,k)\right)\Delta z},\\
C_{hxez}(i,j,k) &= -\frac{2\Delta t}{\left(2\mu_x(i,j,k) + \Delta t\sigma_x^m(i,j,k)\right)\Delta y},\\
C_{hxm}(i,j,k) &= -\frac{2\Delta t}{2\mu_x(i,j,k) + \Delta t\sigma_x^m(i,j,k)}.
\end{aligned}
\tag{1.30}
$$

$$
\begin{aligned}
H_y^{n+\frac{1}{2}}(i,j,k) = {} & C_{hyh}(i,j,k) \times H_y^{n-\frac{1}{2}}(i,j,k) + C_{hyez}(i,j,k)\\
& \times \left(E_z^n(i+1,j,k) - E_z^n(i,j,k)\right) + C_{hyex}(i,j,k)\\
& \times \left(E_x^n(i,j,k+1) - E_x^n(i,j,k)\right) + C_{hym}(i,j,k) \times M_{iy}^n(i,j,k)
\end{aligned}
$$

where

$$
\begin{aligned}
C_{hyh}(i,j,k) &= \frac{2\mu_y(i,j,k) - \Delta t\sigma_y^m(i,j,k)}{2\mu_y(i,j,k) + \Delta t\sigma_y^m(i,j,k)},\\
C_{hyez}(i,j,k) &= \frac{2\Delta t}{\left(2\mu_y(i,j,k) + \Delta t\sigma_y^m(i,j,k)\right)\Delta x},\\
C_{byex}(i,j,k) &= -\frac{2\Delta t}{\left(2\mu_y(i,j,k) + \Delta t\sigma_y^m(i,j,k)\right)\Delta z},\\
C_{hym}(i,j,k) &= -\frac{2\Delta t}{2\mu_y(i,j,k) + \Delta t\sigma_y^m(i,j,k)}.
\end{aligned}
\tag{1.31}
$$

$$
\begin{aligned}
H_z^{n+\frac{1}{2}}(i,j,k) = {} & C_{hzh}(i,j,k) \times H_z^{n-\frac{1}{2}}(i,j,k)\\
& + C_{hzex}(i,j,k) \times \left(E_x^n(i,j+1,k) - E_x^n(i,j,k)\right)\\
& + C_{hzey}(i,j,k) \times \left(E_y^n(i+1,j,k) - E_y^n(i,j,k)\right)\\
& + C_{hzm}(i,j,k) \times M_{iz}^n(i,j,k),
\end{aligned}
$$

where

$$
\begin{aligned}
C_{hzh}(i,j,k) &= \frac{2\mu_y(i,j,k) - \Delta t\sigma_z^m(i,j,k)}{2\mu_z(i,j,k) + \Delta t\sigma_z^m(i,j,k)},\\
C_{hzex}(i,j,k) &= \frac{2\Delta t}{\left(2\mu_z(i,j,k) + \Delta t\sigma_z^m(i,j,k)\right)\Delta y},\\
C_{hzey}(i,j,k) &= -\frac{2\Delta t}{\left(2\mu_z(i,j,k) + \Delta t\sigma_z^m(i,j,k)\right)\Delta x},\\
C_{hzm}(i,j,k) &= -\frac{2\Delta t}{2\mu_z(i,j,k) + \Delta t\sigma_z^m(i,j,k)}.
\end{aligned}
$$

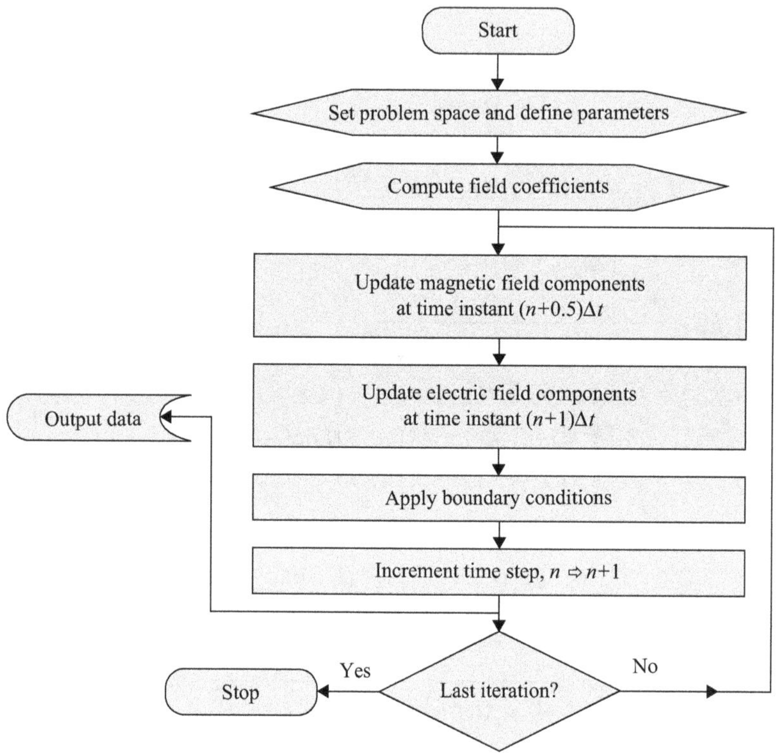

Figure 1.9 Explicit FDTD procedure.

It should be noted that the first two subscripts in each coefficient refer to the corresponding field component being updated. For three subscripts coefficients, the third subscript refers to the type of the field or source (electric or magnetic) that this coefficient is multiplied by. For four subscripts coefficients, the third and fourth subscripts refer to the type of the field that this coefficient is multiplied by.

Having derived the FDTD updating equations, a time-marching algorithm can be constructed as illustrated in Figure 1.9. The first step in this algorithm is setting up the problem space – including the objects, material types, and sources – and defining any other parameters that will be used during the FDTD computation. Then the coefficient terms appearing in (1.26)–(1.31) can be calculated and stored as arrays before the iteration is started. The field components need to be defined as arrays as well and shall be initialized with zeros since the initial values of the fields in the problem space in most cases are zeros, and fields will be induced in the problem space due to sources as the iteration proceeds. At every step of the time-marching iteration the magnetic field components are updated for time instant $(n + 0.5)\Delta t$ using (1.29)–(1.31); then the electric field components are updated for time instant $(n + 1)\Delta t$ using (1.26)–(1.28). The problem space has a finite size, and specific boundary conditions can be enforced on the boundaries of the problem space. Therefore, the field components on the boundaries of the problem space are treated according to the type of boundary conditions during the iteration. The types of boundary conditions and the techniques used to integrate them into the FDTD algorithm are discussed in detail in Chapters 7 and 8. After the fields are updated and boundary conditions are enforced, the current values of any desired field components can be captured and stored as output data, and these data can be used for real-time processing or postprocessing to calculate some other desired parameters. The FDTD iterations can be continued until some stopping criteria are achieved.

1.4 FDTD updating equations for two-dimensional problems

The FDTD updating equations given in (1.26)–(1.31) can be used to solve three-dimensional problems. In the two-dimensional case where there is no variation in the problem geometry and field distributions in one of the dimensions, a simplified set of updating equations can be obtained starting from the Maxwell's curl equations system (1.4). Since there is no variation in one of the dimensions, the derivative terms with respect to that dimension vanish. For instance, if the problem is z dimension independent, equations (1.4) reduce to

$$\frac{\partial E_x}{\partial t} = \frac{1}{\varepsilon_x}\left(\frac{\partial H_z}{\partial y} - \sigma_x^e E_x - J_{ix}\right), \tag{1.32a}$$

$$\frac{\partial E_y}{\partial t} = \frac{1}{\varepsilon_y}\left(-\frac{\partial H_z}{\partial x} - \sigma_y^e E_y - J_{iy}\right), \tag{1.32b}$$

$$\frac{\partial E_z}{\partial t} = \frac{1}{\varepsilon_z}\left(\frac{\partial H_y}{\partial x} - \frac{\partial H_x}{\partial y} - \sigma_z^e E_z - J_{iz}\right), \tag{1.32c}$$

$$\frac{\partial H_x}{\partial t} = \frac{1}{\mu_x}\left(-\frac{\partial E_z}{\partial y} - \sigma_x^m H_x - M_{ix}\right), \tag{1.32d}$$

$$\frac{\partial H_y}{\partial t} = \frac{1}{\mu_y}\left(\frac{\partial E_z}{\partial x} - \sigma_y^m H_y - M_{iy}\right), \tag{1.32e}$$

$$\frac{\partial H_z}{\partial t} = \frac{1}{\mu_z}\left(\frac{\partial E_x}{\partial y} - \frac{\partial E_y}{\partial x} - \sigma_z^m H_z - M_{iz}\right). \tag{1.32f}$$

One should notice that equations (1.32a), (1.32b), and (1.32f) are dependent only on the terms E_x, E_y, and H_z, whereas equations (1.32c), (1.32d), and (1.32e) are dependent only on the terms E_z, H_x, and H_y. Therefore, the six equations (1.32) can be treated as two separate sets of equations. In the first set (1.32a), (1.32b), and (1.32f) – all the electric field components are transverse to the reference dimension z; therefore, this set of equations constitutes the transverse electric to z case – TE_z. In the second set – (1.32c), (1.32d), and (1.32e) – all the magnetic field components are transverse to the reference dimension z; therefore, this set of equations constitutes the transverse magnetic to z case – TM_z. Most two-dimensional problems can be decomposed into two separate problems, each including separate field components that are TE_z and TM_z for the case under consideration. These two problems can be solved separately, and the solution for the main problem can be achieved as the sum of the two solutions.

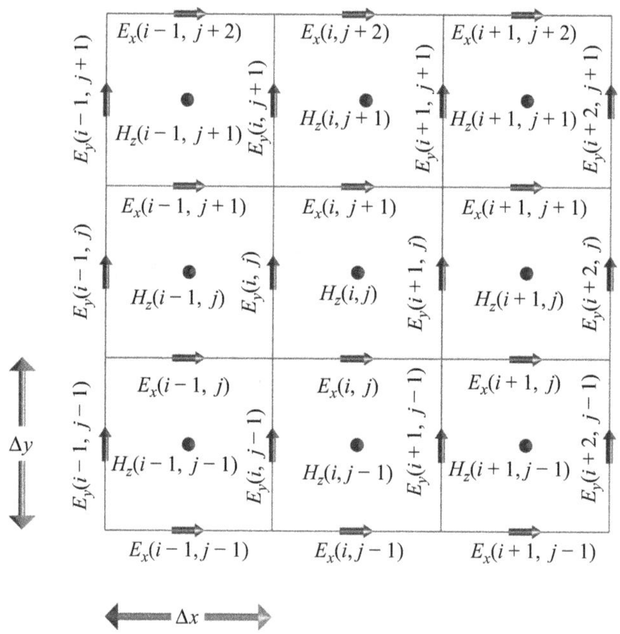

Figure 1.10 Two-dimensional TE_z FDTD field components.

The FDTD updating equations for the TE_z case can be obtained by applying the central difference formula to the equations constituting the TE_z case based on the field positions shown in Figure 1.10, which is obtained by projection of the Yee cells in Figure 1.5 on the xy plane in the z direction. The FDTD updating equations for the TE_z case are therefore obtained as

$$E_x^{n+1}(i,j) = C_{exe}(i,j) \times E_x^n(i,j) + C_{exhz}(i,j) \times \left(H_z^{n+\frac{1}{2}}(i,j) - H_z^{n+\frac{1}{2}}(i,j-1)\right) + C_{exj}(i,j) \times J_{ix}^{n+\frac{1}{2}}(i,j), \tag{1.33}$$

where

$$C_{exe}(i,j) = \frac{2\varepsilon_x(i,j) - \Delta t \sigma_x^e(i,j)}{2\varepsilon_x(i,j) + \Delta t \sigma_x^e(i,j)},$$

$$C_{exhz}(i,j) = \frac{2\Delta t}{\left(2\varepsilon_x(i,j) + \Delta t \sigma_x^e(i,j)\right)\Delta y},$$

$$C_{exj}(i,j) = -\frac{2\Delta t}{2\varepsilon_x(i,j) + \Delta t \sigma_x^e(i,j)}.$$

$$E_y^{n+1}(i,j) = C_{eye}(i,j) \times E_y^n(i,j) + C_{eyhz}(i,j) \times \left(H_z^{n+\frac{1}{2}}(i,j) - H_z^{n+\frac{1}{2}}(i-1,j)\right) + C_{eyj}(i,j) \times J_{iy}^{n+\frac{1}{2}}(i,j), \tag{1.34}$$

where

$$
\begin{aligned}
C_{eye}(i,j) &= \frac{2\varepsilon_y(i,j) - \Delta t\sigma_y^e(i,j)}{2\varepsilon_y(i,j) + \Delta t\sigma_y^e(i,j)}, \\
C_{eyhz}(i,j) &= -\frac{2\Delta t}{\left(2\varepsilon_y(i,j) + \Delta t\sigma_y^e(i,j)\right)\Delta x}, \\
C_{eyj}(i,j) &= -\frac{2\Delta t}{2\varepsilon_y(i,j) + \Delta t\sigma_y^e(i,j)}.
\end{aligned} \tag{1.35}
$$

$$
\begin{aligned}
H_z^{n+\frac{1}{2}}(i,j) &= C_{hzh}(i,j) \times H_z^{n-\frac{1}{2}}(i,j) + C_{hzex}(i,j) \times \left(E_x^n(i,j+1) - E_x^n(i,j)\right) \\
&\quad + C_{hzey}(i,j) \times \left(E_y^n(i+1,j) - E_y^n(i,j)\right) + C_{hzm}(i,j) \times M_{iz}^m(i,j),
\end{aligned}
$$

where

$$
\begin{aligned}
C_{hzh}(i,j) &= \frac{2\mu_z(i,j) - \Delta t\sigma_z^m(i,j)}{2\mu_z(i,j) + \Delta t\sigma_z^m(i,j)}, \\
C_{hzex}(i,j) &= \frac{2\Delta t}{(2\mu_z(i,j) + \Delta t\sigma_z^m(i,j))\Delta y}, \\
C_{hzey}(i,j) &= -\frac{2\Delta t}{(2\mu_z(i,j) + \Delta t\sigma_z^m(i,j))\Delta x}, \\
C_{hzm}(i,j) &= -\frac{2\Delta t}{2\mu_z(i,j) + \Delta t\sigma_z^m(i,j)}.
\end{aligned}
$$

Since the FDTD updating equations for the three-dimensional case are readily available, they can be used to derive (1.33), (1.34), and (1.35) by simply setting the coefficients including $1/\Delta z$ to zero. Hence, the FDTD updating equations for the TM_z case can be obtained by eliminating the term $C_{hxey}(i,j,k)$ in (1.29) and eliminating the term $C_{hyex}(i,j,k)$ in (1.30) based on the field positions shown in Figure 1.11, such that

$$
\begin{aligned}
E_z^{n+1}(i,j) &= C_{eze}(i,j) \times E_z^n(i,j) + C_{ezhy}(i,j) \times \left(H_y^{n+\frac{1}{2}}(i,j) - H_y^{n+\frac{1}{2}}(i-1,j)\right) \\
&\quad + C_{ezhx}(i,j) \times \left(H_x^{n+\frac{1}{2}}(i,j) - H_x^{n+\frac{1}{2}}(i,j-1)\right) + C_{ezj}(i,j) \times J_{iz}^{n+\frac{1}{2}}(i,j),
\end{aligned} \tag{1.36}
$$

where

$$
\begin{aligned}
C_{eze}(i,j) &= \frac{2\varepsilon_z(i,j) - \Delta t\sigma_z^e(i,j)}{2\varepsilon_z(i,j) + \Delta t\sigma_z^e(i,j)}, \\
C_{ezhy}(i,j) &= \frac{2\Delta t}{(2\varepsilon_z(i,j) + \Delta t\sigma_z^e(i,j))\Delta x}, \\
C_{ezhx}(i,j) &= -\frac{2\Delta t}{(2\varepsilon_z(i,j) + \Delta t\sigma_z^e(i,j))\Delta y}, \\
C_{ezj}(i,j) &= -\frac{2\Delta t}{2\varepsilon_z(i,j) + \Delta t\sigma_z^e(i,j)}.
\end{aligned} \tag{1.37}
$$

$$
\begin{aligned}
H_x^{n+\frac{1}{2}}(i,j) &= C_{hxh}(i,j) \times H_x^{n-\frac{1}{2}}(i,j) + C_{hxez}(i,j) \times \left(E_z^n(i,j+1) - E_z^n(i,j)\right) \\
&\quad + C_{hxm}(i,j) \times M_{ix}^n(i,j),
\end{aligned}
$$

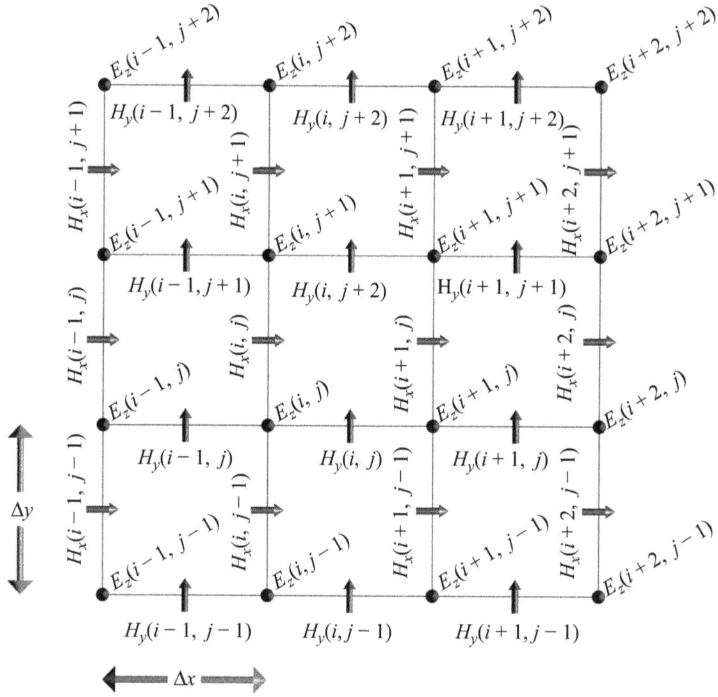

Figure 1.11 Two-dimensional TM_z FDTD field components.

where

$$
\begin{aligned}
C_{hxh}(i,j) &= \frac{2\mu_x(i,j) - \Delta t \sigma_x^m(i,j)}{2\mu_x(i,j) + \Delta t \sigma_x^m(i,j)}, \\
C_{hxez}(i,j) &= -\frac{2\Delta t}{\left(2\mu_x(i,j) + \Delta t \sigma_x^m(i,j)\right)\Delta y}, \\
C_{hxm}(i,j) &= -\frac{2\Delta t}{2\mu_x(i,j) + \Delta t \sigma_x^m(i,j)}.
\end{aligned} \tag{1.38}
$$

$$
\begin{aligned}
H_y^{n+\frac{1}{2}}(i,j) = {} & C_{hyh}(i,j) \times H_y^{n-\frac{1}{2}}(i,j) + C_{hyez}(i,j) \times \left(E_z^n(i+1,j) - E_z^n(i,j)\right) \\
& + C_{hym}(i,j) \times M_{iy}^m(i,j),
\end{aligned}
$$

where

$$
\begin{aligned}
C_{hyh}(i,j) &= \frac{2\mu_y(i,j) - \Delta t \sigma_y^m(i,j)}{2\mu_y(i,j) + \Delta t \sigma_y^m(i,j)}, \\
C_{hyez}(i,j) &= \frac{2\Delta t}{\left(2\mu_y(i,j) + \Delta t \sigma_y^m(i,j)\right)\Delta x}, \\
C_{hym}(i,j) &= -\frac{2\Delta t}{2\mu_y(i,j) + \Delta t \sigma_y^m(i,j)}.
\end{aligned}
$$

1.5 FDTD updating equations for one-dimensional problems

In the one-dimensional case, there is no variation in the problem geometry and field distributions in two of the dimensions. For instance, if the y and z dimensions have no variation, the derivative with respect to the y and z dimensions vanish in Maxwell's curl equations. Therefore, the two-dimensional curl equations (1.32a)–(1.32f) reduce to

$$\frac{\partial E_x}{\partial t} = \frac{1}{\varepsilon_x}\left(-\sigma_x^e E_x - J_{ix}\right), \tag{1.39a}$$

$$\frac{\partial E_y}{\partial t} = \frac{1}{\varepsilon_y}\left(-\frac{\partial H_z}{\partial x} - \sigma_y^e E_y - J_{iy}\right), \tag{1.39b}$$

$$\frac{\partial E_z}{\partial t} = \frac{1}{\varepsilon_z}\left(\frac{\partial H_y}{\partial x} - \sigma_z^e E_z - J_{iz}\right), \tag{1.39c}$$

$$\frac{\partial H_x}{\partial t} = \frac{1}{\mu_x}\left(-\sigma_x^m H_x - M_{ix}\right), \tag{1.39d}$$

$$\frac{\partial H_y}{\partial t} = \frac{1}{\mu_y}\left(\frac{\partial E_z}{\partial x} - \sigma_y^m H_y - M_{iy}\right), \tag{1.39e}$$

$$\frac{\partial H_z}{\partial t} = \frac{1}{\mu_z}\left(-\frac{\partial E_y}{\partial x} - \sigma_z^m H_z - M_{iz}\right). \tag{1.39f}$$

It should be noted that (1.39a) and (1.39d) include time derivatives but not space derivatives. Therefore, these equations do not represent propagating fields, hence the field components E_x and H_x, which only exist in these two equations, do not propagate. The other four equations represent propagating fields, and both the electric and magnetic field components existing in these equations are transverse to the x dimension. Therefore, transverse electric and magnetic to x (TEM_x) fields exist and propagate as plane waves in the one-dimensional case under consideration.

Similar to the two-dimensional case, the one-dimensional case as well can be decomposed into two separate cases, since (1.39b) and (1.39f), which include only the terms E_y and H_z, are decoupled from (1.39c) and (1.39e), which include only the terms E_z and H_y. FDTD updating equations can be obtained for (1.39b) and (1.39f) using the central difference formula based on the field positioning in one-dimensional space as illustrated in Figure 1.12, such that

$$E_y^{n+1}(i) = C_{eye}(i) \times E_y^n(i) + C_{eyhz}(i) \times \left(H_z^{n+\frac{1}{2}}(i) - H_z^{n+\frac{1}{2}}(i-1)\right) + C_{eyj}(i) \times J_{iy}^{n+\frac{1}{2}}(i), \tag{1.40}$$

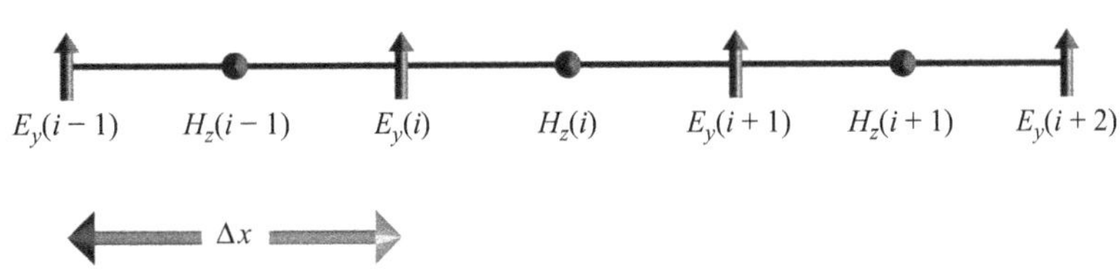

Figure 1.12 One-dimensional FDTD – positions of field components E_y and H_z.

where

$$C_{eye}(i) = \frac{2\varepsilon_y(i) - \Delta t\sigma_y^e(i)}{2\varepsilon_y(i) + \Delta t\sigma_y^e(i)},$$
$$C_{eyhz}(i) = -\frac{2\Delta t}{\left(2\varepsilon_y(i) + \Delta t\sigma_y^e(i)\right)\Delta x},$$
$$C_{eyj}(i) = -\frac{2\Delta t}{2\varepsilon_y(i) + \Delta t\sigma_y^e(i)},$$

and

$$H_z^{n+\frac{1}{2}}(i) = C_{hzh}(i) \times H_z^{n-\frac{1}{2}}(i) + C_{hzey}(i) \times \left(E_y^n(i+1) - E_y^n(i)\right) + C_{hzm}(i) \times M_{iz}^n(i), \quad (1.41)$$

where

$$C_{hzh}(i) = \frac{2\mu_z(i) - \Delta t\sigma_z^m(i)}{2\mu_z(i) + \Delta t\sigma_z^m(i)},$$
$$C_{hzey}(i) = -\frac{2\Delta t}{(2\mu_z(i) + \Delta t\sigma_z^m(i))\Delta x},$$
$$C_{hzm}(i) = -\frac{2\Delta t}{2\mu_z(i) + \Delta t\sigma_z^m(i)}.$$

Similarly, FDTD updating equations can be obtained for (1.39c) and (1.39e) using the central difference formula based on the field positioning in one-dimensional space as illustrated in Figure 1.13, such that

$$E_z^{n+1}(i) = C_{eze}(i) \times E_z^n(i) + C_{ezhy}(i) \times \left(H_y^{n+\frac{1}{2}}(i) - H_y^{n+\frac{1}{2}}(i-1)\right) + C_{ezj}(i) \times J_{iz}^{n+\frac{1}{2}}(i), \quad (1.42)$$

where

$$C_{eze}(i) = \frac{2\varepsilon_z(i) - \Delta t\sigma_z^e(i)}{2\varepsilon_z(i) + \Delta t\sigma_z^e(i)},$$
$$C_{ezhy}(i) = \frac{2\Delta t}{(2\varepsilon_z(i) + \Delta t\sigma_z^e(i))\Delta x},$$
$$C_{ezj}(i) = -\frac{2\Delta t}{2\varepsilon_z(i) + \Delta t\sigma_z^e(i)},$$

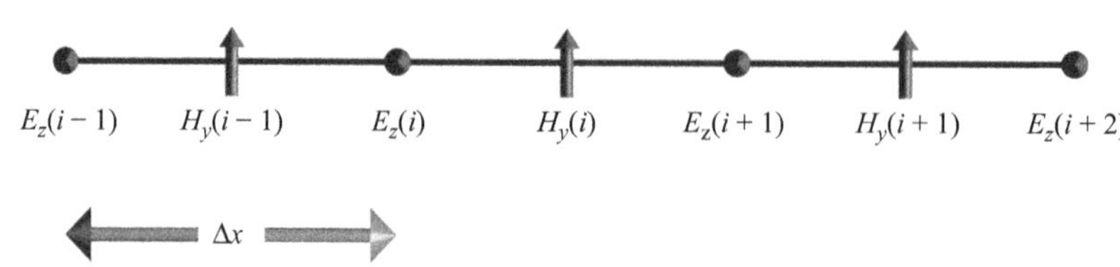

Figure 1.13 One-dimensional FDTD – positions of field components E_z and H_y.

and

$$H_y^{n+\frac{1}{2}}(i) = C_{hyh}(i) \times H_y^{n-\frac{1}{2}}(i) + C_{hyez}(i) \times \left(E_z^n(i+1) - E_z^n(i)\right) + C_{hym}(i) \times M_{iy}^n(i), \quad (1.43)$$

where

$$C_{hyh}(i) = \frac{2\mu_y(i) - \Delta t \sigma_y^m(i)}{2\mu_y(i) + \Delta t \sigma_y^m(i)},$$

$$C_{hyez}(i) = \frac{2\Delta t}{\left(2\mu_y(i) + \Delta t \sigma_y^m(i)\right)\Delta x},$$

$$C_{hym}(i) = -\frac{2\Delta t}{2\mu_y(i) + \Delta t \sigma_y^m(i)}.$$

A MATLAB code is given in Appendix A for a one-dimensional FDTD implementation based on the updating equations (1.42) and (1.43). The code calculates electric and magnetic field components generated by a z-directed current sheet, J_z, placed at the center of a problem space filled with air between two parallel, perfect electric conductor (PEC) plates extending to infinity in the y and z dimensions. Figure 1.14 shows snapshots of E_z and H_y within the FDTD computational domain demonstrating the propagation of the fields and their reflection from the PEC plates at the left and right boundaries.

In the demonstrated problem, the parallel plates are 1 m apart. The one-dimensional problem space is represented by cells having width $\Delta x = 1$ mm. The current sheet at the center excites a current with 1 A/m^2 density and Gaussian waveform

$$J_z(t) = e^{-\left(\frac{t - 2\times 10^{-10}}{5\times 10^{-11}}\right)^2}.$$

The current sheet is excited over a one-cell cross-section (i.e., 1 mm wide), which translates J_z into a surface current density K_z of magnitude 1×10^{-3} A/m. The surface current density causes a discontinuity in magnetic field and generates H_y, which satisfies the boundary condition $\vec{K} = \hat{n} \times \vec{H}$. Therefore, two waves of H_y are generated on both sides of the current sheet, each of which has magnitude 5×10^{-4} A/m. Since the fields are propagating in free space, the magnitude of the generated electric field is $\eta_0 \times 5 \times 10^{-4} \approx 0.1885$ V/m, where η_0 is the intrinsic impedance of free space ($\eta_0 \approx 377$).

Examining the listing of the one-dimensional FDTD code in Appendix A, one immediately notices that the sizes of the arrays associated with E_z and H_y are not the same. The one-dimensional problem space is divided into nx intervals; therefore, there are $nx + 1$ *nodes* in the problem space including the left and right boundary nodes. This code is based on the field positioning scheme given in Figure 1.13, where the E_z components are defined at node positions and the H_y components are defined at the center positions of the intervals. Therefore, arrays associated with E_z have the size $nx + 1$, whereas those associated with H_y have the size nx.

Another point that deserves consideration is the application of the boundary conditions. In this problem the boundaries are PEC; thus, the tangential electric field component (E_z in this case) vanishes on the PEC surface. Therefore, this condition can be enforced on the electric field component on the boundaries, $E_z(1)$ and $E_z(nx + 1)$, as can be seen in the

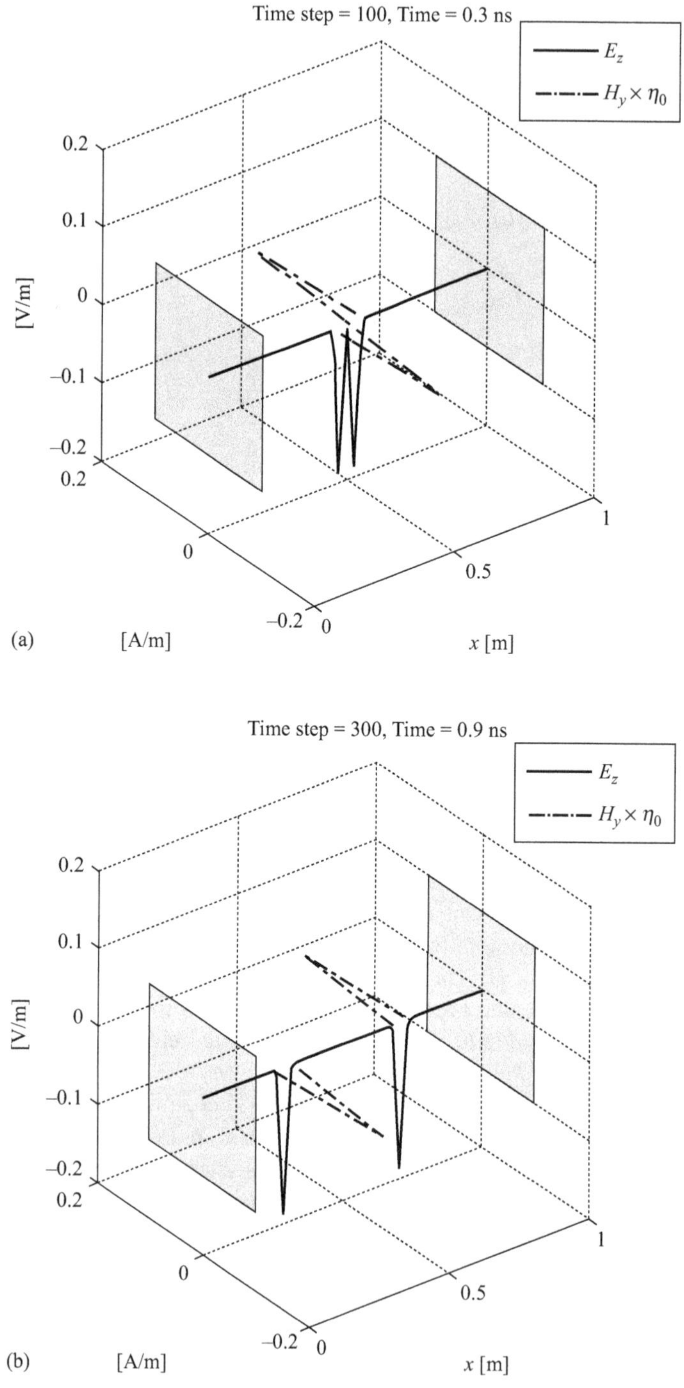

Figure 1.14 Snapshots of a one-dimensional FDTD simulation: (a) fields observed after 100 time steps; (b) fields observed after 300 time steps; (c) fields observed after 615 time steps; and (d) fields observed after 700 time steps.

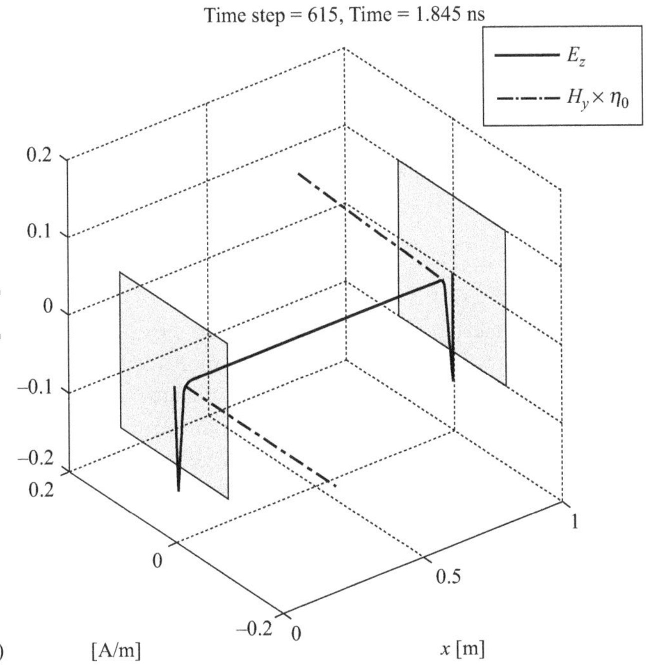

(c)

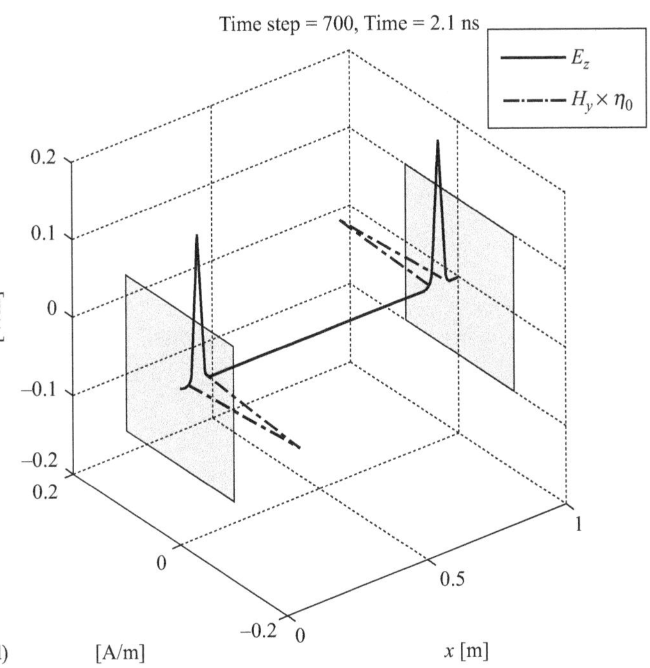

(d)

Figure 1.14 (*Continued*)

code listing. Observing the changes in the fields from time step 650 to 700 in Figure 1.14 reveals the correct behavior of the incident and reflected waves for a PEC plate. This is evident by the reversal of the direction of the peak value of E_z after reflection from the PEC boundaries, which represents a reflection coefficient equal to -1 while the behavior of the reflected H_y component represents the reflection coefficient that is equal to $+1$ as expected.

1.6 Exercises

1.1 Follow the steps presented in Section 1.2 to develop the appropriate approximations for the first derivative of a function with second-order accuracy using the forward and backward difference procedure and fourth-order accuracy using the central difference procedure. Compare your derived expressions with those listed in Table 1.1.

1.2 Update the MATLAB program in Listing 1.1 to regenerate Figure 1.2 using the second-order accurate forward and backward differences and the fourth-order accurate central difference approximations for the same function.

1.3 Update the MATLAB program in Listing 1.1 to generate the corresponding figures to Figure 1.2, but for the second derivatives of the function. Use the approximate expressions in the right column of Table 1.1 while updating the program.

1.4 The steps of the development of the updating equation for the x component of the electric field are demonstrated in Section 1.3. Follow the same procedure to show the steps required to develop the updating equations for the y and z components for the electric field. Pay sufficient attention to the selection of the indices of each individual field component in your expressions. The Yee cell presented in Figure 1.5 should help you in understanding the proper use of these indices.

1.5 Repeat Exercise 1.4, but this time for developing the updating equations for the magnetic field components.

CHAPTER 2

Numerical stability and dispersion

2.1 Numerical stability

The finite-difference time-domain (FDTD) algorithm samples the electric and magnetic fields at discrete points both in time and space. The choice of the period of sampling (Δt in time, Δx, Δy, and Δz in space) must comply with certain restrictions to guarantee the stability of the solution. Furthermore, the choice of these parameters determines the accuracy of the solution. This section focuses on the stability analysis. First, the stability concept is illustrated using a simple partial differential equation (PDE) in space and time domain. Next, the Courant-Friedrichs-Lewy (CFL) condition [3] for the FDTD method is discussed, accompanied by a one-dimensional FDTD example.

2.1.1 Stability in time-domain algorithm

An important issue in designing a time-domain numerical algorithm is the stability condition. To understand the stability concept, let's start with a simple wave equation:

$$\frac{\partial u(x,t)}{\partial t} + \frac{\partial u(x,t)}{\partial x} = 0, \quad u(x,t=0) = u_0(x), \tag{2.1}$$

where $u(x, t)$ is the unknown wave function and $u_0(x)$ is the initial condition at $t = 0$. Using the PDE knowledge, the equation can be analytically solved:

$$u(x,t) = u_0(x-t). \tag{2.2}$$

A time-domain numerical scheme can be developed to solve above wave equation. First, $u(x, t)$ is discretized in both time and space domains:

$$\begin{aligned} x_i &= i\Delta x, \quad i = 0,1,2,\ldots \\ t_n &= n\Delta t, \quad n = 0,1,2,\ldots \\ u_i^n &= u(x_i, t_n). \end{aligned} \tag{2.3}$$

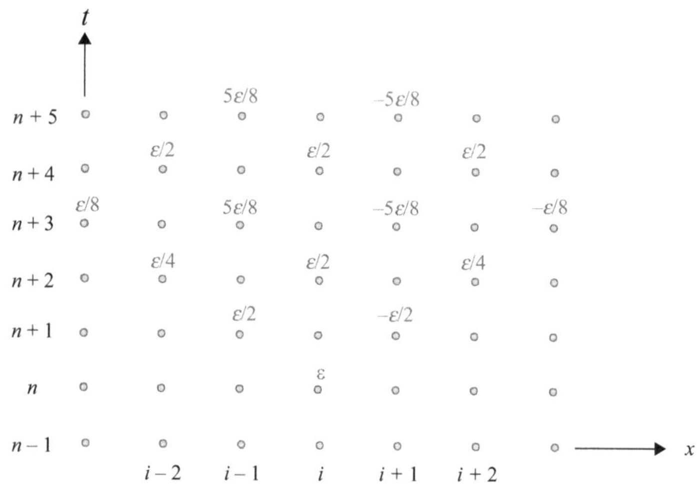

Figure 2.1 The time–space domain grid with error ε propagating with $\lambda = 1/2$.

Here, Δt and Δx are the time and space cell sizes. Then, the finite-difference scheme is used to compute the derivatives, and the following equation is obtained:

$$\frac{u_i^{n+1} - u_i^{n-1}}{2\Delta t} + \frac{u_{i-1}^n - u_{i+1}^n}{2\Delta x} = 0. \tag{2.4}$$

After a simple manipulation, a time-domain numerical scheme can be derived such that

$$u_i^{n+1} = u_i^{n-1} + \lambda(u_{i+1}^n - u_{i-1}^n), \quad \lambda = \frac{\Delta t}{\Delta x}. \tag{2.5}$$

Figure 2.1 shows a time and space domain grid, where the horizontal axis represents the x axis and the vertical axis denotes the t axis. The u values for time index denoted as n and $n-1$ are all assumed to be known. For simplicity we will assume that all these values are zeros. We will also assume that at one x position denoted by the index i and at time denoted by the index n there is a small error represented by the parameter ε. As the time evolves, a u value at $n+1$ is computed from two rows lower in accordance with (2.5).

Now let's analyze the propagation of the assumed numerical error ε in this time-domain algorithm. Note that the error may result from a numerical truncation of a real number. When $\lambda = 1/2$, the errors are shown in Figure 2.1. The error will keep propagating in this time-domain algorithm; however, it can be observed that the errors are always bounded by the original error ε, whereas for the case when $\lambda = 1$, the maximum absolute value of the propagating error is of the same value as the original error. This is clearly obvious from the errors propagating in Figure 2.2. On the contrary, when $\lambda = 2$, the propagation of error as shown in Figure 2.3 will keep increasing as time evolves. Finally, this error will be large enough and will destroy the actual u values. As a result, the time-domain algorithm will not give an accurate result due to a very small initial error. In summary, the numerical scheme in (2.5) is therefore considered conditionally stable. It is stable for small λ values but unstable for large λ values, and the boundary of stability condition is $\lambda = 1$.

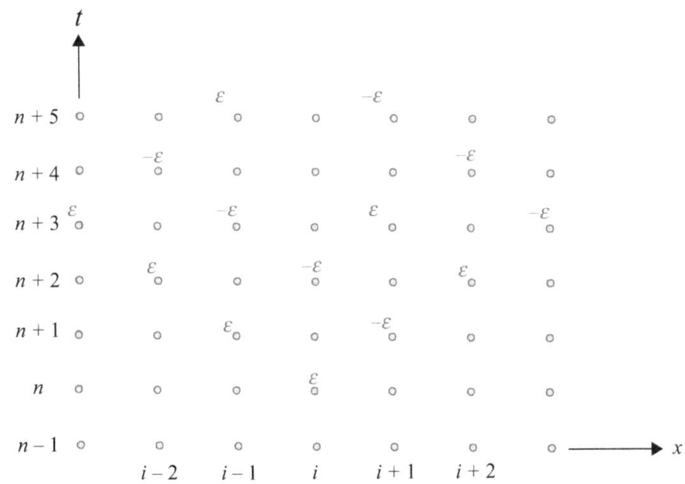

Figure 2.2 The time–space domain grid with error ε propagating with $\lambda = 1$.

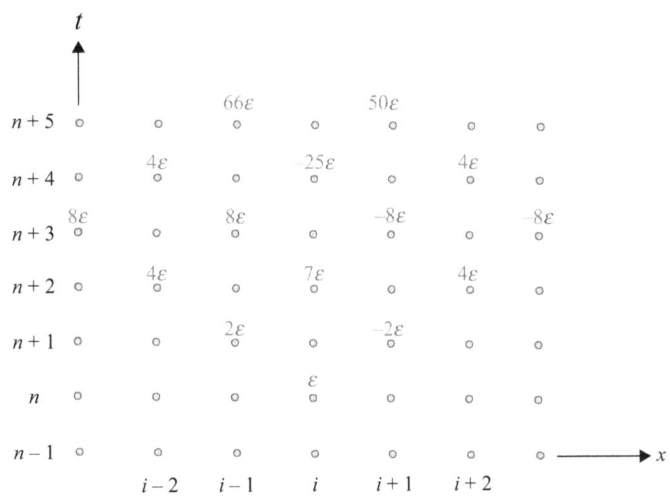

Figure 2.3 The time–space domain grid with error ε propagating with $\lambda = 2$.

2.1.2 CFL condition for the FDTD method

The numerical stability of the FDTD method is determined by the CFL condition, which requires that the time increment Δt has a specific bound relative to the lattice space increments, such that

$$\Delta t \leq \frac{1}{c\sqrt{\frac{1}{(\Delta x)^2} + \frac{1}{(\Delta y)^2} + \frac{1}{(\Delta z)^2}}}, \tag{2.6}$$

where c is the speed of light in free space. Equation (2.6) can be rewritten as

$$c\Delta t\sqrt{\frac{1}{(\Delta x)^2}+\frac{1}{(\Delta y)^2}+\frac{1}{(\Delta z)^2}} \leq 1. \tag{2.7}$$

For a cubical spatial grid where $\Delta x = \Delta y = \Delta z$, the CFL condition reduces to

$$\Delta t \leq \frac{\Delta x}{c\sqrt{3}}. \tag{2.8}$$

One can notice in (2.6) that the smallest value among Δx, Δy, and Δz is the dominant factor controlling the maximum time step and that the maximum time step allowed is always smaller than $\min(\Delta x, \Delta y, \Delta z)/c$.

In the one-dimensional case where $\Delta y \to \infty$ and $\Delta z \to \infty$ in (2.6), the CFL condition reduces to

$$\Delta t \leq \Delta x/c \quad \text{or} \quad c\Delta t \leq \Delta x. \tag{2.9}$$

This equation implies that a wave cannot be allowed to travel more than one cell size in space during one time step.

The existence of instability exposes itself as the development of divergent spurious fields in the problem space as the FDTD iterations proceed. For instance, the one-dimensional FDTD code presented in Chapter 1 simulates a problem space composed of cells having length $\Delta x = 1$ mm. Due to the one-dimensional CFL condition (2.9) the time increment must be chosen as $\Delta t \leq 3.3356$ ps. This code is run with the values of $\Delta t = 3.3356$ ps and $\Delta t = 3.3357$ ps, and the maximum value of electric field magnitude is captured at every time step and plotted in Figure 2.4. The spikes of the electric field magnitude are due to the

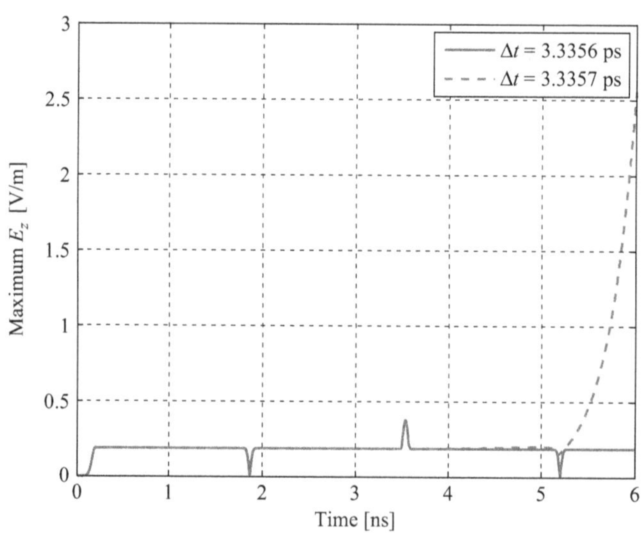

Figure 2.4 Maximum magnitude of E_z in the one-dimensional problem space for simulations with $\Delta t = 3.3356$ ps and $\Delta t = 3.3357$ ps.

additive interference of the fields when fields pass through the center of the problem space, and the nulls are due to the vanishing of electric fields on PEC when the fields hit the PEC boundary walls. However, while the iterations proceed, the electric field magnitude calculated by the simulation running with $\Delta t = 3.3357$ ps starts to diverge. This examination demonstrates how a Δt value larger than the CFL limit gives rise to instability in the FDTD computation.

The CFL stability condition applies to inhomogeneous media as well, since the velocity of propagation in material media is smaller than c. However, numerical stability can be influenced by other factors, such as absorbing boundary conditions, nonuniform spatial grids, and nonlinear materials.

Even if the numerical solution is stable, satisfaction of the CFL condition does not guarantee the numerical accuracy of the solution; it only provides a relationship between the spatial grid size and the time step. One must still satisfy the sampling theory requirements with respect to the highest frequency present in the excitation.

2.2 Numerical dispersion

The FDTD method provides a solution for the behavior of fields that is usually a good approximation to the real physical behavior of the fields. The finite-difference approximation of derivatives of continuous functions introduces an error to the solution. For instance, even in homogeneous free space, the velocity of propagation of the numerical solution for a wave will generally differ from c. Furthermore, it will vary with frequency, the spatial grid size, and direction of propagation of the wave. The differing of phase velocities numerically obtained by the FDTD method from the actual phase velocities is known as numerical dispersion.

For instance, consider a plane wave propagating in free space in the x direction, given by

$$E_z(x,t) = E_0 \cos(k_x x - \omega t), \tag{2.10a}$$

$$H_y(x,t) = H_0 \cos(k_x x - \omega t). \tag{2.10b}$$

Here, $E_z(x, t)$ satisfies the wave equation

$$\frac{\partial^2}{\partial x^2} E_z - \mu_0 \varepsilon_0 \frac{\partial^2}{\partial t^2} E_z = 0. \tag{2.11}$$

Substituting (2.10a) in (2.11) yields the equation

$$k_x^2 = \omega^2 \mu_0 \varepsilon_0 = \left(\frac{\omega}{c}\right)^2, \tag{2.12}$$

which is called the *dispersion relation*. The dispersion relation provides the connection between the spatial frequency k_x and the temporal frequency ω [4]. The dispersion relation (2.12) is analytically exact.

A dispersion relation equation, which is called the numerical dispersion relation, can be obtained based on the finite-difference approximation of Maxwell's curl equations as follows. For the one-dimensional case previously discussed, the plane wave expressions

$E_z(x, t)$ and $H_y(x, t)$ satisfy the Maxwell's one-dimensional curl equations, which are given for the source-free region as

$$\frac{\partial E_z}{\partial t} = \frac{1}{\varepsilon_0}\frac{\partial H_y}{\partial x}, \tag{2.13a}$$

$$\frac{\partial H_y}{\partial t} = \frac{1}{\mu_0}\frac{\partial E_z}{\partial x}. \tag{2.13b}$$

These equations can be rewritten using the central difference formula based on the field positioning scheme in Figure 1.13 as

$$\frac{E_z^{n+1}(i) - E_z^n(i)}{\Delta t} = \frac{1}{\varepsilon_0}\frac{H_y^{n+\frac{1}{2}}(i) - H_y^{n+\frac{1}{2}}(i-1)}{\Delta x}, \tag{2.14a}$$

$$\frac{H_y^{n+\frac{1}{2}}(i) - H_y^{n-\frac{1}{2}}(i)}{\Delta t} = \frac{1}{\mu_0}\frac{E_z^n(i+1) - E_z^n(i)}{\Delta x}. \tag{2.14b}$$

The plane wave equations (2.10) are in continuous time and space, and they can be expressed in discrete time and space with

$$E_z^n(i) = E_0\cos(k_x i\Delta x - \omega n\Delta t), \tag{2.15a}$$

$$E_z^{n+1}(i) = E_0\cos(k_x i\Delta x - \omega(n+1)\Delta t), \tag{2.15b}$$

$$E_z^n(i+1) = E_0\cos(k_x(i+1)\Delta x - \omega n\Delta t), \tag{2.15c}$$

$$H_y^{n+\frac{1}{2}}(i) = H_0\cos(k_x(i+0.5)\Delta x - \omega(n+0.5)\Delta t), \tag{2.15d}$$

$$H_y^{n+\frac{1}{2}}(i-1) = H_0\cos(k_x(i-0.5)\Delta x - \omega(n+0.5)\Delta t), \tag{2.15e}$$

$$H_y^{n-\frac{1}{2}}(i) = H_0\cos(k_x(i+0.5)\Delta x - \omega(n-0.5)\Delta t), \tag{2.15f}$$

The terms (2.15a), (2.15b), (2.15d), and (2.15e) can be used in (2.14a) to obtain

$$\begin{aligned}&\frac{E_0}{\Delta t}\left[\cos(k_x i\Delta x - \omega(n+1)\Delta t) - \cos(k_x i\Delta x - \omega n\Delta t)\right] \\ &= \frac{H_0}{\varepsilon_0\Delta x}\left[\cos(k_x(i+0.5)\Delta x - \omega(n+0.5)\Delta t) - \cos(k_x(i-0.5)\Delta x - \omega(n+0.5)\Delta t)\right].\end{aligned} \tag{2.16}$$

Using the trigonometric identity

$$\cos(u-v) - \cos(u+v) = 2\sin(u)\sin(v)$$

on the left-hand side of (22.16) with $u = k_x i\Delta x - \omega(n + 0.5)\Delta t$ and $v = 0.5\Delta t$, and on the right-hand side with $u = k_x i\Delta x - \omega(n + 0.5)\Delta t$ and $v = -0.5\ \Delta x$, one can obtain

$$\frac{E_0}{\Delta t}\sin(0.5\omega\Delta t) = \frac{-H_0}{\varepsilon_0\Delta x}\sin(0.5k_x\Delta x). \tag{2.17}$$

Similarly, using the terms (2.15a), (2.15c), (2.15d), and (2.15f) in (2.14b) leads to

$$\frac{H_0}{\Delta t}\sin(0.5\omega\Delta t) = \frac{-E_0}{\mu_0 \Delta x}\sin(0.5k_x\Delta x). \tag{2.18}$$

Combining (2.17) and (2.18) one can obtain the numerical dispersion relation for the one-dimensional case as

$$\left[\frac{1}{c\Delta t}\sin\left(\frac{\omega\Delta t}{2}\right)\right]^2 = \left[\frac{1}{\Delta x}\sin\left(\frac{k_x\Delta x}{2}\right)\right]^2. \tag{2.19}$$

One should notice that the numerical dispersion relation (2.19) is different from the ideal dispersion relation (2.12). The difference means that there is a deviation from the actual solution of a problem and, hence, an error introduced by the finite-difference approximations to the numerical solution of the problem. However, it is interesting to note that for the one-dimensional case, if one sets $\Delta t = \Delta x/c$ (2.19) reduces to (2.12), which means that there is no dispersion error for propagation in free space. However, this is of little practical use, since the introduction of a material medium will again create dispersion.

So far, derivation of a numerical dispersion relation has been demonstrated for a one-dimensional case. It is possible to obtain a numerical dispersion relation for the two-dimensional case using a similar approach:

$$\left[\frac{1}{c\Delta t}\sin\left(\frac{\omega\Delta t}{2}\right)\right]^2 = \left[\frac{1}{\Delta x}\sin\left(\frac{k_x\Delta x}{2}\right)\right]^2 + \left[\frac{1}{\Delta y}\sin\left(\frac{k_y\Delta y}{2}\right)\right]^2, \tag{2.20}$$

where there is no variation of fields or geometry in the z dimension. For the particular case where

$$\Delta x = \Delta y = \Delta, \quad \Delta t = \frac{\Delta}{c\sqrt{2}}, \quad \text{and} \quad k_x = k_y,$$

the ideal dispersion relation for the two-dimensional case can be recovered as

$$k_x^2 + k_y^2 = \left(\frac{\omega}{c}\right)^2. \tag{2.21}$$

The extension to the three-dimensional case is straightforward but tedious, yielding

$$\left[\frac{1}{c\Delta t}\sin\left(\frac{\omega\Delta t}{2}\right)\right]^2 = \left[\frac{1}{\Delta x}\sin\left(\frac{k_x\Delta x}{2}\right)\right]^2 + \left[\frac{1}{\Delta y}\sin\left(\frac{k_y\Delta y}{2}\right)\right]^2 + \left[\frac{1}{\Delta z}\sin\left(\frac{k_z\Delta z}{2}\right)\right]^2. \tag{2.22}$$

Similar to the one- and two-dimensional cases, it is possible to recover the three-dimensional ideal dispersion relation

$$k_x^2 + k_y^2 + k_z^2 = \left(\frac{\omega}{c}\right)^2 \tag{2.23}$$

for specific choices of Δt, Δx, Δy, Δz, and angle of propagation.

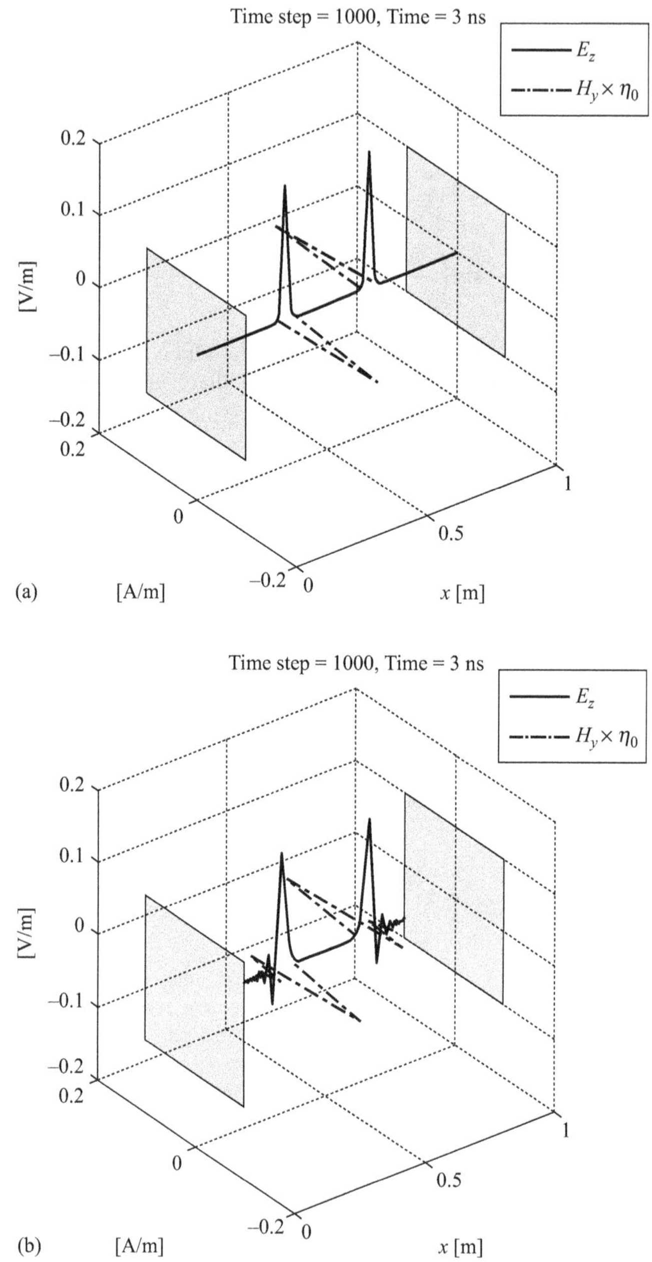

Figure 2.5 (a) Maximum magnitude of E_z in the one-dimensional problem space: simulation with $\Delta x = 1$ mm; (b) maximum magnitude of E_z in the one-dimensional problem space: simulation with $\Delta x = 4$ mm.

Let's rewrite (2.22) in the following form:

$$\left[\frac{\omega}{2c}\frac{\sin(\omega\Delta t/2)}{(\omega\Delta t/2)}\right]^2 = \left[\frac{k_x}{2}\frac{\sin(k_x\Delta x/2)}{(k_x\Delta x/2)}\right]^2 + \left[\frac{k_y}{2}\frac{\sin(k_y\Delta y/2)}{(k_y\Delta y/2)}\right]^2 + \left[\frac{k_z}{2}\frac{\sin(k_z\Delta z/2)}{(k_z\Delta z/2)}\right]^2. \quad (2.24)$$

Since $\lim_{x\to 0}(\sin(x)/x) = 1$ (2.24) reduces to the ideal dispersion relation (2.23) when $\Delta t \to 0$, $\Delta x \to 0$, $\Delta y \to 0$, and $\Delta z \to 0$. This is an expected result since when the sampling periods approach zero, the discrete approximation turns into the continuous case. This also indicates that if the temporal and spatial sampling periods Δt, Δx, Δy, and Δz are taken smaller, then the numerical dispersion error reduces.

So far, we have discussed the numerical dispersion in the context of waves propagating in free space. A more general discussion of numerical stability and dispersion, including the derivation of two- and three-dimensional numerical dispersion relations, other factors affecting the numerical dispersion, and strategies to reduce the associated errors, can be found in [1]. We conclude our discussion with an example demonstrating the numerical dispersion.

Due to numerical dispersion, waves with different frequencies propagate with different phase velocities. Any waveform is a sum of sinusoidal waves with different frequencies, and a waveform does not maintain its shape while propagating since its sinusoidal components do not propagate with the same velocity due to dispersion. Therefore, the existence of numerical dispersion exposes itself as distortion of the waveform. For instance, Figure 2.5 shows the electric and magnetic field in the one-dimensional problem space calculated by the one-dimensional FDTD code presented in Chapter 1. The problem space is 1 m wide, and $\Delta t = 3$ ps. Figure 2.5(a) shows the field distribution at time instant 3 ns as calculated by the program when $\Delta x = 1$ mm and $J_z = 1$ A/m^2. Similarly, Figure 2.5(b) shows the field distribution at time instant 3 ns when $\Delta x = 4$ mm. In this second simulation the magnitude of J_z is taken as 0.25 A/m^2 to maintain the magnitude of the surface current density K_z as 1×10^{-3} A/m; hence, the magnitudes of electric and magnetic fields are almost the same. Figure 2.5(a) shows that the Gaussian pulse is not distorted; even though there is numerical dispersion, it is not significant. However, in Figure 2.5(b) the Gaussian pulse is distorted due to use of a larger cell size, and the error introduced by numerical dispersion is more pronounced.

2.3 Exercises

2.1 Using the one-dimensional MATLAB® FDTD program listed in Appendix A, verify the results presented in Figure 2.5.

2.2 Update the one-dimensional FDTD MATLAB program such that the Gaussian pulse propagates in a medium characterized with electric losses instead of free space. Observe the decay of the amplitude of the propagating pulse and the effect of losses on the dispersion shown in Figure 2.5(b) at time = 3 ns.

2.3 Repeat Exercise 2.2, but with the introduction of magnetic losses in the medium, instead of free space. Record your observations related to the pulse propagation at time = 3 ns.

CHAPTER 3

Building objects in the Yee grid

In Chapter 1 we discussed the derivation of the finite-difference time-domain (FDTD) updating equations based on the staircased grid composed of Yee cells in Cartesian coordinates. We defined material components ε, μ, σ^e, and σ^m associated with the field components such that they represent linear, anisotropic, and nondispersive media. Due to the staircase gridding, the objects in an FDTD problem space can be represented in terms of the size of the cells building the grid. Therefore, the staircased gridding imposes some restrictions on the accurate representation of the objects in the problem space. Some advanced modeling techniques called *local subcell models* are available for defining objects with fine features, and some other FDTD formulations based on nonorthogonal and unstructured grids have been introduced for problems that contain objects that do not conform with the staircased grid [1]. However, the staircased FDTD model can provide sufficiently accurate results for many practical problems. In this chapter, we discuss construction of objects in the staircased FDTD grid through some MATLAB® code examples.

3.1 Definition of objects

Throughout this book, while describing the concepts of the FDTD method we illustrate the construction of an FDTD program as well. After completing this book, the reader should be able to construct an FDTD program that can be used to solve three-dimensional electromagnetics problems of moderate difficulty. In this chapter, we start to provide the blocks of a three-dimensional FDTD program.

We name the main MATLAB program file that runs an FDTD calculation as ***fdtd_solve***. The MATLAB code in Listing 3.1 shows the contents of ***fdtd_solve***, in which the program routines are named by their functions. Usually, it is a good practice to divide a program into functionally separable modules; this will facilitate the readability of the programs and will simplify the debugging process. This file is not in the final form for a complete FDTD program; while we proceed with the chapters we will add more routines with additional functions, will describe the functions of the routines, and will provide partial code sections to illustrate the programming of the respective concepts that have been described in the text. The program structures and codes presented throughout this book are not necessarily samples of the most efficient ways of programming the FDTD method; there

Listing 3.1 fdtd_solve.m

```
% initialize the matlab workspace
clear all; close all; clc;

% define the problem
define_problem_space_parameters;
define_geometry;
define_sources_and_lumped_elements;
define_output_parameters;

% initialize the problem space and parameters
initialize_fdtd_material_grid;
display_problem_space;
display_material_mesh;
if run_simulation
    initialize_fdtd_parameters_and_arrays;
    initialize_sources_and_lumped_elements;
    initialize_updating_coefficients;
    initialize_boundary_conditions;
    initialize_output_parameters;
    initialize_display_parameters;

    % FDTD time marching loop
    run_fdtd_time_marching_loop;

    % display simulation results
    post_process_and_display_results;
end
```

are many ways of translating a concept into a program, and readers can develop program structures or algorithms that they think are more efficient while writing their own code. However, we tried to be as explicit as possible while linking the FDTD concepts with the respective codes.

In the FDTD algorithm, before the time-marching iterations are performed the problem should be set up as the first step as indicated in the concise flow chart in Figure 1.9. Problem setup can be performed in two sets of routines: (1) routines that *define* the problem; and (2) routines that *initialize* the problem space and FDTD parameters. The problem definition process consists of construction of data structures that store the necessary information about the electromagnetic problem to be solved. This information would include the geometries of the objects in the problem space, their material types, electromagnetic properties of these material types, types of sources and waveforms, types of simulation results sought from the FDTD computation, and some other simulation parameters specific to the FDTD algorithm, such as the number of time steps, and types of boundaries of the problem space. The initialization process translates the data structures into an FDTD material grid and constructs and initializes data structures and arrays representing the FDTD updating coefficients, field components, and any other data structures that would be used during and after the FDTD iterations. In other words, it prepares the structural framework for the FDTD computation and postprocessing.

3.1.1 Defining the problem space parameters

Listing 3.1 starts with the initialization of MATLAB workspace. The MATLAB command **clear all** removes all items from the current workspace and frees the memory, **close all** deletes all open figures, and **clc** clears the command window. After the MATLAB workspace initialization, there are four subroutines listed in Listing 3.1 for definition of the problem and other subroutines for initialization of the FDTD procedure. In this chapter, we implement the subroutines ***define_problem_space_parameters***, ***define_geometry***, and ***initialize_fdtd_material_grid***.

The contents of the subroutine ***define_problem_space_parameters*** are given in Listing 3.2. Listing 3.2 starts with the MATLAB command ***disp***, which displays the string in its argument on MATLAB command window. Here the string argument of ***disp*** indicates that the program is

Listing 3.2 define_problem_space_parameters.m

```
disp('defining the problem space parameters');

% maximum number of time steps to run FDTD simulation
number_of_time_steps = 700;

% A factor that determines duration of a time step
% wrt CFL limit
courant_factor = 0.9;

% A factor determining the accuracy limit of FDTD results
number_of_cells_per_wavelength = 20;

% Dimensions of a unit cell in x, y, and z directions (meters)
dx=2.4e-3;
dy=2.0e-3;
dz=2.2e-3;

% ==<boundary conditions>========
% Here we define the boundary conditions parameters
% 'pec' : perfect electric conductor
boundary.type_xp = 'pec';
boundary.air_buffer_number_of_cells_xp = 5;

boundary.type_xn = 'pec';
boundary.air_buffer_number_of_cells_xn = 5;

boundary.type_yp = 'pec';
boundary.air_buffer_number_of_cells_yp = 10;

boundary.type_yn = 'pec';
boundary.air_buffer_number_of_cells_yn = 5;

boundary.type_zp = 'pec';
boundary.air_buffer_number_of_cells_zp = 5;

boundary.type_zn = 'pec';
boundary.air_buffer_number_of_cells_zn = 0;
```

```
39 % ===<material types>============
   % Here we define and initialize the arrays of material types
41 % eps_r   : relative permittivity
   % mu_r    : relative permeability
43 % sigma_e : electric conductivity
   % sigma_m : magnetic conductivity
45
   % air
47 material_types(1).eps_r   = 1;
   material_types(1).mu_r    = 1;
49 material_types(1).sigma_e = 0;
   material_types(1).sigma_m = 0;
51 material_types(1).color   = [1 1 1];

53 % PEC : perfect electric conductor
   material_types(2).eps_r   = 1;
55 material_types(2).mu_r    = 1;
   material_types(2).sigma_e = 1e10;
57 material_types(2).sigma_m = 0;
   material_types(2).color   = [1 0 0];
59
   % PMC : perfect magnetic conductor
61 material_types(3).eps_r   = 1;
   material_types(3).mu_r    = 1;
63 material_types(3).sigma_e = 0;
   material_types(3).sigma_m = 1e10;
65 material_types(3).color   = [0 1 0];

67 % a dielectric
   material_types(4).eps_r   = 2.2;
69 material_types(4).mu_r    = 1;
   material_types(4).sigma_e = 0;
71 material_types(4).sigma_m = 0.2;
   material_types(4).color   = [0 0 1];
73
   % a dielectric
75 material_types(5).eps_r   = 3.2;
   material_types(5).mu_r    = 1.4;
77 material_types(5).sigma_e = 0.5;
   material_types(5).sigma_m = 0.3;
79 material_types(5).color   = [1 1 0];

81 % indices of material types defining air, PEC, and PMC
   material_type_index_air = 1;
83 material_type_index_pec = 2;
   material_type_index_pmc = 3;
```

executing the routine in which the problem space parameters are defined. Displaying this kind of informative statement at various stages of the program indicates the progress of the execution as well as helps for debugging any sources of errors if they exist.

The total number of time steps that the FDTD time-marching iterations will run for is defined in line 4 of Listing 3.2 with the parameter **number_of_time_steps**. Line 8 defines a parameter named **courant_factor**: a factor by which the duration of a time step is determined with respect to the CFL stability limit as described in Chapter 2. In line 11 another parameter is defined with the name **number_of_cells_per_wavelength**. This is a parameter by which the highest frequency in the Fourier spectrum of a source waveform is determined for a certain accuracy level. This parameter is described in more detail in Chapter 5, where the source waveforms are discussed. On lines 14, 15, and 16 the dimensions of a unit cell constructing the uniform FDTD problem grid are defined as parameters **dx**, **dy**, and **dz**, respectively.

One of the important concepts in the FDTD technique is the treatment of boundaries of the problem space. The treatment of perfect electric conductor (PEC) boundaries was demonstrated in Chapter 1 through a one-dimensional FDTD code. However, some types of problems may require other types of boundaries. For instance, an antenna problem requires that the boundaries simulate radiated fields propagating outside the finite problem space to infinity. Unlike a PEC boundary, this type of boundary requires advanced algorithms, which is the subject of a separate chapter. However, for now, we limit our discussion to PEC boundaries. A conventional FDTD problem space composed of uniform Yee cells is a rectangular box having six faces. On lines 21–37 of Listing 3.2 the boundary types of these six faces are defined using a data structure called **boundary**. The structure **boundary** has two types of fields: **type** and **air_buffer_number_of_cells**. These field names have extensions indicating the specific face of the domain under consideration. For instance, the extension **_xn** refers to the face of the problem space for which the normal vector is directed in the negative x direction, where **n** stands for *negative direction*. Similarly, the extension **_zp** refers to the face of the problem space for which the normal vector is directed in the positive z direction, where **p** stands for *positive direction*. The other four extensions refer to four other faces. The field **type** defines the type of the boundary under consideration, and in this case all boundaries are defined as PEC by the keyword ***pec***. Other types of boundaries are identified with other keywords.

Some types of boundary conditions require that the boundaries be placed at a distance from the objects. Most of the time this space is filled with air, and the distance in between is given in terms of the number of cells. Hence, the parameter **air_buffer_number_of_cells** defines the distance of the respective boundary faces of the rectangular problem space from the objects placed in the problem space. If the value of this parameter is zero, it implies that the boundary touches the objects.

Finally, we define different types of materials that the objects in the problem space are made of. Since more than one type of material may be used to construct the objects, an array structure with name **material_types** is used as shown in lines 47–79 of Listing 3.2. An index i in **material_types(i)** refers to the ith material type. Later on, every object will be assigned a material type through the index of the material type. An isotropic and homogeneous material type can be defined by its electromagnetic parameters: relative permittivity ε_r, relative permeability μ_r, electric conductivity σ^e, and magnetic conductivity σ^m; hence, fields **eps_r**, **mu_r**, **sigma_e**, and **sigma_m** are used with parameter **material_types(i)**, respectively, to define the electromagnetic parameters of ith material type. In the given example code, a very high value is assigned to **material_types(2).sigma_e** to construct a material type simulating a PEC, and a very high value is assigned to **material_types(3).sigma_m** to construct a material type simulating a perfect magnetic conductor (PMC). A fifth field **color** in **material_types(i)** is optional and is used to assign a color to each material type, which will help in distinguishing different material types when displaying the objects in the problem space on a MATLAB figure. The field **color** is a vector of three values corresponding to red, green, and blue intensities of the red, green, and blue (RGB) color scheme, respectively, with each color scaled between 0 and 1.

While defining the **material_types** it is a good practice to reserve some indices for some common material types. Some of these material types are air, PEC, and PMC, and these material types are indexed as 1, 2, and 3 in the **material_types** structure array, respectively. Then we can define additional parameters **material_type_index_air**, **material_type_index_pec**, and **material_type_index_pmc**, which store the indices of these materials and can be used to identify them in the program.

3.1.2 Defining the objects in the problem space

In the previous section, we discussed the definition of some problem space parameters including the boundary types and material types. In this section we discuss the implementation of the routine ***define_geometry*** contents, which are given in Listing 3.3. Different types of three-dimensional objects can be placed in a problem geometry. We show example implementations using *prisms* and *spheres* due to their geometrical simplicity, and it is possible to construct more complicated shapes by using a combination of these simple objects. Here we will use the term *brick* to indicate a *prism*. A brick, which has its faces parallel to the Cartesian coordinate axes, can be represented by two corner points: (1) the point with lower x, y, and z coordinates; and (2) the point with upper x, y, and z coordinates as illustrated in Figure 3.1(a). Therefore, a structure array called **bricks** is used in Listing 3.3 together with the fields **min_x**, **min_y**, **min_z**, **max_x**, **max_y**, and **max_z** corresponding to their respective positions in Cartesian coordinates. Each element of the array **bricks** is a brick object indexed by i and is referred to by **bricks(i)**. Another field of the parameter **bricks(i)** is the parameter **material_type**, which refers to the index of the material type of the ith brick. In Listing 3.3 **bricks(1).material_type** is 4, indicating that brick 1 is made of **material_types(4)**, which is defined in Listing 3.2 as a dielectric with $\varepsilon_r = 2.2$ and $\sigma^m = 0.2$. Similarly, **bricks(2).material_type** is 2, indicating that brick 2 is made of **material_types(2)**, which is defined as a PEC.

A sphere can be defined by its center coordinates and radius as illustrated in Figure 3.1(b). Therefore, a structure array **spheres** is used in Listing 3.3 together with the fields **center_x**, **center_y**, **center_z**, and **radius** which correspond to the respective parameters in Cartesian coordinates. Similar to the brick case, the field **material_type** is used together with **spheres(i)** to define the material type of the ith sphere. For example, in Listing 3.3 two spheres are defined with coinciding centers. Sphere 1 is a dielectric of **material_types(5)** with a radius of 20 mm, and sphere 2 is a dielectric of **material_types(1)** with a radius of 15 mm. If these two spheres are created in the problem space in sequence (sphere 1 first and sphere 2 second), their combination is going to form a hollow shell of thickness 5 mm since the material type of the inner sphere **material_types(1)** is air. One should notice that Listing 3.3 starts with two lines defining two arrays, **bricks** and **spheres**, and initializes them by null. Even though we are not going to define an object corresponding to some of these arrays, we still define them and initialize them as empty arrays. Later, when the program executes, these arrays will be accessed by some routines of the program.

Listing 3.3 define_geometry.m

```
disp('defining the problem geometry');

bricks  = [];
spheres = [];

% define a brick with material type 4
bricks(1).min_x = 0;
bricks(1).min_y = 0;
bricks(1).min_z = 0;
bricks(1).max_x = 24e-3;
bricks(1).max_y = 20e-3;
bricks(1).max_z = 11e-3;
bricks(1).material_type = 4;

% define a brick with material type 2
bricks(2).min_x = -20e-3;
bricks(2).min_y = -20e-3;
bricks(2).min_z = -11e-3;
bricks(2).max_x = 0;
bricks(2).max_y = 0;
bricks(2).max_z = 0;
bricks(2).material_type = 2;

% define a sphere with material type 5
spheres(1).radius   = 20e-3;
spheres(1).center_x = 0;
spheres(1).center_y = 0;
spheres(1).center_z = 40e-3;
spheres(1).material_type = 5;

% define a sphere with material type 1
spheres(2).radius   = 15e-3;
spheres(2).center_x = 0;
spheres(2).center_y = 0;
spheres(2).center_z = 40e-3;
spheres(2).material_type = 1;
```

If MATLAB tries to access a parameter that does not exist in its workspace, it is going to fail and stop the execution of the program. To prevent such a runtime error we define these arrays even though they are empty.

So far we have defined an FDTD problem space and defined some objects existing in it. Combining the definitions in Listings 3.2 and 3.3 we obtained a problem space, which is plotted in Figure 3.2. The dimensions of the cells in the grids in Figure 3.2 are the same as the dimensions of the unit cell making the FDTD grid. Therefore, it is evident that the boundaries are away from the objects by the distances defined in Listing 3.2. Furthermore, the objects are placed in the three-dimensional space with the positions and dimensions as defined in Listing 3.3, and their colors are set as their corresponding material type colors.

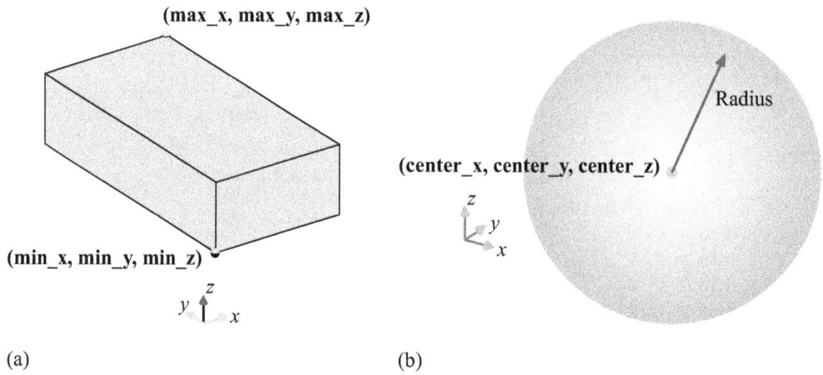

Figure 3.1 Parameters defining a brick and a sphere in Cartesian coordinates: (a) parameters defining a brick and (b) parameters defining a sphere.

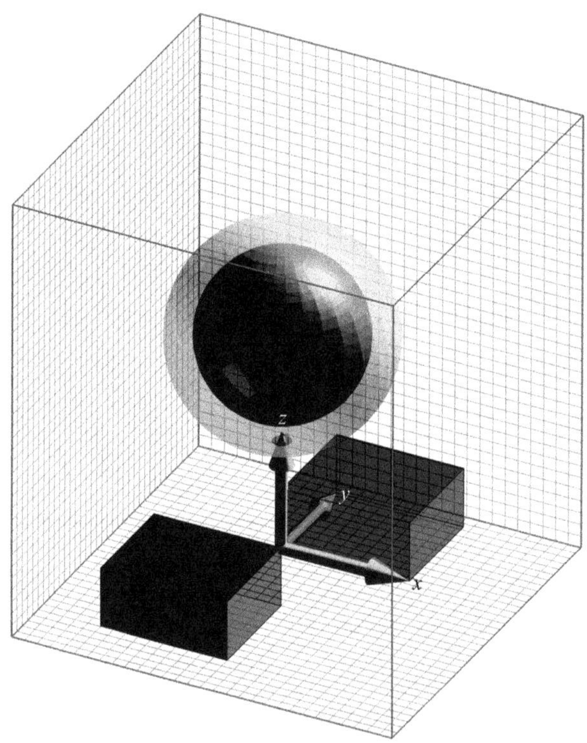

Figure 3.2 An FDTD problem space and the objects defined in it.

3.2 Material approximations

In a three-dimensional FDTD problem space, the field components are defined at discrete locations, and the associated material components are defined at the same locations as illustrated in Figures 1.5 and 1.6. The material components need to be assigned appropriate

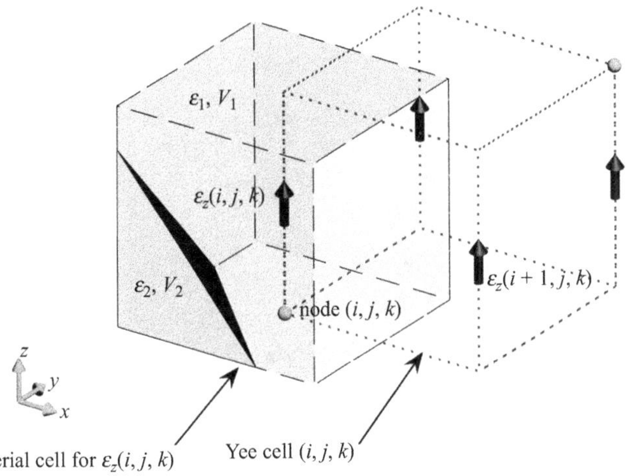

Figure 3.3 A cell around material component $\varepsilon_z(i, j, k)$ partially filled with two media.

values representing the media existing at the respective positions. However, the medium around a material component may not be homogeneous, and some approximation strategies need to be adopted.

One of the strategies is to assume that each material component is residing at the center of a cell. These cells are offset from the Yee cells, and we refer to them as *material cells*. For instance, consider the z-component of permittivity shown in Figure 3.3, which can be imagined as being at the center of a material cell partially filled with two different media denoted by subscripts 1 and 2, with permittivity values ε_1 and ε_2. The simplest approach is to assume that the cell is completely filled with medium 1 since the material component $\varepsilon_z(i, j, k)$ resides in the medium 1. Therefore, $\varepsilon_z(i, j, k)$ can be assigned ε_1; $\varepsilon_z(i, j, k) = \varepsilon_1$.

A better approach would include the effect of ε_2 by employing an averaging scheme. If it is possible to obtain the volume of each medium filling the material cell, a simple weighted volume averaging can be used as employed in [5]. In the example case shown in Figure 3.3, $\varepsilon_z(i, j, k)$ can be calculated as $\varepsilon_z(i, j, k) = (V_1 \times \varepsilon_1 + V_2 \times \varepsilon_2)/(V_1 + V_2)$, where V_1 and V_2 are the volumes of medium 1 and medium 2 in the material cell, respectively.

In many cases, calculation of the volume of each media in a cell may require tedious calculations, whereas it may be much simpler to determine the medium in which a point resides. A crude approximation to the weighted volume averaging scheme is proposed in [6], which may be useful for such cases. In this approach, every cell in the Yee grid is divided into eight subcells, and each material component is located in between eight subcells, as illustrated in Figure 3.4. For every subcell center point the respective medium can be determined and every subcell can be assigned the respective medium material property. Then an effective average of the eight subcells surrounding a material component can be calculated and assigned to the respective material component. For instance, $\varepsilon_z(i, j, k)$ in Figure 3.4, for uniform cell sizes in the x, y, and z directions, can be calculated as

$$\varepsilon_z(i, j, k) = \frac{\varepsilon_1 + \varepsilon_2 + \varepsilon_3 + \varepsilon_4 + \varepsilon_5 + \varepsilon_6 + \varepsilon_7 + \varepsilon_8}{8}. \tag{3.1}$$

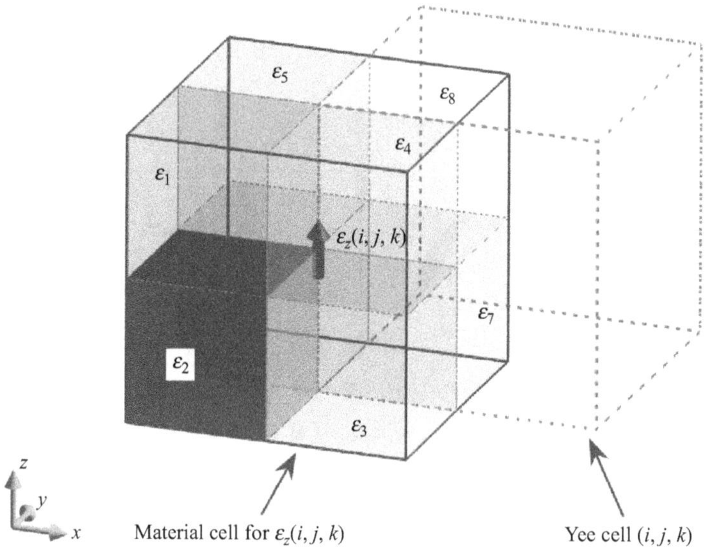

Figure 3.4 A cell around material component $\varepsilon_z(i, j, k)$ divided into eight subcells.

The advantage of this method is that objects may be modeled with eight times the geometrical resolution (two times in each direction) without increasing the size of the staggered grid and memory requirements.

3.3 Subcell averaging schemes for tangential and normal components

The methods discussed thus far do not account for the orientation of the field component associated with the material component under consideration with respect to the boundary interface between two different media. For instance, Figures 3.5 and 3.6 show two such cases where a material cell is partially filled with two different types of media. In the first case the material component is parallel to the boundary of the two media, whereas in the second case the material component is normal to the boundary. Two different averaging schemes can be developed to obtain an equivalent medium type to be assigned to the material component.

For instance, Ampere's law can be expressed in integral form based on the configuration in Figure 3.5 as

$$\oint \vec{H} \cdot d\bar{l} = \frac{d}{dt} \int \vec{D} \cdot d\bar{s} = \frac{d}{dt} D_z \times (A_1 + A_2),$$

where D_z is the electric displacement vector component in the z direction, which is assumed to be uniform in cross-section, and A_1 and A_2 are the cross-section areas of the two media types in the plane normal to D_z. The electric field components in the two media are E_{z1} and E_{z2}. The total electric flux through the cell cross-section can be expressed as

$$D_z \times (A_1 + A_2) = \varepsilon_1 E_{z1} A_1 + \varepsilon_2 E_{z2} A_2 = \varepsilon_{eff} E_z \times (A_1 + A_2),$$

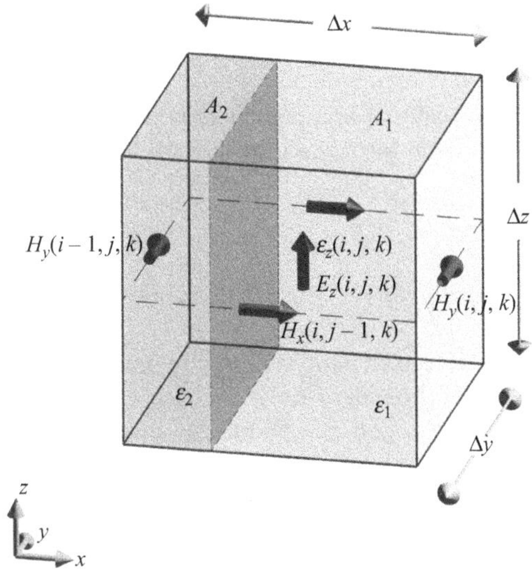

Figure 3.5 Material component $\varepsilon_z(i, j, k)$ parallel to the boundary of two different media partially filling a material cell.

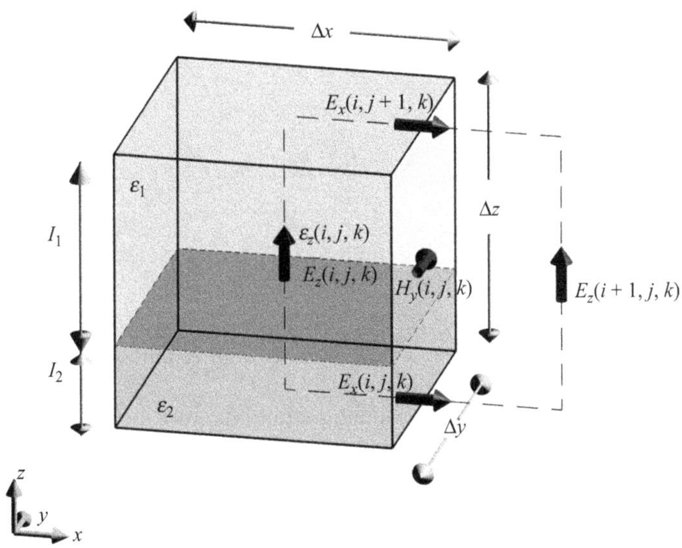

Figure 3.6 Material component $\varepsilon_z(i, j, k)$ normal to the boundary of two different media partially filling a material cell.

where E_z is the electric field component (i, j, k), which is assumed to be uniform in the cross-section of the cell and associated with the permittivity ε_{eff} that is equivalent to $\varepsilon_z(i, j, k)$. On the boundary of the media the tangential components of the electric field are continuous such that $E_{z1} = E_{z2} = E_z$; therefore, one can write

$$\varepsilon_1 A_1 + \varepsilon_2 A_2 = \varepsilon_{eff} \times (A_1 + A_2) \Rightarrow \varepsilon_{eff} = \varepsilon_z(i, j, k) = \frac{\varepsilon_1 A_1 + \varepsilon_2 A_2}{(A_1 + A_2)}. \tag{3.2}$$

Hence, an equivalent permittivity can be calculated for a material component parallel to the interface of two different material types of permittivity ε_1 and ε_2 using (3.2), by which the electric flux conservation is maintained.

For the case where the material component is normal to the media boundary we can employ Faradays's law in integral form based on the configuration in Figure 3.6, which reads

$$\oint \vec{E} \cdot d\vec{l} = -\frac{d}{dt}\int \vec{B} \cdot d\vec{s}$$
$$= -\Delta x E_x(i, j, k) + \Delta x E_x(i, j+1, k) + \Delta z E_z(i, j, k) - \Delta z E_z(i+1, j, k),$$

where E_z is the equivalent electric field vector component in the z direction, which is assumed to be uniform in the boundary cross-section. However, in reality there are two different electric field values, E_{z1} and E_{z2}, due to the different material types. We want to obtain E_z to represent the equivalent effect of E_{z1} and E_{z2} and ε_{eff} to represent the equivalent effect of ε_1 and ε_2. The electric field terms shall satisfy

$$\Delta z E_z(i, j, k) = (l_1 + l_2) \times E_z(i, j, k) = l_1 E_{z1} + l_2 E_{z2}. \tag{3.3}$$

On the boundary of the media the normal components of the electric flux $\vec{D}$ are continuous such that $D_{z1} = D_{z2} = D_z(i, j, k)$; therefore, one can write

$$D_z(i, j, k) = \varepsilon_1 E_{z1} = \varepsilon_2 E_{z2} = \varepsilon_{eff} E_z(i, j, k), \tag{3.4}$$

which can be expressed in the form

$$E_{z1} = \frac{D_z(i, j, k)}{\varepsilon_1}, \quad E_{z2} = \frac{D_z(i, j, k)}{\varepsilon_2}, \quad E_z(i,j,k) = \frac{D_z(i, j, k)}{\varepsilon_{eff}}. \tag{3.5}$$

Using (3.5) in (3.3) and eliminating the terms $D_z(i, j, k)$ yields

$$\frac{l_1 + l_2}{\varepsilon_{eff}} = \frac{l_1}{\varepsilon_1} + \frac{l_2}{\varepsilon_2} \Rightarrow \varepsilon_{eff} = \varepsilon_z(i, j, k) = \frac{\varepsilon_1 \varepsilon_2 \times (l_1 + l_2)}{(l_1 \varepsilon_2 + l_2 \varepsilon_1)}. \tag{3.6}$$

Hence, an expression for the equivalent permittivity has been obtained in (3.6), by which the magnetic flux conservation is maintained.

The averaging schemes given by (3.2) and (3.6) are specific cases of the more general approaches introduced in [7], [8], and [9].

It can be shown that these schemes can be employed for other material parameters as well. For instance, the equivalent permeability μ can be written as

$$\mu_{eff} = \frac{\mu_1 A_1 + \mu_2 A_2}{(A_1 + A_2)}, \quad \mu \text{ parallel to media boundary}, \tag{3.7a}$$

$$\mu_{eff} = \frac{\mu_1 \mu_2 \times (l_1 + l_2)}{(l_1 \mu_2 + l_2 \mu_1)}, \quad \mu \text{ normal to media boundary}. \tag{3.7b}$$

These schemes serve as good approximations for low values of σ^e and σ^m as well, where they can be written as

$$\sigma^e_{eff} = \frac{\sigma^e_1 A_1 + \sigma^e_2 A_2}{(A_1 + A_2)}, \quad \sigma^e \text{ parallel to media boundary,} \tag{3.8a}$$

$$\sigma^e_{eff} = \frac{\sigma^e_1 \sigma^e_2 \times (l_1 + l_2)}{(l_1 \sigma^e_2 + l_2 \sigma^e_1)}, \quad \sigma^e \text{ normal to media boundary,} \tag{3.8b}$$

and

$$\sigma^m_{eff} = \frac{\sigma^m_1 A_1 + \sigma^m_2 A_2}{(A_1 + A_2)}, \quad \sigma^m \text{ parallel to media boundary,} \tag{3.9a}$$

$$\sigma^m_{eff} = \frac{\sigma^m_1 \sigma^m_2 \times (l_1 + l_2)}{(l_1 \sigma^m_2 + l_2 \sigma^m_1)}, \quad \sigma^m \text{ normal to media boundary,} \tag{3.9b}$$

3.4 Defining objects snapped to the Yee grid

The averaging schemes provided in the previous section are usually discussed in the context of subcell modeling techniques in which effective material parameter values are introduced while the staircased gridding scheme is maintained. Although there are some other more advanced subcell modeling techniques that have been introduced for modeling fine features in the FDTD method, in this section we discuss modeling of three-dimensional objects assuming that the objects are conforming to the staircased Yee grid. Therefore, each Yee cell is filled with a single material type. Although this assumption gives a crude model compared with the subcell modeling techniques discussed before, we consider this case since it does not require complicated programming effort and since it is sufficient to obtain reliable results for many problems. Furthermore, we still employ the aforementioned subcell averaging schemes for obtaining effective values for the material components lying on the boundaries of cells filled with different material types.

For instance, consider Figure 3.7, where the material component $\varepsilon_z(i, j, k)$ is located in between four Yee cells, each of which is filled with a material type. Every Yee cell is indexed by a respective node. Since every cell is filled with a material type, three-dimensional

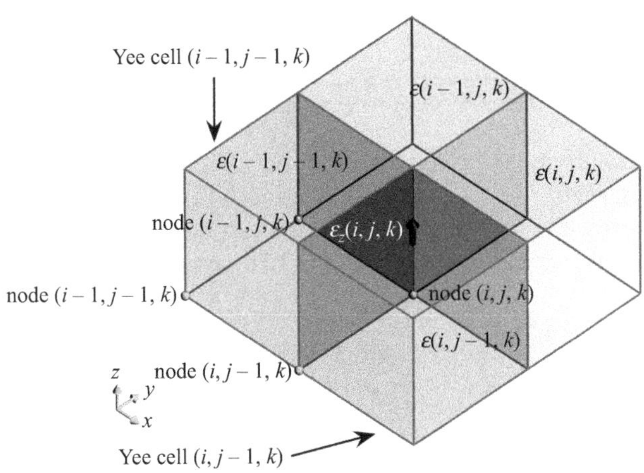

Figure 3.7 Material component $\varepsilon_z(i, j, k)$ located between four Yee cells filled with four different material types.

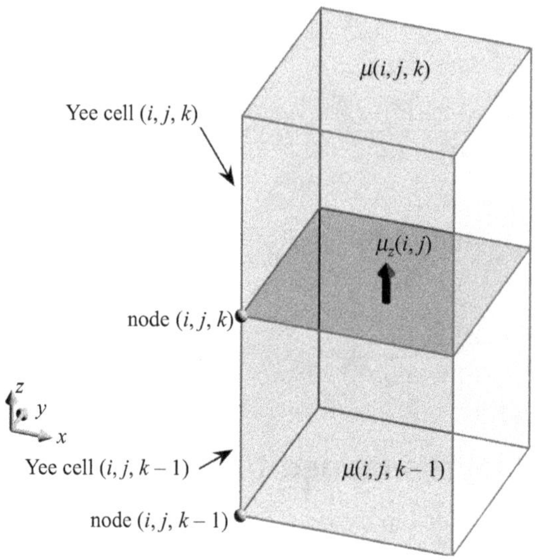

Figure 3.8 Material component $\mu_z(i, j, k)$ located between two Yee cells filled with two different material types.

material arrays can be utilized, each element of which corresponds to the material type filling the respective cell. For instance, $\varepsilon(i-1, j-1, k)$ holds the permittivity value of the material filling the cell $(i-1, j-1, k)$. Consider the permittivity parameters that are indexed as $\varepsilon(i-1, j-1, k)$, $\varepsilon(i-1, j, k)$, $\varepsilon(i, j-1, k)$, and $\varepsilon(i, j, k)$ in Figure 3.7. The material component $\varepsilon_z(i, j, k)$ is tangential to the four surrounding media types; therefore, we can employ (3.2) to get the equivalent permittivity $\varepsilon_z(i, j, k)$ as

$$\varepsilon_z(i, j, k) = \frac{\varepsilon(i, j, k) + \varepsilon(i-1, j, k) + \varepsilon(i, j-1, k) + \varepsilon(i-1, j-1, k)}{4}. \tag{3.10}$$

Since the electric conductivity material components are defined at the same positions as the permittivity components, the same averaging scheme can be employed to get the equivalent electric conductivity $\sigma_z^e(i, j, k)$ as

$$\sigma_z^e(i, j, k) = \frac{\sigma^e(i, j, k) + \sigma^e(i-1, j, k) + \sigma^e(i, j-1, k) + \sigma^e(i-1, j-1, k)}{4}. \tag{3.11}$$

Unlike the permittivity and electric conductivity components, the material components associated with magnetic field components, permeability and magnetic conductivity, are located between two cells and are oriented normal to the cell boundaries as illustrated in Figure 3.8. In this case we can use the relation (3.7b) to get $\mu_z(i, j, k)$ as

$$\mu_z(i, j, k) = \frac{2 \times \mu(i, j, k) \times \mu(i, j, k-1)}{\mu(i, j, k) + \mu(i, j, k-1)}. \tag{3.12}$$

Similarly, we can write $\sigma_z^m(i, j, k)$ as

$$\sigma_z^m(i, j, k) = \frac{2 \times \sigma^m(i, j, k) \times \sigma^m(i, j, k-1)}{\sigma^m(i, j, k) + \sigma^m(i, j, k-1)}. \tag{3.13}$$

3.4.1 Defining zero-thickness PEC objects

In many electromagnetics problems, especially in planar circuits, some very thin objects exist, most of the time in the form of microstrip or stripline structures. In these cases, when constructing an FDTD simulation, it may not be feasible to choose the unit cell size as small as the thickness of the strip, since this will lead to a very large number of cells and, hence to an excessive amount of computer memory that prevents practical simulations with current computer resources. Then if the cell size is larger than the strip thickness, subcell modeling techniques, some of which are discussed in the previous sections, can be employed for accurate modeling of the thin strips in the FDTD method. Usually the subcell modeling of thin strips is studied in the context of *thin-sheet* modeling techniques. Here we discuss a simple modeling technique that can be used to obtain results with a reasonable accuracy for the majority of problems including thin strips where the thin strips are PEC. Here we assume that the thin-strip thickness is zero, which is a reasonable approximation since most of the time the strip thicknesses are much smaller than the cell sizes. Furthermore, we assume that the strips are conforming to the faces of the Yee cells in the problem space.

Consider the PEC plate with zero thickness as shown in Figure 3.9, which conforms to the face of the Yee cell with index (i, j, k). This plate is surrounded by four electric conductivity material components, namely, $\sigma_x^e(i, j, k)$, $\sigma_x^e(i, j+1, k)$, $\sigma_y^e(i, j, k)$, and $\sigma_y^e(i+1, j, k)$. We can simply assign the conductivity of PEC, σ_{pec}^e, to these material components, such that

$$\sigma_x^e(i, j, k) = \sigma_{pec}^e, \qquad \sigma_x^e(i, j+1, k) = \sigma_{pec}^e,$$
$$\sigma_y^e(i, j, k) = \sigma_{pec}^e, \quad \text{and} \quad \sigma_y^e(i+1, j, k) = \sigma_{pec}^e.$$

Therefore, any electric conductivity components surrounding and coinciding with PEC plates can be assigned the conductivity of PEC to model zero-thickness PEC plates in the FDTD method.

Similarly, if a PEC object with thickness smaller than a cell size in two dimensions, such as a thin conductor wire, needs to be modeled in the FDTD method, it is sufficient to assign the conductivity of PEC to the electric conductivity components coinciding with the object.

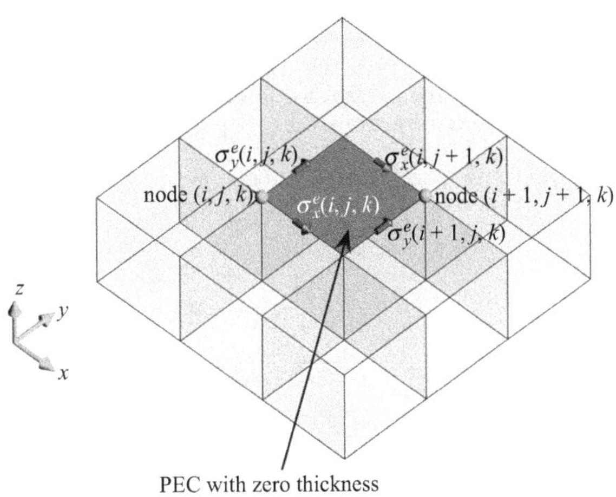

Figure 3.9 A PEC plate with zero thickness surrounded by four σ^e material components.

This approach gives a reasonable approximation to a thin wire in the FDTD method. However, since the field variation around a thin wire is large, the sampling of the fields at the discrete points around the wire may not be accurate. Therefore, it is more appropriate to use more advanced *thin-wire modeling* techniques, which take into account the wire thickness, for better accuracy. A thin-wire modeling technique is discussed in Chapter 10.

3.5 Creation of the material grid

At the beginning of this chapter we discussed how we can define the objects using data structures in MATLAB. We use these structures to initialize the FDTD material grid; we create and initialize three-dimensional arrays representing the components of the material parameters ε, μ, σ^e, and σ^m. Then we assign appropriate values to these arrays using the averaging schemes discussed in Section 3.4. Later on these material arrays will be used to calculate the FDTD updating coefficients.

The routine ***initialize_fdtd_material_grid***, which is used to initialize the material arrays, is given in Listing 3.4. The material array initialization process is composed of multiple stages; therefore, ***initialize_fdtd_material_grid*** is divided into functional subroutines. We describe the implementation of these subroutines and demonstrate how the concepts described in Section 3.4 are coded.

The first subroutine in Listing 3.4 is **calculate_domain_size**, the implementation of which is given in Listing 3.5. In this subroutine the dimensions of the FDTD problem space are determined, including the number of cells (*Nx*, *Ny*, and *Nz*) making the problem space. Once the number of cells in the x, y, and z dimensions are determined, they are assigned to

Listing 3.4 initialize_fdtd_material_grid.m

```
disp('initializing FDTD material grid');

% calculate problem space size based on the object
% locations and boundary conditions
calculate_domain_size;

% Array to store material type indices for every cell
% in the problem space. By default the space is filled
% with air by initializing the array by ones
material_3d_space = ones(nx, ny, nz);

% Create the 3D objects in the problem space by
% assigning indices of material types in the cells
% to material_3d_space

% create spheres
create_spheres;

% create bricks
create_bricks;

% Material component arrays for a problem space
% composed of (nx, ny, nz) cells
eps_r_x    = ones (nx   , nyp1  , nzp1);
eps_r_y    = ones (nxp1 , ny    , nzp1);
eps_r_z    = ones (nxp1 , nyp1  , nz);
```

```
mu_r_x      = ones (nxp1, ny   , nz);
mu_r_y      = ones (nx  , nyp1 , nz);
mu_r_z      = ones (nx  , ny   , nzp1);
sigma_e_x   = zeros(nx  , nyp1 , nzp1);
sigma_e_y   = zeros(nxp1, ny   , nzp1);
sigma_e_z   = zeros(nxp1, nyp1 , nz);
sigma_m_x   = zeros(nxp1, ny   , nz);
sigma_m_y   = zeros(nx  , nyp1 , nz);
sigma_m_z   = zeros(nx  , ny   , nzp1);

% calculate material component values by averaging
calculate_material_component_values;

% create zero thickness PEC plates
create_PEC_plates;
```

Listing 3.5 calculate_domain_size.m

```
disp('calculating the number of cells in the problem space');

number_of_spheres = size(spheres,2);
number_of_bricks  = size(bricks,2);

% find the minimum and maximum coordinates of a
% box encapsulating the objects
number_of_objects = 1;
for i=1:number_of_spheres
    min_x(number_of_objects) = spheres(i).center_x - spheres(i).radius;
    min_y(number_of_objects) = spheres(i).center_y - spheres(i).radius;
    min_z(number_of_objects) = spheres(i).center_z - spheres(i).radius;
    max_x(number_of_objects) = spheres(i).center_x + spheres(i).radius;
    max_y(number_of_objects) = spheres(i).center_y + spheres(i).radius;
    max_z(number_of_objects) = spheres(i).center_z + spheres(i).radius;
    number_of_objects = number_of_objects + 1;
end
for i=1:number_of_bricks
    min_x(number_of_objects) = bricks(i).min_x;
    min_y(number_of_objects) = bricks(i).min_y;
    min_z(number_of_objects) = bricks(i).min_z;
    max_x(number_of_objects) = bricks(i).max_x;
    max_y(number_of_objects) = bricks(i).max_y;
    max_z(number_of_objects) = bricks(i).max_z;
    number_of_objects = number_of_objects + 1;
end

fdtd_domain.min_x = min(min_x);
fdtd_domain.min_y = min(min_y);
fdtd_domain.min_z = min(min_z);
fdtd_domain.max_x = max(max_x);
fdtd_domain.max_y = max(max_y);
fdtd_domain.max_z = max(max_z);

```

```
% Determine the problem space boundaries including air buffers
fdtd_domain.min_x = fdtd_domain.min_x ...
    - dx * boundary.air_buffer_number_of_cells_xn;
fdtd_domain.min_y = fdtd_domain.min_y ...
    - dy * boundary.air_buffer_number_of_cells_yn;
fdtd_domain.min_z = fdtd_domain.min_z ...
    - dz * boundary.air_buffer_number_of_cells_zn;
fdtd_domain.max_x = fdtd_domain.max_x ...
    + dx * boundary.air_buffer_number_of_cells_xp;
fdtd_domain.max_y = fdtd_domain.max_y ...
    + dy * boundary.air_buffer_number_of_cells_yp;
fdtd_domain.max_z = fdtd_domain.max_z ...
    + dz * boundary.air_buffer_number_of_cells_zp;

% Determining the problem space size
fdtd_domain.size_x = fdtd_domain.max_x - fdtd_domain.min_x;
fdtd_domain.size_y = fdtd_domain.max_y - fdtd_domain.min_y;
fdtd_domain.size_z = fdtd_domain.max_z - fdtd_domain.min_z;

% number of cells in x, y, and z directions
nx = round(fdtd_domain.size_x/dx);
ny = round(fdtd_domain.size_y/dy);
nz = round(fdtd_domain.size_z/dz);

% adjust domain size by snapping to cells
fdtd_domain.size_x = nx * dx;
fdtd_domain.size_y = ny * dy;
fdtd_domain.size_z = nz * dz;

fdtd_domain.max_x = fdtd_domain.min_x + fdtd_domain.size_x;
fdtd_domain.max_y = fdtd_domain.min_y + fdtd_domain.size_y;
fdtd_domain.max_z = fdtd_domain.min_z + fdtd_domain.size_z;

% some frequently used auxiliary parameters
nxp1 = nx+1;    nyp1 = ny+1;    nzp1 = nz+1;
nxm1 = nx-1;    nxm2 = nx-2;    nym1 = ny-1;
nym2 = ny-2;    nzm1 = nz-1;    nzm2 = nz-2;

% create arrays storing the center coordinates of the cells
fdtd_domain.cell_center_coordinates_x = zeros(nx,ny,nz);
fdtd_domain.cell_center_coordinates_y = zeros(nx,ny,nz);
fdtd_domain.cell_center_coordinates_z = zeros(nx,ny,nz);
for ind = 1:nx
    fdtd_domain.cell_center_coordinates_x(ind,:,:) = ...
        (ind - 0.5) * dx + fdtd_domain.min_x;
end
for ind = 1:ny
    fdtd_domain.cell_center_coordinates_y(:,ind,:) = ...
       (ind - 0.5) * dy + fdtd_domain.min_y;
end
for ind = 1:nz
    fdtd_domain.cell_center_coordinates_z(:,:,ind) = ...
        (ind - 0.5) * dz + fdtd_domain.min_z;
end
```

parameters **nx**, **ny**, and **nz**. Based on the discussion in Section 3.4, we assume that every Yee cell is filled with a material type. Then in line 10 of Listing 3.4 a three-dimensional array **material_3d_space** with size ($Nx \times Ny \times Nz$) is initialized with ones. In this array each element will store the index of the material type filling the corresponding cell in the problem space. Since **material_3d_space** is initialized by the MATLAB function **ones**, all of its elements have the value 1. Referring to Listing 3.2 one can see that **material_types(1)** is reserved for *air*; therefore, the problem space is initially filled with air.

After this point we find the cells overlapping with the objects defined in ***define_geometry*** and assign the indices of the material types of the objects to the corresponding elements of the array **material_3d_space**. This procedure is performed in the subroutines ***create_spheres***, which is shown in Listing 3.6, and ***create_bricks***, which is shown in Listing 3.7. With the given implementations, first the spheres are created in the problem space, and then the bricks are created. Furthermore, the objects will be created in the sequence as they are defined in

Listing 3.6 create_spheres.m

```
disp('creating spheres');

cx = fdtd_domain.cell_center_coordinates_x;
cy = fdtd_domain.cell_center_coordinates_y;
cz = fdtd_domain.cell_center_coordinates_z;

for ind=1:number_of_spheres
% distance of the centers of the cells from the center of the sphere
    distance = sqrt((spheres(ind).center_x - cx).^2 ...
             + (spheres(ind).center_y - cy).^2 ...
             + (spheres(ind).center_z - cz).^2);
    I = find(distance<=spheres(ind).radius);
    material_3d_space(I) = spheres(ind).material_type;
end
clear cx cy cz;
```

Listing 3.7 create_bricks.m

```
disp('creating bricks');

for ind = 1:number_of_bricks
    % convert brick end coordinates to node indices
    blx = round((bricks(ind).min_x - fdtd_domain.min_x)/dx) + 1;
    bly = round((bricks(ind).min_y - fdtd_domain.min_y)/dy) + 1;
    blz = round((bricks(ind).min_z - fdtd_domain.min_z)/dz) + 1;

    bux = round((bricks(ind).max_x - fdtd_domain.min_x)/dx)+1;
    buy = round((bricks(ind).max_y - fdtd_domain.min_y)/dy)+1;
    buz = round((bricks(ind).max_z - fdtd_domain.min_z)/dz)+1;

    % assign material type of the brick to the cells
    material_3d_space (blx:bux-1, bly:buy-1, blz:buz-1) ...
        = bricks(ind).material_type;
end
```

define_geometry. This means that if more than one object overlaps in the same cell, the object defined last will overwrite the objects defined before. So one should define the objects accordingly. These two subroutines (**create_spheres** and **create_bricks**) can be replaced by a more advanced algorithm in which the objects can be assigned priorities in case of overlapping, and changing the priorities would be sufficient to have the objects created in the desired sequence in the problem space.

For example, the array **material_3d_space** is filled with the material parameter indices of the objects that were defined in Listing 3.3 and displayed in Figure 3.2. The cells represented by **material_3d_space** are plotted in Figure 3.10, which clearly shows the staircased approximations of the objects as generated by the subroutine ***display_problem_space***, which is called in ***fdtd_solve***.

Then in ***initialize_fdtd_material_grid*** we create three-dimensional arrays representing the material parameter components ε_x, ε_y, ε_z, μ_x, μ_y, μ_z, σ_x^e, σ_y^e, σ_z^e, σ_x^m, σ_y^m, and σ_z^m. Here the parameters **eps_r_x, eps_r_y, eps_r_z, mu_r_x, mu_r_y,** and **mu_r_z** are relative permittivity

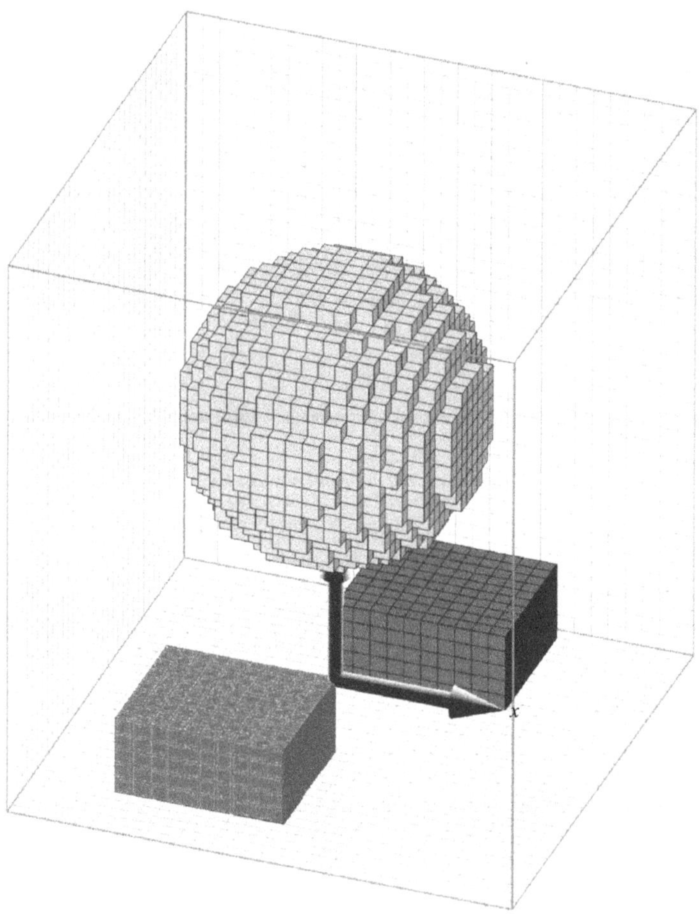

Figure 3.10 An FDTD problem space and the objects approximated by snapping to cells.

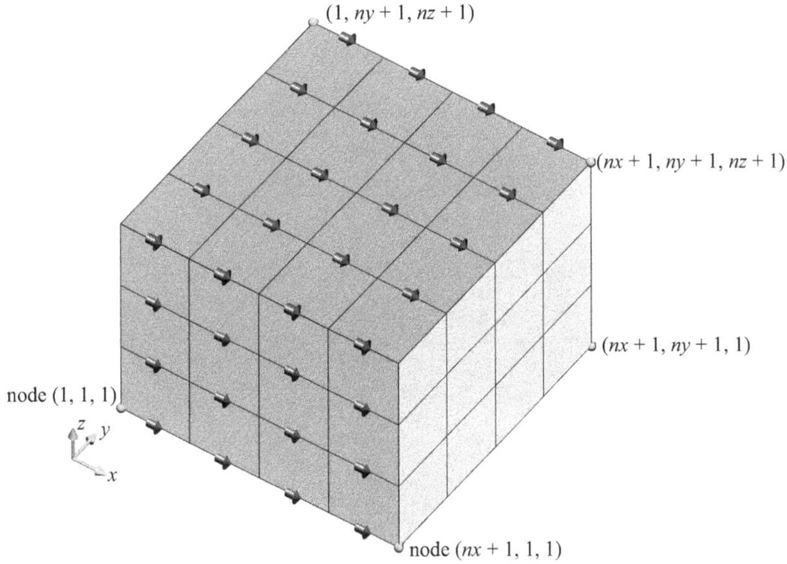

Figure 3.11 Positions of the E_x components on the Yee grid for a problem space composed of $(Nx \times Ny \times Nz)$ cells.

and relative permeability arrays, and they are initialized with 1. The electric and magnetic conductivity arrays are initialized with 0. One can notice that the sizes of these arrays are not the same as they correspond to the size of the associated field component. This is due to the specific positioning scheme of the field components on the Yee cell as shown in Figure 1.5. For instance, the distribution of the x components of the electric field, E_x, in a problem space composed of $(Nx \times Ny \times Nz)$ cells is illustrated in Figure 3.11. It can be observed that there exist $(Nx \times Ny + 1 \times Nz + 1)$ E_x components in the problem space. Therefore, in the program the three-dimensional arrays representing E_x and the material components associated with E_x shall be constructed with size $(Nx \times Ny + 1 \times Nz + 1)$. Hence, in Listing 3.4 the parameters **eps_r_x** and **sigma_e_x** are constructed with size (*nx*, *nyp* 1, *nzp* 1), where *nyp* 1 is $Ny + 1$, and *nzp* 1 is $Nz + 1$. Similarly, it can be observed in Figure 3.12 that there exist $(Nx + 1 \times Ny \times Nz)$ H_x components in the problem space. Therefore, in Listing 3.4 the parameters **mu_r_x** and **sigma_m_x** are constructed with size (*nxp* 1, *ny*, *nz*), where *nxp* 1 is $Nx + 1$.

Since we have created and filled the array **material_3d_space** as the staircased representation of the FDTD problem space and have initialized material component arrays, we can proceed with assigning appropriate parameter values to the material component array elements based on the averaging schemes discussed in Section 3.4 to form the material grid. This is done in the subroutine ***calculate_material_component_values***; partial implementation of this subroutine is shown in Listing 3.8. In Listing 3.8 calculations for the x components of the material component arrays are illustrated; implementation of other components is left to the reader. Here one can notice that **eps_r_x**, **sigma_e_x**, **mu_r_x**, and **sigma_m_x** are calculated using the forms of equations (3.10), (3.11), (3.12), and (3.13), respectively.

The last step in constructing the material grid is to create the zero-thickness PEC plates in the material grid. The subroutine ***create_PEC_plates*** shown in Listing 3.9 is implemented for

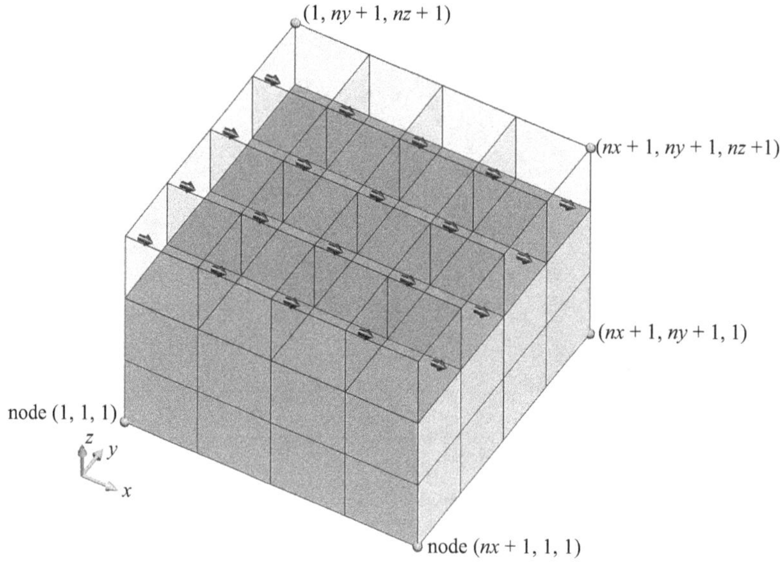

Figure 3.12 Positions of the H_x components on the Yee grid for a problem space composed of $(Nx \times Ny \times Nz)$ cells.

Listing 3.8 calculate_material_component_values.m

```
disp('filling material components arrays');

% creating temporary 1D arrays for storing
% parameter values of material types
for ind = 1:size(material_types,2)
    t_eps_r(ind)   = material_types(ind).eps_r;
    t_mu_r(ind)    = material_types(ind).mu_r;
    t_sigma_e(ind) = material_types(ind).sigma_e;
    t_sigma_m(ind) = material_types(ind).sigma_m;
end

% assign negligibly small values to t_mu_r and t_sigma_m where they are
% zero in order to prevent division by zero error
t_mu_r(find(t_mu_r==0)) = 1e-20;
t_sigma_m(find(t_sigma_m==0)) = 1e-20;

disp('Calculating eps_r_x');
% eps_r_x(i,j,k) is average of four cells
% (i,j,k),(i,j-1,k), (i,j,k-1), (i,j-1,k-1)
eps_r_x(1:nx,2:ny,2:nz) = ...
    0.25 * (t_eps_r(material_3d_space(1:nx,2:ny,2:nz)) ...
    + t_eps_r(material_3d_space(1:nx,1:ny-1,2:nz)) ...
    + t_eps_r(material_3d_space(1:nx,2:ny,1:nz-1)) ...
    + t_eps_r(material_3d_space(1:nx,1:ny-1,1:nz-1)));
disp('Calculating sigma_e_x');
% sigma_e_x(i,j,k) is average of four cells
% (i,j,k),(i,j-1,k), (i,j,k-1), (i,j-1,k-1)
```

```
sigma_e_x(1:nx,2:ny,2:nz) = ...
    0.25 * (t_sigma_e(material_3d_space(1:nx,2:ny,2:nz)) ...
    + t_sigma_e(material_3d_space(1:nx,1:ny-1,2:nz)) ...
    + t_sigma_e(material_3d_space(1:nx,2:ny,1:nz-1)) ...
    + t_sigma_e(material_3d_space(1:nx,1:ny-1,1:nz-1)));
disp('Calculating mu_r_x');
% mu_r_x(i,j,k) is average of two cells (i,j,k),(i-1,j,k)
mu_r_x(2:nx,1:ny,1:nz) = ...
    2 * (t_mu_r(material_3d_space(2:nx,1:ny,1:nz)) ...
    .* t_mu_r(material_3d_space(1:nx-1,1:ny,1:nz))) ...
    ./(t_mu_r(material_3d_space(2:nx,1:ny,1:nz)) ...
    + t_mu_r(material_3d_space(1:nx-1,1:ny,1:nz)));
disp('Calculating sigma_m_x');
% sigma_m_x(i,j,k) is average of two cells (i,j,k),(i-1,j,k)
sigma_m_x(2:nx,1:ny,1:nz) = ...
    2 * (t_sigma_m(material_3d_space(2:nx,1:ny,1:nz)) ...
    .* t_sigma_m(material_3d_space(1:nx-1,1:ny,1:nz))) ...
    ./(t_sigma_m(material_3d_space(2:nx,1:ny,1:nz)) ...
    + t_sigma_m(material_3d_space(1:nx-1,1:ny,1:nz)));
```

Listing 3.9 create_PEC_plates.m

```
disp('creating PEC plates on the material grid');

for ind = 1:number_of_bricks

    mtype = bricks(ind).material_type;
    sigma_pec = material_types(mtype).sigma_e;

    % convert coordinates to node indices on the FDTD grid
    blx = round((bricks(ind).min_x - fdtd_domain.min_x)/dx)+1;
    bly = round((bricks(ind).min_y - fdtd_domain.min_y)/dy)+1;
    blz = round((bricks(ind).min_z - fdtd_domain.min_z)/dz)+1;

    bux = round((bricks(ind).max_x - fdtd_domain.min_x)/dx)+1;
    buy = round((bricks(ind).max_y - fdtd_domain.min_y)/dy)+1;
    buz = round((bricks(ind).max_z - fdtd_domain.min_z)/dz)+1;

    % find the zero thickness bricks
    if (blx == bux)
        sigma_e_y(blx, bly:buy-1,blz:buz) = sigma_pec;
        sigma_e_z(blx, bly:buy,blz:buz-1) = sigma_pec;
    end
    if (bly == buy)
        sigma_e_z(blx:bux, bly, blz:buz-1) = sigma_pec;
        sigma_e_x(blx:bux-1, bly, blz:buz) = sigma_pec;
    end
    if (blz == buz)
        sigma_e_x(blx:bux-1,bly:buy, blz) = sigma_pec;
        sigma_e_y(blx:bux,bly:buy-1,blz)  = sigma_pec;
    end
end
```

this purpose. The code checks all of the defined bricks for zero thickness and assigns their material's conductivity values to the electric conductivity components overlapping with them.

Using the code sections described in this section, the material grid for the material cell distribution in Figure 3.10 is obtained and plotted on three plane cuts for relative permittivity distribution in Figure 3.13(a) and for relative permeability distribution in Figure 3.13(b). The effects of averaging can be clearly observed on the object boundaries.

3.6 Improved eight-subcell averaging

In Section 3.2 we described how any cell can be divided into eight subcells and how every material parameter component is located in between eight surrounding cells as illustrated in Figure 3.4. A simple averaging scheme as (3.1) was used to find an equivalent value for a given material component. We can combine the averaging schemes discussed in Section 3.3 with the eight-subcell modeling scheme to obtain an improved approach for modeling the material components in the FDTD grid.

Consider the eight subcells in Figure 3.4, each of which is filled with a material type. We can assume an equivalent permittivity ε_p for the four subcells with permittivities ε_1, ε_4, ε_5, and ε_8, which can be calculated as $\varepsilon_p = 0.25 \times (\varepsilon_1 + \varepsilon_4 + \varepsilon_5 + \varepsilon_8)$. Similarly, we can assume an equivalent permittivity ε_n for the four subcells with permittivities ε_2, ε_3, ε_6, and ε_7 that can be calculated as $\varepsilon_n = 0.25 \times (\varepsilon_2 + \varepsilon_3 + \varepsilon_6 + \varepsilon_7)$. The permittivity component $\varepsilon_z(i, j, k)$ is located between the half-cells filled with materials of permittivities ε_p and ε_n, and it is oriented normal to the boundaries of these half-cells. Therefore, using (3.6) we can write

$$\varepsilon_z(i, j, k) = \frac{2 \times \varepsilon_p \times \varepsilon_n}{\varepsilon_p + \varepsilon_n}. \tag{3.14}$$

The same approach can be used for parameter components permeability and for electric and magnetic conductivities as well. Furthermore, it should be noted that the improved eight-subcell averaging scheme presented here is based on the application of the more general approach for eight subcells [8].

3.7 Exercises

3.1 The file **fdtd_solve** is the main program used to run FDTD simulations. Open this subroutine and examine its contents. It calls several subroutines that are not implemented yet. In the code disable the lines 7, 8, and 14–27 by "commenting" these lines. Simply insert "%" sign in front of these lines. Now the code is ready to run with its current functionality.

Open the subroutine ***define_problem_space-parameters***, and define the problem space parameters. Set the cell size as 1 mm on a side, boundaries as "***pec***," and air gap between the objects and boundaries as five cells on all sides. Define a material type with index 4 having the relative permittivity as 4, relative permeability as 2, electric conductivity as 0.2, and magnetic conductivity as 0.1. Define another material type with index 5 having relative permittivity as 6, relative permeability as 3, electric conductivity as 0.4, and magnetic conductivity as 0.2. Open the subroutine ***define_geometry***, and define the problem geometry. Define a brick with size

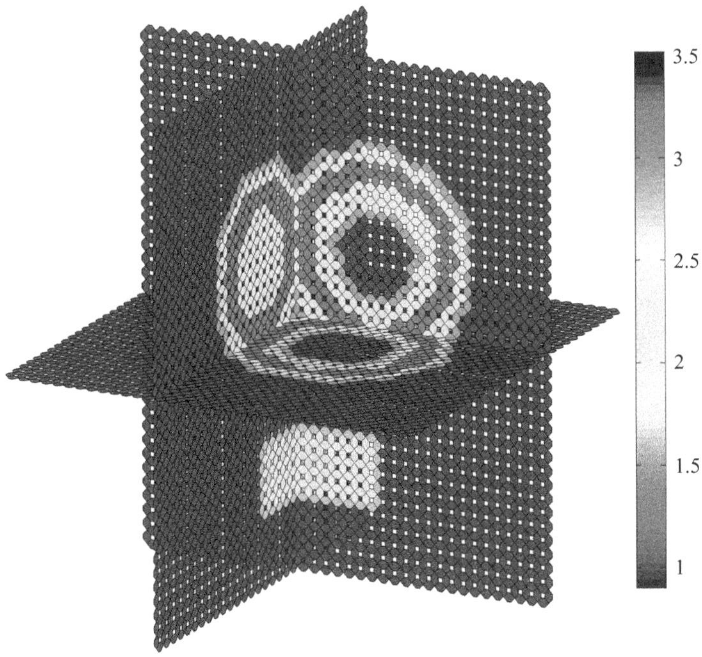

(a)

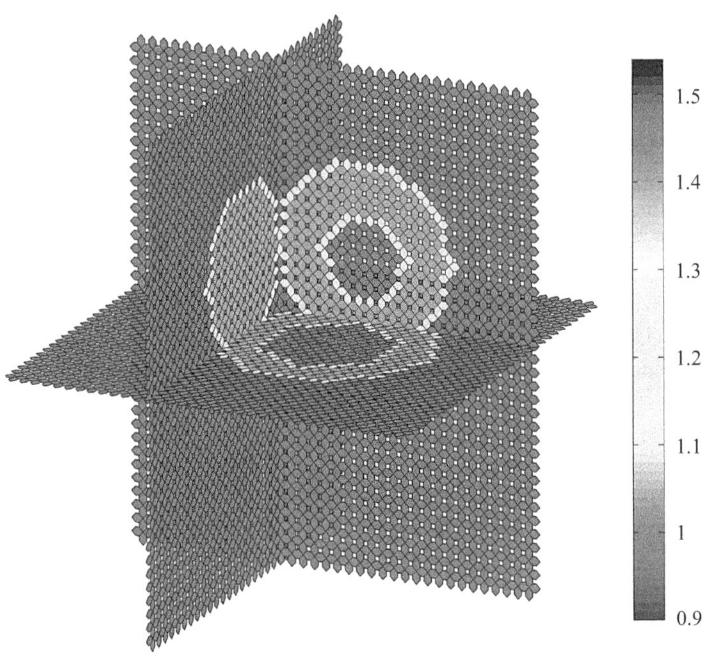

(b)

Figure 3.13 Material grid on three plane cuts: (a) relative permittivity components and (b) relative permeability components.

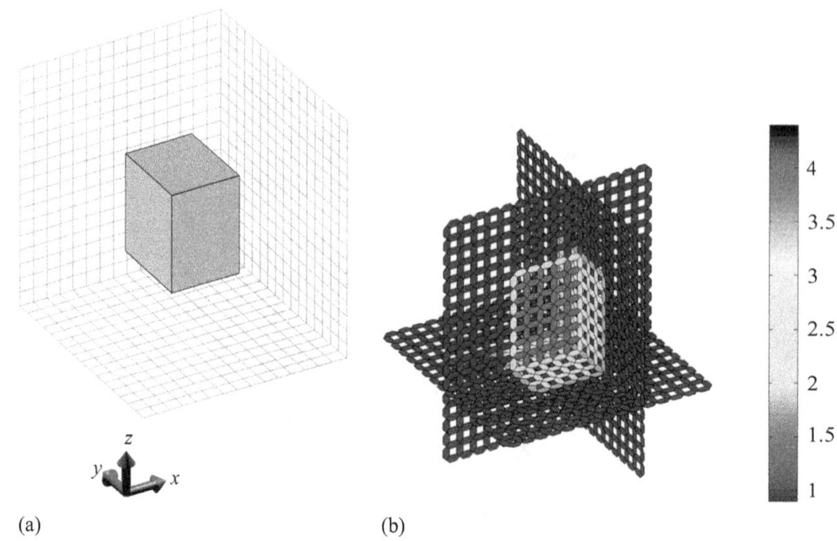

Figure 3.14 The FDTD problem space of Exercise 3.1: (a) geometry of the problem and (b) relative permittivity distribution.

5 mm × 6 mm × 7 mm, and assign material type index 4 as its material type. Figure 3.14(a) shows the geometry of the problem in consideration.

In the directory including your code files there are some additional files prepared as plotting routines that you can use together with your code. These routines are called in ***fdtd_solve*** after the definition routines. Insert the command "show_problem_space =true;" in line 9 of fdtd_solve temporarily to enable three-dimensional geometry display. Run the program ***fdtd_solve***. The program will open a figure including a three-dimensional view of the geometry you have defined and another program named as "Display Material Mesh" that helps you examine the material mesh created for the defined geometry. For instance, Figure 3.14(b) shows the relative permittivity distribution of the FDTD problem space in three plane cuts generated by the program "Display Material Mesh." Examine the geometry and the material mesh, and verify that the material parameter values are averaged on the boundaries as explained in Chapter 3.

3.2 Consider the problem you constructed in Exercise 3.1. Now define another brick with size 3 mm × 4 mm × 5 mm, assign material type index 5 as its material type, and place it at the center of the first brick. Run ***fdtd_solve***, and then examine the geometry and the material mesh and verify that the material parameter values are averaged on the boundaries as explained in Chapter 3.

Now change the definition sequence of these two bricks in ***define_geometry***; define the second brick first with index 1 and the first brick as second with its index 2. Run ***fdtd_solve***, and examine the geometry and the material mesh. Verify that the smaller brick has disappeared in the material mesh. With the given implementation of the program, the sequence of defining the objects determines the way the FDTD material mesh is created.

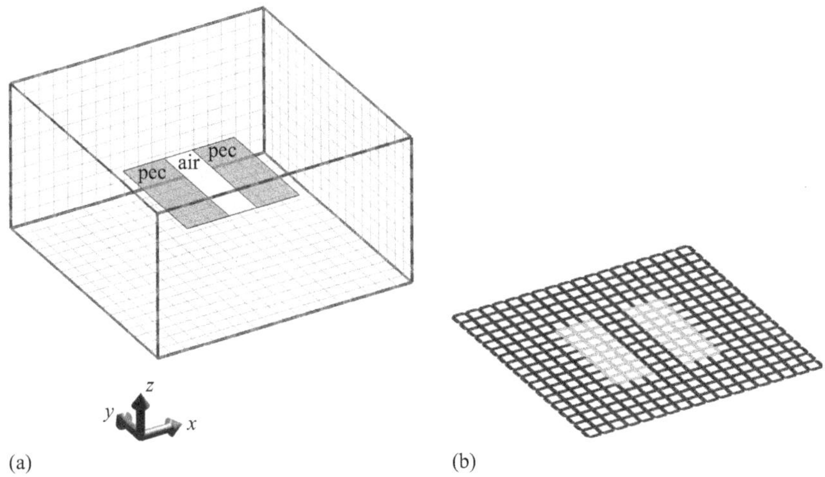

Figure 3.15 The FDTD problem space of Exercise 3.3: (a) geometry of the problem and (b) electric conductivity distribution.

3.3 Construct a problem space with the cell size as 1 mm on a side, boundaries as "***pec***," and air gap between the objects and boundaries as five cells on all sides. Define a brick as a PEC plate with zero thickness placed between the coordinates given in millimeters as (0, 0, 0) and (8, 8, 0). Define another brick as a plate with zero thickness, material type air, and placed between the coordinates (3, 0, 0) and (5, 8, 0) as illustrated in Figure 3.15(a), which shows the geometry of the problem in consideration. Run ***fdtd_solve***, and then examine the geometry and the material mesh. Examine the electric conductivity distribution, and verify that on the boundary between the plates the material components are air as shown in Figure 3.15(b).

3.4 Construct a problem space with the cell size as 1 mm on a side, boundaries as "***pec***," and air gap between the objects and boundaries as five cells on all sides. Define a brick as a PEC plate with zero thickness placed between the coordinates given in millimeters as (0, 0, 0) and (3, 8, 0). Define another brick as PEC plate and placed between the coordinates (5, 0, 0) and (8, 8, 0). Run ***fdtd_solve***, and then examine the geometry and the material mesh. Examine the electric conductivity distribution, and verify that on the boundaries of the plates facing each other the material components are PEC.

Although the geometry constructed in this example is physically the same as the one in Exercise 3.3, the material meshes are different since the ways the geometries are defined are different.

CHAPTER 4

Active and passive lumped elements

4.1 FDTD updating equations for lumped elements

Many practical electromagnetics applications require inclusion of lumped circuit elements. The lumped elements may be active sources in the form of voltage and current sources or passive in the form of resistors, inductors, and capacitors. Nonlinear circuit elements such as diodes and transistors are also required to be integrated in the numerical simulation of antennas and microwave devices. The electric current flowing through these circuit elements can be represented by the impressed current density term, $\vec{J}_i$, in Maxwell's curl equation

$$\nabla \times \vec{H} = \varepsilon \frac{\partial \vec{E}}{\partial t} + \sigma^e \vec{E} + \vec{J}_i. \tag{4.1}$$

Impressed currents are used to represent sources or known quantities. In this sense, they are the sources that cause the electric and magnetic fields in the computational domain [10]. A lumped element component placed between two nodes is characterized by the relationship between the voltage V across and the current I flowing between these two nodes. This relationship can be incorporated into Maxwell's curl equation (4.1) by expressing $\vec{E}$ in terms of V using

$$\vec{E} = -\nabla V \tag{4.2}$$

and by expressing $\vec{J}$ in terms of I using the relation

$$I = \int_S \vec{J} \cdot d\bar{s}, \tag{4.3}$$

where S is the cross-sectional area of a unit cell normal to the flow of the current I. These equations can be implemented in discrete time and space and can be incorporated into (4.1), which are expressed in terms of finite differences. Then, the finite-difference time-domain (FDTD) updating equations can be obtained that simulate the respective lumped element characteristics.

In this chapter we discuss the construction of the FDTD updating equations for lumped element components, and we demonstrate MATLAB® implementation of *definition*, *initialization*, and *simulation* of these elements.

4.1.1 Voltage source

In any electromagnetics simulation, one of the necessary components is the inclusion of sources. Types of sources vary depending on the problem type; scattering problems require incident fields from far zone sources, such as plane waves, to excite objects in a problem space, whereas many other problems require near zone sources, which are usually in the forms of voltage or current sources. In this section we derive updating equations that simulate the effects of a voltage source present in the problem space.

Consider the scalar curl equation (1.4c) repeated here for convenience:

$$\frac{\partial E_z}{\partial t} = \frac{1}{\varepsilon_z}\left(\frac{\partial H_y}{\partial x} - \frac{\partial H_x}{\partial y} - \sigma_z^e E_z - J_{iz}\right). \tag{4.4}$$

This equation constitutes the relation between the current density J_{iz} flowing in the z direction and the electric and magnetic field vector components. Application of the central difference formula to the time and space derivatives based on the field positioning scheme illustrated in Figure 4.1 yields

$$\begin{aligned}\frac{E_z^{n+1}(i,j,k) - E_z^n(i,j,k)}{\Delta t} = &\frac{1}{\varepsilon_z(i,j,k)} \frac{H_y^{n+\frac{1}{2}}(i,j,k) - H_y^{n+\frac{1}{2}}(i-1,j,k)}{\Delta x} \\ &- \frac{1}{\varepsilon_z(i,j,k)} \frac{H_x^{n+\frac{1}{2}}(i,j,k) - H_x^{n+\frac{1}{2}}(i,j-1,k)}{\Delta y} \\ &- \frac{\sigma_z^e(i,j,k)}{2\varepsilon_z(i,j,k)}\left(E_z^{n+1}(i,j,k) + E_z^n(i,j,k)\right) \\ &- \frac{1}{\varepsilon_z(i,j,k)} J_{iz}^{n+\frac{1}{2}}(i,j,k)\end{aligned} \tag{4.5}$$

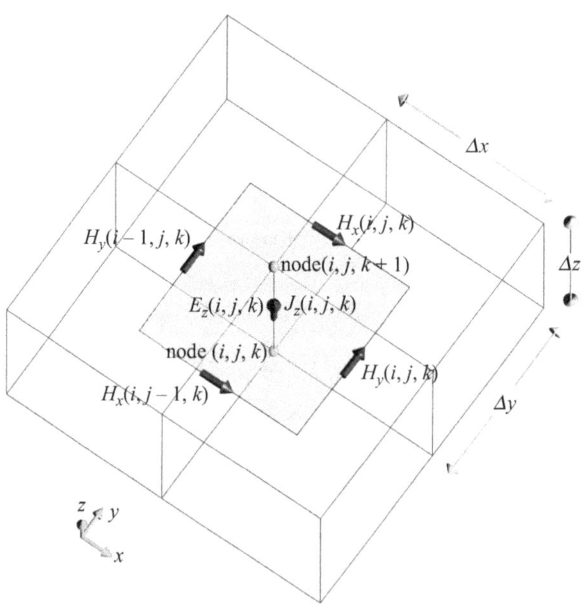

Figure 4.1 Field components around $E_z(i, j, k)$.

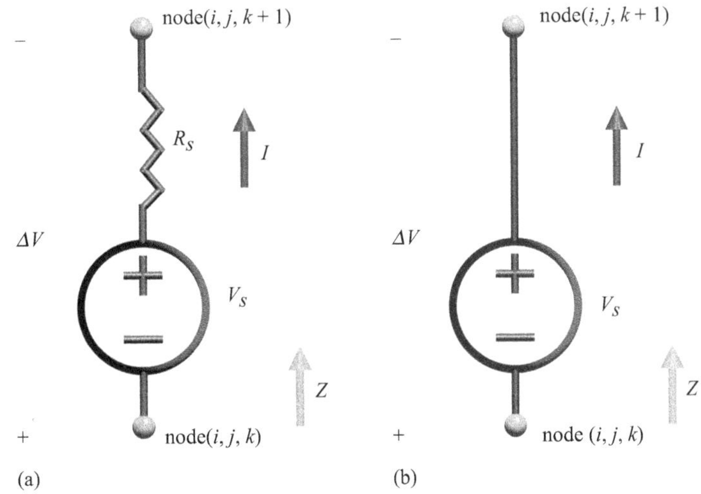

Figure 4.2 Voltage sources placed between nodes (i, j, k) and $(i, j, k+1)$: (a) voltage source with internal resistance (soft source) and (b) voltage source without internal resistance (hard source).

We want to place a voltage source with V_s volts magnitude and R_s ohms internal resistance between nodes (i, j, k) and $(i, j, k+1)$ as illustrated in Figure 4.2(a), where V_s is a time-varying function with a predetermined waveform. The voltage–current relation for this circuit can be written as

$$I = \frac{\Delta V + V_s}{R_s}, \tag{4.6}$$

where ΔV is the potential difference between nodes (i, j, k) and $(i, j, k+1)$. The term ΔV can be expressed in terms of E_z using (4.2), which translates to

$$\Delta V = \Delta z \times E_z^{n+\frac{1}{2}}(i, j, k) = \Delta z \times \frac{E_z^{n+1}(i, j, k) + E_z^{n}(i, j, k)}{2}, \tag{4.7}$$

in discrete form at time instant $(n + 0.5)\Delta t$. Current I is the current flowing through the surface enclosed by the magnetic field components in Figure 4.1, which can be expressed in terms of J_{iz} using (4.3) as

$$I^{n+\frac{1}{2}} = \Delta x \Delta y\, J_{iz}^{n+\frac{1}{2}}(i, j, k). \tag{4.8}$$

One should notice that ΔV in (4.7) is evaluated at time instant $(n + 0.5)\Delta t$, which corresponds to the required time instant of I and J dictated by (4.5). Inserting (4.7) and (4.8) in (4.6) one can obtain

$$J_{iz}^{n+\frac{1}{2}}(i, j, k) = \frac{\Delta z}{2\Delta x \Delta y R_s} \times \left(E_z^{n+1}(i, j, k) + E_z^{n}(i, j, k)\right) + \frac{1}{\Delta x \Delta y R_s} \times V_s^{n+\frac{1}{2}}. \tag{4.9}$$

Equation (4.9) includes the voltage–current relation for a voltage source tying V_s and R_s to electric field components in the discrete space and time. One can use (4.9) in (4.5) and rearrange the terms such that the future value of the electric field component E_z^{n+1} can be calculated using other terms, which yields our standard form updating equations such that

$$\begin{aligned} E_z^{n+1}(i,j,k) &= C_{eze}(i,j,k) \times E_z^n(i,j,k) \\ &+ C_{ezhy}(i,j,k) \times \left(H_y^{n+\frac{1}{2}}(i,j,k) - H_y^{n+\frac{1}{2}}(i-1,j,k) \right) \\ &+ C_{ezhx}(i,j,k) \times \left(H_x^{n+\frac{1}{2}}(i,j,k) - H_x^{n+\frac{1}{2}}(i,j-1,k) \right) \\ &+ C_{ezs}(i,j,k) \times V_s^{n+\frac{1}{2}}(i,j,k), \end{aligned} \tag{4.10}$$

where

$$C_{eze}(i,j,k) = \frac{2\varepsilon_z(i,j,k) - \Delta t \sigma_z^e(i,j,k) - \dfrac{\Delta t \Delta z}{R_s \Delta x \Delta y}}{2\varepsilon_z(i,j,k) + \Delta t \sigma_z^e(i,j,k) + \dfrac{\Delta t \Delta z}{R_s \Delta x \Delta y}},$$

$$C_{ezhy}(i,j,k) = \frac{2\Delta t}{\left(2\varepsilon_z(i,j,k) + \Delta t \sigma_z^e(i,j,k) + \dfrac{\Delta t \Delta z}{R_s \Delta x \Delta y} \right) \Delta x},$$

$$C_{ezhx}(i,j,k) = -\frac{2\Delta t}{\left(2\varepsilon_z(i,j,k) + \Delta t \sigma_z^e(i,j,k) + \dfrac{\Delta t \Delta z}{R_s \Delta x \Delta y} \right) \Delta y},$$

$$C_{ezs}(i,j,k) = -\frac{2\Delta t}{\left(2\varepsilon_z(i,j,k) + \Delta t \sigma_z^e(i,j,k) + \dfrac{\Delta t \Delta z}{R_s \Delta x \Delta y} \right) (R_s \Delta x \Delta y)},$$

Equation (4.10) is the FDTD updating equation modeling a voltage source placed between the nodes (i, j, k) and $(i, j, k + 1)$, which is oriented in the z direction. The FDTD updating equations for voltage sources oriented in other directions can easily be obtained following the same steps as just illustrated.

The FDTD updating equation (4.10) is given for a voltage source with a polarity in the positive z direction as indicated in Figure 4.2(a). To model a voltage source with the opposite polarity, one needs only to use the inverse of the voltage magnitude waveform by changing V_s to $-V_s$.

4.1.2 Hard voltage source

In some applications it may be desired that a voltage difference is enforced between two points in the problem space. This can be achieved by using a voltage source without any internal resistances, which is called a *hard voltage source*. For instance, Figure 4.2(b) illustrates such a scenario, where a voltage source with V_s volts magnitude is placed between the nodes (i, j, k) and $(i, j, k + 1)$. The FDTD updating equation for this voltage source can simply be obtained by letting $R_s \to 0$ in (4.10), which can be written as

$$E_z^{n+1}(i,j,k) = -E_z^n(i,j,k) - \frac{2}{\Delta z} V_s^{n+\frac{1}{2}}(i,j,k). \tag{4.11}$$

To conform with the general form of the FDTD updating equations, (4.11) can be expressed as

$$\begin{aligned} E_z^{n+1}(i,j,k) = {} & C_{eze}(i,j,k) \times E_z^n(i,j,k) \\ & + C_{ezhy}(i,j,k) \times \left(H_y^{n+\frac{1}{2}}(i,j,k) - H_y^{n+\frac{1}{2}}(i-1,j,k)\right) \\ & + C_{ezhx}(i,j,k) \times \left(H_x^{n+\frac{1}{2}}(i,j,k) - H_x^{n+\frac{1}{2}}(i,j-1,k)\right) \\ & + C_{ezs}(i,j,k) \times V_s^{n+\frac{1}{2}}(i,j,k), \end{aligned} \tag{4.12}$$

where

$$C_{eze}(i,j,k) = -1, \quad C_{ezhy}(i,j,k) = 0, \quad C_{ezhx}(i,j,k) = 0, \quad C_{ezs}(i,j,k) = -\frac{2}{\Delta z}.$$

4.1.3 Current source

Figure 4.3(a) illustrates a current source with I_s amperes magnitude and R_s internal resistance, where I_s is a function of time with a time-varying waveform. The voltage–current relation for the current source can be written as

$$I = I_s + \frac{\Delta V}{R_s}. \tag{4.13}$$

Expressions of ΔV in (4.7) and $I^{n+\frac{1}{2}}$ in (4.8) can be used in (4.13), which yields in discrete time and space

$$J_{iz}^{n+\frac{1}{2}}(i,j,k) = \frac{\Delta z}{2\Delta x \Delta y R_s} \times \left(E_z^{n+1}(i,j,k) + E_z^n(i,j,k)\right) + \frac{1}{\Delta x \Delta y} \times I_s^{n+\frac{1}{2}}. \tag{4.14}$$

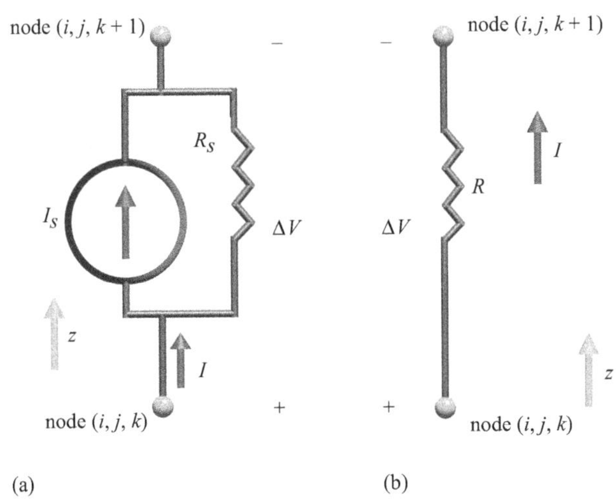

Figure 4.3 Lumped elements placed between nodes (i, j, k) and $(i, j, k+1)$: (a) current source with internal resistance and (b) resistor.

One can use (4.14) in (4.5) and rearrange the terms such that E_z^{n+1} can be calculated using other terms, which yields

$$\begin{aligned}E_z^{n+1}(i,j,k) = C_{eze}(i,j,k) \times E_z^{n}(i,j,k) \\ + C_{ezhy}(i,j,k) \times \left(H_y^{n+\frac{1}{2}}(i,j,k) - H_y^{n+\frac{1}{2}}(i-1,j,k)\right) \\ + C_{ezhx}(i,j,k) \times \left(H_x^{n+\frac{1}{2}}(i,j,k) - H_x^{n+\frac{1}{2}}(i,j-1,k)\right) \\ + C_{ezs}(i,j,k) \times I_s^{n+\frac{1}{2}}(i,j,k),\end{aligned} \tag{4.15}$$

where

$$C_{eze}(i,j,k) = \frac{2\varepsilon_z(i,j,k) - \Delta t\sigma_z^e(i,j,k) - \dfrac{\Delta t\Delta z}{R_s\Delta x\Delta y}}{2\varepsilon_z(i,j,k) + \Delta t\sigma_z^e(i,j,k) + \dfrac{\Delta t\Delta z}{R_s\Delta x\Delta y}},$$

$$C_{ezhy}(i,j,k) = \frac{2\Delta t}{\left(2\varepsilon_z(i,j,k) + \Delta t\sigma_z^e(i,j,k) + \dfrac{\Delta t\Delta z}{R_s\Delta x\Delta y}\right)\Delta x},$$

$$C_{ezhx}(i,j,k) = -\frac{2\Delta t}{\left(2\varepsilon_z(i,j,k) + \Delta t\sigma_z^e(i,j,k) + \dfrac{\Delta t\Delta z}{R_s\Delta x\Delta y}\right)\Delta y},$$

$$C_{ezs}(i,j,k) = -\frac{2\Delta t}{\left(2\varepsilon_z(i,j,k) + \Delta t\sigma_z^e(i,j,k) + \dfrac{\Delta t\Delta z}{R_s\Delta x\Delta y}\right)\Delta x\Delta y},$$

Equation (4.15) is the updating equation modeling a current source. One should notice that (4.15) is the same as (4.10) except for the source term; that is, replacing the $V_s^{n+\frac{1}{2}}(i,j,k)$ term in (4.10) by $R_s \times I_s^{n+\frac{1}{2}}(i,j,k)$ yields (4.15).

The FDTD updating equation (4.15) is given for a current source with a current flowing in the positive z direction as indicated in Figure 4.3(a). To model a current source with the opposite flow direction, one needs only to use the inverse of the current magnitude waveform by changing I_s to $-I_s$.

4.1.4 Resistor

Having derived the updating equations for a voltage source or a current source with internal resistance, it is straightforward to derive updating equations for a resistor. For instance, eliminating the current source in Figure 4.3(a) yields the resistor in Figure 4.3(b). Hence, setting the source term $I_s^{n+\frac{1}{2}}(i,j,k)$ in (4.15) to zero results the updating equation for a resistor as

$$\begin{aligned}E_z^{n+1}(i,j,k) = C_{eze}(i,j,k) \times E_z^{n}(i,j,k) \\ + C_{ezhy}(i,j,k) \times \left(H_y^{n+\frac{1}{2}}(i,j,k) - H_y^{n+\frac{1}{2}}(i-1,j,k)\right) \\ + C_{ezhx}(i,j,k) \times \left(H_x^{n+\frac{1}{2}}(i,j,k) - H_x^{n+\frac{1}{2}}(i,j-1,k)\right),\end{aligned} \tag{4.16}$$

where

$$C_{eze}(i,j,k) = \frac{2\varepsilon_z(i,j,k) - \Delta t\sigma_z^e(i,j,k) - \dfrac{\Delta t\Delta z}{R\Delta x\Delta y}}{2\varepsilon_z(i,j,k) + \Delta t\sigma_z^e(i,j,k) + \dfrac{\Delta t\Delta z}{R\Delta x\Delta y}},$$

$$C_{ezhy}(i,j,k) = \frac{2\Delta t}{\left(2\varepsilon_z(i,j,k) + \Delta t\sigma_z^e(i,j,k) + \dfrac{\Delta t\Delta z}{R\Delta x\Delta y}\right)\Delta x},$$

$$C_{ezhx}(i,j,k) = -\frac{2\Delta t}{\left(2\varepsilon_z(i,j,k) + \Delta t\sigma_z^e(i,j,k) + \dfrac{\Delta t\Delta z}{R\Delta x\Delta y}\right)\Delta y}.$$

4.1.5 Capacitor

Figure 4.4(a) illustrates a capacitor with C farads capacitance. The voltage–current relation for the capacitor can be written as

$$I = C\frac{d\Delta V}{dt}. \tag{4.17}$$

This relation can be expressed in discrete time and space as

$$I^{n+\frac{1}{2}} = C\frac{\Delta V^{n+1} - \Delta V^n}{\Delta t}. \tag{4.18}$$

Using ΔV in (4.7) and $I^{n+\frac{1}{2}}$ in (4.8) together with (4.18), one can obtain

$$J_{iz}^{n+\frac{1}{2}}(i,j,k) = \frac{C\Delta z}{\Delta t\Delta x\Delta y} \times \left(E_z^{n+1}(i,j,k) - E_z^n(i,j,k)\right). \tag{4.19}$$

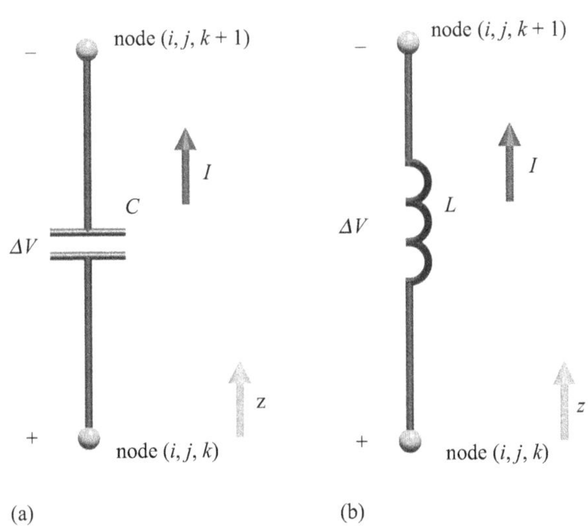

Figure 4.4 Lumped elements placed between nodes (i, j, k) and $(i, j, k+1)$: (a) capacitor and (b) inductor.

One can use (4.19) in (4.5) and rearrange the terms such that E_z^{n+1} can be calculated using other terms, which yields the updating equation for a capacitor as

$$\begin{aligned}E_z^{n+1}(i,j,k) &= C_{eze}(i,j,k)\times E_z^n(i,j,k)\\&+C_{ezhy}(i,j,k)\times\left(H_y^{n+\frac{1}{2}}(i,j,k)-H_y^{n+\frac{1}{2}}(i-1,j,k)\right)\\&+C_{ezhx}(i,j,k)\times\left(H_x^{n+\frac{1}{2}}(i,j,k)-H_x^{n+\frac{1}{2}}(i,j-1,k)\right),\end{aligned}\tag{4.20}$$

where

$$C_{eze}(i,j,k)=\frac{2\varepsilon_z(i,j,k)-\Delta t\sigma_z^e(i,j,k)+\dfrac{2C\Delta z}{\Delta x\Delta y}}{2\varepsilon_z(i,j,k)+\Delta t\sigma_z^e(i,j,k)+\dfrac{2C\Delta z}{\Delta x\Delta y}},$$

$$C_{ezhy}(i,j,k)=\frac{2\Delta t}{\left(2\varepsilon_z(i,j,k)+\Delta t\sigma_z^e(i,j,k)+\dfrac{2C\Delta z}{\Delta x\Delta y}\right)\Delta x},$$

$$C_{ezhx}(i,j,k)=-\frac{2\Delta t}{\left(2\varepsilon_z(i,j,k)+\Delta t\sigma_z^e(i,j,k)+\dfrac{2C\Delta z}{\Delta x\Delta y}\right)\Delta y}.$$

4.1.6 Inductor

Figure 4.4(b) illustrates an inductor with L henrys inductance. The inductor is characterized by the voltage–current relation

$$V=L\frac{dI}{dt}.\tag{4.21}$$

This relation can be expressed in discrete time and space using the central difference formula at time instant $n\Delta t$ as

$$\Delta V^n=\frac{L}{\Delta t}\left(I^{n+\frac{1}{2}}-I^{n-\frac{1}{2}}\right).\tag{4.22}$$

Using the discrete domain relation (4.7) and (4.8) one can obtain

$$J_{iz}^{n+\frac{1}{2}}(i,j,k)=J_{iz}^{n-\frac{1}{2}}(i,j,k)+\frac{\Delta t\Delta z}{L\Delta x\Delta y}E_z^n(i,j,k).\tag{4.23}$$

Since we were able to obtain an expression for $J_{iz}^{n+\frac{1}{2}}(i,j,k)$ in terms of the previous value of the electric field component $E_z^n(i,j,k)$ and the previous value of the impressed current density component $J_{iz}^{n-\frac{1}{2}}(i,j,k)$, we can use the general form of the FDTD updating equation (1.28) without any modification, which is rewritten here for convenience as

$$\begin{aligned} E_z^{n+1}(i,j,k) &= C_{eze}(i,j,k) \times E_z^n(i,j,k) \\ &+ C_{ezhy}(i,j,k) \times \left(H_y^{n+\frac{1}{2}}(i,j,k) - H_y^{n+\frac{1}{2}}(i-1,j,k)\right) \\ &+ C_{ezhx}(i,j,k) \times \left(H_x^{n+\frac{1}{2}}(i,j,k) - H_x^{n+\frac{1}{2}}(i,j-1,k)\right) \\ &+ C_{ezj}(i,j,k) \times J_{iz}^{n+\frac{1}{2}}(i,j,k), \end{aligned} \tag{4.24}$$

where

$$\begin{aligned} C_{eze}(i,j,k) &= \frac{2\varepsilon_z(i,j,k) - \Delta t\sigma_z^e(i,j,k)}{2\varepsilon_z(i,j,k) + \Delta t\sigma_z^e(i,j,k)}, \\ C_{ezhy}(i,j,k) &= \frac{2\Delta t}{\left(2\varepsilon_z(i,j,k) + \Delta t\sigma_z^e(i,j,k)\right)\Delta x}, \\ C_{ezhx}(i,j,k) &= -\frac{2\Delta t}{\left(2\varepsilon_z(i,j,k) + \Delta t\sigma_z^e(i,j,k)\right)\Delta y}, \\ C_{ezj}(i,j,k) &= -\frac{2\Delta t}{2\varepsilon_z(i,j,k) + \Delta t\sigma_z^e(i,j,k)}. \end{aligned}$$

One should notice that during the FDTD time-marching iteration, at every time step the new value of $J_{iz}^{n+\frac{1}{2}}(i,j,k)$ should be calculated by (4.23) using $E_z^n(i,j,k)$ and $J_{iz}^{n-\frac{1}{2}}(i,j,k)$ before updating $E_z^{n+1}(i,j,k)$ by (4.24).

4.1.7 Lumped elements distributed over a surface or within a volume

In previous sections, we have demonstrated the derivation of FDTD updating equations for common lumped element circuit components. These components were assumed to be placed between two neighboring nodes along the edge of a cell oriented in a certain direction (x, y, or z). However, in some applications it may be desired to simulate a lumped element component behavior over a surface or within a volume, which extends to a number of cells in the discrete space.

For instance, consider the voltage source in Figure 4.5. Such a source can be used to feed a strip with a uniform potential V_s across its cross-section at its edge. Similarly, a resistor is illustrated in Figure 4.5, which is distributed over a volume and maintains a resistance R between the top and bottom strips. Similarly, the other types of lumped elements as well can be distributed over a surface or a volume. In such cases, the lumped element behavior extends over a number of cells, and the surfaces or volumes of the elements can be indicated by the indices of the nodes at the lower points and upper points of the cell groups. For instance, the voltage source is indicated by the nodes (*vis, vjs, vks*) and (*vie, vje, vke*), whereas the resistor is indicated by the nodes (*ris, rjs, rks*) and (*rie, rje, rke*), as illustrated in Figure 4.6.

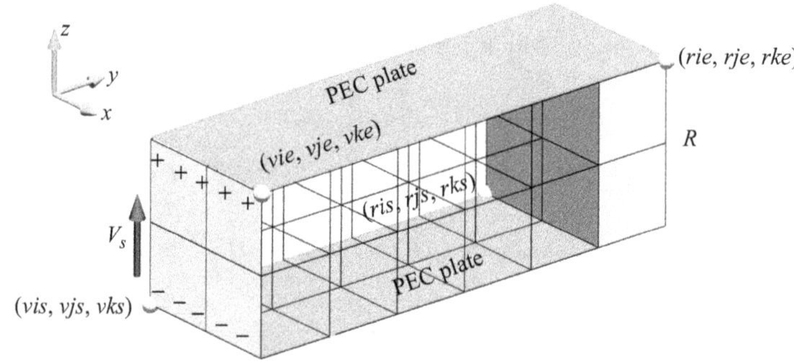

Figure 4.5 Parallel PEC plates excited by a voltage source at one end and terminated by a resistor at the other end.

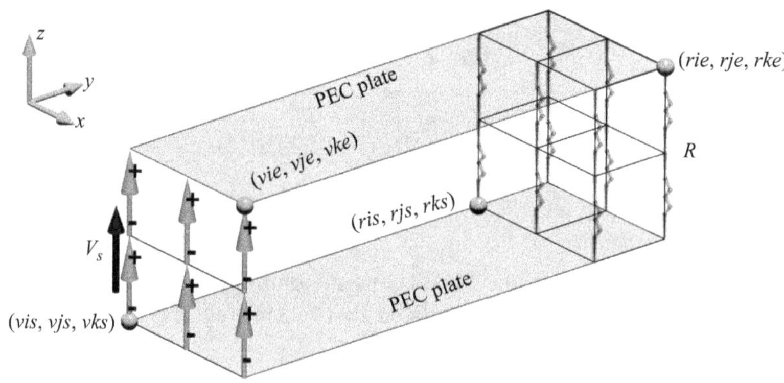

Figure 4.6 A voltage source distributed over a surface and a resistor distributed over a volume.

The lumped element updating equations derived in previous sections still can be used to model the elements distributed over a number of cells, but a modification of parameter values is required to maintain the respective values of the lumped elements. For instance, consider the voltage source in Figure 4.5. To simulate this source, the electric field components $E_z(vis : vie, vjs : vje, vks : vke - 1)$ need to be updated as shown in Figure 4.6. Hence, the total number of field components needing to be updated is $(vie - vis + 1) \times (vje - vjs + 1) \times (vke - vks)$. Therefore, the main voltage source can be replaced by multiple voltage sources, each of which is associated with a respective field component. Each individual voltage source in Figure 4.6 then should be assigned a voltage value V'_s,

$$V'_s = \frac{V_s}{(vke - vks)}, \tag{4.25}$$

since the potential difference between the ends of the main voltage source in the z direction shall be kept the same. If the main voltage source has an internal resistance R_s, it should be maintained as well. Then the main resistance R_s will be represented by a network of $(vie - vis + 1) \times (vje - vjs + 1)$ resistors connected in parallel where each element in this parallel combination is made of $(vke - vks)$ series resistors. Therefore, the resistance for each source component R_s' shall be set as

$$R_s' = R_s \times \frac{(vie - vis + 1) \times (vje - vjs + 1)}{(vke - vks)}. \tag{4.26}$$

Having applied these modifications to V_s and R_s, the electric field components $E_z(vis : vie, vjs : vje, vks : vke - 1)$ can be updated using (4.10).

The same procedure holds for each individual resistor R' at the other end of the parallel PEC plates. Thus, the resistance R' can be given as

$$R' = R \times \frac{(rie - ris + 1) \times (rje - rjs + 1)}{(rke - rks)}, \tag{4.27}$$

and (4.16) can be used to update the electric field components $E_z(ris : rie, rjs : rje, rks : rke - 1)$.

Similarly, for an inductor with inductance L extending between the nodes (is, js, ks) and (ie, je, ke), the individual inductance L' can be defined as

$$L' = L \times \frac{(ie - is + 1) \times (je - js + 1)}{(ke - ks)}, \tag{4.28}$$

and (4.24) can be used to update the electric field components $E_z(is : ie, js : je, ks : ke - 1)$.

For a capacitor with capacitance C extending between the nodes (is, js, ks) and (ie, je, ke), each individual capacitance C' can be defined as

$$C' = C \times \frac{(ke - ks)}{(ie - is + 1) \times (je - js + 1)}, \tag{4.29}$$

and (4.19) can be used to update the electric field components $E_z(is : ie, js : je, ks : ke - 1)$.

A current source with magnitude I_s and resistance R_s extending between the nodes (is, js, ks) and (ie, je, ke) can be modeled by updating $E_z(is : ie, js : je, ks : ke - 1)$ using (4.15) after modifying I_s and resistance R_s such that

$$I_s' = \frac{I_s}{(ie - is + 1) \times (je - js + 1)}, \tag{4.30}$$

and

$$R_s' = R_s \times \frac{(vie - vis + 1) \times (vje - vjs + 1)}{(vke - vks)}. \tag{4.31}$$

One should notice that the equations given for modifying the lumped values are all for the elements oriented in the z direction. The indexing scheme should be reflected appropriately to obtain equations for lumped elements oriented in other directions.

4.1.8 Diode

Derivation of the FDTD updating equations of some lumped element circuit components have been presented in previous sections. These circuit components are characterized by

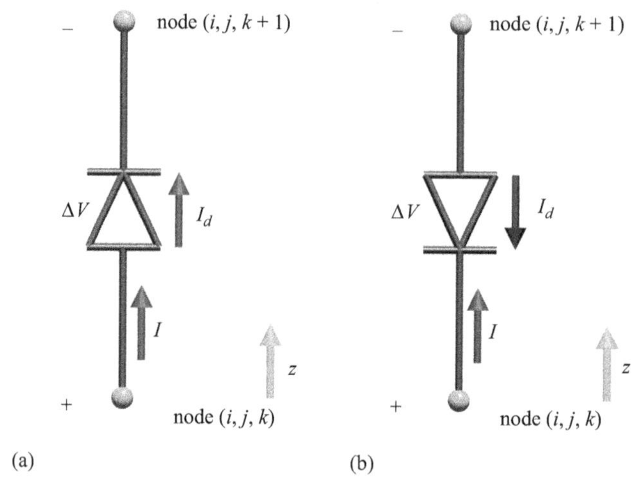

Figure 4.7 Diodes placed between nodes (i, j, k) and $(i, j, k + 1)$: (a) diode oriented in the positive z direction and (b) diode oriented in the negative z direction.

linear voltage–current relations, and it is straightforward to model their behavior in FDTD by properly expressing their voltage–current relations in discrete time and space. However, one of the strengths of the FDTD method is that it is possible to model nonlinear components as well. In this section we present the derivation of the updating equations for modeling a diode.

Figure 4.7(a) illustrates a diode placed between nodes (i, j, k) and $(i, j, k + 1)$ in an FDTD problem space. This diode allows currents flowing only in the positive z direction, as indicated by direction of the current I_d and characterized by the voltage–current relation

$$I = I_d = I_0 \left[e^{(qV_d/kT)} - 1 \right], \tag{4.32}$$

where q is the absolute value of electron charge in coulombs, k is Boltzmann's constant, and T is the absolute temperature in kelvins. This equation can be expressed in discrete form as

$$J_{iz}^{n+\frac{1}{2}}(i, j, k) = \frac{I_0}{\Delta x \Delta y} \left[e^{(q\Delta z/2kT)(E_z^{n+1}(i,j,k)+E_z^n(i,j,k))} - 1 \right], \tag{4.33}$$

where the relation $V_d = \Delta V = \Delta z E_z$ is used. Equation 4.33 can be used in (4.5), which yields

$$\begin{aligned}
\frac{E_z^{n+1}(i, j, k) - E_z^n(i, j, k)}{\Delta t} = {} & \frac{1}{\varepsilon_z(i, j, k)} \frac{H_y^{n+\frac{1}{2}}(i, j, k) - H_y^{n+\frac{1}{2}}(i-1, j, k)}{\Delta x} \\
& - \frac{1}{\varepsilon_z(i, j, k)} \frac{H_x^{n+\frac{1}{2}}(i, j, k) - H_x^{n+\frac{1}{2}}(i, j-1, k)}{\Delta y} \\
& - \frac{\sigma_z^e(i, j, k)}{2\varepsilon_z(i, j, k)} \left(E_z^{n+1}(i, j, k) + E_z^n(i, j, k) \right) \\
& - \frac{I_0}{\varepsilon_z(i, j, k)\Delta x \Delta y} \left[e^{(q\Delta z/2kT)\left(E_z^{n+1}(i, j, k)+E_z^n(i, j, k)\right)} - 1 \right].
\end{aligned} \tag{4.34}$$

This equation can be expanded as

$$
\begin{aligned}
E_z^{n+1}(i,j,k) + \frac{\sigma_z^e(i,j,k)\Delta t}{2\varepsilon_z(i,j,k)} E_z^{n+1}(i,j,k) = E_z^n(i,j,k) - \frac{\sigma_z^e(i,j,k)\Delta t}{2\varepsilon_z(i,j,k)} E_z^n(i,j,k) \\
+ \frac{\Delta t}{\varepsilon_z(i,j,k)} \frac{H_y^{n+\frac{1}{2}}(i,j,k) - H_y^{n+\frac{1}{2}}(i-1,j,k)}{\Delta x} \\
- \frac{\Delta t}{\varepsilon_z(i,j,k)} \frac{H_x^{n+\frac{1}{2}}(i,j,k) - H_x^{n+\frac{1}{2}}(i,j-1,k)}{\Delta y} \\
- \frac{I_0 \Delta t}{\varepsilon_z(i,j,k)\Delta x \Delta y} e^{(q\Delta z/2kT)E_z^n(i,j,k)} e^{(q\Delta z/2kT)E_z^{n+1}(i,j,k)} \\
+ \frac{I_0 \Delta t}{\varepsilon_z(i,j,k)\Delta x \Delta y}
\end{aligned}
\tag{4.35}
$$

and can be arranged in the following form

$$
Ae^{Bx} + x + C = 0, \tag{4.36}
$$

where

$$
\begin{aligned}
& x = E_z^{n+1}(i,j,k), \quad A = -C_{ezd}(i,j,k)e^{B \times E_z^n(i,j,k)}, \quad B = (q\Delta z/2kT), \\
& C = C_{eze}(i,j,k) \times E_z^n(i,j,k) + C_{ezhy}(i,j,k) \times \left(H_y^{n+\frac{1}{2}}(i,j,k) - H_y^{n+\frac{1}{2}}(i-1,j,k)\right) \\
& \quad + C_{ezhx}(i,j,k) \times \left(H_x^{n+\frac{1}{2}}(i,j,k) - H_x^{n+\frac{1}{2}}(i,j-1,k)\right) + C_{ezd}(i,j,k), \\
& C_{eze}(i,j,k) = -\frac{2\varepsilon_z(i,j,k) - \Delta t \sigma_z^e(i,j,k)}{2\varepsilon_z(i,j,k) + \Delta t \sigma_z^e(i,j,k)}, \\
& C_{ezhy}(i,j,k) = -\frac{2\Delta t}{(2\varepsilon_z(i,j,k) + \Delta t \sigma_z^e(i,j,k))\Delta x}, \\
& C_{ezhx}(i,j,k) = \frac{2\Delta t}{(2\varepsilon_z(i,j,k) + \Delta t \sigma_z^e(i,j,k))\Delta y}, \\
& C_{ezd}(i,j,k) = -\frac{2\Delta t I_0}{2\varepsilon_z(i,j,k)\Delta x \Delta y + \Delta t \sigma_z^e(i,j,k)\Delta x \Delta y}.
\end{aligned}
$$

Here the parameters A and C are dependent on $E_z^n(i,j,k)$. Therefore, at every time step the terms A and C must be calculated using the past values of magnetic and electric field components, and then (4.36) must be solved to obtain $E_z^{n+1}(i,j,k)$.

Equation (4.36) can easily be solved numerically by using the Newton–Raphson method, which is one of the most widely used methods for finding roots of a nonlinear function. It is an iterative process that starts from an initial point in the proximity of a root and approaches to the root by the use of the derivative of the function. For instance, Figure 4.8 shows a function for which we want to find the value of x that satisfies $f(x) = 0$. We can start with an initial guess x_0 as shown in the figure. The derivative of the function $f(x)$ at point x_0 is

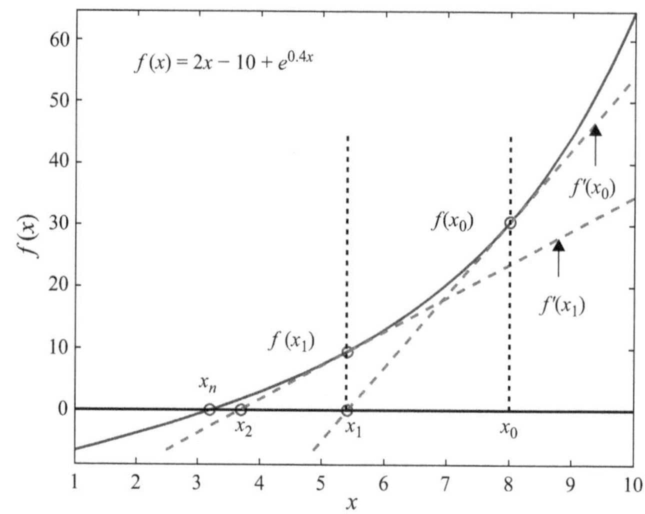

Figure 4.8 A function $f(x)$ and the points approaching to the root of the function iteratively calculated using the Newton–Raphson method.

the slope of the tangential line passing through the point $f(x_0)$ and intersecting the x axis at point x_1. We can easily calculate point x_1 as

$$x_1 = x_0 - \frac{f(x_0)}{f'(x_0)}. \tag{4.37}$$

Then x_1 can be used as the reference point, and a second point x_2 can be obtained similarly by

$$x_2 = x_1 - \frac{f(x_1)}{f'(x_1)}. \tag{4.38}$$

As can be followed from Figure 4.8, the new calculated x leads to the target point where $f(x)$ intersects with the x axis. Due to the high convergence rate of this procedure a point x_n can be achieved after couple of iterations, which is in the very close proximity of the exact root of the function $f(x)$. This iterative procedure can be continued until a stopping criteria or a maximum number of iterations is reached. Such a stopping criteria can be given by

$$|f(x_n)| < \varepsilon, \tag{4.39}$$

where ε is a very small positive number accepted as the error tolerance accepted for $f(x)$ approximately equals to zero.

It is very convenient to use the Newton–Raphson method to solve the diode equation (4.36), since the nature of this equation does not allow local minima or maxima, which can be an obstacle for the convergence of this method. Furthermore, the value of the electric field component $E_z^n(i, j, k)$ can be used as the initial guess for $E_z^{n+1}(i, j, k)$. Since the value of E_z (i, j, k) will change only a small amount between the consecutive time steps, the initial guess will be fairly close to the target value, and it would take only a couple of Newton–Raphson iterations to reach the desired solution.

The diode equation given by (4.36) is for a positive z-directed diode illustrated in Figure 4.7(a), which is characterized by the voltage–current relation (4.32). A diode oriented in the negative z direction, as shown in 4.7(b), can be characterized by the voltage–current relation

$$I = -I_d = -I_0\left[e^{(qV_d/kT)} - 1\right] \tag{4.40}$$

and $V_d = -\Delta V$ due to the inversion of the diode polarity. Therefore, one can reconstruct (4.34) and (4.36) considering the reversed polarities of V_d and I_d, which yields the updating equation for a negative z-directed diode as

$$Ae^{Bx} + x + C = 0, \tag{4.41}$$

where

$$x = E_z^{n+1}(i,j,k), \quad A = -C_{ezd}(i,j,k)e^{B\times E_z^n(i,j,k)}, \quad B = -(q\Delta z/2kT),$$

$$C = C_{eze}(i,j,k) \times E_z^n(i,j,k) + C_{ezhy}(i,j,k) \times \left(H_y^{n+\frac{1}{2}}(i,j,k) - H_y^{n+\frac{1}{2}}(i-1,j,k)\right)$$

$$+ C_{ezhx}(i,j,k) \times \left(H_x^{n+\frac{1}{2}}(i,j,k) - H_x^{n+\frac{1}{2}}(i,j-1,k)\right) + C_{ezd}(i,j,k),$$

$$C_{eze}(i,j,k) = -\frac{2\varepsilon_z(i,j,k) - \Delta t\sigma_z^e(i,j,k)}{2\varepsilon_z(i,j,k) + \Delta t\sigma_z^e(i,j,k)},$$

$$C_{ezhy}(i,j,k) = -\frac{2\Delta t}{(2\varepsilon_z(i,j,k) + \Delta t\sigma_z^e(i,j,k))\Delta x},$$

$$C_{ezhx}(i,j,k) = \frac{2\Delta t}{(2\varepsilon_z(i,j,k) + \Delta t\sigma_z^e(i,j,k))\Delta y},$$

$$C_{ezd}(i,j,k) = \frac{2\Delta t I_0}{2\varepsilon_z(i,j,k)\Delta x\Delta y + \Delta t\sigma_z^e(i,j,k)\Delta x\Delta y}.$$

Here, one should notice that only the coefficients B and $C_{ezd}(i, j, k)$ are reversed when compared with the respective coefficients of (4.36), and the rest of the terms are the same.

Due to the nonlinear nature of the diode voltage–current relation, it is not convenient to define a diode extending over a number of cells in the FDTD problem grid. If a diode is going to be placed between two nodes that are apart from each other more than one cell size, it is more convenient to place the diode between two neighboring nodes and then to establish connection between the other nodes by other means, (e.g., by thin wires). The thin-wires FDTD updating procedure and equations are discussed in Chapter 10.

4.1.9 Summary

In this chapter, we have provided the derivation of FDTD updating equations for modeling some common lumped element circuit components. We have shown the construction of updating equations for components placed between two neighboring nodes and have illustrated how these components can be modeled if they extend over a number of cells on the problem grid.

It is possible to obtain updating equations for the circuits composed of combinations of these lumped elements; one only needs to obtain the appropriate voltage–current relations, express them in discrete time and space, and establish the relation between the impressed current densities and the field components. Then impressed current density terms can be used in the general form of the updating equation to obtain the specific updating equations that model the lumped element circuit under consideration.

4.2 Definition, initialization, and simulation of lumped elements

We discussed modeling of several types of lumped element components in the FDTD method in previous sections. In this section we demonstrate the implementation of the aforementioned concepts in MATLAB. Furthermore, we show the implementation of other routines that initialize the FDTD problem space, including some auxiliary parameters, field arrays, and updating coefficient arrays. Then the MATLAB workspace will be ready to run an FDTD simulation, and we show how the FDTD time-marching loop can be implemented and some sample results of interest can be obtained.

4.2.1 Definition of lumped elements

First we show the implementation of the definition of the lumped element components. As discussed before, these components can be defined as prism-like objects distributed over a volume in the FDTD problem space. Any prism-like object can be defined with its lower and upper coordinates in the Cartesian coordinate system. Therefore, we define the positions of the lumped components the same as the brick object, as demonstrated in Section 3.1. Referring to the FDTD solver main program ***fdtd_solve***, which is shown in Listing 3.1, the implementation of the definition of lumped element components is done in the subroutine ***define_sources_and_lumped_elements***, a sample content of which is given as a template in Listing 4.1. As can be seen in Listing 4.1, the structure arrays **voltage_sources**, **current_sources**, **resistors**, **inductors**, **capacitors**, and **diodes** are defined to store the properties of the respective types of lumped components, and these arrays are initialized as empty arrays. Similar to a brick, the positions and dimensions of these components are indicated by the parameters **min_x**, **min_y**, **min_z**, **max_x**, **max_y**, and **max_z**. One should be careful when defining the coordinates of diodes; as discussed in Section 4.8, we assume diodes as objects having zero thickness in two dimensions.

Besides the parameters defining the positioning, some additional parameters as well are needed to specify the properties of these components. The lumped element components are modeled in FDTD by the way the respective electric field components are updated due to the voltage–current relations. These field components that need specific updates are determined by the functional directions of these components. For voltage sources, current sources, and diodes there are six directions in which they can be defined in the staircased FDTD grid: "***xn***," "***xp***," "***yn***," "***yp***," "***zn***," or "***zp***." Here "***p***" refers to positive direction, whereas "***n***" refers to negative direction. The other lumped elements (resistors, capacitors, and inductors) can be defined with the directions "*x*," "*y*," or "*z*." The other parameters needed to specify the lumped components are **magnitude**, **resistance**, **inductance**, and **capacitance**, which represent the characteristics of the lumped elements as shown in Listing 4.1.

Listing 4.1 define_sources_and_lumped_elements.m

```
disp('defining sources and lumped element components');

voltage_sources = [];
current_sources = [];
diodes = [];
resistors = [];
inductors = [];
capacitors = [];

% define source waveform types and parameters
waveforms.sinusoidal(1).frequency = 1e9;
waveforms.sinusoidal(2).frequency = 5e8;
waveforms.unit_step(1).start_time_step = 50;

% voltage sources
% direction: 'xp', 'xn', 'yp', 'yn', 'zp', or 'zn'
% resistance : ohms, magitude   : volts
voltage_sources(1).min_x = 0;
voltage_sources(1).min_y = 0;
voltage_sources(1).min_z = 0;
voltage_sources(1).max_x = 1.0e-3;
voltage_sources(1).max_y = 2.0e-3;
voltage_sources(1).max_z = 4.0e-3;
voltage_sources(1).direction = 'zp';
voltage_sources(1).resistance = 50;
voltage_sources(1).magnitude = 1;
voltage_sources(1).waveform_type = 'sinusoidal';
voltage_sources(1).waveform_index = 2;

% current sources
% direction: 'xp', 'xn', 'yp', 'yn', 'zp', or 'zn'
% resistance : ohms, magitude   : amperes
current_sources(1).min_x = 30*dx;
current_sources(1).min_y = 10*dy;
current_sources(1).min_z = 10*dz;
current_sources(1).max_x = 36*dx;
current_sources(1).max_y = 10*dy;
current_sources(1).max_z = 13*dz;
current_sources(1).direction = 'xp';
current_sources(1).resistance = 50;
current_sources(1).magnitude = 1;
current_sources(1).waveform_type = 'unit_step';
current_sources(1).waveform_index = 1;

% resistors
% direction: 'x', 'y', or 'z'
% resistance : ohms
resistors(1).min_x = 7.0e-3;
resistors(1).min_y = 0;
resistors(1).min_z = 0;
```

```
resistors(1).max_x = 8.0e-3;
resistors(1).max_y = 2.0e-3;
resistors(1).max_z = 4.0e-3;
resistors(1).direction = 'z';
resistors(1).resistance = 50;

% inductors
% direction: 'x', 'y', or 'z'
% inductance : henrys
inductors(1).min_x = 30*dx;
inductors(1).min_y = 10*dy;
inductors(1).min_z = 10*dz;
inductors(1).max_x = 36*dx;
inductors(1).max_y = 10*dy;
inductors(1).max_z = 13*dz;
inductors(1).direction = 'x';
inductors(1).inductance = 1e-9;

% capacitors
% direction: 'x', 'y', or 'z'
% capacitance : farads
capacitors(1).min_x = 30*dx;
capacitors(1).min_y = 10*dy;
capacitors(1).min_z = 10*dz;
capacitors(1).max_x = 36*dx;
capacitors(1).max_y = 10*dy;
capacitors(1).max_z = 13*dz;
capacitors(1).direction = 'x';
capacitors(1).capacitance = 1e-12;

% diodes
% direction: 'xp', 'xn', 'yp', 'yn', 'zp', or 'zn'
diodes(1).min_x = 30*dx;
diodes(1).min_y = 10*dy;
diodes(1).min_z = 10*dz;
diodes(1).max_x = 36*dx;
diodes(1).max_y = 10*dy;
diodes(1).max_z = 13*dz;
diodes(1).direction = 'xp';
```

Another type of parameter that is required for time-domain simulation of sources is the type of waveforms that the voltages and current sources generate as a function of time. For each source, the voltage or current value generated at every time step can be stored as the waveform in a one-dimensional array with the size of **number_of_time_steps**. However, before associating a waveform to a source, we can define the parameters specific to various types of waveforms in a structure named **waveforms** as illustrated in Listing 4.1. There are various options for constructing a waveform in FDTD. Some of the most common types of waveforms are Gaussian, sinusoidal, cosine modulated Gaussian, and unit step. The definition and construction of these waveforms will be the subject of Chapter 5. However, we provide a template here showing how the waveform parameters can be defined. In Listing 4.1 two fields are defined in the parameter **waveforms**: **sinusoidal** and **unit_step**. Later on we can add new parameters representing other types of waveforms. The parameters representing the waveform types are arrays of structures as illustrated with the indices **waveforms.sinuoidal(i)** and **waveforms. unit_step(i)**. Each waveform type has parameters representing its specific characteristics.

Here **frequency** is a parameter for the sinusoidal waveform **waveforms.sinusoidal(i)**, whereas **start_time_step** is a parameter for the waveform **waveforms.unit_step(i)**.

Having defined the types of waveforms and their parameters, we can associate them to the sources by referring to them with the **waveform_type** and **waveform_index** in the source parameters **voltage_sources(i)** and **current_sources(i)**. In the example code, we assign the type of the waveform "***sinusoidal***" to **voltage_sources(1).waveform_type**, and the index of the sinusoidal waveform **waveforms.sinusoidal(2)**, which is 2, to **voltage_sources(1).waveform_index**. Similarly, the type of the waveform "***unit_step***" is assigned to **current_sources(1).waveform_type** and the index of the waveform **waveforms.unit_step(1)**, which is 1, is assigned to **current_sources(1).waveform_index**.

4.2.2 Initialization of FDTD parameters and arrays

Some constant parameters and auxiliary parameters are needed in various stages of an FDTD program. These parameters can be defined in the subroutine ***initialize_fdtd_parameters_and_arrays***, which is shown in Listing 4.2. The most frequently used ones of these parameters are permittivity, permeability of free space, and speed of light in free space, which are represented in the code with the parameters **eps_0**, **mu_0**, and **c**, respectively. Duration of a time step is denoted by **dt** and is calculated based on the CFL stability limit (2.7) and the **courant_factor** described in Section 3.1.1. Once the duration of a time step is calculated, an auxiliary one-dimensional array, **time**, of size **number_of_time_steps** can be constructed to store the time instant values at the middle of the FDTD time steps. When a new simulation starts, the initial values of electric and magnetic fields are zero. Therefore, the three-dimensional arrays representing the vector components of the fields can be created and initialized using the MATLAB function **zeros** as illustrated in Listing 4.2. One can

Listing 4.2 initialize_fdtd_parameters_and_arrays.m

```
disp('initializing FDTD parameters and arrays');

% constant parameters
eps_0 = 8.854187817e-12; % permittivity of free space
mu_0  = 4*pi*1e-7; % permeability of free space
c = 1/sqrt(mu_0*eps_0); % speed of light in free space

% Duration of a time step in seconds
dt = 1/(c*sqrt((1/dx^2)+(1/dy^2)+(1/dz^2)));
dt = courant_factor*dt;

% time array
time = ([1:number_of_time_steps]-0.5)*dt;

% Create and initialize field and current arrays
disp('creating field arrays');

Hx = zeros(nxp1,ny,nz);
Hy = zeros(nx,nyp1,nz);
Hz = zeros(nx,ny,nzp1);
Ex = zeros(nx,nyp1,nzp1);
Ey = zeros(nxp1,ny,nzp1);
Ez = zeros(nxp1,nyp1,nz);
```

notice that the sizes of the arrays are not the same due to the positioning scheme of field components on the Yee grid as discussed in Section 3.5.

At this point one can ask if there are any arrays that need to be defined to represent impressed currents. The answer is that we do not need to define any arrays representing the impressed currents unless they are explicitly needed in an FDTD simulation. Throughout Chapter 4 we have shown that the concept of impressed currents is used to establish the voltage–current relations for the lumped element components and the final updating equations do not include the impressed current terms except for the inductor. Therefore, we do not need to create three-dimensional arrays expanded to the problem space to represent impressed current components. However, for the case of the inductor, there exists an impressed electric current term. Since an inductor would be expanding only over a couple of cells and only a small number of field components need to be updated using the inductor updating equations, it is sufficient to create auxiliary arrays representing these impressed currents that have the limited size based on the number of field components associated with the respective inductors. The creation and initialization of these arrays are performed in the subroutine where the updating coefficients of the lumped element components are initialized.

4.2.3 Initialization of lumped element components

We have shown templates for defining some types of source waveforms, voltage and current sources, and other lumped element components. The initialization of these components is performed in ***initialize_sources_and_lumped_elements*** as shown in Listing 4.3.

Listing 4.3 initialize_sources_and_lumped_elements.m

```
disp('initializing sources and lumped element components');

number_of_voltage_sources   = size(voltage_sources,2);
number_of_current_sources   = size(current_sources,2);
number_of_resistors   = size(resistors,2);
number_of_inductors   = size(inductors,2);
number_of_capacitors   = size(capacitors,2);
number_of_diodes   = size(diodes,2);

% initialize waveforms
initialize_waveforms;

% voltage sources
for ind = 1:number_of_voltage_sources
    is = round((voltage_sources(ind).min_x - fdtd_domain.min_x)/dx)+1;
    js = round((voltage_sources(ind).min_y - fdtd_domain.min_y)/dy)+1;
    ks = round((voltage_sources(ind).min_z - fdtd_domain.min_z)/dz)+1;
    ie = round((voltage_sources(ind).max_x - fdtd_domain.min_x)/dx)+1;
    je = round((voltage_sources(ind).max_y - fdtd_domain.min_y)/dy)+1;
    ke = round((voltage_sources(ind).max_z - fdtd_domain.min_z)/dz)+1;
    voltage_sources(ind).is = is;
```

```
    voltage_sources(ind).js = js;
    voltage_sources(ind).ks = ks;
    voltage_sources(ind).ie = ie;
    voltage_sources(ind).je = je;
    voltage_sources(ind).ke = ke;

    switch (voltage_sources(ind).direction(1))
    case 'x'
        n_fields = ie - is;
        r_magnitude_factor = (1 + je - js) * (1 + ke - ks) / (ie - is);
    case 'y'
        n_fields = je - js;
        r_magnitude_factor = (1 + ie - is) * (1 + ke - ks) / (je - js);
    case 'z'
        n_fields = ke - ks;
        r_magnitude_factor = (1 + ie - is) * (1 + je - js) / (ke - ks);
    end
    if strcmp(voltage_sources(ind).direction(2), 'n')
        v_magnitude_factor = ...
            -1*voltage_sources(ind).magnitude/n_fields;
    else
        v_magnitude_factor = ...
            1*voltage_sources(ind).magnitude/n_fields;
    end
    voltage_sources(ind).resistance_per_component = ...
        r_magnitude_factor * voltage_sources(ind).resistance;

    % copy waveform of the waveform type to waveform of the source
    wt_str = voltage_sources(ind).waveform_type;
    wi_str = num2str(voltage_sources(ind).waveform_index);
    eval_str = ['a_waveform = waveforms.' ...
        wt_str '(' wi_str ').waveform;'];
    eval(eval_str);
    voltage_sources(ind).voltage_per_e_field = ...
        v_magnitude_factor * a_waveform;
    voltage_sources(ind).waveform = ...
        v_magnitude_factor * a_waveform * n_fields;
end

% current sources
for ind = 1:number_of_current_sources
    is = round((current_sources(ind).min_x - fdtd_domain.min_x)/dx)+1;
    js = round((current_sources(ind).min_y - fdtd_domain.min_y)/dy)+1;
    ks = round((current_sources(ind).min_z - fdtd_domain.min_z)/dz)+1;
    ie = round((current_sources(ind).max_x - fdtd_domain.min_x)/dx)+1;
    je = round((current_sources(ind).max_y - fdtd_domain.min_y)/dy)+1;
    ke = round((current_sources(ind).max_z - fdtd_domain.min_z)/dz)+1;
    current_sources(ind).is = is;
    current_sources(ind).js = js;
    current_sources(ind).ks = ks;
    current_sources(ind).ie = ie;
    current_sources(ind).je = je;
    current_sources(ind).ke = ke;
```

```
        switch (current_sources(ind).direction(1))
        case 'x'
            n_fields = (1 + je - js) * (1 + ke - ks);
            r_magnitude_factor = (1 + je - js) * (1 + ke - ks) / (ie - is);
        case 'y'
            n_fields = (1 + ie - is) * (1 + ke - ks);
            r_magnitude_factor = (1 + ie - is) * (1 + ke - ks) / (je - js);
        case 'z'
            n_fields = (1 + ie - is) * (1 + je - js);
            r_magnitude_factor = (1 + ie - is) * (1 + je - js) / (ke - ks);
        end
        if strcmp(current_sources(ind).direction(2),'n')
            i_magnitude_factor = ...
                -1*current_sources(ind).magnitude/n_fields;
        else
            i_magnitude_factor =  ...
                1*current_sources(ind).magnitude/n_fields;
        end
        current_sources(ind).resistance_per_component = ...
            r_magnitude_factor * current_sources(ind).resistance;

        % copy waveform of the waveform type to waveform of the source
        wt_str = current_sources(ind).waveform_type;
        wi_str = num2str(current_sources(ind).waveform_index);
        eval_str = ['a_waveform␣=␣waveforms.' ...
            wt_str '(' wi_str ').waveform;'];
        eval(eval_str);
        current_sources(ind).current_per_e_field = ...
            i_magnitude_factor * a_waveform;
        current_sources(ind).waveform = ...
            i_magnitude_factor * a_waveform * n_fields;
    end

    % resistors
    for ind = 1:number_of_resistors
        is = round((resistors(ind).min_x - fdtd_domain.min_x)/dx)+1;
        js = round((resistors(ind).min_y - fdtd_domain.min_y)/dy)+1;
        ks = round((resistors(ind).min_z - fdtd_domain.min_z)/dz)+1;
        ie = round((resistors(ind).max_x - fdtd_domain.min_x)/dx)+1;
        je = round((resistors(ind).max_y - fdtd_domain.min_y)/dy)+1;
        ke = round((resistors(ind).max_z - fdtd_domain.min_z)/dz)+1;
        resistors(ind).is = is;
        resistors(ind).js = js;
        resistors(ind).ks = ks;
        resistors(ind).ie = ie;
        resistors(ind).je = je;
        resistors(ind).ke = ke;
        switch (resistors(ind).direction)
        case 'x'
            r_magnitude_factor = (1 + je - js) * (1 + ke - ks) / (ie - is);

        case 'y'
            r_magnitude_factor = (1 + ie - is) * (1 + ke - ks) / (je - js);
        case 'z'
            r_magnitude_factor = (1 + ie - is) * (1 + je - js) / (ke - ks);
        end
        resistors(ind).resistance_per_component = ...
            r_magnitude_factor * resistors(ind).resistance;
    end
```

```
% inductors
for ind = 1:number_of_inductors
    is = round((inductors(ind).min_x - fdtd_domain.min_x)/dx)+1;
    js = round((inductors(ind).min_y - fdtd_domain.min_y)/dy)+1;
    ks = round((inductors(ind).min_z - fdtd_domain.min_z)/dz)+1;
    ie = round((inductors(ind).max_x - fdtd_domain.min_x)/dx)+1;
    je = round((inductors(ind).max_y - fdtd_domain.min_y)/dy)+1;
    ke = round((inductors(ind).max_z - fdtd_domain.min_z)/dz)+1;
    inductors(ind).is = is;
    inductors(ind).js = js;
    inductors(ind).ks = ks;
    inductors(ind).ie = ie;
    inductors(ind).je = je;
    inductors(ind).ke = ke;
    switch (inductors(ind).direction)
    case 'x'
        l_magnitude_factor = (1 + je - js) * (1 + ke - ks) / (ie - is);
    case 'y'
        l_magnitude_factor = (1 + ie - is) * (1 + ke - ks) / (je - js);
    case 'z'
        l_magnitude_factor = (1 + ie - is) * (1 + je - js) / (ke - ks);
    end
    inductors(ind).inductance_per_component = ...
        l_magnitude_factor * inductors(ind).inductance;
end

% capacitors
for ind = 1:number_of_capacitors
    is = round((capacitors(ind).min_x - fdtd_domain.min_x)/dx)+1;
    js = round((capacitors(ind).min_y - fdtd_domain.min_y)/dy)+1;
    ks = round((capacitors(ind).min_z - fdtd_domain.min_z)/dz)+1;
    ie = round((capacitors(ind).max_x - fdtd_domain.min_x)/dx)+1;
    je = round((capacitors(ind).max_y - fdtd_domain.min_y)/dy)+1;
    ke = round((capacitors(ind).max_z - fdtd_domain.min_z)/dz)+1;
    capacitors(ind).is = is;
    capacitors(ind).js = js;
    capacitors(ind).ks = ks;
    capacitors(ind).ie = ie;
    capacitors(ind).je = je;
    capacitors(ind).ke = ke;
    switch (capacitors(ind).direction)
    case 'x'
        c_magnitude_factor = (ie - is) ...

            / ((1 + je - js) * (1 + ke - ks));
    case 'y'
        c_magnitude_factor = (je - js) ...
            / ((1 + ie - is) * (1 + ke - ks));
    case 'z'
        c_magnitude_factor = (ke - ks) ...
            / ((1 + ie - is) * (1 + je - js));
    end
    capacitors(ind).capacitance_per_component = ...
        c_magnitude_factor * capacitors(ind).capacitance;
end

sigma_pec = material_types(material_type_index_pec).sigma_e;
```

```
% diodes
for ind = 1:number_of_diodes
    is = round((diodes(ind).min_x - fdtd_domain.min_x)/dx)+1;
    js = round((diodes(ind).min_y - fdtd_domain.min_y)/dy)+1;
    ks = round((diodes(ind).min_z - fdtd_domain.min_z)/dz)+1;
    ie = round((diodes(ind).max_x - fdtd_domain.min_x)/dx)+1;
    je = round((diodes(ind).max_y - fdtd_domain.min_y)/dy)+1;
    ke = round((diodes(ind).max_z - fdtd_domain.min_z)/dz)+1;
    diodes(ind).is = is;
    diodes(ind).js = js;
    diodes(ind).ks = ks;
    diodes(ind).ie = ie;
    diodes(ind).je = je;
    diodes(ind).ke = ke;

    switch (diodes(ind).direction(1))
        case 'x'
            sigma_e_x(is+1:ie-1,js,ks) = sigma_pec;
        case 'y'
            sigma_e_y(is,js+1:je-1,ks) = sigma_pec;
        case 'z'
            sigma_e_z(is,js,ks+1:ke-1) = sigma_pec;
    end
end
```

Listing 4.4 initialize_waveforms.m

```
disp('initializing source waveforms');

% initialize sinusoidal waveforms
for ind=1:size(waveforms.sinusoidal,2)
    waveforms.sinusoidal(ind).waveform = ...
        sin(2 * pi * waveforms.sinusoidal(ind).frequency * time);
end

% initialize unit step waveforms
for ind=1:size(waveforms.unit_step,2)
    start_index = waveforms.unit_step(ind).start_time_step;
    waveforms.unit_step(ind).waveform(1:number_of_time_steps) = 1;
    waveforms.unit_step(ind).waveform(1:start_index-1) = 0;
end
```

Before initializing the sources, the source waveforms need to be initialized. Therefore, a subroutine, ***initialize_waveforms***, is called in Listing 4.3 before initialization of lumped components. The sample implementation of ***initialize_waveforms*** is given in Listing 4.4, and this subroutine will be expanded as new source waveform types are discussed in the following chapter. In subroutine ***initialize_waveforms*** for every waveform type the respective waveforms are calculated as functions of time based on the waveform type specific parameters. Then the calculated waveforms are stored in the arrays with the associated name **waveform**. For instance, in Listing 4.4 the parameter **waveforms.sinusoidal(i).waveform** is constructed as a sinusoidal waveform using $sin(2\pi \times f \times t)$, where f is the frequency defined with the parameter **waveforms.sinusoidal(i).frequency** and t is the discrete time array **time** storing the time instants of the time steps.

Once the waveform arrays are constructed for the defined waveform types they can be copied to **waveform** fields of the structures of sources associated with them as demonstrated in Listing 4.3. For example, consider the code section in the loop initializing the **voltage_sources**. The type of source waveform is stored in **voltage_sources(ind).waveform_type** as a string, and the index of the source waveform is stored in **voltage_sources(ind).waveform_index** as an integer number. Here we employ the MATLAB function **eval**, which executes a string containing MATLAB expression. We construct a string using the following expression:

```
wt_str = voltage_sources(ind).waveform_type;
wi_str = num2str(voltage_sources(ind).waveform_index);
eval_str = ['a_waveform = waveforms.' wt_str '(' wi_str ').waveform;'];
```

which constructs the string **eval_str** in the MATLAB workspace for the waveforms and the voltage source **voltage_sources(1)** defined in Listing 4.1 as

```
eval_str = 'a_waveform = waveforms.sinusoidal(2).waveform;'
```

Executing this string using ***eval*** as

```
eval(eval_str);
```

creates a parameter **a_waveform** in the MATLAB workspace, which has a copy of the waveform that the voltage source **voltage_sources(1)** is associated with. Then we can assign **a_waveform** to **voltage_sources(1).waveform** after multiplying with a factor. This multiplication factor is explained in the following paragraphs.

As discussed in Section 4.7, all the lumped element components except the diodes are assumed to be distributed over surfaces or volumes extending over a number of cells. Therefore, a number of field components require special updating due to the lumped elements associated with them. We assume that the surfaces or volumes that the lumped elements are coinciding with are conforming to the cells composing the FDTD problem space. Then we can identify the field components that are associated with lumped components by the indices of the *start* node, (is, js, ks), coinciding with the lower coordinates of the lumped component, and the indices of the *end* node, (ie, je, ke), coinciding with the upper coordinates of the lumped component, as illustrated in Figure 4.6. Therefore, for every lumped element, the start and end node indices are calculated and stored in the parameters **is**, **js**, **ks**, **ie**, **je**, and **ke** as shown in Listing 4.3.

Based on the discussion in Section 4.7, if a lumped element is associated with more than one field component, the value of the lumped element should be assigned to the field components after a modification. Therefore, new parameters are defined in the structures representing the lumped elements, which store the value of the lumped element per field component. For instance, **resistance_per_component** is defined in **voltage_sources(i)**, which stores the resistance of the voltage source per field component as calculated using the form of (4.26) by taking into account the direction of the component. Similarly, for a resistor **resistance_per_component** is defined and calculated by (4.27). For a current source **resistance_per_component** is defined and calculated by (4.31). The parameter **inductance_per_component** is calculated by (4.28) for an inductor, and **capacitance_per_component** is calculated by (4.29) for a capacitor.

Similarly, the **voltage_sources(i).voltage_per_e_field** is scaled using (4.25), and **current_sources(i).current_per_e_field** is scaled using (4.30). Furthermore, directions of these components as well are taken into account in the scaling factors; if the direction is *negative* the scaling factor is multiplied by -1. Meanwhile, parameters **voltage_sources(i).waveform** and **current_sources(i).waveform** are constructed to store the main voltages and currents of the lumped sources rather than the values used to update the associated field components.

We assume that a diode is defined as a one-dimensional object, such as a line section, extending over a number of electric field components on the cell edges as illustrated in Figure 4.9. Here the diode is extending between the nodes (is, js, ks) and (ie, je, ke) – hence the field components $E_z(is, js, ks : ke - 1)$, where $ie = is$ and $je = js$. However, as discussed in Section 4.8, we update only one of these field components, in this case $E_z(is, js, ks)$, using the diode updating equations, and we establish an electrical connection between other nodes $(is, js, ks + 1)$ and (ie, je, ke). The easiest way of establishing the electrical connection is considering a PEC line between these nodes. Therefore, we can assign conductivity of PEC to the material components associated with the electric field components $E_z(is, js, ks + 1 : ke - 1)$, which are $\sigma_z^e(is, js, ks + 1 : ke - 1)$. Therefore, the necessary material

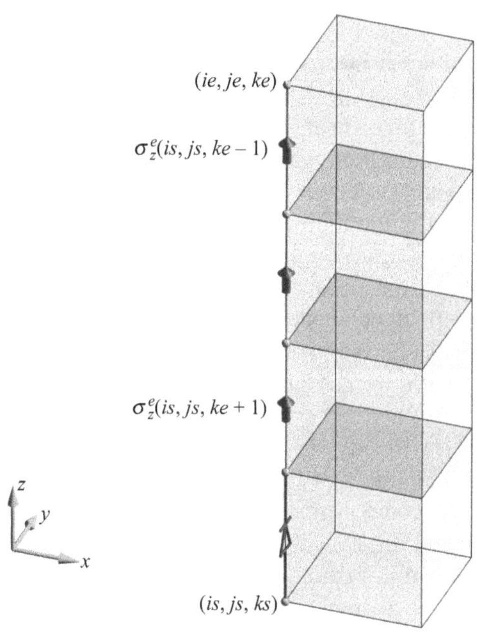

Figure 4.9 A diode defined between the nodes (is, js, ks) and (ie, je, ke).

components are assigned the conductivity of PEC in Listing 4.3 while initializing the **diodes**. A more accurate representation for a PEC line is presented Chapter 10 where thin PEC wire updating equations will be developed.

4.2.4 Initialization of updating coefficients

After initializing the lumped element components by determining the indices of nodes identifying their positions in the FDTD problem space grid and by setting other parameters necessary to characterize them, we can construct the FDTD updating coefficients. We calculate and store updating coefficients in three-dimensional arrays before the FDTD time-marching loop begins and use them while updating the fields during the time-marching loop. The initialization of the updating coefficients is done in the subroutine ***initialize_updating_coeffcients***, which is given in Listing 4.5.

Listing 4.5 initialize_updating_coefficients.m

```
disp('initializing general updating coefficients');

% General electric field updating coefficients
% Coeffiecients updating Ex
Cexe  =  (2*eps_r_x*eps_0 - dt*sigma_e_x) ...
    ./(2*eps_r_x*eps_0 + dt*sigma_e_x);
Cexhz =  (2*dt/dy)./(2*eps_r_x*eps_0 + dt*sigma_e_x);
Cexhy = -(2*dt/dz)./(2*eps_r_x*eps_0 + dt*sigma_e_x);

% Coeffiecients updating Ey
Ceye  =  (2*eps_r_y*eps_0 - dt*sigma_e_y) ...
    ./(2*eps_r_y*eps_0 + dt*sigma_e_y);
Ceyhx =  (2*dt/dz)./(2*eps_r_y*eps_0 + dt*sigma_e_y);
Ceyhz = -(2*dt/dx)./(2*eps_r_y*eps_0 + dt*sigma_e_y);

% Coeffiecients updating Ez
Ceze  =  (2*eps_r_z*eps_0 - dt*sigma_e_z) ...
    ./(2*eps_r_z*eps_0 + dt*sigma_e_z);
Cezhy =  (2*dt/dx)./(2*eps_r_z*eps_0 + dt*sigma_e_z);
Cezhx = -(2*dt/dy)./(2*eps_r_z*eps_0 + dt*sigma_e_z);

% General magnetic field updating coefficients
% Coeffiecients updating Hx
Chxh  =  (2*mu_r_x*mu_0 - dt*sigma_m_x) ...
    ./(2*mu_r_x*mu_0 + dt*sigma_m_x);
Chxez = -(2*dt/dy)./(2*mu_r_x*mu_0 + dt*sigma_m_x);
Chxey =  (2*dt/dz)./(2*mu_r_x*mu_0 + dt*sigma_m_x);

% Coeffiecients updating Hy
Chyh  =  (2*mu_r_y*mu_0 - dt*sigma_m_y) ...
    ./(2*mu_r_y*mu_0 + dt*sigma_m_y);
Chyex = -(2*dt/dz)./(2*mu_r_y*mu_0 + dt*sigma_m_y);
Chyez =  (2*dt/dx)./(2*mu_r_y*mu_0 + dt*sigma_m_y);

% Coeffiecients updating Hz
Chzh  =  (2*mu_r_z*mu_0 - dt*sigma_m_z) ...
```

```
    ./(2* mu_r_z*mu_0 + dt*sigma_m_z);
Chzey = -(2*dt/dx)./(2* mu_r_z*mu_0 + dt*sigma_m_z);
Chzex =  (2*dt/dy)./(2* mu_r_z*mu_0 + dt*sigma_m_z);

% Initialize coeffiecients for lumped element components
initialize_voltage_source_updating_coefficients;
initialize_current_source_updating_coefficients;
initialize_resistor_updating_coefficients;
initialize_capacitor_updating_coefficients;
initialize_inductor_updating_coefficients;
initialize_diode_updating_coefficients;
```

In Chapter 1 we obtained the set of equations (1.26)–(1.31) as the general form of the updating equations. One can notice that the form of the updating equations obtained for the lumped elements is the same as the general form. The only difference is that for some types of lumped elements the terms associated with the impressed currents are replaced by some other terms; for instance, for a voltage source the impressed current term is replaced by a voltage source term in the updating equation. Furthermore, as discussed in Section 4.2.2, the impressed current terms in (1.26)–(1.31) are used to establish the voltage–current relations for the lumped elements, and they vanish in updating equations modeling media that does not include any lumped components.

Therefore, in an FDTD simulation before the time-marching iteration starts the updating coefficients can be calculated for the whole problem space using the general updating equation coefficients appearing in (1.26)–(1.31) and excluding the impressed current terms. Then the updating coefficients can be recalculated only for the position indices where the lumped components exist. The additional coefficients required for the lumped elements (e.g., voltage sources, current sources, and inductors), can be calculated and stored in separate arrays associated with these components. Then during the FDTD time-marching iterations, at every time step, the electric field components can be updated using the general updating equations (1.26)–(1.31) excluding the impressed current terms, and then the additional terms can be added to the electric field components at the respective position indices where the lumped elements exist. In this section we illustrate the implementation of the previous discussion in the MATLAB program. The vector type operation in MATLAB is being used to efficiently generate these arrays.

In Listing 4.5 the updating coefficients, except for the impressed current terms, defined in the general updating equations (1.26)–(1.31) are constructed using the material component arrays that have been initialized in the subroutine ***initialize_fdtd_material_grid***. The sizes of the coefficient arrays are the same as the sizes of the material component arrays that have been used to construct them. Then the updating coefficients related to the lumped element components are constructed in respective subroutines as shown in Listing 4.5.

4.2.4.1 Initialization of voltage source updating coefficients

The updating coefficients related to the voltage sources are constructed in the subroutine ***initialize_voltage_source_updating_coefficients*** as implemented in Listing 4.6. The indices of the nodes identifying the positions of the voltage sources, **is**, **js**, **ks**, **ie**, **je**, and **ke**, were

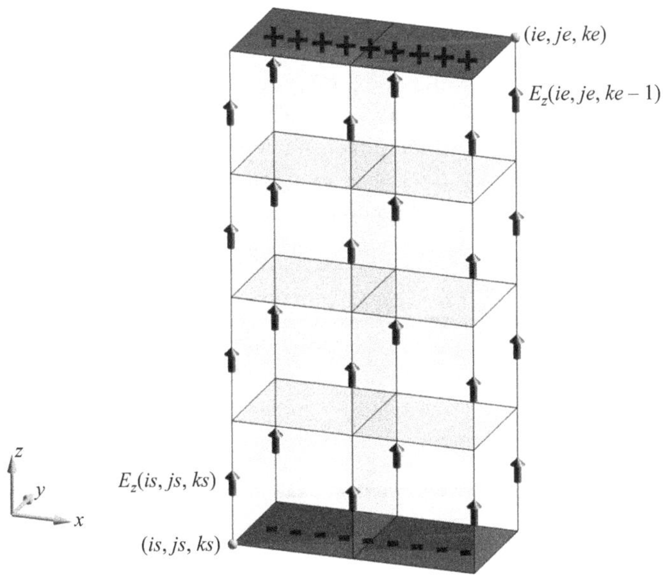

Figure 4.10 A z-directed voltage source defined between the nodes (is, js, ks) and (ie, je, ke).

determined while initializing the voltage sources as shown in Listing 4.3. For instance, consider the z-directed voltage source defined between the nodes (is, js, ks) and (ie, je, ke) and the associated electric field components as illustrated in Figure 4.10. These indices of the nodes are used to identify the indices of the field components that need the special updates due to the voltage sources. The indices of the field components are determined by taking into account the directions of the voltage sources. Since these field components will be accessed frequently, they can be stored in a parameter, **field_indices**, associated with the voltage sources for an easy access. The field components under consideration are usually accessed using three index values, such as

```
Ceze(is:ie, js:je, ks:ke-1)
```

since the field arrays are three-dimensional. An alternative way of accessing an element of a multidimensional array is to use the nice feature of MATLAB called *linear indexing*. Using linear indexing the three indices of a three-dimensional array can be mapped to a single index. Therefore, a single index would be sufficient. For the case where a number of elements need to be indexed, a vector can be constructed in which each value holds the index of the respective element in the three-dimensional array. Therefore, as shown in Listing 4.6, a temporary array of indices, **fi**, is constructed using

```
fi = create_linear_index_list(eps_r_z, is:ie, js:je, ks:ke-1);
```

Here ***create_linear_index_list*** is a function that takes three parameters indicating the range of indices of a subsection of a three-dimensional array and converts it to a one-dimensional linear index array. While doing this conversion, it uses the MATLAB function ***sub2ind***, which needs the size of the three-dimensional array. Therefore, the first argument of ***create_linear_index_list*** is the three-dimensional array under consideration. The implementation of ***create_linear_index_list*** is given in Listing 4.7. After the linear indices array is constructed, instead of accessing a subsection of an array by

```
Ceze(is:ie,js:je,ks:ke-1)
```

one can simply implement

```
Ceze(fi)
```

Listing 4.6 initialize_voltage_source_updating_coefficients.m

```
disp('initializing voltage source updating coefficients');

for ind = 1:number_of_voltage_sources
    is = voltage_sources(ind).is;
    js = voltage_sources(ind).js;
    ks = voltage_sources(ind).ks;
    ie = voltage_sources(ind).ie;
    je = voltage_sources(ind).je;
    ke = voltage_sources(ind).ke;
    R = voltage_sources(ind).resistance_per_component;
    if (R == 0) R = 1e-20; end

    switch (voltage_sources(ind).direction(1))
    case 'x'
        fi = create_linear_index_list(eps_r_x,is:ie-1,js:je,ks:ke);
        a_term = (dt*dx)/(R*dy*dz);
        Cexe(fi) = ...
            (2*eps_0*eps_r_x(fi)-dt*sigma_e_x(fi)-a_term) ...
                ./ (2*eps_0*eps_r_x(fi)+dt*sigma_e_x(fi)+a_term);
        Cexhz(fi)= (2*dt/dy)...
            ./ (2*eps_r_x(fi)*eps_0+dt*sigma_e_x(fi)+a_term);
        Cexhy(fi)= -(2*dt/dz) ...
            ./ (2*eps_r_x(fi)*eps_0 + dt*sigma_e_x(fi)+a_term);
        voltage_sources(ind).Cexs = -(2*dt/(R*dy*dz)) ...
            ./(2*eps_r_x(fi)*eps_0+dt*sigma_e_x(fi)+a_term);
    case 'y'
        fi = create_linear_index_list(eps_r_y,is:ie,js:je-1,ks:ke);
        a_term = (dt*dy)/(R*dz*dx);
        Ceye(fi) = ...
            (2*eps_0*eps_r_y(fi)-dt*sigma_e_y(fi)-a_term) ...
                ./ (2*eps_0*eps_r_y(fi)+dt*sigma_e_y(fi)+a_term);
```

```
        Ceyhx(fi)= (2*dt/dz)...
            ./ (2*eps_r_y(fi)*eps_0+dt*sigma_e_y(fi)+a_term);
        Ceyhz(fi)= -(2*dt/dx) ...
            ./ (2*eps_r_y(fi)*eps_0 + dt*sigma_e_y(fi)+a_term);
        voltage_sources(ind).Ceys = -(2*dt/(R*dz*dx)) ...
            ./(2*eps_r_y(fi)*eps_0+dt*sigma_e_y(fi)+a_term);
    case 'z'
        fi = create_linear_index_list(eps_r_z,is:ie,js:je,ks:ke-1);
        a_term = (dt*dz)/(R*dx*dy);
        Ceze(fi) = ...
            (2*eps_0*eps_r_z(fi)-dt*sigma_e_z(fi)-a_term) ...
                ./ (2*eps_0*eps_r_z(fi)+dt*sigma_e_z(fi)+a_term);
        Cezhy(fi)= (2*dt/dx)...
            ./ (2*eps_r_z(fi)*eps_0+dt*sigma_e_z(fi)+a_term);
        Cezhx(fi)= -(2*dt/dy) ...
            ./ (2*eps_r_z(fi)*eps_0+dt*sigma_e_z(fi)+a_term);
        voltage_sources(ind).Cezs = -(2*dt/(R*dx*dy)) ...
            ./(2*eps_r_z(fi)*eps_0+dt*sigma_e_z(fi)+a_term);
        end
        voltage_sources(ind).field_indices = fi;
end
```

Listing 4.7 create_linear_index_list.m

```
% a function that generates a list of linear indices
function [fi] = create_linear_index_list(array_3d,i_list,j_list,k_list)

i_size = size(i_list,2);
j_size = size(j_list,2);
k_size = size(k_list,2);
number_of_indices = i_size * j_size * k_size;
I = zeros(number_of_indices,1);
J = zeros(number_of_indices,1);
K = zeros(number_of_indices,1);
ind = 1;
for mk = k_list(1):k_list(k_size)
    for mj = j_list(1):j_list(j_size)
        for mi = i_list(1):i_list(i_size)
            I(ind) = mi;
            J(ind) = mj;
            K(ind) = mk;
            ind = ind + 1;
        end
    end
end
fi = sub2ind(size(array_3d), I, J, K);
```

Once the field component indices needing special updates are determined as **fi** in Listing 4.6, the respective components of the general updating coefficients are *recalculated* based on the voltage source updating equations as given in (4.10).

One can notice that there exists an additional term in (4.10), given by $C_{ezs}(i,j,k) \times V_s^{n+\frac{1}{2}}(i,j,k)$, for updating $E_z^{n+1}(i,j,k)$. The term $V_s^{n+\frac{1}{2}}(i,j,k)$ is the value of the voltage source waveform at time instant $\left(n+\frac{1}{2}\right) \times \Delta t$, and this value is stored in the parameter **voltage_sources(i).waveform**. The other term C_{ezs} is the additional coefficient term for the voltage source. This term is constructed based on (4.10) as the parameter **Cezs** and associated with the respective voltage source as **voltage_sources(i).Cezs**. Similarly, if the voltage source updates the electric field components E_x the parameter **voltage_sources(i).Cexs** can be constructed, whereas if the voltage source updates E_y the parameter **voltage_sources(i).Ceys** can be constructed, based on the direction of the voltage sources as implemented in Listing 4.6. Finally, the parameter **voltage_sources(i).field_indices** is constructed from **fi**. Then the parameter **field_indices** are used to access the respective field components while updating the electric field components during the time-marching loop.

4.2.4.2 Initialization of current source updating coefficients

The initialization of current source updating coefficients is performed in the subroutine ***initialize_current_source_updating_coefficients***, which is shown in Listing 4.8. The initialization of current source coefficients is the same as the initialization of voltage source coefficients and follows the updating equation given in (4.15).

Listing 4.8 initialize_current_source_updating_coefficients.m

```
disp('initializing current source updating coefficients');

for ind = 1:number_of_current_sources
    is = current_sources(ind).is;
    js = current_sources(ind).js;
    ks = current_sources(ind).ks;
    ie = current_sources(ind).ie;
    je = current_sources(ind).je;
    ke = current_sources(ind).ke;

    R = current_sources(ind).resistance_per_component;

    switch (current_sources(ind).direction(1))
    case 'x'
        fi = create_linear_index_list(eps_r_x, is:ie-1,js:je,ks:ke);
        a_term = (dt*dx)/(R*dy*dz);
        Cexe(fi) = ...
            (2*eps_0*eps_r_x(fi)-dt*sigma_e_x(fi)-a_term) ...
                ./ (2*eps_0*eps_r_x(fi)+dt*sigma_e_x(fi)+a_term);
        Cexhz(fi)= (2*dt/dy)...
            ./ (2*eps_r_x(fi)*eps_0+dt*sigma_e_x(fi)+a_term);
        Cexhy(fi)= -(2*dt/dz) ...
            ./ (2*eps_r_x(fi)*eps_0 + dt*sigma_e_x(fi)+a_term);
        current_sources(ind).Cexs = -(2*dt/(dy*dz)) ...
            ./(2*eps_r_x(fi)*eps_0+dt*sigma_e_x(fi)+a_term);
```

```
    case 'y'
        fi = create_linear_index_list(eps_r_y,is:ie,js:je-1,ks:ke);
        a_term = (dt*dy)/(R*dz*dx);
        Ceye(fi) = ...
            (2*eps_0*eps_r_y(fi)-dt*sigma_e_y(fi)-a_term) ...
                ./ (2*eps_0*eps_r_y(fi)+dt*sigma_e_y(fi)+a_term);
        Ceyhx(fi)= (2*dt/dz)...
            ./ (2*eps_r_y(fi)*eps_0+dt*sigma_e_y(fi)+a_term);
        Ceyhz(fi)= -(2*dt/dx) ...
            ./ (2*eps_r_y(fi)*eps_0 + dt*sigma_e_y(fi)+a_term);
        current_sources(ind).Ceys = -(2*dt/(dz*dx)) ...
            ./(2*eps_r_y(fi)*eps_0+dt*sigma_e_y(fi)+a_term);
    case 'z'
        fi = create_linear_index_list(eps_r_z,is:ie,js:je,ks:ke-1);
        a_term = (dt*dz)/(R*dx*dy);
        Ceze(fi) = ...
            (2*eps_0*eps_r_z(fi)-dt*sigma_e_z(fi)-a_term) ...
                ./ (2*eps_0*eps_r_z(fi)+dt*sigma_e_z(fi)+a_term);
        Cezhy(fi)= (2*dt/dx)...
            ./ (2*eps_r_z(fi)*eps_0+dt*sigma_e_z(fi)+a_term);
        Cezhx(fi)= -(2*dt/dy) ...
            ./ (2*eps_r_z(fi)*eps_0+dt*sigma_e_z(fi)+a_term);
        current_sources(ind).Cezs = -(2*dt/(dx*dy)) ...
            ./(2*eps_r_z(fi)*eps_0+dt*sigma_e_z(fi)+a_term);
    end
    current_sources(ind).field_indices = fi;
end
```

4.2.4.3 Initialization of inductor updating coefficients

The updating coefficients for modeling inductors are initialized in the subroutine ***initialize_inductor_updating_coefficients***, which is shown in Listing 4.9, based on the updating equation (4.29). Comparing (4.24) with (1.28) one can observe that the form of the updating equation modeling an inductor is the same as the form of the general updating equation. Therefore, there is no need to recalculate the coefficients that have been initialized as general updating coefficients. However, a coefficient term C_{ezj} has not been defined in the implementation of the general updating equations. Therefore, the coefficient term C_{ezj} associated with the inductor is defined as **inductors(i).Cezj** in Listing 4.9. Furthermore, the J_{iz} term is also associated with the inductor, and it is also defined as **inductors(i).Jiz** and is initialized with zeros.

As discussed in Section 4.1.6, during the FDTD time-marching iteration, at every time step the new value of $J_{iz}^{n+\frac{1}{2}}(i, j, k)$ is to be calculated by (4.23) using $E_z^n(i, j, k)$ and $J_{iz}^{n-\frac{1}{2}}(i, j, k)$ before updating $E_z^{n+1}(i, j, k)$ using (4.24). Equation (4.23) can be rewritten as

$$\begin{aligned} J_{iz}^{n+\frac{1}{2}}(i, j, k) &= J_{iz}^{n-\frac{1}{2}}(i, j, k) + \frac{\Delta t \Delta z}{L \Delta x \Delta y} E_z^n(i, j, k) \\ &= J_{iz}^{n-\frac{1}{2}}(i, j, k) + C_{jez}(i, j, k) \times E_z^n(i, j, k), \end{aligned}$$

Listing 4.9 initialize_inductor_updating_coefficients.m

```
disp('initializing inductor updating coefficients');

for ind = 1:number_of_inductors
    is = inductors(ind).is;
    js = inductors(ind).js;
    ks = inductors(ind).ks;
    ie = inductors(ind).ie;
    je = inductors(ind).je;
    ke = inductors(ind).ke;

    L = inductors(ind).inductance_per_component;

    switch (inductors(ind).direction(1))
    case 'x'
        fi = create_linear_index_list(eps_r_x,is:ie-1,js:je,ks:ke);
        inductors(ind).Cexj = -(2*dt) ...
            ./ (2*eps_r_x(fi)*eps_0+dt*sigma_e_x(fi));
        inductors(ind).Jix = zeros(size(fi));
        inductors(ind).Cjex = (dt*dx)/ (L*dy*dz);
    case 'y'
        fi = create_linear_index_list(eps_r_y,is:ie,js:je-1,ks:ke);
        inductors(ind).Ceyj = -(2*dt) ...
            ./ (2*eps_r_y(fi)*eps_0+dt*sigma_e_y(fi));
        inductors(ind).Jiy = zeros(size(fi));
        inductors(ind).Cjey = (dt*dy)/ (L*dz*dx);
    case 'z'
        fi = create_linear_index_list(eps_r_z,is:ie,js:je,ks:ke-1);
        inductors(ind).Cezj = -(2*dt) ...
            ./ (2*eps_r_z(fi)*eps_0+dt*sigma_e_z(fi));
        inductors(ind).Jiz = zeros(size(fi));
        inductors(ind).Cjez = (dt*dz)/ (L*dx*dy);
    end
    inductors(ind).field_indices = fi;
end
```

where

$$C_{jez}(i,j,k) = \frac{\Delta t \Delta z}{L \Delta x \Delta y}.$$

Therefore, C_{jez} is an additional coefficient that can be initialized before the time-marching loop. Hence, the parameter **inductors(i).Cjez** is constructed as shown in Listing 4.9.

4.2.4.4 Initialization of resistor updating coefficients

The initialization of updating coefficients related to resistors is performed in the subroutine ***initialize_resistor_updating_coeffcients***, which is shown in Listing 4.10. The initialization

Listing 4.10 initialize_resistor_updating_coefficients.m

```
disp('initializing resistor updating coefficients');

for ind = 1:number_of_resistors
    is = resistors(ind).is;
    js = resistors(ind).js;
    ks = resistors(ind).ks;
    ie = resistors(ind).ie;
    je = resistors(ind).je;
    ke = resistors(ind).ke;

    R = resistors(ind).resistance_per_component;

    switch (resistors(ind).direction(1))
    case 'x'
        fi = create_linear_index_list(eps_r_x, is:ie-1,js:je,ks:ke);
        a_term = (dt*dx)/(R*dy*dz);
        Cexe(fi) = ...
            (2*eps_0*eps_r_x(fi)-dt*sigma_e_x(fi)-a_term) ...
                ./ (2*eps_0*eps_r_x(fi)+dt*sigma_e_x(fi)+a_term);
        Cexhz(fi)= (2*dt/dy)...
            ./ (2*eps_r_x(fi)*eps_0+dt*sigma_e_x(fi)+a_term);
        Cexhy(fi)= -(2*dt/dz) ...
            ./ (2*eps_r_x(fi)*eps_0 + dt*sigma_e_x(fi)+a_term);
    case 'y'
        fi = create_linear_index_list(eps_r_y, is:ie,js:je-1,ks:ke);
        a_term = (dt*dy)/(R*dz*dx);
        Ceye(fi) = ...
            (2*eps_0*eps_r_y(fi)-dt*sigma_e_y(fi)-a_term) ...
                ./ (2*eps_0*eps_r_y(fi)+dt*sigma_e_y(fi)+a_term);
        Ceyhx(fi)= (2*dt/dz)...
            ./ (2*eps_r_y(fi)*eps_0+dt*sigma_e_y(fi)+a_term);
        Ceyhz(fi)= -(2*dt/dx) ...
            ./ (2*eps_r_y(fi)*eps_0 + dt*sigma_e_y(fi)+a_term);
    case 'z'
        fi = create_linear_index_list(eps_r_z, is:ie,js:je,ks:ke-1);
        a_term = (dt*dz)/(R*dx*dy);
        Ceze(fi) = ...
            (2*eps_0*eps_r_z(fi)-dt*sigma_e_z(fi)-a_term) ...
                ./ (2*eps_0*eps_r_z(fi)+dt*sigma_e_z(fi)+a_term);
        Cezhy(fi)= (2*dt/dx)...
            ./ (2*eps_r_z(fi)*eps_0+dt*sigma_e_z(fi)+a_term);
        Cezhx(fi)= -(2*dt/dy) ...
            ./ (2*eps_r_z(fi)*eps_0+dt*sigma_e_z(fi)+a_term);
    end
    resistors(ind).field_indices = fi;
end
```

of resistor coefficients is straightforward; the indices of field components associated with a resistor are determined in the temporary parameter **fi**, and the elements of the general coefficient arrays are accessed through **fi** and are recalculated based on the resistor updating equation (4.16).

4.2.4.5 Initialization of capacitor updating coefficients

The initialization of updating coefficients related to capacitors is performed in the subroutine ***initialize_capacitor_updating_coefficients***, which is shown in Listing 4.11. The initialization of capacitor coefficients is similar to the initialization of resistor coefficients and is based on the updating equation (4.20).

Listing 4.11 initialize_capacitor_updating_coefficients.m

```
disp('initializing capacitor updating coefficients');

for ind = 1:number_of_capacitors
    is = capacitors(ind).is;
    js = capacitors(ind).js;
    ks = capacitors(ind).ks;
    ie = capacitors(ind).ie;
    je = capacitors(ind).je;
    ke = capacitors(ind).ke;

    C = capacitors(ind).capacitance_per_component;

    switch (capacitors(ind).direction(1))
        case 'x'
            fi = create_linear_index_list(eps_r_x,is:ie-1,js:je,ks:ke);
            a_term = (2*C*dx)/(dy*dz);
            Cexe(fi) = ...
                (2*eps_0*eps_r_x(fi)-dt*sigma_e_x(fi)+a_term) ...
                    ./ (2*eps_0*eps_r_x(fi)+dt*sigma_e_x(fi)+a_term);
            Cexhz(fi)= (2*dt/dy)...
                ./ (2*eps_r_x(fi)*eps_0+dt*sigma_e_x(fi)+a_term);
            Cexhy(fi)= -(2*dt/dz) ...
                ./ (2*eps_r_x(fi)*eps_0 + dt*sigma_e_x(fi)+a_term);
        case 'y'
            fi = create_linear_index_list(eps_r_y,is:ie,js:je-1,ks:ke);
            a_term = (2*C*dy)/(dz*dx);
            Ceye(fi) = ...
                (2*eps_0*eps_r_y(fi)-dt*sigma_e_y(fi)+a_term) ...
                    ./ (2*eps_0*eps_r_y(fi)+dt*sigma_e_y(fi)+a_term);
            Ceyhx(fi)= (2*dt/dz)...
                ./ (2*eps_r_y(fi)*eps_0+dt*sigma_e_y(fi)+a_term);
            Ceyhz(fi)= -(2*dt/dx) ...
                ./ (2*eps_r_y(fi)*eps_0 + dt*sigma_e_y(fi)+a_term);
        case 'z'
            fi = create_linear_index_list(eps_r_z,is:ie,js:je,ks:ke-1);
            a_term = (2*C*dz)/(dx*dy);
            Ceze(fi) = ...
                (2*eps_0*eps_r_z(fi)-dt*sigma_e_z(fi)+a_term) ...
                    ./ (2*eps_0*eps_r_z(fi)+dt*sigma_e_z(fi)+a_term);
            Cezhy(fi)= (2*dt/dx)...
                ./ (2*eps_r_z(fi)*eps_0+dt*sigma_e_z(fi)+a_term);
            Cezhx(fi)= -(2*dt/dy) ...
                ./ (2*eps_r_z(fi)*eps_0+dt*sigma_e_z(fi)+a_term);
    end
    capacitors(ind).field_indices = fi;
end
```

4.2.4.6 Initialization of diode updating coefficients

The algorithm for modeling diodes is significantly different compared with other lumped element components, and it requires the initialization of a different set of parameters. The updating coefficients for modeling diodes are initialized in ***initialize_diode_updating_coefficients***, and implementation of this subroutine is shown in Listing 4.12.

Listing 4.12 initialize_diode_updating_coefficients.m

```
disp('initializing diode updating coefficients');

q = 1.602*1e-19; % charge of an electron
k = 1.38066e-23; % Boltzman constant, joule/kelvin
T = 273+27;      % Kelvin; room temperature
I_0 = 1e-14;     % saturation current

for ind = 1:number_of_diodes
    is = diodes(ind).is;
    js = diodes(ind).js;
    ks = diodes(ind).ks;
    ie = diodes(ind).ie;
    je = diodes(ind).je;
    ke = diodes(ind).ke;

    if strcmp(diodes(ind).direction(2),'n')
        sgn = -1;
    else
        sgn = 1;
    end
    switch (diodes(ind).direction(1))
        case 'x'
            fi = create_linear_index_list(eps_r_x,is,js,ks);
            diodes(ind).B = sgn*q*dx/(2*k*T);
            diodes(ind).Cexd = ...
                -sgn*(2*dt*I_0/(dy*dz)) ...
                ./(2*eps_r_x(fi)*eps_0 + dt*sigma_e_x(fi));
            diodes(ind).Exn = 0;
        case 'y'
            fi = create_linear_index_list(eps_r_y,is,js,ks);
            diodes(ind).B = sgn*q*dy/(2*k*T);
            diodes(ind).Ceyd = ...
                -sgn*(2*dt*I_0/(dz*dx)) ...
                ./(2*eps_r_y(fi)*eps_0 + dt*sigma_e_y(fi));
            diodes(ind).Eyn = 0;
        case 'z'
            fi = create_linear_index_list(eps_r_z,is,js,ks);
            diodes(ind).B = sgn*q*dz/(2*k*T);
            diodes(ind).Cezd = ...
                -sgn*(2*dt*I_0/(dx*dy)) ...
                ./(2*eps_r_z(fi)*eps_0 + dt*sigma_e_z(fi));
            diodes(ind).Ezn = 0;
    end
    diodes(ind).field_indices = fi;
end
```

As mentioned in Section 4.2.3, only one electric field component is associated with a diode; therefore, it needs a special updating based on the equations in Section 4.1.8. For example, for a diode oriented in the positive z direction and defined between the nodes (is, js, ks) and (ie, je, ke), only the field component $E_z(is, js, ks)$ is updated as illustrated in Figure 4.9.

As discussed in Section 4.1.8, the new value of the electric field component associated with a diode is obtained by the solution of the equation

$$Ae^{Bx} + x + C = 0, \tag{4.42}$$

where

$$x = E_z^{n+1}(i, j, k), \quad A = -C_{ezd}(i, j, k)e^{B \times E_z^n(i, j, k)}, \quad B = (q\Delta z/2kT),$$

$$C = C_{eze}(i, j, k) \times E_z^n(i, j, k) + C_{ezhy}(i, j, k) \times \left(H_y^{n+\frac{1}{2}}(i, j, k) - H_y^{n+\frac{1}{2}}(i-1, j, k)\right)$$

$$+ C_{ezhx}(i, j, k) \times \left(H_x^{n+\frac{1}{2}}(i, j, k) - H_x^{n+\frac{1}{2}}(i, j-1, k)\right) + C_{ezd}(i, j, k).$$

Here the $E_z^{n+1}(i, j, k)$ is the electric field component that is sought. The term B is a constant that can be constructed before the time-marching loop starts and stored as parameter **diodes(i).B**. However, the parameters A and C are expressions that have to be recalculated at every time step of the time-marching iteration using the previous values of electric and magnetic field components.

Comparing the first three terms on the right-hand side of the equation expressing C in (4.34) with the updating equation for general media (1.28), one can see that their forms are the same and that the coefficients C_{eze}, C_{ezhy}, and C_{ezhx} are different only by a minus sign. These coefficients were initialized as **Ceze**, **Cezhy**, and **Cezhx** in the subroutine ***initialize_updating_coeffcients***, and there is no need to redefine them. We only need to keep in mind that the components of these parameters with the indices (is, js, ks) – **Ceze(is,js,ks)**, **Cezhy(is,js,ks)**, and **Cezhz(is,js,ks)** – are the negatives of the coefficients that are used to recalculate C while updating $E_z^{n+1}(is, js, ks)$ as associated with a diode.

However, the fourth term in the expression of C includes a coefficient, $C_{ezd}(i, j, k)$, which has not been defined before. Therefore, a parameter, **diodes(i).Cezd**, is defined to store the value of $C_{ezd}(is, js, ks)$ in Listing 4.12.

An additional parameter that is defined and initialized with zero in Listing 4.12 is **diodes (i).Ezn**. This parameter stores $E_z^n(is, js, ks)$ which is the value of electric field component at the previous time step. This parameter is used to recalculate the coefficient C at every time step, and it should be available when recalculating C. Therefore, at every time step the value of $E_z^n(is, js, ks)$ is stored in the parameter **diodes(i).Ezn** to have it available at the next time step for the calculation of C.

4.2.5 Sampling electric and magnetic fields, voltages, and currents

The reason for performing an FDTD simulation is to obtain some results that characterize the response of an electromagnetics problem. The types of results that can be obtained from an

FDTD simulation vary based on the type of the electromagnetics problem at hand. However, the fundamental result that can be obtained from an FDTD simulation is the behavior of electric and magnetic fields in time due to a source exciting the problem space, since these are the fundamental quantities that the FDTD algorithm is based on. Other types of available results can be obtained only by the translation of transient electric and magnetic fields into other parameters. For instance, in a microwave circuit simulation, once the transient electric and magnetic fields are obtained, it is easy to calculate transient voltages and currents. Then the transient voltages and currents can be translated into the frequency domain by using the Fourier transform, and scattering parameters (S-parameters) of the microwave circuit can be obtained. Similarly, radiation patterns of an antenna or radar cross-section of a scatterer can be obtained by employing the appropriate transforms.

4.2.5.1 Calculation of sampled voltages

The voltages and currents that will be sampled are represented similar to lumped element components by prism-like volumes and their directions. The voltages across and the currents flowing through these volumes will be calculated using the electric field components across and the magnetic field components around these volumes, respectively, during the time-marching loop.

For instance, consider the volume defined between the nodes (is, js, ks) and (ie, je, ke). This volume is placed between two parallel PEC plates. An average value of voltage can be calculated between the PEC plates across the volume. Here we employ the integral form of (4.2), which can be given as

$$V = -\int \vec{E} \cdot d\vec{l}. \tag{4.43}$$

This integration is expressed as a summation in the discrete space. For the configuration in Figure 4.11, the voltage between the upper and lower PEC plates can be expressed in discrete form as

$$V = -dz \times \sum_{k=ks}^{ke-1} E_z(is, js, k). \tag{4.44}$$

However, it can be seen in Figure 4.11 that there are multiple paths for performing the discrete summation and that the voltages calculated through these paths should be very close to each other. Therefore, an average of these voltage values can be calculated by

$$V = \frac{-dz}{(ie - is + 1) \times (je - js + 1)} \times \sum_{i=is}^{ie} \sum_{j=js}^{je} \sum_{k=ks}^{ke-1} E_z(i, j, k). \tag{4.45}$$

This expression can be written in the form

$$V = C_{suf} \times \sum_{i=is}^{ie} \sum_{j=js}^{je} \sum_{k=ks}^{ke-1} E_z(i, j, k), \tag{4.46}$$

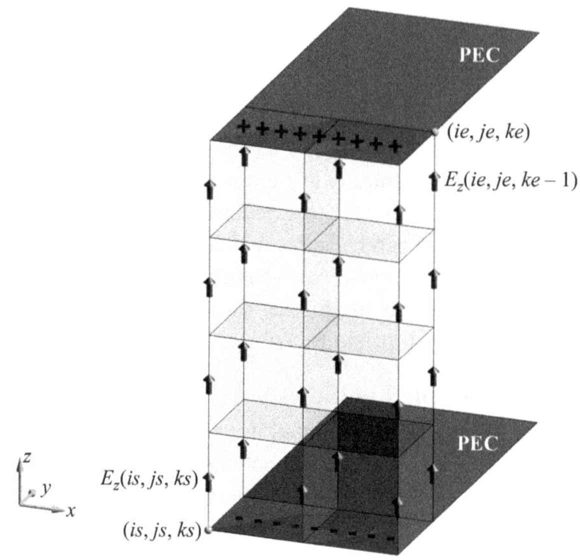

Figure 4.11 A volume between PEC plates where a z-directed sampled voltage is required.

where

$$C_{suf} = \frac{-dz}{(ie - is + 1) \times (je - js + 1)}. \tag{4.47}$$

The coefficient C_{svf} is a scaling factor multiplied by the sum of the field components $E_z(is:ie, js:je, ks:ke-1)$ to obtain the average voltage.

The aforementioned equations still hold for the cases where is is equal to ie or js is equal to je. In those cases the voltage is sampled along a line segment or across a surface.

4.2.5.2 Calculation of sampled currents

To calculate the current flowing through a surface we can employ Ampere's law in integral form, which is given by

$$I_{free} = \oint \vec{H} \cdot d\bar{l}, \tag{4.48}$$

where I_{free} is the total free current. We need to describe the integral in (4.48) as a summation of magnetic field components in the discrete space. For instance, consider the case shown in Figure 4.12. The figure shows magnetic field components enclosing a surface. The positions of the fields are given with respect to the Yee cell, and indices of the fields are given with respect to the nodes (is, js, ks) and (ie, je, ke). These fields can be used to calculate the enclosed total free current by

$$\begin{aligned} I_z(ke-1) = {} & dx \times \sum_{i=is}^{ie} H_x(i, js-1, ke-1) + dy \times \sum_{j=js}^{je} H_y(ie, j, ke-1) \\ & - dx \times \sum_{i=is}^{ie} H_x(i, je, ke-1) - dy \times \sum_{j=js}^{je} H_y(is-1, j, ke-1). \end{aligned} \tag{4.49}$$

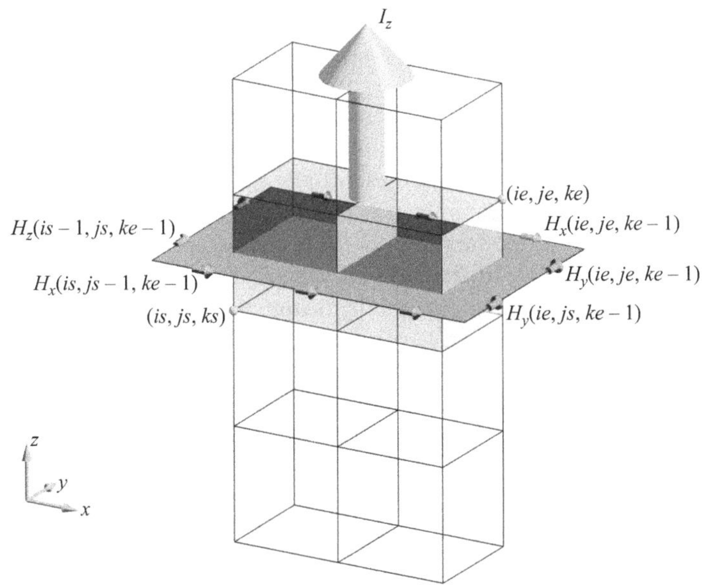

Figure 4.12 A surface enclosed by magnetic field components and z-directed current flowing through it.

Here we indexed I_z by $(ke-1)$ since the magnetic field components with the index $(ke-1)$ in the z direction are used to calculate I_z.

As discussed before, we can determine the position of the current sampling by a volume identified by the node indices (is, js, ks) and (ie, je, ke). Then, a surface intersecting with the cross-section of this volume and enclosing it can be used to determine the magnetic field components that are used to calculate the current as illustrated in Figure 4.12. It is possible that the volume identified by the nodes (is, js, ks) and (ie, je, ke) reduces to a surface or a line segment or even to a point in the case where $ie = is, je = js$, and $ke = ks$. The expression in (4.49) still can be used to calculate the current in those cases. However, one should notice that the surface represented by the magnetic field components indexed by $(ke - 1)$ is not coinciding with the nodes; it instead crosses the centers of the cells. Therefore, if sampling of both the voltage and the current is required at the same position, one should sample one of these quantities over multiple positions and should take the average to obtain both the voltage and current effectively at the same position.

Furthermore, another point that should be kept in mind is that the voltages and currents are sampled at different time instants and a half time step apart from each other, since the voltages are calculated from electric field components and currents are calculated from magnetic field components, which are offset in time due to the leap-frog Yee algorithm. This time difference can also be compensated for as pointed out in [11].

4.2.6 Definition and initialization of output parameters

Definition of types of results sought from a simulation is part of the construction of an electromagnetics problem in the FDTD method. This can be done in the *definition* stage of the main FDTD program ***fdtd_solve*** as shown in Listing 3.1, in the subroutine ***define_output_parameters***. In this section, we demonstrate how the electric field, magnetic field, voltages, and currents can be defined as output parameters in Listing 4.13. We also demonstrate how to set up some auxiliary parameters that would be used to display the three-dimensional view of the objects in the problem space, the material mesh of the problem space, and a runtime animation of fields on some plane cuts while the simulation is running. The definition of other types of parameters is discussed in the subsequent chapters and their implementation are added to the subroutine ***define_output_parameters***.

Listing 4.13 starts with the definition and initialization of the empty structure arrays **sampled_electric_fields**, **sampled_magnetic_fields**, **sampled-voltages**, and **sampled_currents** for representing the respective parameters to be sampled at every time step of the FDTD loop as functions of time. As these parameters are captured during the FDTD loop, they can be plotted on MATLAB figures to display the progress of the sampled parameters. However, it is not meaningful to display the progress of these parameters at every time step, as it may slow down the simulation considerably. It is better to refresh the plots at a rate defined with the parameter **plotting_step**. For instance, a value of 10 implies that the figures will be refreshed once every 10 time steps.

Listing 4.13 define_output_parameters.m

```
disp('defining output parameters');

sampled_electric_fields = [];
sampled_magnetic_fields = [];
sampled_voltages = [];
sampled_currents = [];

% figure refresh rate
plotting_step = 10;

% mode of operation
run_simulation = true;
show_material_mesh = true;
show_problem_space = true;

% define sampled electric fields
% component: vector component 'x','y','z', or magnitude 'm'
% display_plot = true, in order to plot field during simulation
sampled_electric_fields(1).x = 30*dx;
sampled_electric_fields(1).y = 30*dy;
sampled_electric_fields(1).z = 10*dz;
sampled_electric_fields(1).component = 'x';
sampled_electric_fields(1).display_plot = true;

% define sampled magnetic fields
% component: vector component 'x','y','z', or magnitude 'm'
% display_plot = true, in order to plot field during simulation
sampled_magnetic_fields(1).x = 30*dx;
sampled_magnetic_fields(1).y = 30*dy;
sampled_magnetic_fields(1).z = 10*dz;
sampled_magnetic_fields(1).component = 'm';
sampled_magnetic_fields(1).display_plot = true;
```

```
% define sampled voltages
sampled_voltages(1).min_x = 5.0e-3;
sampled_voltages(1).min_y = 0;
sampled_voltages(1).min_z = 0;
sampled_voltages(1).max_x = 5.0e-3;
sampled_voltages(1).max_y = 2.0e-3;
sampled_voltages(1).max_z = 4.0e-3;
sampled_voltages(1).direction = 'zp';
sampled_voltages(1).display_plot = true;

% define sampled currents
sampled_currents(1).min_x = 5.0e-3;
sampled_currents(1).min_y = 0;
sampled_currents(1).min_z = 4.0e-3;
sampled_currents(1).max_x = 5.0e-3;
sampled_currents(1).max_y = 2.0e-3;
sampled_currents(1).max_z = 4.0e-3;
sampled_currents(1).direction = 'xp';
sampled_currents(1).display_plot = true;

% display problem space parameters
problem_space_display.labels = true;
problem_space_display.axis_at_origin = false;
problem_space_display.axis_outside_domain = true;
problem_space_display.grid_xn = false;
problem_space_display.grid_xp = true;
problem_space_display.grid_yn = false;
problem_space_display.grid_yp = true;
problem_space_display.grid_zn = true;
problem_space_display.grid_zp = false;
problem_space_display.outer_boundaries = true;
problem_space_display.cpml_boundaries = true;

% define animation
% field_type shall be 'e' or 'h'
% plane cut shall be 'xy', yz, or zx
% component shall be 'x', 'y', 'z', or 'm;
animation(1).field_type = 'e';
animation(1).component = 'm';
animation(1).plane_cut(1).type = 'xy';
animation(1).plane_cut(1).position = 0;
animation(1).plane_cut(2).type = 'yz';
animation(1).plane_cut(2).position = 0;
animation(1).plane_cut(3).type = 'zx';
animation(1).plane_cut(3).position = 0;
animation(1).enable = true;
animation(1).display_grid = false;
animation(1).display_objects = true;

animation(2).field_type = 'h';
animation(2).component = 'x';
animation(2).plane_cut(1).type = 'xy';
animation(2).plane_cut(1).position = -5;
```

```
animation(2).plane_cut(2).type = 'xy';
animation(2).plane_cut(2).position  = 5;
animation(2).enable = true;
animation(2).display_grid = true;
animation(2).display_objects = true;
```

Usually it is a good practice to examine the positioning of the objects in the three-dimensional problem space before running the simulation. Furthermore, examining the distribution of the material parameters on the FDTD grid will also reveal if the simulation is set up correctly. In these cases it is better to run the simulation up to a point where the three-dimensional view of the problem space and the material mesh can be displayed. Therefore, a parameter named **run_simulation** is defined. If this parameter is ***true***, then the program proceeds with running the simulation; otherwise, it stops after calling the subroutines ***display_problem_space*** and ***display_material_mesh*** as shown in Listing 3.1. The ***display_problem_space*** is a subroutine that displays the three-dimensional view of the problem space, and it is controlled by a logical parameter named **show_problem_space**. A set of parameters that can be used to set the view of the problem space is listed at the end of Listing 4.13. Similarly, ***display_material_mesh*** is a subroutine that launches a utility program with a graphical user interface, ***display_material_mesh_gui***, and it is controlled by a parameter named **show_material_mesh**. For instance, the material grid in Figure 3.13 is obtained using ***display_material_mesh_gui***.

During the FDTD time-marching loop, the electric and magnetic fields are calculated all over the problem space at their respective positions on the Yee grid. Since all of the field components are available, any of them can be captured, plotted, or stored. For instance, some field components on some plane cuts can be captured and plotted on figures so as to demonstrate the real-time progression of fields on these plane cuts. Or some fields can be captured and stored for postprocessing. Here we limit our implementation to field components sampled at nodes of the FDTD grid. To define a node, we need its position in the Cartesian coordinates. Therefore, the parameters **x**, **y**, and **z** are defined in the structures **sampled_electric_fields(i)** and **sampled_magnetic_fields(i)**. Since the fields are vectors, their x, y, and z components or magnitudes can be sampled. Hence, the parameter **component**, which can take one of the values "***x***," "***y***," "***z***," or "***m***," is defined with **sampled_electric_fields(i)** and **sampled_magnetic_fields(i)**. If the **component** takes the value "***m***" then the magnitude of the field is calculated using

$$E_m = \sqrt{E_x^2 + E_y^2 + E_z^2}, \quad H_m = \sqrt{H_x^2 + H_y^2 + H_z^2}.$$

One additional parameter is **display_plot**, which takes one of the values ***true*** or ***false*** and determines whether this field component will be plotted while the simulation is running or not.

Meanwhile, the volumes representing the sampled voltages and currents are identified by the parameters **min_x**, **min_y**, **min_z**, **max_x**, **max_y**, and **max_z**. The directions of the sampled voltages and currents are defined by the parameter **direction**, which can take one of the values "***xp***," "***yp***," "***zp***," "***xn***," "***yn***," or "***zn***."

As discussed already, since all of the field components are available at every time step of the simulation, they can be captured on desired plane cuts and plotted; thus, a propagation of fields on these plane cuts can be simulated. The definition of animation as an output is done using a parameter **animation**. The parameter **animation** is an array so each element in the array generates a figure that displays the fields. An element **animation(i)** has the following subfields that are used to define the properties of the animation. The subfield **field_type** can take a value either "*e*" or "*h*," determining whether the electric field or magnetic field will be animated, respectively. Then the subfield **component** is used to define which component of the field to display, and it takes one of the values "*x*," "*y*," "*z*," or "*m*." Then several plane cuts can be defined to be displayed on a figure using the subfield **plane_cut(j)**. The parameter **plane_cut(j).type** can take one of the values "*xy*," "*yz*," or "*zx*," describing which principal plane the plane cut is parallel to. Then the parameter **plane_cut(j).position** determines the position of the plane cut. For instance, if **type** is "*yz*," and **position** is 5, then the plane cut is the plane at $x = 5$. Finally, three other subfields of **animation(i)** are the logical parameters **enable**, **display_grid**, and **display_objects**. The animation can be disabled by simply setting **enable** "*false*," the FDTD grid can be displayed during the simulation by setting **display_grid** "*true*," and the objects in the problem space can be displayed by a wire grid by setting **display_objects** "*true*."

4.2.6.1 Initialization of output parameters

The initialization of output parameters is implemented in the subroutine ***initialize_output_parameters***, as shown in Listing 4.14. Here the parameter **sampled_electric_fields** represents the sampled electric field components, and **sampled_magnetic_fields** represents the sampled magnetic field components. Similarly, the parameters **sampled_voltages** and **sampled_currents** represent sampled voltages and currents, respectively. As indicated before, the electric and magnetic field components are sampled at specified positions, which coincide with respective nodes in the Yee grid. The indices of the sampling nodes are determined in Listing 4.14 and are stored in the parameters **is**, **js**, and **ks**. Furthermore, a one-dimensional array with size **number_of_time_steps** is constructed as **sampled_value** and initialized with zeros. This parameter is used to store the new computed values of the sampled component while the FDTD iterations proceed. Another one-dimensional array with size **number_of_time_steps** is **time**, which is used to store the time instants of the respective captured values. One should notice that there is a half time step duration ($dt/2$) shift between the electric field and magnetic field **time** arrays.

Listing 4.14 initialize_output_parameters.m

```
disp('initializing the output parameters');

number_of_sampled_electric_fields  = size(sampled_electric_fields,2);
number_of_sampled_magnetic_fields  = size(sampled_magnetic_fields,2);
number_of_sampled_voltages  = size(sampled_voltages,2);
number_of_sampled_currents  = size(sampled_currents,2);

% initialize sampled electric field terms
for ind=1:number_of_sampled_electric_fields
    is = round((sampled_electric_fields(ind).x ...
        - fdtd_domain.min_x)/dx)+1;
```

```
    js = round((sampled_electric_fields(ind).y ...
        - fdtd_domain.min_y)/dy)+1;
    ks = round((sampled_electric_fields(ind).z ...
        - fdtd_domain.min_z)/dz)+1;
    sampled_electric_fields(ind).is = is;
    sampled_electric_fields(ind).js = js;
    sampled_electric_fields(ind).ks = ks;
    sampled_electric_fields(ind).sampled_value = ...
        zeros(1, number_of_time_steps);
    sampled_electric_fields(ind).time = ...
        ([1:number_of_time_steps])*dt;
end

% initialize sampled magnetic field terms
for ind=1:number_of_sampled_magnetic_fields
    is = round((sampled_magnetic_fields(ind).x ...
        - fdtd_domain.min_x)/dx)+1;
    js = round((sampled_magnetic_fields(ind).y ...
        - fdtd_domain.min_y)/dy)+1;
    ks = round((sampled_magnetic_fields(ind).z ...
        - fdtd_domain.min_z)/dz)+1;
    sampled_magnetic_fields(ind).is = is;
    sampled_magnetic_fields(ind).js = js;
    sampled_magnetic_fields(ind).ks = ks;
    sampled_magnetic_fields(ind).sampled_value = ...
        zeros(1, number_of_time_steps);
    sampled_magnetic_fields(ind).time = ...
        ([1:number_of_time_steps]-0.5)*dt;
end

% initialize sampled voltage terms
for ind=1:number_of_sampled_voltages
    is = round((sampled_voltages(ind).min_x - fdtd_domain.min_x)/dx)
    js = round((sampled_voltages(ind).min_y - fdtd_domain.min_y)/dy)+1;
    ks = round((sampled_voltages(ind).min_z - fdtd_domain.min_z)/dz)+1;
    ie = round((sampled_voltages(ind).max_x - fdtd_domain.min_x)/dx)+1;
    je = round((sampled_voltages(ind).max_y - fdtd_domain.min_y)/dy)+1;
    ke = round((sampled_voltages(ind).max_z - fdtd_domain.min_z)/dz)+1;
    sampled_voltages(ind).is = is;
    sampled_voltages(ind).js = js;
    sampled_voltages(ind).ks = ks;
    sampled_voltages(ind).ie = ie;
    sampled_voltages(ind).je = je;
    sampled_voltages(ind).ke = ke;
    sampled_voltages(ind).sampled_value = ...
                        zeros(1, number_of_time_steps);

    switch (sampled_voltages(ind).direction(1))
    case 'x'
        fi = create_linear_index_list(Ex,is:ie-1,js:je,ks:ke);
        sampled_voltages(ind).Csvf = -dx/((je-js+1)*(ke-ks+1));
    case 'y'
        fi = create_linear_index_list(Ey,is:ie,js:je-1,ks:ke);
        sampled_voltages(ind).Csvf = -dy/((ke-ks+1)*(ie-is+1));
```

```
    case 'z'
        fi = create_linear_index_list(Ez,is:ie,js:je,ks:ke-1);
        sampled_voltages(ind).Csvf = -dz/((ie-is+1)*(je-js+1));
    end
    if strcmp(sampled_voltages(ind).direction(2),'n')
        sampled_voltages(ind).Csvf = ...
            -1 * sampled_voltages(ind).Csvf;
    end
    sampled_voltages(ind).field_indices = fi;
    sampled_voltages(ind).time = ([1:number_of_time_steps])*dt;
end

% initialize sampled current terms
for ind=1:number_of_sampled_currents
    is = round((sampled_currents(ind).min_x - fdtd_domain.min_x)/dx)+1;
    js = round((sampled_currents(ind).min_y - fdtd_domain.min_y)/dy)+1;
    ks = round((sampled_currents(ind).min_z - fdtd_domain.min_z)/dz)+1;
    ie = round((sampled_currents(ind).max_x - fdtd_domain.min_x)/dx)+1;
    je = round((sampled_currents(ind).max_y - fdtd_domain.min_y)/dy)+1;
    ke = round((sampled_currents(ind).max_z - fdtd_domain.min_z)/dz)+1;
    sampled_currents(ind).is = is;
    sampled_currents(ind).js = js;
    sampled_currents(ind).ks = ks;
    sampled_currents(ind).ie = ie;
    sampled_currents(ind).je = je;
    sampled_currents(ind).ke = ke;
    sampled_currents(ind).sampled_value = ...
                        zeros(1, number_of_time_steps);
    sampled_currents(ind).time =([1:number_of_time_steps]-0.5)*dt;
end
```

The indices of the nodes indicating the positions of the sampled voltages are calculated and assigned to the parameters **is**, **js**, **ks**, **ie**, **je**, and **ke**. The indices of the fields that are used to calculate sampled voltages are determined and assigned to the parameter **field_indices**. Then coefficient C_{svf}, which is a scaling factor as described in Section 4.2.5, is calculated using (4.47) and is assigned to the parameter **Csvf**.

4.2.6.2 Initialization of figures for runtime display

So far we have defined four types of outputs for an FDTD simulation: namely, sampled electric and magnetic fields, voltages, and currents. While the FDTD simulation is running, the progress of these sampled parameters can be displayed on MATLAB figures. In the subroutine ***define_output_parameters***, while defining the sampled components, a parameter **display_plot** is associated to each defined component indicating whether the real-time progress of this component displayed. If there are any components defined for runtime display, figures displaying these components can be opened and their properties can be set before the FDTD time-marching loop starts. This figure initialization process can be performed in the subroutine ***initialize_display_parameters***, which is given in Listing 4.15. Here figures are launched for every sampled component that is displayed, and their axis labels are set. Furthermore, the number of the figure associated with each sampled component is stored in the parameter **figure_number**.

Listing 4.15 initialize_display_parameters.m

```
disp('initializing display parameters');

% open figures for sampled electric fields
for ind=1:number_of_sampled_electric_fields
    if sampled_electric_fields(ind).display_plot == true
        sampled_electric_fields(ind).figure_number = figure;
        sampled_electric_fields(ind).plot_handle = plot(0,0,'b-');
        xlabel('time (ns)','fontsize',12);
        ylabel('(volt/meter)','fontsize',12);
        title(['sampled electric field [' ...
            num2str(ind) ']'],'fontsize',12);
        grid on;
        hold on;
    end
end

% open figures for sampled magnetic fields
for ind=1:number_of_sampled_magnetic_fields
    if sampled_magnetic_fields(ind).display_plot == true
        sampled_magnetic_fields(ind).figure_number = figure;
        sampled_magnetic_fields(ind).plot_handle = plot(0,0,'b-');
        xlabel('time (ns)','fontsize',12);
        ylabel('(ampere/meter)','fontsize',12);
        title(['sampled magnetic field [' num2str(ind) ']'], ...
            'fontsize',12);
        grid on;
        hold on;
    end
end

% initialize figures for sampled voltages
for ind=1:number_of_sampled_voltages
    if sampled_voltages(ind).display_plot == true
        sampled_voltages(ind).figure_number = figure;
        sampled_voltages(ind).plot_handle = plot(0,0,'b-');
        xlabel('time (ns)','fontsize',12);
        ylabel('(volt)','fontsize',12);
        title(['sampled voltage [' num2str(ind) ']'], ...
            'fontsize',12);
        grid on;
        hold on;
    end
end

% initialize figures for sampled currents
for ind=1:number_of_sampled_currents
    if sampled_currents(ind).display_plot == true
        sampled_currents(ind).figure_number = figure;
        sampled_currents(ind).plot_handle = plot(0,0,'b-');
        xlabel('time (ns)','fontsize',12);
        ylabel('(ampere)','fontsize',12);
        title(['sampled current [' num2str(ind) ']'], ...
            'fontsize',12);
```

```
        grid on;
        hold on;
    end
end

% initialize field animation parameters
initialize_animation_parameters;
```

Finally, a subroutine ***initialize_animation_parameters*** is called, which performs the initialization of parameters for field animation on the plane cuts defined in ***define_output_parameters***.

Furthermore, figures and display parameters for other types of sampled outputs can be initialized in this routine.

4.2.7 Running an FDTD simulation: The time-marching loop

So far we have discussed all *definition* and *initialization* routines in ***fdtd_solve*** as given in Listing 3.1 except the subroutines ***initialize_boundary_conditions*** and ***run_fdtd_time_marching_loop***. The only boundary type we have discussed is the PEC boundary. We only need to ensure that the tangential electric field components on the boundaries of the problem space are zeros to satisfy the PEC boundary conditions. Therefore, there is no need for special initialization routines for PEC boundaries. The subroutine ***run_fdtd_time_marching_loop*** is reserved for other types of boundaries we have not discussed yet and are the subject of another chapter.

Therefore, so far we have constructed the necessary components for starting an FDTD simulation for a problem space assuming PEC boundaries and including objects of isotropic and linear materials and lumped element components. Now we are ready to implement the subroutine ***run_fdtd_time_marching_loop***, which includes the steps of the leap-frog time-marching algorithm of the FDTD method. The implementation of ***run_fdtd_time_marching_loop*** is given in Listing 4.16. We discuss the functions and provide the implementation of subroutines of these steps in this section.

Listing 4.16 starts with the definition and initialization of a parameter **current_time**, which is used to include the current value of time instant during the FDTD loop. Then the MATLAB function ***cputime*** is used to capture the time on the computer system. This is first called to capture the start time of the FDTD loop and is stored in **start_time**. The function **cputime** is called one more time to capture the time on the system after the FDTD loop is completed. The captured value is copied to **end_time**. Hence, the difference between the start and end times is used to calculate the total simulation time spent for the time-marching iterations.

4.2.7.1 Updating magnetic fields

The time-marching iterations are performed **number_of_time_steps** times. At every iteration, first the magnetic field components are updated in the subroutine ***update_magnetic_fields***,

Listing 4.16 run_fdtd_time_marching_loop.m

```
disp (['Starting the time marching loop']);
disp(['Total number of time steps : ' ...
    num2str(number_of_time_steps)]);

start_time = cputime;
current_time = 0;

for time_step = 1:number_of_time_steps
    update_magnetic_fields;
    capture_sampled_magnetic_fields;
    capture_sampled_currents;
    update_electric_fields;
    update_voltage_sources;
    update_current_sources;
    update_inductors;
    update_diodes;
    capture_sampled_electric_fields;
    capture_sampled_voltages;
    display_sampled_parameters;
end

end_time = cputime;
total_time_in_minutes = (end_time - start_time)/60;
disp(['Total simulation time is ' ...
    num2str(total_time_in_minutes) ' minutes.']);
```

Listing 4.17 update_magnetic_fields.m

```
% update magnetic fields

current_time  = current_time + dt/2;

Hx = Chxh.*Hx+Chxey.*(Ey(1:nxp1,1:ny,2:nzp1)-Ey(1:nxp1,1:ny,1:nz)) ...
    + Chxez.*(Ez(1:nxp1,2:nyp1,1:nz)-Ez(1:nxp1,1:ny,1:nz));

Hy = Chyh.*Hy+Chyez.*(Ez(2:nxp1,1:nyp1,1:nz)-Ez(1:nx,1:nyp1,1:nz)) ...
    + Chyex.*(Ex(1:nx,1:nyp1,2:nzp1)-Ex(1:nx,1:nyp1,1:nz));

Hz = Chzh.*Hz+Chzex.*(Ex(1:nx,2:nyp1,1:nzp1)-Ex(1:nx,1:ny,1:nzp1)) ...
    + Chzey.*(Ey(2:nxp1,1:ny,1:nzp1)-Ey(1:nx,1:ny,1:nzp1));
```

as given in Listing 4.17. Here the new values of the magnetic field components are calculated all over the problem space based on the general updating equations (1.29)–(1.31), excluding the impressed magnetic current terms since there are no impressed magnetic currents defined.

4.2.7.2 Updating electric fields

Then the next important step in an FDTD iteration is the update of electric fields, which is performed in the subroutine ***update_electric_fields*** and is shown in Listing 4.18. The electric field components are updated using the general updating equations given in (1.26)–(1.28), except for the impressed electric current terms. While discussing the lumped element components, it has been shown that impressed current terms are needed for modeling inductors. Since only a small number of electric field components need to be updated for modeling the inductors; these components are recalculated by addition of impressed current related terms in the subroutine ***update_inductors***.

One should notice that in Listing 4.18, the electric field components that are tangential to the boundaries are excluded in the updating. The reason is the specific arrangement of field positions on the Yee cell. While updating an electric field component, the magnetic field components encircling it are required. However, for an electric field component resting tangential to the boundary of the problem space, at least one of the required magnetic field components would be out of the problem space and thus would be undefined. Therefore, the electric field components tangential to the boundaries cannot be updated using the general updating equations. However, if these components are not updated, they maintain their initial values, which are zeros, thus naturally simulating the PEC boundaries. Therefore, there is no need for any additional effort to enforce PEC boundaries. However, some other types of boundaries may require special updates for these electric field components.

The updating coefficient values that model the lumped element components were assigned to the respective positions in the updating coefficient arrays; therefore, the updating of the electric field in ***update_electric_fields*** completes the modeling of capacitors and

Listing 4.18 update_electric_fields.m

```
% update electric fields except the tangential components
% on the boundaries

current_time = current_time + dt/2;

Ex(1:nx,2:ny,2:nz) = Cexe(1:nx,2:ny,2:nz).*Ex(1:nx,2:ny,2:nz) ...
                     + Cexhz(1:nx,2:ny,2:nz).*...
                     (Hz(1:nx,2:ny,2:nz)-Hz(1:nx,1:ny-1,2:nz)) ...
                     + Cexhy(1:nx,2:ny,2:nz).*...
                     (Hy(1:nx,2:ny,2:nz)-Hy(1:nx,2:ny,1:nz-1));

Ey(2:nx,1:ny,2:nz)=Ceye(2:nx,1:ny,2:nz).*Ey(2:nx,1:ny,2:nz) ...
                      + Ceyhx(2:nx,1:ny,2:nz).* ...
                      (Hx(2:nx,1:ny,2:nz)-Hx(2:nx,1:ny,1:nz-1)) ...
                      + Ceyhz(2:nx,1:ny,2:nz).* ...
                      (Hz(2:nx,1:ny,2:nz)-Hz(1:nx-1,1:ny,2:nz));

Ez(2:nx,2:ny,1:nz)=Ceze(2:nx,2:ny,1:nz).*Ez(2:nx,2:ny,1:nz) ...
                      + Cezhy(2:nx,2:ny,1:nz).* ...
                      (Hy(2:nx,2:ny,1:nz)-Hy(1:nx-1,2:ny,1:nz)) ...
                      + Cezhx(2:nx,2:ny,1:nz).*...
                      (Hx(2:nx,2:ny,1:nz)-Hx(2:nx,1:ny-1,1:nz));
```

resistors. The modeling of other lumped elements requires addition of some other terms to the electric fields at their respective positions. The following sections elaborate on updates for these other lumped elements.

4.2.7.3 Updating electric fields associated with voltage sources

The updating equations modeling the voltage sources are derived and represented by (4.10). This equation includes an additional voltage source specific term,

$$C_{ezs}(i, j, k) \times V_s^{n+\frac{1}{2}}(i, j, k), \tag{4.50}$$

which was omitted in the implementation of ***update_electric_fields*** in Listing 4.18. This term can be added to the electric field components that are associated with the voltage sources. The indices of the electric field components requiring an additional update are stored in the parameter **voltage_sources(i).field_indices**. The coefficient terms are constructed as **Cexs**, **Ceys**, and **Cezs**, based on the direction of the voltage source. The value of the voltage at the current time step can be retrieved from the waveform array **voltage_sources(i).waveform**. The additional update of electric field components due to voltage sources is performed in ***update_voltage_sources*** as shown in Listing 4.19.

4.2.7.4 Updating electric fields associated with current sources

Updating of the field components associated with the current sources is similar to that for the voltage sources. The updating equations modeling the current sources are derived and given by (4.15). This equation includes an additional current source specific term,

$$C_{ezs}(i, j, k) \times I_s^{n+\frac{1}{2}}(i, j, k). \tag{4.51}$$

The additional update of electric field components due to current sources is performed in ***update_current_sources*** as shown in Listing 4.20.

Listing 4.19 update_voltage_sources.m

```
% updating electric field components
% associated with the voltage sources

for ind = 1:number_of_voltage_sources
    fi = voltage_sources(ind).field_indices;
    switch (voltage_sources(ind).direction(1))
    case 'x'
        Ex(fi) = Ex(fi) + voltage_sources(ind).Cexs ...
            * voltage_sources(ind).voltage_per_e_field(time_step);
    case 'y'
        Ey(fi) = Ey(fi) + voltage_sources(ind).Ceys ...
            * voltage_sources(ind).voltage_per_e_field(time_step);
    case 'z'
        Ez(fi) = Ez(fi) + voltage_sources(ind).Cezs ...
            * voltage_sources(ind).voltage_per_e_field(time_step);
    end
end
```

Listing 4.20 update_current_sources.m

```
% updating electric field components
% associated with the current sources

for ind = 1:number_of_current_sources
    fi = current_sources(ind).field_indices;
    switch (current_sources(ind).direction(1))
    case 'x'
        Ex(fi) = Ex(fi) + current_sources(ind).Cexs ...
            * current_sources(ind).current_per_e_field(time_step);
    case 'y'
        Ey(fi) = Ey(fi) + current_sources(ind).Ceys ...
            * current_sources(ind).current_per_e_field(time_step);
    case 'z'
        Ez(fi) = Ez(fi) + current_sources(ind).Cezs ...
            * current_sources(ind).current_per_e_field(time_step);
    end
end
```

4.2.7.5 Updating electric fields associated with inductors

The updating equations modeling the inductors are derived as in equation (4.24), which includes an additional impressed electric current source term

$$C_{ezj}(i, j, k) \times J_{iz}^{n+\frac{1}{2}}(i, j, k). \tag{4.52}$$

The coefficients in the additional terms have been constructed in Listing 4.9 as **Cexj**, **Ceyj**, and **Cezj**. However, the impressed current term $J_{iz}^{n+\frac{1}{2}}$ should be recalculated at every time step using the previous value of the impressed current, $J_{iz}^{n-\frac{1}{2}}$, and the associated electric field component, E_z^n, based on (4.23), before adding the terms in (4.52) to E_z^{n+1}. However, since ***update_electric_fields*** overwrites $E_z^n, J_{iz}^{n+\frac{1}{2}}$ must be calculated before ***update_electric_fields***. Therefore, $J_{iz}^{n+\frac{1}{2}}$ can be recalculated after updating the electric fields due to inductors for use in the following time step. The algorithm for updating electric field components and calculating the impressed currents can be followed in Listing 4.21, which shows the implementation of ***update_inductors***.

4.2.7.6 Updating electric fields associated with diodes

Update of electric field components for modeling diodes is more complicated compared to other types of lumped elements. As discussed in Section 4.1.8, the new value of the electric field component associated with a diode is obtained by the solution of the equation

$$Ae^{Bx} + x + C = 0, \tag{4.53}$$

where

$$\begin{aligned}
x &= E_z^{n+1}(i, j, k), \quad A = -C_{ezd}(i, j, k)e^{B\times E_z^n(i,j,k)}, \quad B = (q\Delta z/2kT), \\
C &= C_{eze}(i, j, k) \times E_z^n(i, j, k) + C_{ezhy}(i, j, k) \times \left(H_y^{n+\frac{1}{2}}(i, j, k) - H_y^{n+\frac{1}{2}}(i-1, j, k)\right) \\
&\quad + C_{ezhx}(i, j, k) \times \left(H_x^{n+\frac{1}{2}}(i, j, k) - H_x^{n+\frac{1}{2}}(i, j-1, k)\right) + C_{ezd}(i, j, k).
\end{aligned} \tag{4.54}$$

Listing 4.21 update_inductors.m

```
% updating electric field components
% associated with the inductors

for ind = 1:number_of_inductors
    fi = inductors(ind).field_indices;
    switch (inductors(ind).direction(1))
    case 'x'
        Ex(fi) = Ex(fi) + inductors(ind).Cexj ...
            .* inductors(ind).Jix;
        inductors(ind).Jix = inductors(ind).Jix ...
            + inductors(ind).Cjex .* Ex(fi);
    case 'y'
        Ey(fi) = Ey(fi) + inductors(ind).Ceyj ...
            .* inductors(ind).Jiy;
        inductors(ind).Jiy = inductors(ind).Jiy ...
            + inductors(ind).Cjey .* Ey(fi);
    case 'z'
        Ez(fi) = Ez(fi) + inductors(ind).Cezj ...
            .* inductors(ind).Jiz;
        inductors(ind).Jiz = inductors(ind).Jiz ...
            + inductors(ind).Cjez .* Ez(fi);
    end
end
```

The coefficients B and C_{ezd} are constructed as parameters **diodes(i).B** and **diodes(i).Cezd** in the subroutine ***initialize_diode_updating_coeffcients***. However, the coefficients A and C must be recalculated at every time step using the equations just shown. As discussed before, the coefficients C_{eze}, C_{ezhy}, and C_{ezhx} are negatives of the corresponding coefficients appearing in the general updating equations, which were executed in ***update_electric_fields***. Therefore, after execution of ***update_electric_fields*** the electric field component associated with the diode, E_z^{n+1}, would be holding the negative value of the sum of first three terms on the right-hand side of (4.54).

Then the coefficient C can be expressed as

$$C = -E_z^{n+1}(i, j, k) + C_{ezd}(i, j, k). \tag{4.55}$$

At this point, to calculate C we need the previous time step value of the electric field component, $E_z^n(i, j, k)$. Since this component is overwritten in ***update_electric_fields***, we copy it to the parameter **diodes(ind).Ezn** before executing ***update_electrie_fields***. Therefore, at every time step after the value of the electric field component associated with a diode is obtained, it is copied to **diodes(ind).Ezn** for use in the following time step. Similarly, the new value of A can be calculated using $E_z^n(i, j, k)$.

Once the parameters A and C are recalculated, the new value of electric field component E_z^{n+1} can be calculated by solving (4.53). As discussed in Section 4.8, this equation can easily be solved numerically by using the Newton–Raphson method. The updating procedure

Listing 4.22 update_diodes.m

```
% updating electric field components
% associated with the diodes

for ind = 1:number_of_diodes
    fi = diodes(ind).field_indices;
    B  = diodes(ind).B;
    switch (diodes(ind).direction(1))
    case 'x'
        E = diodes(ind).Exn;
        C = -Ex(fi) + diodes(ind).Cexd;
        A = -diodes(ind).Cexd * exp(B * E);
        E = solve_diode_equation(A, B, C, E);
        Ex(fi) = E;
        diodes(ind).Exn = E;
    case 'y'
        E = diodes(ind).Eyn;
        C = -Ey(fi) + diodes(ind).Ceyd;
        A = -diodes(ind).Ceyd * exp(B * E);
        E = solve_diode_equation(A, B, C, E);
        Ey(fi) = E;
        diodes(ind).Eyn = E;
    case 'z'
        E = diodes(ind).Ezn;
        C = -Ez(fi) + diodes(ind).Cezd;
        A = -diodes(ind).Cezd * exp(B * E);
        E = solve_diode_equation(A, B, C, E);
        Ez(fi) = E;
        diodes(ind).Ezn = E;
    end
end
```

of electric field components for modeling diodes is implemented in the subroutine ***update_diodes***, which is shown in Listing 4.22. Comparing the given implementation with the previous discussion helps in understanding this complicated updating process. Finally, a function with name ***solve_diode_equation*** is implemented for solving (4.53) based on the Newton–Raphson method, and its implementation is given in Listing 4.23.

4.2.7.7 Capturing the sampled magnetic field components

After the magnetic field components are updated by ***update_magnetic_fields***, they can be sampled for the current time step. As mentioned before, although all of the field components in the problem space can be captured, for now we limit our discussion with sampling the magnetic field values at predefined positions coinciding with the nodes of the problem space. The indices of the nodes for which the magnetic field will be sampled are stored in the parameters **is**, **js**, and **ks**. However, examining the field positioning scheme of the Yee cell reveals that the magnetic field components are not defined at the node positions. But it is possible to take the average of the four surrounding field components as the field value of the

Listing 4.23 solve_diode_equation.m

```
function [x] = solve_diode_equation(A, B, C, x)
% Function used to solve the diode equation
% which is in the form Ae^{Bx}+x+C=0
% using the Newton-Raphson method

tolerance = 1e-25;
max_iter = 50;
iter = 0;
f = A * exp(B*x) + x + C;
while ((iter < max_iter) && (abs(f) > tolerance))
    fp = A * B * exp(B*x) + 1;
    x = x - f/fp;
    f = A * exp(B*x) + x + C;
    iter = iter + 1;
end
```

Listing 4.24 capture_sampled_magnetic_fields.m

```
% Capturing magnetic fields

for ind=1:number_of_sampled_magnetic_fields
    is = sampled_magnetic_fields(ind).is;
    js = sampled_magnetic_fields(ind).js;
    ks = sampled_magnetic_fields(ind).ks;

    switch (sampled_magnetic_fields(ind).component)
        case 'x'
            sampled_value = 0.25 * sum(sum(Hx(is,js-1:js,ks-1:ks)));
        case 'y'
            sampled_value = 0.25 * sum(sum(Hy(is-1:is,js,ks-1:ks)));
        case 'z'
            sampled_value = 0.25 * sum(sum(Hz(is-1:is,js-1:js,ks)));
        case 'm'
            svx = 0.25 * sum(sum(Hx(is,js-1:js,ks-1:ks)));
            svy = 0.25 * sum(sum(Hy(is-1:is,js,ks-1:ks)));
            svz = 0.25 * sum(sum(Hz(is-1:is,js-1:js,ks)));
            sampled_value = sqrt(svx^2 + svy^2 + svz^2);
    end
    sampled_magnetic_fields(ind).sampled_value(time_step) = ...
        sampled_value;
end
```

node. This procedure is illustrated in Listing 4.24, which shows the implementation of the subroutine ***capture_sampled_magnetic_fields***. For instance, to sample the y component of the magnetic field at node (is, js, ks), an average of the components $H_y(is, js, ks)$, $H_y(is-1, js, ks)$, $H_y(is, js, ks-1)$, and $H_y(is-1, js, ks-1)$ can be used as the sampled values as illustrated in Figure 4.13(a).

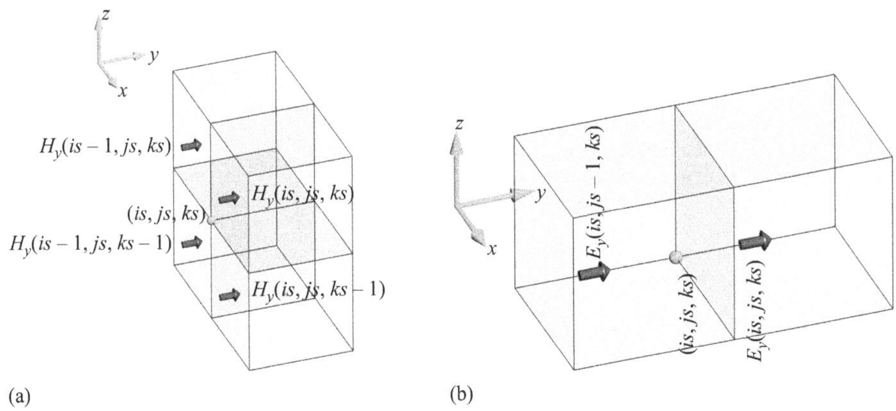

Figure 4.13 The y components of the magnetic and electric fields around the node (is, js, ks): (a) magnetic field components and (b) electric field components.

If the magnitude of the field vector is sought, the x, y, and z components of the field vector can be sampled, and then their vector magnitude can be calculated as shown in Listing 4.24.

4.2.7.8 Capturing the sampled currents

As discussed before, the currents flowing through a given surface can be calculated using the magnetic fields components circling around the surface. Therefore, after the magnetic field components are updated by ***update_magnetic_fields***, the currents can be calculated for the current time step. The procedure for capturing the electric current is implemented in the subroutine ***capture_sampled_currents*** as shown in Listing 4.25.

4.2.7.9 Capturing the sampled electric field components

After the electric field components are updated by ***update_electric_fields***, they can be sampled for the current time step. Since there is not an electric field component defined at the position of the sampling node (is, js, ks), an average of the two field components around the node can be used as the sampled value. For instance, to sample the y component of the electric field, an average of the components $E_y(is, js, ks)$ and $E_y(is, js - 1, ks)$ can be used as the sampled value as illustrated in Figure 4.13(b). The sampled electric field components are captured in the subroutine ***update_electric_fields***, which is shown in Listing 4.26.

4.2.7.10 Capturing the sampled voltages

The indices of the field components that are required for calculating the voltage across a volume indicated by the nodes (is, js, ks) and (ie, je, ke) are determined in the subroutine ***initialize_output_parameters*** and are stored in the parameter **field_indices**. Furthermore, another parameter **Csvf** is defined as a coefficient that would transform the sum of the electric field components to a voltage value as discussed in Section 4.2.5. Then a procedure for calculating the sampled voltages is implemented in the subroutine **capture_sampled_voltages** shown in Listing 4.27.

Listing 4.25 capture_sampled_currents.m

```
% Capturing sampled currents

for ind=1:number_of_sampled_currents
        is = sampled_currents(ind).is;
        js = sampled_currents(ind).js;
        ks = sampled_currents(ind).ks;
        ie = sampled_currents(ind).ie;
        je = sampled_currents(ind).je;
        ke = sampled_currents(ind).ke;

    switch (sampled_currents(ind).direction(1))
    case 'x'
        sampled_value = ...
        + dy * sum(sum(sum(Hy(ie-1,js:je,ks-1))))  ...
        + dz * sum(sum(sum(Hz(ie-1,je,ks:ke))))...
        - dy * sum(sum(sum(Hy(ie-1,js:je,ke))))...
        - dz * sum(sum(sum(Hz(ie-1,js-1,ks:ke))));
    case 'y'
        sampled_value = ...
        + dz * sum(sum(sum(Hz(is-1,je-1,ks:ke))))  ...
        + dx * sum(sum(sum(Hx(is:ie,je-1,ke))))...
        - dz * sum(sum(sum(Hz(ie,je-1,ks:ke))))...
        - dx * sum(sum(sum(Hx(is:ie,je-1,ks-1))));
    case 'z'
        sampled_value = ...
        + dx * sum(sum(sum(Hx(is:ie,js-1,ke-1))))  ...
        + dy * sum(sum(sum(Hy(ie,js:je,ke-1))))...
        - dx * sum(sum(sum(Hx(is:ie,je,ke-1))))...
        - dy * sum(sum(sum(Hy(is-1,js:je,ke-1))));
    end
    if strcmp(sampled_currents(ind).direction(2),'n')
        sampled_value = -1 * sampled_value;
    end
    sampled_currents(ind).sampled_value(time_step) = sampled_value;
end
```

Listing 4.26 capture_sampled_electric_fields.m

```
% Capturing electric fields

for ind=1:number_of_sampled_electric_fields
    is = sampled_electric_fields(ind).is;
    js = sampled_electric_fields(ind).js;
    ks = sampled_electric_fields(ind).ks;

    switch (sampled_electric_fields(ind).component)
        case 'x'
            sampled_value = 0.5 * sum(Ex(is-1:is,js,ks));
        case 'y'
            sampled_value = 0.5 * sum(Ey(is,js-1:js,ks));
        case 'z'
```

```
                sampled_value = 0.5 * sum(Ez(is,js,ks-1:ks));
            case 'm'
                svx = 0.5 * sum(Ex(is-1:is,js,ks));
                svy = 0.5 * sum(Ey(is,js-1:js,ks));
                svz = 0.5 * sum(Ez(is,js,ks-1:ks));
                sampled_value = sqrt(svx^2 + svy^2 + svz^2);
        end
        sampled_electric_fields(ind).sampled_value(time_step) = sampled_value;
end
```

Listing 4.27 capture_sampled_voltages.m

```
% Capturing sampled voltages

for ind=1:number_of_sampled_voltages
    fi   = sampled_voltages(ind).field_indices;
    Csvf = sampled_voltages(ind).Csvf;
    switch (sampled_voltages(ind).direction(1))
    case 'x'
        sampled_value = Csvf * sum(Ex(fi));
    case 'y'
        sampled_value = Csvf * sum(Ey(fi));
    case 'z'
        sampled_value = Csvf * sum(Ez(fi));
    end
    sampled_voltages(ind).sampled_value(time_step) = sampled_value;
end
```

4.2.7.11 Displaying the sampled parameters

After all fields are updated and the sampled parameters are captured, the sampled parameters defined for runtime display can be plotted using the subroutine ***displayj_sampled_parameters***, which is shown in Listing 4.28.

One should notice that in Listing 4.28 the instructions are executed once every **plotting_step** time steps. Then for each sampled parameter requiring display, the respective figure is activated using the respective figure number. Then the sampled parameter is plotted from the first time step to current time step using MATLAB ***plot*** function.

Finally, at the end of ***display_sampled_parameters*** another subroutine ***display_animation*** is called to display the fields captured on the predefined plane cuts.

4.2.8 Displaying FDTD simulation results

After the FDTD time-marching loop is completed, the output parameters can be displayed. Furthermore, the output parameters may be postprocessed to obtain some other types of outputs. The postprocessing and display of the simulation results are performed in the subroutine ***post_proces_and_display_results*** following ***run_fdtd_time_marching_loop***

Listing 4.28 display_sampled_parameters.m

```
% displaying sampled parameters

if mod(time_step, plotting_step) ~= 0
    return;
end

remaining_time = (number_of_time_steps-time_step) ...
    *(cputime-start_time)/(60*time_step);
disp([num2str(time_step) ' of ' ...
    num2str(number_of_time_steps) ' is completed, ' ...
    num2str(remaining_time) ' minutes remaining']);

% display sampled electric fields
for ind = 1:number_of_sampled_electric_fields
    if sampled_electric_fields(ind).display_plot == true
    sampled_time = ...
        sampled_electric_fields(ind).time(1:time_step)*1e9;
    sampled_value = ...
        sampled_electric_fields(ind).sampled_value(1:time_step);
    figure(sampled_electric_fields(ind).figure_number);
    delete(sampled_electric_fields(ind).plot_handle);
    sampled_electric_fields(ind).plot_handle = ...
    plot(sampled_time, sampled_value(1:time_step),'b-', ...
    'linewidth',1.5);
    drawnow;
    end
end

% display sampled magnetic fields
for ind = 1:number_of_sampled_magnetic_fields
    if sampled_magnetic_fields(ind).display_plot == true
    sampled_time = ...
        sampled_magnetic_fields(ind).time(1:time_step)*1e9;
    sampled_value = ...
        sampled_magnetic_fields(ind).sampled_value(1:time_step);
    figure(sampled_magnetic_fields(ind).figure_number);
    delete(sampled_magnetic_fields(ind).plot_handle);
    sampled_magnetic_fields(ind).plot_handle = ...
    plot(sampled_time, sampled_value(1:time_step),'b-', ...
    'linewidth',1.5);
    drawnow;
    end
end

% display sampled voltages
for ind = 1:number_of_sampled_voltages
    if sampled_voltages(ind).display_plot == true
    sampled_time = ...
        sampled_voltages(ind).time(1:time_step)*1e9;
    sampled_value = ...
```

```
            sampled_voltages(ind).sampled_value(1:time_step);
        figure(sampled_voltages(ind).figure_number);
        delete(sampled_voltages(ind).plot_handle);
        sampled_voltages(ind).plot_handle = ...
        plot(sampled_time, sampled_value(1:time_step),'b-', ...
        'linewidth',1.5);
        drawnow;
        end
end

% display sampled currents
for ind = 1:number_of_sampled_currents
        if sampled_currents(ind).display_plot == true
        sampled_time = ...
            sampled_currents(ind).time(1:time_step)*1e9;
        sampled_value = ...
            sampled_currents(ind).sampled_value(1:time_step);
        figure(sampled_currents(ind).figure_number);
        delete(sampled_currents(ind).plot_handle);
        sampled_currents(ind).plot_handle = ...
        plot(sampled_time, sampled_value(1:time_step),'b-',...
        'linewidth',1.5);
        drawnow;
        end
end


% display animated fields
display_animation;
```

Listing 4.29 post_process_and_display_results.m

```
disp('displaying simulation results');

display_transient_parameters;
```

in ***fdtd_solve***. Listing 4.29 shows the contents of this subroutine. So far we have defined four types of outputs: sampled electric and magnetic fields, voltages, and currents. All these outputs are captured during the simulation and are stored in the arrays **sampled_value**. The variations of these components are obtained as functions of time; hence, they are called the transient parameters.

Listing 4.29 includes a subroutine ***display_transient_parameters***, which is used to plot these sampled transient parameters. As other types of output parameters are defined for the FDTD solution, new subroutines that implement the postprocessing and displaying of these output parameters can be added to ***post_process_and_display_results***.

Listing 4.30 display_transient_parameters.m

```
disp('plotting the transient parameters');

% figures for sampled electric fields
for ind=1:number_of_sampled_electric_fields
    if sampled_electric_fields(ind).display_plot == false
        sampled_electric_fields(ind).figure_number = figure;
        xlabel('time (ns)','fontsize',12);
        ylabel('(volt/meter)','fontsize',12);
        title(['sampled electric field [' ...
            num2str(ind) ']'],'fontsize',12);
        grid on; hold on;
    else
        figure(sampled_electric_fields(ind).figure_number);
        delete(sampled_electric_fields(ind).plot_handle);
    end
    sampled_time = ...
        sampled_electric_fields(ind).time(1:time_step)*1e9;
    sampled_value = ...
        sampled_electric_fields(ind).sampled_value(1:time_step);
    plot(sampled_time, sampled_value(1:time_step), 'b-', ...
        'linewidth',1.5);
    drawnow;
end
```

The partial implementation of ***display_transient_parameters*** is given in Listing 4.30. Listing 4.30 shows the instructions for plotting only the sampled electric field components. The other parameters are plotted similarly. One can notice that, if a parameter has not been displayed during the time-marching loop, a new figure is launched and initialized for it. Then the sampled data are plotted.

4.3 Simulation examples

So far we have discussed all the basic components for performing an FDTD simulation, including the definition of the problem, initialization of the problem workspace, running an FDTD time-marching loop, and capturing and displaying the outputs. In this section we provide examples of some FDTD simulations. We provide the definitions of the problems and show the simulation results.

4.3.1 A resistor excited by a sinusoidal voltage source

The first example is a simulation of a resistor excited by a voltage source. The geometry of the problem is shown in Figure 4.14, and the partial code sections for the definition of the problem are given in Listings 4.31–4.34. The code sections given next are not complete and can be completed by the reader referring to the code listings given before.

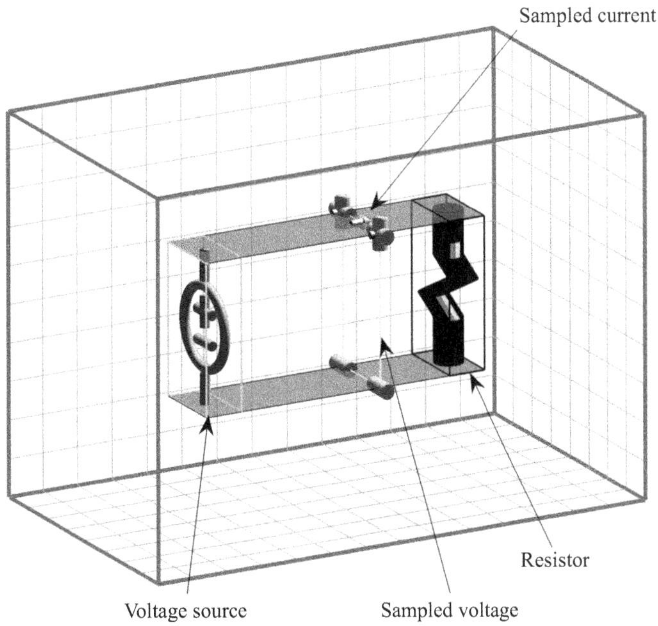

Figure 4.14 A voltage source terminated by a resistor.

Listing 4.31 define_problem_space_parameters.m

```
% maximum number of time steps to run FDTD simulation
number_of_time_steps = 3000;

% A factor that determines duration of a time step
% wrt CFL limit
courant_factor = 0.9;

% A factor determining the accuracy limit of FDTD results
number_of_cells_per_wavelength = 20;

% Dimensions of a unit cell in x, y, and z directions (meters)
dx=1.0e-3;
dy=1.0e-3;
dz=1.0e-3;

% ==<boundary conditions>========
% Here we define the boundary conditions parameters
% 'pec' : perfect electric conductor
boundary.type_xp = 'pec';
boundary.air_buffer_number_of_cells_xp = 3;

% PEC : perfect electric conductor
material_types(2).eps_r   = 1;
material_types(2).mu_r    = 1;
material_types(2).sigma_e = 1e10;
material_types(2).sigma_m = 0;
material_types(2).color   = [1 0 0];
```

Listing 4.32 define_geometry.m

```
% define a PEC plate
bricks(1).min_x = 0;
bricks(1).min_y = 0;
bricks(1).min_z = 0;
bricks(1).max_x = 8e-3;
bricks(1).max_y = 2e-3;
bricks(1).max_z = 0;
bricks(1).material_type = 2;

% define a PEC plate
bricks(2).min_x = 0;
bricks(2).min_y = 0;
bricks(2).min_z = 4e-3;
bricks(2).max_x = 8e-3;
bricks(2).max_y = 2e-3;
bricks(2).max_z = 4e-3;
bricks(2).material_type = 2;
```

Listing 4.33 define_source_and_lumped_elements.m

```
% define source waveform types and parameters
waveforms.sinusoidal(1).frequency = 1e9;
waveforms.sinusoidal(2).frequency = 5e8;
waveforms.unit_step(1).start_time_step = 50;

% voltage sources
% direction: 'xp', 'xn', 'yp', 'yn', 'zp', or 'zn'
% resistance : ohms, magitude   : volts
voltage_sources(1).min_x = 0;
voltage_sources(1).min_y = 0;
voltage_sources(1).min_z = 0;
voltage_sources(1).max_x = 1.0e-3;
voltage_sources(1).max_y = 2.0e-3;
voltage_sources(1).max_z = 4.0e-3;
voltage_sources(1).direction = 'zp';
voltage_sources(1).resistance = 50;
voltage_sources(1).magnitude = 1;
voltage_sources(1).waveform_type = 'sinusoidal';
voltage_sources(1).waveform_index = 2;

% resistors
% direction: 'x', 'y', or 'z'
% resistance : ohms
resistors(1).min_x = 7.0e-3;
resistors(1).min_y = 0;
resistors(1).min_z = 0;
resistors(1).max_x = 8.0e-3;
resistors(1).max_y = 2.0e-3;
resistors(1).max_z = 4.0e-3;
resistors(1).direction = 'z';
resistors(1).resistance = 50;
```

Listing 4.34 define_output_parameters.m

```
% figure refresh rate
plotting_step = 100;

% mode of operation
run_simulation = true;
show_material_mesh = true;
show_problem_space = true;

% define sampled voltages
sampled_voltages(1).min_x = 5.0e-3;
sampled_voltages(1).min_y = 0;
sampled_voltages(1).min_z = 0;
sampled_voltages(1).max_x = 5.0e-3;
sampled_voltages(1).max_y = 2.0e-3;
sampled_voltages(1).max_z = 4.0e-3;
sampled_voltages(1).direction = 'zp';
sampled_voltages(1).display_plot = true;

% define sampled currents
sampled_currents(1).min_x = 5.0e-3;
sampled_currents(1).min_y = 0;
sampled_currents(1).min_z = 4.0e-3;
sampled_currents(1).max_x = 5.0e-3;
sampled_currents(1).max_y = 2.0e-3;
sampled_currents(1).max_z = 4.0e-3;
sampled_currents(1).direction = 'xp';
sampled_currents(1).display_plot = true;
```

In Listing 4.31, the size of a unit cell is set to 1 mm on each side. The number of time steps to run the simulation is 3000. In Listing 4.32, two parallel PEC plates are defined as 4 mm apart. Between these two plates a voltage source is placed at one end and a resistor is placed at the other end. Then in Listing 4.33 sinusoidal and unit step waveforms are defined. A voltage source with 50 Ω internal resistance is defined, and the sinusoidal source waveform with frequency 500 MHz is assigned to the voltage source. Then a resistor with 50 Ω resistance is defined. Listing 4.34 shows the outputs defined for this simulation. A sampled voltage is defined between the parallel PEC plates. Furthermore, a sampled current is defined for the upper PEC plate.

Figure 4.15 shows the result of this simulation. Figure 4.15(a) shows a comparison of the source voltage and the sampled voltage. As can be seen, the sampled voltage is half the source voltage as expected. Furthermore, the sampled current shown in 4.15(b) confirms that the sum of the resistances is 100 Ω.

4.3.2 A diode excited by a sinusoidal voltage source

The second example is the simulation of a diode excited by a sinusoidal voltage source. The geometry of the problem is given in Figure 4.16, and the partial code sections for the definition of the problem are given in Listings 4.35–4.37.

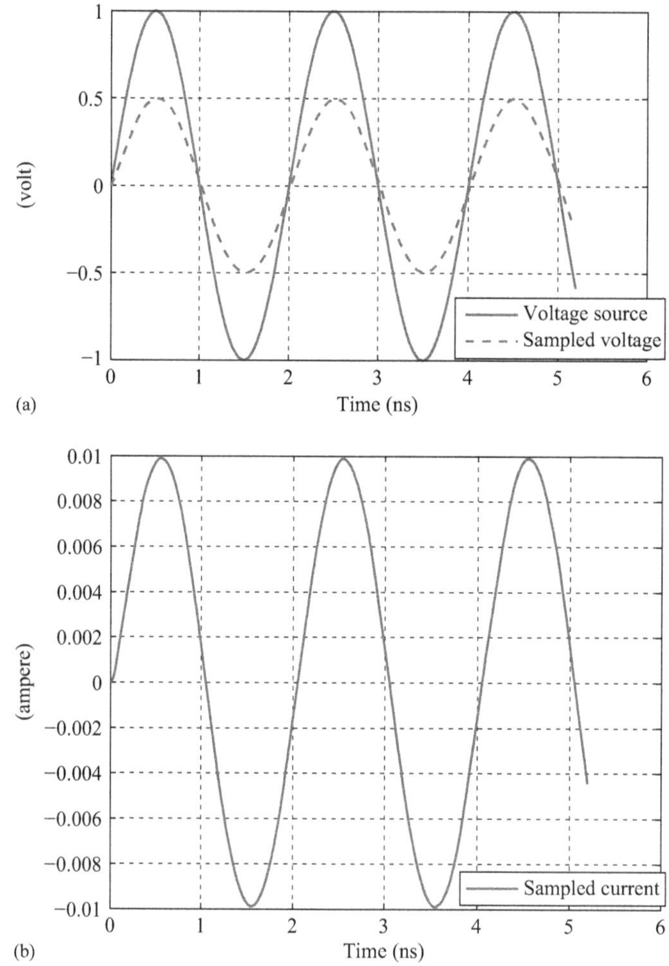

Figure 4.15 Sinusoidal voltage source waveform with 500 MHz frequency and sampled voltage and current: (a) source waveform and sampled voltage and (b) sampled current.

In Listing 4.35, two parallel PEC plates are defined 1 mm apart. Between these two plates a voltage source is placed at one end and a diode is placed at the other end. Then in Listing 4.36 sinusoidal and unit step waveforms are defined. A voltage source with 50 Ω internal resistance is defined, and the sinusoidal source waveform with frequency 500 MHz is assigned to the voltage source. Then a diode in the negative z direction is defined. Listing 4.37 shows the output defined for this simulation; a sampled voltage is defined across the diode.

Figure 4.16(b) shows the result of this simulation, where the source voltage and the sampled voltage are compared. As can be seen, the sampled voltage demonstrates the response of a diode; when the source voltage is larger than 0.7 V, the sampled voltage on the diode is maintained as 0.7 V.

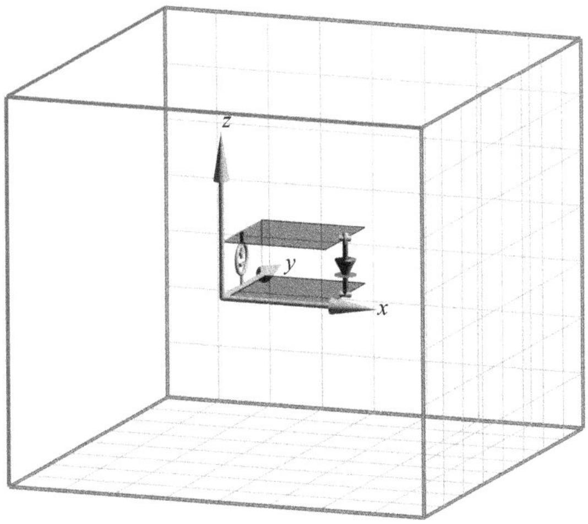

(a)

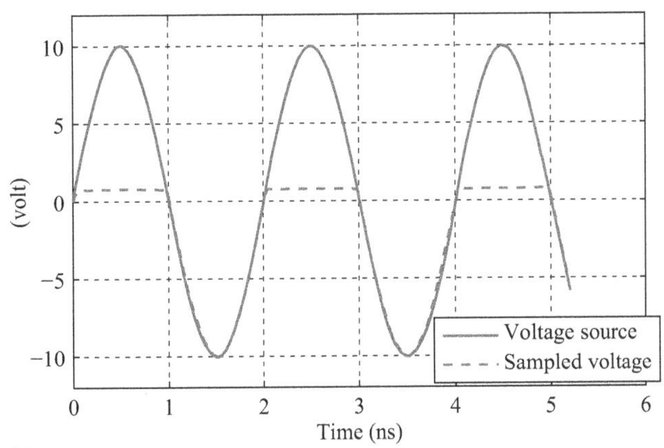

(b)

Figure 4.16 Sinusoidal voltage source waveform with 500 MHz frequency and sampled voltage: (a) a voltage source terminated by a diode and (b) source waveform and sampled voltage.

4.3.3 A capacitor excited by a unit-step voltage source

The third example is the simulation of a capacitor excited by a unit step voltage source. The geometry of the problem is given in Figure 4.17, and the partial code sections for the definition of the problem are given in Listings 4.38 and 4.39.

In Listing 4.38, two parallel PEC plates are defined 1 mm apart. A voltage source with 50 Ω internal resistances is places at one end and a capacitor with 10 pF capacitance is placed at the other end between these two plates as shown in Listing 4.39. A unit step waveform is assigned to the voltage source. The unit step function is excited 50 time steps after the start of simulation.

Listing 4.35 define_geometry.m

```
% define a PEC plate
bricks(1).min_x = 0;
bricks(1).min_y = 0;
bricks(1).min_z = 0;
bricks(1).max_x = 2e-3;
bricks(1).max_y = 2e-3;
bricks(1).max_z = 0;
bricks(1).material_type = 2;

% define a PEC plate
bricks(2).min_x = 0;
bricks(2).min_y = 0;
bricks(2).min_z = 1e-3;
bricks(2).max_x = 2e-3;
bricks(2).max_y = 2e-3;
bricks(2).max_z = 1e-3;
bricks(2).material_type = 2;
```

Listing 4.36 define_sources_and_lumped_elements.m

```
% define source waveform types and parameters
waveforms.sinusoidal(1).frequency = 1e9;
waveforms.sinusoidal(2).frequency = 5e8;
waveforms.unit_step(1).start_time_step = 50;

% voltage sources
% direction: 'xp', 'xn', 'yp', 'yn', 'zp', or 'zn'
% resistance : ohms, magitude   : volts
voltage_sources(1).min_x = 0;
voltage_sources(1).min_y = 0;
voltage_sources(1).min_z = 0;
voltage_sources(1).max_x = 0.0e-3;
voltage_sources(1).max_y = 2.0e-3;
voltage_sources(1).max_z = 1.0e-3;
voltage_sources(1).direction = 'zp';
voltage_sources(1).resistance = 50;
voltage_sources(1).magnitude = 10;
voltage_sources(1).waveform_type = 'sinusoidal';
voltage_sources(1).waveform_index = 2;

% diodes
% direction: 'xp', 'xn', 'yp', 'yn', 'zp', or 'zn'
diodes(1).min_x = 2.0e-3;
diodes(1).min_y = 1.0e-3;
diodes(1).min_z = 0.0e-3;
diodes(1).max_x = 2.0e-3;
diodes(1).max_y = 1.0e-3;
diodes(1).max_z = 1.0e-3;
diodes(1).direction = 'zn';
```

Listing 4.37 define_output_parameters.m

```
% figure refresh rate
plotting_step = 10;

% mode of operation
run_simulation = true;
show_material_mesh = true;
show_problem_space = true;

% define sampled voltages
sampled_voltages(1).min_x = 2.0e-3;
sampled_voltages(1).min_y = 1.0e-3;
sampled_voltages(1).min_z = 0;
sampled_voltages(1).max_x = 2.0e-3;
sampled_voltages(1).max_y = 1.0e-3;
sampled_voltages(1).max_z = 1.0e-3;
sampled_voltages(1).direction = 'zp';
sampled_voltages(1).display_plot = true;
```

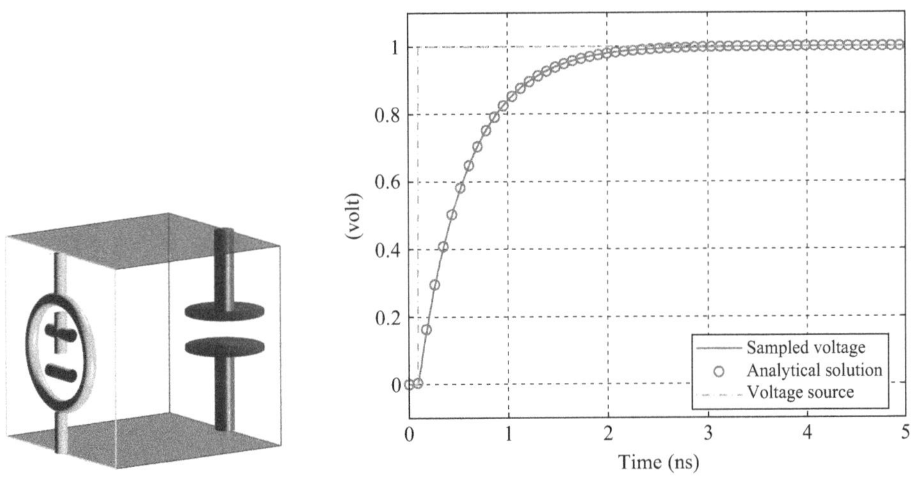

(a) (b)

Figure 4.17 Unit step voltage source waveform and sampled voltage: (a) a voltage source terminated by a capacitor and (b) sampled voltage and analytical solution.

Figure 4.17(b) shows the result of this simulation, where the sampled voltage across the capacitor is compared with the analytical solution of the equivalent lumped circuit problem. A simple circuit analysis reveals that the voltage across a capacitor excited by a unit step function can be calculated from

$$v_C(t) = 1 - e^{-\frac{t-t_0}{RC}}, \tag{4.56}$$

where t_0 is time delay, which is represented in this example by 50 time steps, and R is 50 Ω.

Listing 4.38 define_geometry.m

```
% define a PEC plate
bricks(1).min_x = 0;
bricks(1).min_y = 0;
bricks(1).min_z = 0;
bricks(1).max_x = 1e-3;
bricks(1).max_y = 1e-3;
bricks(1).max_z = 0;
bricks(1).material_type = 2;

% define a PEC plate
bricks(2).min_x = 0;
bricks(2).min_y = 0;
bricks(2).min_z = 1e-3;
bricks(2).max_x = 1e-3;
bricks(2).max_y = 1e-3;
bricks(2).max_z = 1e-3;
bricks(2).material_type = 2;
```

Listing 4.39 define_sources_and_lumped_elements.m

```
% define source waveform types and parameters
waveforms.sinusoidal(1).frequency = 1e9;
waveforms.sinusoidal(2).frequency = 5e8;
waveforms.unit_step(1).start_time_step = 50;

% voltage sources
% direction: 'xp', 'xn', 'yp', 'yn', 'zp', or 'zn'
% resistance : ohms, magitude   : volts
voltage_sources(1).min_x = 0;
voltage_sources(1).min_y = 0;
voltage_sources(1).min_z = 0;
voltage_sources(1).max_x = 0.0e-3;
voltage_sources(1).max_y = 1.0e-3;
voltage_sources(1).max_z = 1.0e-3;
voltage_sources(1).direction = 'zp';
voltage_sources(1).resistance = 50;
voltage_sources(1).magnitude = 1;
voltage_sources(1).waveform_type = 'unit_step';
voltage_sources(1).waveform_index = 1;

% capacitors
% direction: 'x', 'y', or 'z'
% capacitance : farads
capacitors(1).min_x = 1.0e-3;
capacitors(1).min_y = 0.0;
capacitors(1).min_z = 0.0;
capacitors(1).max_x = 1.0e-3;
capacitors(1).max_y = 1.0e-3;
capacitors(1).max_z = 1.0e-3;
capacitors(1).direction = 'z';
capacitors(1).capacitance = 10e-12;
```

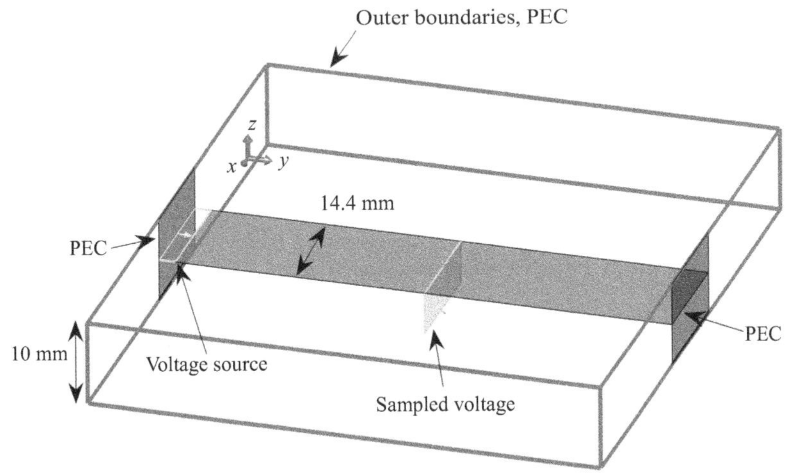

Figure 4.18 Dimensions of 50 Ω stripline.

4.4 Exercises

4.1 A stripline structure that has width/height ratio of 1.44 and an air substrate is forming a transmission line with 50 Ω characteristic impedance. Construct such a 50 Ω stripline structure, and keep all boundaries as PEC. The geometry and dimensions of such a stripline is shown in Figure 4.18 as a reference. Feed the stripline with a voltage source from one end, and let the other end touch the problem space boundary. Capture the voltage on the stripline around the center of the stripline. Show that if the voltage source is excited by a unit step waveform with 1 volt magnitude, the sampled voltage will be a pulse with 0.5 volts.

4.2 Consider the stripline in Exercise 4.1. Place a resistor with 50 Ω between the end of stripline and the problem space boundary as shown in Figure 4.19. Show that with the same excitation, the sampled voltage is going to be 0.5 volts unit step function.

4.3 Construct a problem space with the cell size as 1 mm on a side, boundaries as "***pec***," and air gap between the objects and boundaries as five cells on all sides. Connect a sinusoidal voltage source with 5 Ω internal resistance at one end between two parallel PEC plates. At the other end place a 50 pF capacitor and a diode connected in series between the plates. The diode is oriented in the positive z direction. The geometry of the circuit is illustrated in Figure 4.20. Run the simulation with the voltage source having 2 volts magnitude and 1 GHz frequency. Capture the voltage across the capacitor. Run the same simulation using a circuit simulator, and verify your simulation result.

4.4 Construct a problem space with the cell size as 1 mm on a side, boundaries as "***pec***," and air gap between the objects and boundaries as five cells on all sides. Connect a sinusoidal voltage source with 10 Ω internal resistance at one end between two parallel PEC plates. At the other end place an inductor, a capacitor, and a resistor connected in series between the plates. For these components use the values L = 10 nH,

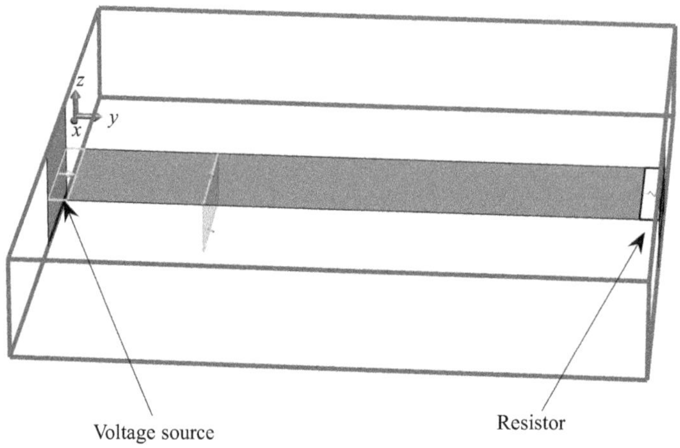

Figure 4.19 A 50 Ω stripline terminated by a resistor.

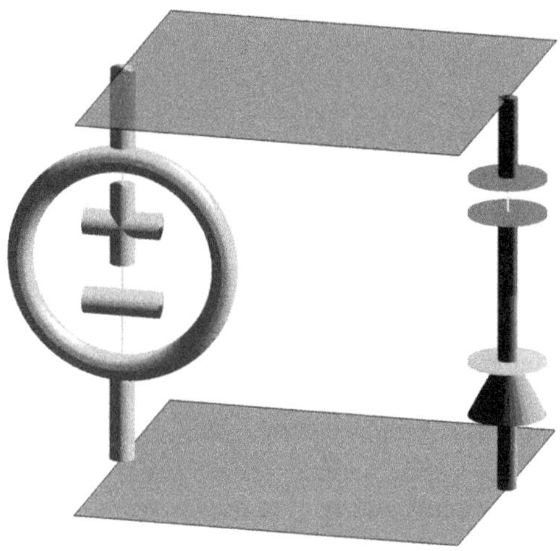

Figure 4.20 A circuit composed of a voltage source, a capacitor, and a diode.

C = 253.3 pF, and R = 10 Ω, respectively. At 100 MHz the series RLC circuit is in resonance. Excite the voltage source with a sinusoidal waveform with 100 MHz frequency, and capture the voltage between the plates approximately at the middle of the PEC plates. Show that the magnitude of the captured voltage waveform is half the magnitude of the source waveform when it reaches steady state. Repeat the same simulation with another frequency, and verify the magnitude of the observed voltage by calculating the exact expected value.

CHAPTER 5

Source waveforms and time to frequency domain transformation

Sources are necessary components of a finite-difference time-domain (FDTD) simulation and their types vary depending on the type of problem under consideration. Usually sources are of two types: (1) *near*, such as the voltage and current sources described in Chapter 4; and (2) *far*, such as the incident fields appearing in scattering problems. In any case, a source excites electric and magnetic fields with a waveform as a function of time. The type of waveform can be selected specific to the problem under consideration. However, some limitations of the FDTD method should be kept in mind while constructing the source waveforms to obtain a valid and accurate simulation result.

One of the considerations for the source waveform construction is the spectrum of the frequency components of the waveform. A temporal waveform is the sum of time-harmonic waveforms with a spectrum of frequencies that can be obtained using the Fourier transform. The temporal waveform is said to be in *time domain*, and its Fourier transformed form is said to be in *frequency domain*. The Fourier transform of the source waveforms and other output parameters can be used to obtain some useful results in the frequency domain. In this chapter, we discuss selected types of waveforms and then introduce the numerical algorithms for calculating the Fourier transform of temporal functions in discrete time.

5.1 Common source waveforms for FDTD simulations

A source waveform should be chosen such that its frequency spectrum includes all the frequencies of interest for the simulation, and it should have a smooth turn-on and turn-off to minimize the undesired effects of high-frequency components. A sine or a cosine function is a single-frequency waveform, whereas other waveforms such as Gaussian pulse, time derivative of Gaussian pulse, cosine-modulated Gaussian pulse, and Blackman-Harris window are multiple frequency waveforms. The type and parameters of the waveforms can be chosen based on the frequency spectrum of the waveform of interest.

5.1.1 Sinusoidal waveform

Definition and construction of a sinusoidal waveform was presented in Chapter 4. As mentioned before, a sinusoidal waveform is ideally a single-frequency waveform. However, in an FDTD simulation the waveform can be excited for a limited duration, and the turn-on and turn-off of the finite duration of the waveform is going to add other frequency components to the spectrum. The initial conditions for the sources are zero before the simulation is started in an FDTD simulation, and the sources are active during the duration of simulation. Therefore, if a sinusoidal signal ($\sin(2\pi t)$, $0 < t < 4$) is used for a source its duration is finite as shown in Figure 5.1(a).

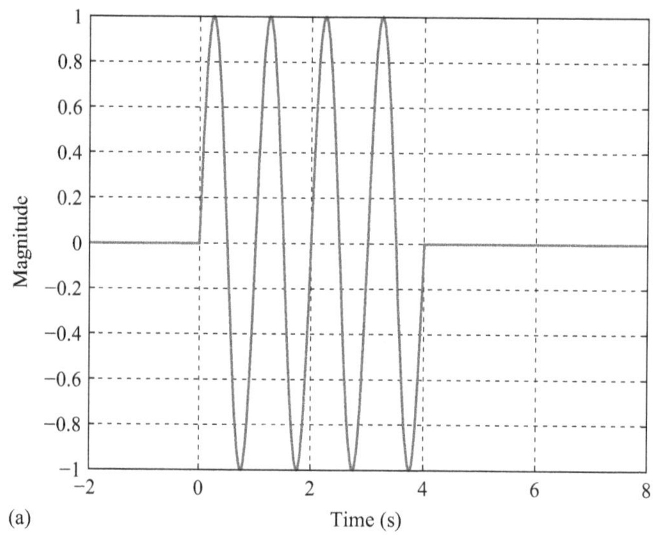

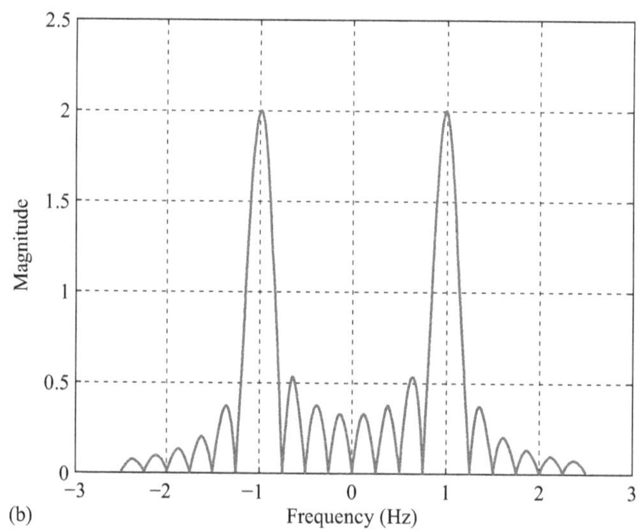

Figure 5.1 A sinusoidal waveform excited for 4 s: (a) $x(t)$ and (b) $X(\omega)$ magnitude.

The Fourier transform of a continuous time function $x(t)$ is $X(\omega)$ given by

$$X(\omega) = \int_{-\infty}^{+\infty} x(t)e^{-j\omega t}dt, \tag{5.1}$$

while the inverse Fourier transform is

$$x(t) = \frac{1}{2\pi}\int_{-\infty}^{+\infty} X(\omega)e^{j\omega t}d\omega. \tag{5.2}$$

The Fourier transform of the signal in Figure 5.1(a) is a complex function, and its magnitude is plotted in Figure 5.1(b). The frequency spectrum of the signal in Figure 5.1(a) covers a range of frequencies while being maximum at 1 Hz, which is the frequency of the finite sinusoidal function. If the simulation runs for a longer duration as shown in Figure 5.2(a), the Fourier transform of the waveform becomes more pronounced at 1 Hz as shown in Figure 5.2(b). In Figure 5.1(b) the ratio of the highest harmonic to the main signal at 1 Hz is $0.53/2 \equiv -11.53$ dB, while in Figure 5.2(b) this ratio is $0.96/4 \equiv -12.39$ dB. In this case, running the simulation twice as long in time reduced the effect of the highest harmonic.

If it is intended to observe the response of an electromagnetic problem due to a sinusoidal excitation, the simulation should be run long enough such that the transient response due to the turn on of the sources die out and only the sinusoidal response persists.

5.1.2 Gaussian waveform

Running an FDTD simulation will yield numerical results for a dominant frequency as well as for other frequencies in the spectrum of the finite sinusoidal excitation. However, the sinusoidal waveform is not an appropriate choice if simulation results for a wideband of frequencies are sought.

The frequency spectrum of the source waveform determines the range of frequencies for which valid and accurate results can be obtained. As the frequency increases, the wavelength decreases and becomes comparable to the cell size of the problem space. If the cell size is too large compared with a fraction of a wavelength, the signal at that frequency cannot be sampled accurately in space. Therefore, the highest frequency in the source waveform spectrum should be chosen such that the cell size is not larger than a fraction of the highest-frequency wavelength. For many applications, setting the highest-frequency wavelength larger than 20 cells size is sufficient for a reasonable FDTD simulation. A Gaussian waveform is the best choice for a source waveform, since it can be constructed to contain all frequencies up to a highest frequency that is tied to a cell size by a factor. This factor, which is the proportion of the highest frequency wavelength to the unit cell size, is referred as *number of cells per wavelength* n_c.

A Gaussian waveform can be written as a function of time as

$$g(t) = e^{-\frac{t^2}{\tau^2}}, \tag{5.3}$$

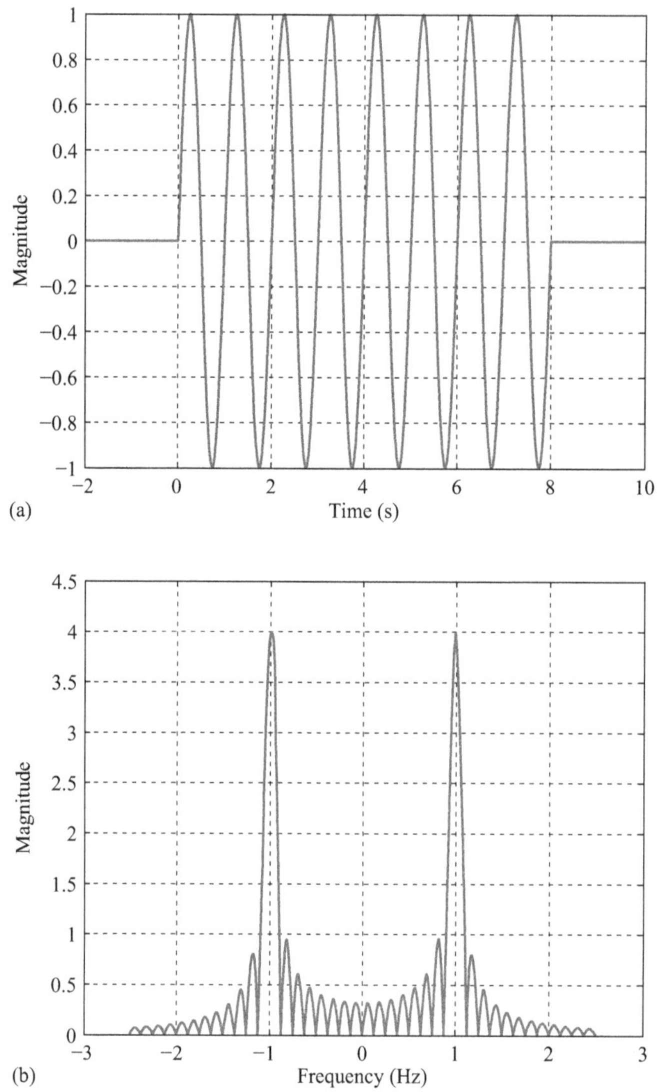

Figure 5.2 A sinusoidal waveform excited for 8 s: (a) $x(t)$ and (b) $X(\omega)$ magnitude.

where τ is a parameter that determines the width of the Gaussian pulse both in the time domain and the frequency domain. The Fourier transform of a Gaussian waveform is also a Gaussian waveform, which can be expressed as a function of frequency as

$$G(w) = \tau\sqrt{\pi}e^{-\frac{\tau^2\omega^2}{4}}. \tag{5.4}$$

The highest frequency that is available out of an FDTD calculation can be determined by the accuracy parameter *number of cells per wavelength* such that

$$f_{\max} = \frac{c}{\lambda_{\min}} = \frac{c}{n_c \Delta s_{\max}}, \tag{5.5}$$

where c is the speed of light in free space, Δs_{max} is the maximum of the cell dimensions (Δx, Δy, or Δz), and λ_{min} is the wavelength of the highest frequency in free space. Once the highest frequency in the spectrum of the frequency domain Gaussian waveform is determined, it is possible to find a τ that constructs the corresponding time-domain Gaussian waveform. Following is a procedure that determines τ, so that one can construct the Gaussian waveform in the time domain before the simulation starts with a good estimate of the highest frequency at which the simulation results will be acceptable.

Consider the Gaussian waveform plotted in Figure 5.3(a) and the magnitude of its Fourier transform pair plotted in Figure 5.3(b) using (5.4). The function (5.4) is real, and its phase

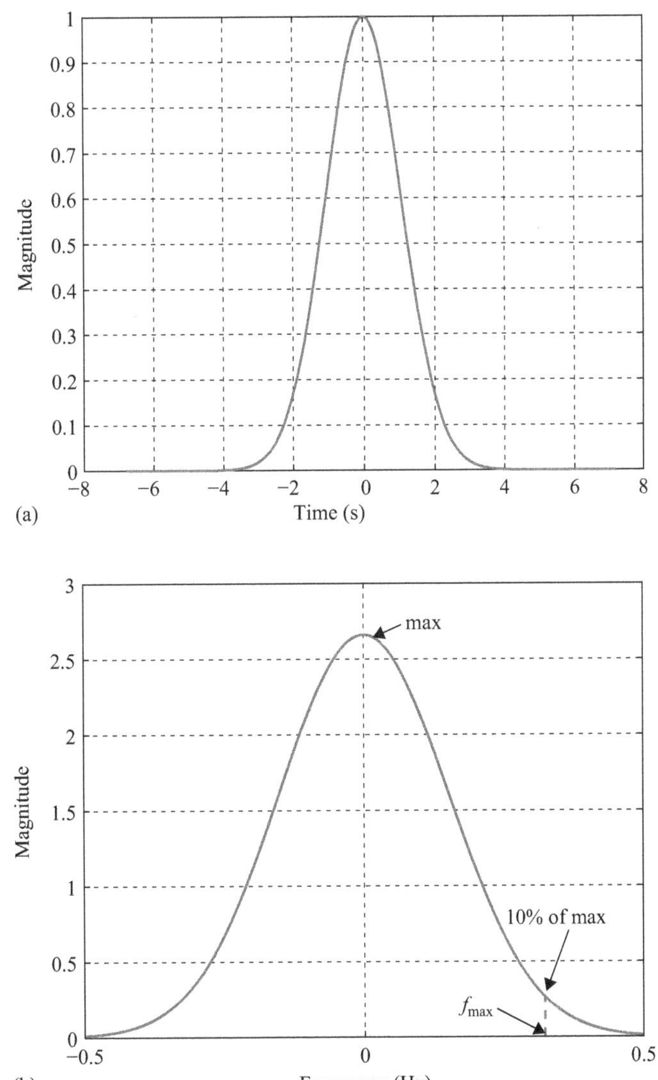

Figure 5.3 A Gaussian waveform and its Fourier transform: (a) $g(t)$ and (b) $G(\omega)$ magnitude.

vanishes; hence, the phase of the Fourier transform is not plotted here. We consider the maximum valid frequency of the Gaussian spectrum as the frequency where the magnitude of $G(\omega)$ is 10% of its maximum magnitude as illustrated in Figure 5.3(b). Therefore, by solving the equation

$$0.1 = e^{-\frac{\tau^2 \omega_{max}^2}{4}},$$

we can find a relation between τ and f_{max}, such that

$$\tau = \frac{\sqrt{2.3}}{\pi f_{max}}. \tag{5.6}$$

Using (5.5) in (5.6), we can tie n_c and Δs_{max} to τ as

$$\tau = \frac{\sqrt{2.3} n_c \Delta s_{max}}{\pi c} \cong \frac{n_c \Delta s_{max}}{2c}. \tag{5.7}$$

Once the parameters n_c and Δs_{max} are defined, the parameter τ can be calculated and can be used in (5.3) to construct the Gaussian waveform spectrum that allows one to obtain a valid frequency response up to f_{max}.

Determination of the parameter τ is not enough to construct the Gaussian waveform that can be directly used in an FDTD simulation. One should notice in Figure 5.3(a) that the Gaussian waveform's maximum coincides with time instant zero. However, in an FDTD simulation the initial conditions of the fields are zero; hence, the sources must also be zero. This can be achieved by shifting the Gaussian waveform in time such that at time instant zero the waveform value is zero or is (practically) negligible. Furthermore, time shifting preserves the frequency content of the waveform.

With time shifting the Gaussian waveform takes the form

$$g(t) = e^{-\frac{(t-t_0)^2}{\tau^2}}, \tag{5.8}$$

where t_0 is the amount of time shift. We can find t_0 as the time instant in (5.3) where $g(t = 0) = e^{-20}$, which is a negligible value and is good for more than 16 digits of numerical accuracy. Solving (5.8) with $g(0) = e^{-20}$ we can find

$$t_0 = \sqrt{20}\tau \cong 4.5\tau. \tag{5.9}$$

Using the time shift t_0, we obtain a Gaussian waveform as illustrated in Figure 5.4, which can be used as a source waveform in FDTD simulations to obtain acceptable results for frequencies up to f_{max}.

5.1.3 Normalized derivative of a Gaussian waveform

In some applications it may be desired to excite the problem space with a pulse that has a wide frequency spectrum but does not include zero frequency or very low-frequency

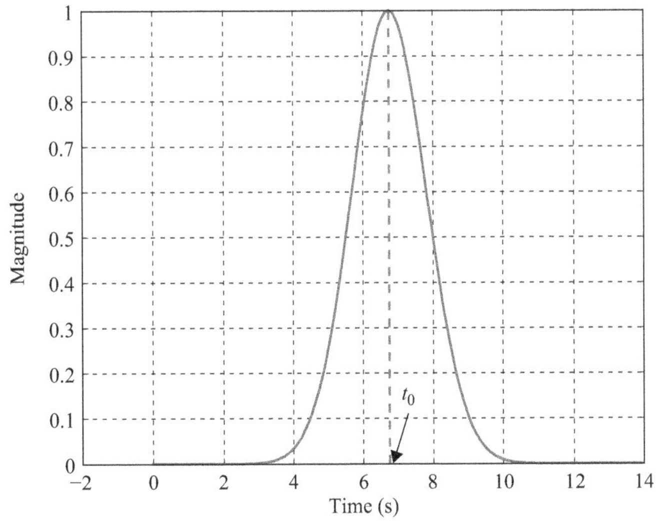

Figure 5.4 Gaussian waveform shifted by $t_0 = 4.5\tau$ in time.

components. Furthermore, if it is not necessary, it is better to avoid the simulation of low-frequency components since simulation of these components need long simulation time. For such cases the derivative of the Gaussian waveform is a suitable choice. The derivative of a Gaussian waveform normalized to the maximum can be found as

$$g(t) = \frac{\sqrt{2e}}{\tau} t e^{-\frac{t^2}{\tau^2}}. \tag{5.10}$$

The Fourier transform of this function can be determined as

$$G(\omega) = \frac{j\omega\tau^2\sqrt{\pi e}}{\sqrt{2}} e^{-\frac{\tau^2\omega^2}{4}}. \tag{5.11}$$

The normalized derivative of a Gaussian waveform and its Fourier transform are illustrated in Figure 5.5(a) and 5.5(b), respectively. One can notice that the function in the frequency domain vanishes at zero frequency. The form of the function suggests that there is a minimum frequency, f_{min}, and a maximum frequency, f_{max}, that determine the bounds of the valid frequency spectrum for an FDTD simulation considering 10% of the maximum as the limit. Having determined the maximum usable frequency from the *number of cells per wavelength* and maximum cell size using (5.5), one can find the parameter τ that sets $G(\omega)$ to 10% of its maximum value at f_{max} and can use τ in (5.10) to construct a time-domain waveform for the FDTD simulation. However, this procedure requires solution of a nonlinear equation and is not convenient. The equations readily available for a Gaussian waveform can be used instead for convenience. For instance, τ can be calculated using (5.7). Similar to the Gaussian waveform, the derivative of a Gaussian waveform is also centered at time instant zero as it appears in (5.10) and as shown in Figure 5.5(a). Therefore, it also needs a time shift

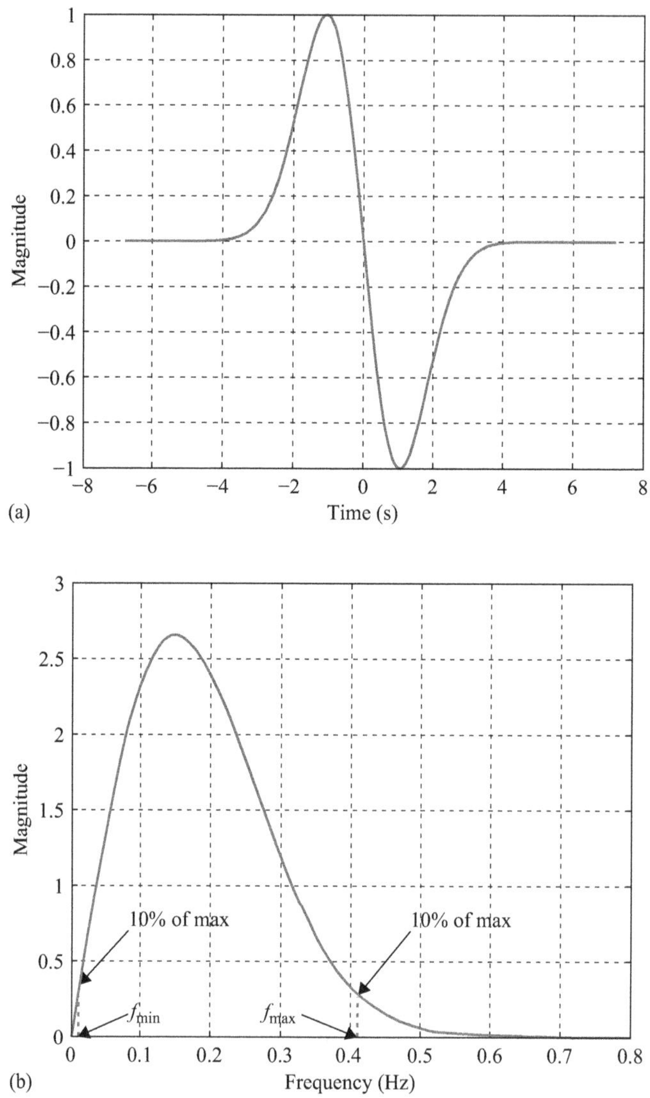

Figure 5.5 Normalized derivative of a Gaussian waveform and its Fourier transform: (a) $g(t)$ and (b) $G(\omega)$ magnitude.

to have practically zero value at time instant zero; hence, it can be written as

$$g(t) = \frac{\sqrt{2e}}{\tau}(t - t_0)e^{-\frac{(t-t_0)^2}{\tau^2}}. \tag{5.12}$$

The time shift value t_0 can be obtained as well using the readily available equation (5.9).

5.1.4 Cosine-modulated Gaussian waveform

Another common waveform used in FDTD applications is the cosine-modulated Gaussian waveform. A Gaussian waveform includes frequency components over a band centered at zero frequency as illustrated in Figure 5.3(b). Modulating a signal with a cosine function having frequency f_c corresponds to shifting the frequency spectrum of the modulated signal by f_c in the frequency domain. Therefore, if it is desired to perform a simulation to obtain results for a frequency band centered at a frequency f_c, the cosine-modulated Gaussian pulse is a suitable choice. One can first construct a Gaussian waveform with a desired bandwidth and then can multiply the Gaussian pulse with a cosine function with frequency f_c. Then the corresponding waveform can be expressed as

$$g(t) = \cos(\omega_c t)e^{-\frac{t^2}{\tau^2}}, \tag{5.13}$$

while its Fourier transform can be written as

$$G(\omega) = \frac{\tau\sqrt{\pi}}{2}e^{-\frac{\tau^2(\omega-\omega_c)^2}{4}} + \frac{\tau\sqrt{\pi}}{2}e^{-\frac{\tau^2(\omega+\omega_c)^2}{4}}, \tag{5.14}$$

where $\omega_c = 2\pi f_c$. A cosine-modulated Gaussian waveform and its Fourier transform are illustrated in Figure 5.6. To construct a cosine-modulated Gaussian waveform with spectrum having Δf bandwidth centered at f_c, first we find τ in (5.14) that satisfies the Δf bandwidth requirement. We can utilize (5.6) to find τ such that

$$\tau = \frac{2\sqrt{2.3}}{\pi\Delta f} \cong \frac{0.966}{\Delta f}. \tag{5.15}$$

Then τ can be used in (5.13) to construct the intended waveform. However, we still need to apply a time shift to the waveform such that initial value of the excitation is zero. We can use (5.9) to find the required time shift value, t_0. Then the final equation for the cosine-modulated Gaussian waveform takes the form

$$g(t) = \cos(\omega_c(t - t_0)) \times e^{-\frac{(t-t_0)^2}{\tau^2}}. \tag{5.16}$$

5.2 Definition and initialization of source waveforms for FDTD simulations

In the previous section, we discussed some common types of source waveforms that can be used to excite an FDTD problem space, and we derived equations for waveform function parameters to construct waveforms for desired frequency spectrum properties. In this section, we discuss the definition and implementation of these source waveforms in an FDTD MATLAB® code.

In Section 4.2.1, we showed that we can define the parameters specific to various types of waveforms in a structure named waveforms. We have shown how a unit step function and a sinusoidal waveform can be defined as subfields of the structure **waveforms**. We can add new

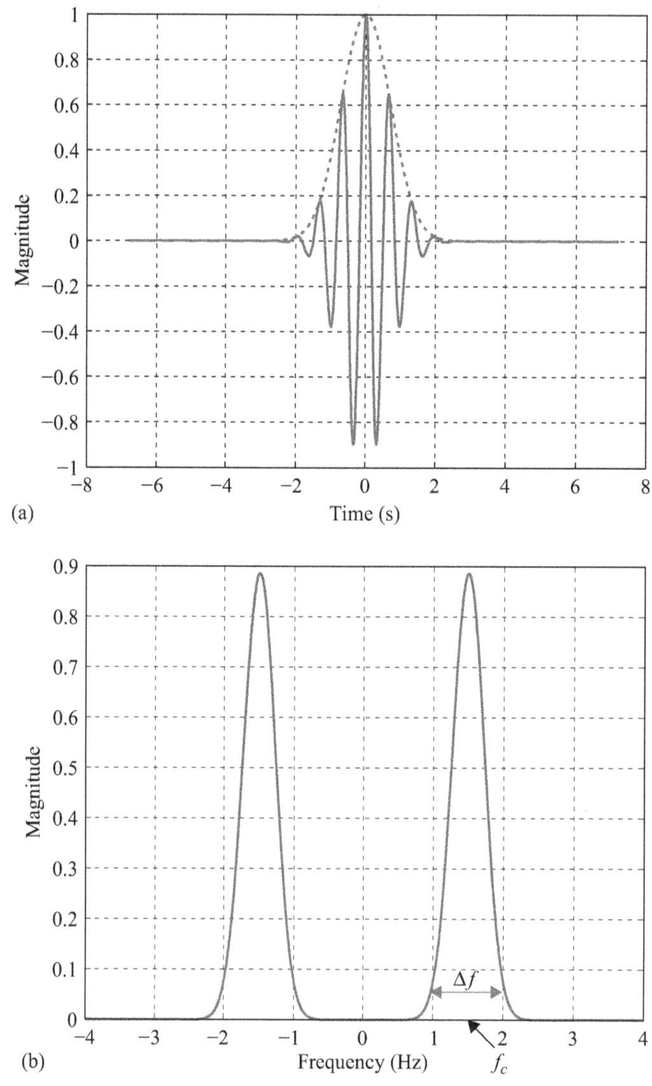

Figure 5.6 Cosine-modulated Gaussian waveform and its Fourier transform: (a) $g(t)$ and (b) $G(\omega)$ magnitude.

subfields to the structure **waveforms** to define the new types of source waveforms. For instance, consider the Listing 5.1, where a sample section of the subroutine ***define_sources_and_lumped_elements*** illustrates the definition of the source waveforms. In addition to the unit step function and sinusoidal waveform, the Gaussian waveform is defined as the subfield **waveforms.gaussian**, the derivative of a Gaussian waveform is defined as **waveforms.**

derivative_gaussian, and the cosine-modulated Gaussian waveform is defined as **waveforms.cosine_modulated_gaussian**. Each of these subfields is an array, so more than one waveform of the same type can be defined in the same structure **waveforms**. For instance, in this listing two sinusoidal waveforms and two Gaussian waveforms are defined. The Gaussian and the derivative of Gaussian waveforms have the parameter **number_of_cells_per_wavelength**, which is the accuracy parameter n_c described in the previous section used to determine the maximum frequency of the spectrum of the Gaussian waveform. If the **number_of_cells_per_wavelength** is assigned zero, then the **number_of_cells_per_wavelength** that is defined in the subroutine ***define_problems_space_parameters*** will be used as the default value.

The cosine-modulated Gaussian waveform has the parameter **bandwidth**, which specifies the width of the frequency band Δf. One more parameter is **modulation_frequency**, which is the center frequency of the band, f_c.

Listing 5.1 define_sources_and_lumped_elements.m

```
disp('defining sources and lumped element components');

voltage_sources = [];
current_sources = [];
diodes = [];
resistors = [];
inductors = [];
capacitors = [];

% define source waveform types and parameters
waveforms.sinusoidal(1).frequency = 1e9;
waveforms.sinusoidal(2).frequency = 5e8;
waveforms.unit_step(1).start_time_step = 50;
waveforms.gaussian(1).number_of_cells_per_wavelength = 0;
waveforms.gaussian(2).number_of_cells_per_wavelength = 15;
waveforms.derivative_gaussian(1).number_of_cells_per_wavelength = 20;
waveforms.cosine_modulated_gaussian(1).bandwidth = 1e9;
waveforms.cosine_modulated_gaussian(1).modulation_frequency = 2e9;

% voltage sources
% direction: 'xp', 'xn', 'yp', 'yn', 'zp', or 'zn'
% resistance : ohms, magitude   : volts
voltage_sources(1).min_x = 0;
voltage_sources(1).min_y = 0;
voltage_sources(1).min_z = 0;
voltage_sources(1).max_x = 1.0e-3;
voltage_sources(1).max_y = 2.0e-3;
voltage_sources(1).max_z = 4.0e-3;
voltage_sources(1).direction = 'zp';
voltage_sources(1).resistance = 50;
voltage_sources(1).magnitude = 1;
voltage_sources(1).waveform_type = 'gaussian';
voltage_sources(1).waveform_index = 2;
```

Once the waveform types are defined, the desired waveform type can be assigned to the sources as illustrated in Listing 5.1. For instance, for the voltage source **voltage_sources(1)** the parameter **waveform_type** is assigned the value "***gaussian***," and the parameter **waveform_index** is assigned the value 2, indicating that the waveform **waveforms.gaussian(2)** is the waveform of the voltage source.

The initialization of the waveforms is performed in the subroutine ***initialize_waveforms***, which is called in ***initialize_sources_and_lumped_elements*** before the initialization of lumped sources as discussed in Section 4.2.3. The implementation of ***initialize_waveforms*** is extended to include the Gaussian, derivative of Gaussian, and cosine-modulated Gaussian waveforms based on the discussions in Section 5 as shown in Listing 5.2. After the waveforms are initialized, the constructed waveforms are copied to appropriate subfields in the source structures in ***initialize_sources_and_lumped_elements***.

Listing 5.2 initialize_waveforms.m

```
disp('initializing source waveforms');

% initialize sinusoidal waveforms
if isfield(waveforms,'sinusoidal')
    for ind=1:size(waveforms.sinusoidal,2)
        waveforms.sinusoidal(ind).waveform = ...
            sin(2 * pi * waveforms.sinusoidal(ind).frequency * time);
        waveforms.sinusoidal(ind).t_0 = 0;
    end
end

% initialize unit step waveforms
if isfield(waveforms,'unit_step')
    for ind=1:size(waveforms.unit_step,2)
        start_index = waveforms.unit_step(ind).start_time_step;
        waveforms.unit_step(ind).waveform(1:number_of_time_steps) = 1;
        waveforms.unit_step(ind).waveform(1:start_index-1) = 0;
        waveforms.unit_step(ind).t_0 = 0;
    end
end

% initialize Gaussian waveforms
if isfield(waveforms,'gaussian')
    for ind=1:size(waveforms.gaussian,2)
        if waveforms.gaussian(ind).number_of_cells_per_wavelength == 0
            nc = number_of_cells_per_wavelength;
        else
            nc = waveforms.gaussian(ind).number_of_cells_per_wavelength;
        end
        waveforms.gaussian(ind).maximum_frequency = ...
            c/(nc*max([dx,dy,dz]));
        tau = (nc*max([dx,dy,dz]))/(2*c);
        waveforms.gaussian(ind).tau = tau;
        t_0 = 4.5 * waveforms.gaussian(ind).tau;
```

```
            waveforms.gaussian(ind).t_0 = t_0;
            waveforms.gaussian(ind).waveform = exp(-((time - t_0)/tau).^2);
        end
end

% initialize derivative of Gaussian waveforms
if isfield(waveforms,'derivative_gaussian')
    for ind=1:size(waveforms.derivative_gaussian,2)
    wfrm = waveforms.derivative_gaussian(ind);
    if wfrm.number_of_cells_per_wavelength == 0
        nc = number_of_cells_per_wavelength;
    else
        nc = ...
        waveforms.derivative_gaussian(ind).number_of_cells_per_wavelength;
    end
        waveforms.derivative_gaussian(ind).maximum_frequency = ...
            c/(nc*max([dx,dy,dz]));
        tau = (nc*max([dx,dy,dz]))/(2*c);
        waveforms.derivative_gaussian(ind).tau = tau;
        t_0 = 4.5 * waveforms.derivative_gaussian(ind).tau;
        waveforms.derivative_gaussian(ind).t_0 = t_0;
        waveforms.derivative_gaussian(ind).waveform = ...
            -(sqrt(2*exp(1))/tau)*(time - t_0).*exp(-((time - t_0)/tau).^2);
        end
end

% initialize cosine modulated Gaussian waveforms
if isfield(waveforms,'cosine_modulated_gaussian')
        for ind=1:size(waveforms.cosine_modulated_gaussian,2)
        frequency = ...
        waveforms.cosine_modulated_gaussian(ind).modulation_frequency;
        tau = 0.966/waveforms.cosine_modulated_gaussian(ind).bandwidth;
        waveforms.cosine_modulated_gaussian(ind).tau = tau;
        t_0 = 4.5 * waveforms.cosine_modulated_gaussian(ind).tau;
        waveforms.cosine_modulated_gaussian(ind).t_0 = t_0;
        waveforms.cosine_modulated_gaussian(ind).waveform = ...
        cos(2*pi*frequency*(time - t_0)).*exp(-((time - t_0)/tau).^2);
        end
end
```

5.3 Transformation from time domain to frequency domain

In the previous sections, we discussed construction of different types of source waveforms for an FDTD simulation. These waveforms are functions of time, and any primary output obtained from an FDTD simulation is in the time domain. The input–output relationship is available in the time domain after an FDTD simulation is completed. The input and output time functions can be transformed to the frequency domain by the use of the Fourier transform to obtain the response of the system in the case of time-harmonic excitations. The Fourier transform is an integration applied to a continuous time function. In the FDTD method the values are sampled at discrete time instants; therefore, the continuous time integration can be approximated by a discrete time summation. The Fourier transform of a continuous time function $x(t)$ is given as $X(\omega)$ using (5.1). In FDTD the time function is sampled at a period Δt; therefore, the values $x(n\Delta t)$ are known. Then the Fourier transform (5.1) can be expressed for discrete samples $x(n\Delta t)$ as

$$X(\omega) = \Delta t \sum_{n=1}^{n=N_{steps}} x(n\Delta t) e^{-j\omega n\Delta t}, \tag{5.17}$$

where N_{steps} is the number of time steps. It is easy to implement a function based on (5.17) that calculates the Fourier transform of the discrete samples $x(n\Delta t)$ for a list of frequencies. Implementation of such a function, ***time_to_frequency_domain***, is shown in Listing 5.3. This function accepts a parameter **x**, which is an array of sampled time-domain values of a function. The parameter **frequency_array** includes list of frequencies for which the transform is to be performed. The output parameter of the function **X** is an array including the transform at the respective frequencies.

Another input parameter that needs consideration is **time_shift**. As discussed in previous chapters, the electric field-related values and magnetic field-related values are sampled at different time instants during the FDTD time-marching scheme. There is a half time step duration in between, and the function ***time_to_frequency_domain*** accepts a value **time_shift** to account for this time shift in the Fourier transform. For electric field-related values (e.g., sampled voltages), **time_shift** can take the value 0, whereas for the magnetic field-related values (e.g., sampled currents), **time_shift** can take the value $-dt/2$.

The function implementation given in Listing 5.3 is intended for an easy understanding of the time-to-frequency-domain transform action, and is not necessarily the best or optimum algorithm for this purpose. More efficient discrete Fourier transform implementations, such as fast Fourier transforms, can be used as well.

Listing 5.3 time_to_frequency_domain.m

```
function [X] = time_to_frequency_domain(x,dt,frequency_array,time_shift)
% x  : array including the sampled values at discrete time steps
% dt : sampling period, duration of an FDTD time step
% frequency_array : list of frequencies for which transform is performed
% time_shift : a value in order to account for the time shift between
% electric and magnetic fields

number_of_time_steps  = size(x,1);
number_of_frequencies = size(frequency_array,2);
X = zeros(1, number_of_frequencies);
w = 2 * pi * frequency_array;
for n = 1:number_of_time_steps
    t = n * dt + time_shift;
    X = X + x(n) * exp(-j*w*t);
end
X = X * dt;
```

Similarly, the inverse Fourier transform can also be implemented. Consider a band-limited frequency domain function $X(\omega)$ sampled at uniformly distributed discrete frequency points with sampling period $\Delta\omega$, where the sampling frequencies include zero and positive frequencies. Then the inverse Fourier transform (5.2) can be expressed as

$$x(t) = \frac{\Delta\omega}{2\pi}\left(X(0) + \sum_{m=1}^{m=M-1}\left[X(m\Delta\omega)e^{j\omega t} + X^*(m\Delta\omega)e^{-j\omega t}\right]\right), \tag{5.18}$$

where X^* is the complex conjugate of X, $X(0)$ is the value of X at zero frequency, and M is the number of sampled frequency points. We have assumed that X includes the samples at zero and positive frequencies. It is evident from (5.2) that the integration includes the negative frequencies as well. However, for a causal time function, the Fourier transform at a negative frequency is the complex conjugate of its positive-frequency counterpart. So, if the Fourier transform is known for the positive frequencies, the transform at the negative frequencies can be obtained by the conjugate. Therefore, (5.18) includes summation at the negative frequencies as well. Since the zero frequency appears once, it is added to the discrete summation separately. Then a function named as ***frequency_to_time_domain*** is constructed based on (5.18) as shown in Listing 5.4.

Listing 5.4 frequency_to_time_domain.m

```
function [x] = frequency_to_time_domain(X, df, time_array)
% X  : array including the sampled values at discrete frequency steps
% df : sampling period in frequency domain
% time_array : list of time steps for which
%              inverse transform is performed

number_of_frequencies  = size(X,2);
number_of_time_points = size(time_array,2);
x = zeros(1, number_of_time_points);
dw = 2 * pi * df;
x = X(1); % zero frequency component
for m = 2:number_of_frequencies
    w = (m-1) * dw;
    x = x + X(m)* exp(j*w*time_array)+ conj(X(m)) ...
        * exp(-j*w*time_array);
end
x = x * df;
```

In general, we can define a list of frequencies for which we seek the frequency-domain response in the ***define_output_parameters*** subroutine, which is illustrated in Listing 5.5. Here a structure **frequency_domain** is defined to store the frequency-domain-specific parameters. The subfields **start**, **end**, and **step** correspond to the respective values of a uniformly sampled list of frequencies.

The desired list of frequencies is calculated and assigned to an array frequency_domain. frequencies in the initialize_output_parameters subroutine as shown in Listing 5.6. The initialization of output parameters is implemented in the subroutine **initialize_output_ parameters**, as shown in Listing 5.6.

Listing 5.5 define_output_parameters.m

```
% frequency domain parameters
frequency_domain.start = 1e7;
frequency_domain.end   = 1e9;
frequency_domain.step  = 1e7;
```

Listing 5.6 initialize_output_parameters.m

```
% intialize frequency domain parameters
frequency_domain.frequencies = [frequency_domain.start: ...
    frequency_domain.step:frequency_domain.end];
frequency_domain.number_of_frequencies = ...
    size(frequency_domain.frequencies,2);
```

Listing 5.7 post_process_and_display_results.m

```
disp('displaying simulation results');

display_transient_parameters;
calculate_frequency_domain_outputs;
display_frequency_domain_outputs;
```

The post processing and display of the simulation results are performed in the subroutine ***post_process_and_display_results*** following ***run_fdtd_time_marching_loop*** in **fdtd_solve**. We can add two new subroutines to ***post_process_and_display_results*** as illustrated in Listing 5.7: ***calculate_frequency_domain_outputs*** and ***display_frequency_domain_outputs***. The time-to-frequency-domain transformations can be performed in ***calculate_frequency_domain_outputs***, for which the sample code is shown in Listing 5.8. In this subroutine the time-domain arrays of sampled values are transformed to frequency domain using the function **time_to_frequency_domain**, and the calculated frequency-domain arrays are assigned to the **frequency_domain_value** subfield of the respective structures. Finally, the subroutine ***display_frequency_domain_outputs***, a partial section of which is listed in Listing 5.9, can be called to display the frequency-domain outputs.

5.4 Simulation examples

So far we have discussed some source waveform types that can be used to excite an FDTD problem space. We considered the relationship between the time- and frequency-domain representations of the waveforms and provided equations that can be used to construct a temporal waveform having specific frequency spectrum characteristics. Then we provided functions that perform time-to-frequency-domain transformation, which can be used to obtain simulation results in the frequency domain after an FDTD simulation is completed. In this section we provide examples that utilize source waveforms and transformation functions.

Listing 5.8 calculate_frequency_domain_outputs.m

```
disp('generating frequency domain outputs');

frequency_array = frequency_domain.frequencies;

% sampled electric fields in frequency domain
for ind=1:number_of_sampled_electric_fields
    x = sampled_electric_fields(ind).sampled_value;
    time_shift = 0;
    [X] = time_to_frequency_domain(x, dt, frequency_array, time_shift);
    sampled_electric_fields(ind).frequency_domain_value = X;
    sampled_electric_fields(ind).frequencies = frequency_array;
end

% sampled currents in frequency domain
for ind=1:number_of_sampled_currents
    x = sampled_currents(ind).sampled_value;
    time_shift = -dt/2;
    [X] = time_to_frequency_domain(x, dt, frequency_array, time_shift);
    sampled_currents(ind).frequency_domain_value = X;
    sampled_currents(ind).frequencies = frequency_array;
end

% voltage sources in frequency domain
for ind=1:number_of_voltage_sources
    x = voltage_sources(ind).waveform;
    time_shift = 0;
    [X] = time_to_frequency_domain(x, dt, frequency_array, time_shift);
    voltage_sources(ind).frequency_domain_value = X;
    voltage_sources(ind).frequencies = frequency_array;
end
```

Listing 5.9 display_frequency_domain_outputs.m

```
% figures for sampled voltages
for ind=1:number_of_sampled_voltages
    frequencies = sampled_voltages(ind).frequencies*1e-9;
    fd_value = sampled_voltages(ind).frequency_domain_value;
    figure;
    title(['sampled voltage [' num2str(ind) ']'],'fontsize',12);
    subplot(2,1,1);
    plot(frequencies, abs(fd_value),'b-','linewidth',1.5);
    xlabel('frequency (GHz)','fontsize',12);
    ylabel('magnitude','fontsize',12);
    grid on;
    subplot(2,1,2);
    plot(frequencies, angle(fd_value)*180/pi,'r-','linewidth',1.5);
    xlabel('frequency (GHz)','fontsize',12);
    ylabel('phase (degrees)','fontsize',12);
    grid on;
    drawnow;
end
```

5.4.1 Recovering a time waveform from its Fourier transform

The first example illustrates how the functions ***time_to_frequency_domain*** and ***frequency_to_time_domain*** can be used. Consider the program ***recover_a_time_waveform*** given in Listing 5.10.

Listing 5.10 recover_a_time_waveform.m

```
clc; close all; clear all;
% Construct a Gaussian Waveform in time, frequency spectrum of which
% has its magnitude at 1 GHz as 10% of the maximum.
maximum_frequency = 1e9;
tau = sqrt(2.3)/(pi*maximum_frequency);
t_0 = 4.5 * tau;
time_array = [1:1000]*1e-11;
g = exp(-((time_array - t_0)/tau).^2);
figure(1);
plot(time_array*1e9, g,'b-','linewidth',1.5);
title('g(t)=e^{-((t-t_0)/\tau)^2}','fontsize',14);
xlabel('time (ns)','fontsize',12);
ylabel('magnitude','fontsize',12);
set(gca,'fontsize',12);
grid on;

% Perform time to frequency domain transform
frequency_array = [0:1000]*2e6;
dt = time_array(2)-time_array(1);
G = time_to_frequency_domain(g, dt, frequency_array, 0);

figure(2);
subplot(2,1,1);
plot(frequency_array*1e-9, abs(G),'b-','linewidth',1.5);
title('G(\omega) = F(g(t))','fontsize',12);
xlabel('frequency (GHz)','fontsize',12);
ylabel('magnitude','fontsize',12);
set(gca,'fontsize',12);
grid on;
subplot(2,1,2);
plot(frequency_array*1e-9, angle(G)*180/pi,'r-','linewidth',1.5);
xlabel('frequency (GHz)','fontsize',12);
ylabel('phase (degrees)','fontsize',12);
set(gca,'fontsize',12);
grid on;
drawnow;

% Perform frequency to time domain transform
df = frequency_array(2)-frequency_array(1);
g2 = frequency_to_time_domain(G, df, time_array);

figure(3);
plot(time_array*1e9, abs(g2),'b-','linewidth',1.5);
title('g(t)= F^{-1}(G(\omega))','fontsize',14);
xlabel('time (ns)','fontsize',12);
ylabel('magnitude','fontsize',12);
set(gca,'fontsize',12);
grid on;
```

First, a Gaussian waveform, $g(t)$, for which the frequency spectrum is 1 GHz wide, is constructed and is time shifted as shown in Figure 5.7(a). The Fourier transform of $g(t)$ is obtained as $G(\omega)$, as plotted in Figure 5.7(b), using ***time_to_frequency_domain***. One should notice that the value of $G(\omega)$ is 10% of the maximum at 1 GHz. Furthermore, the phase vanishes for the Fourier transform of a time-domain Gaussian waveform having its center at time instant zero. However, the phase shown in Figure 5.7(b) is nonzero. This is due to the time shift introduced to the Gaussian waveform in the time domain. Finally, the time-domain Gaussian waveform is reconstructed from $G(\omega)$ using the function ***frequency_to_time_domain*** and is plotted in Figure 5.7(c).

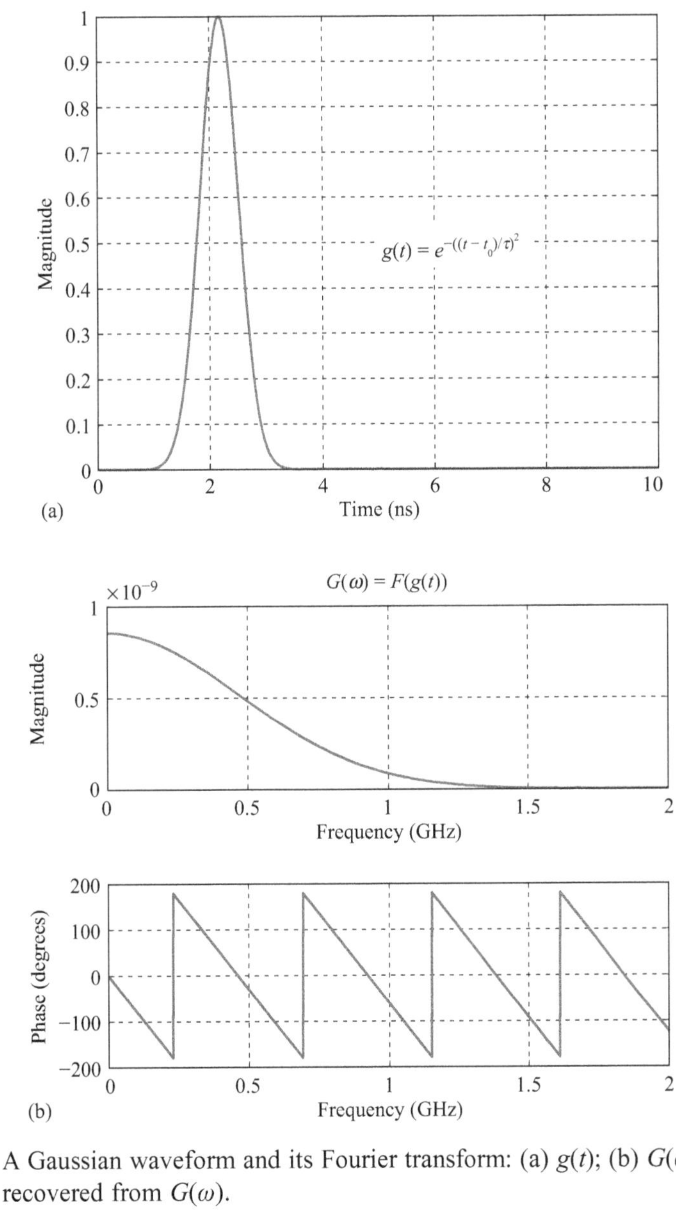

Figure 5.7 A Gaussian waveform and its Fourier transform: (a) $g(t)$; (b) $G(\omega)$; and (c) $g(t)$ recovered from $G(\omega)$.

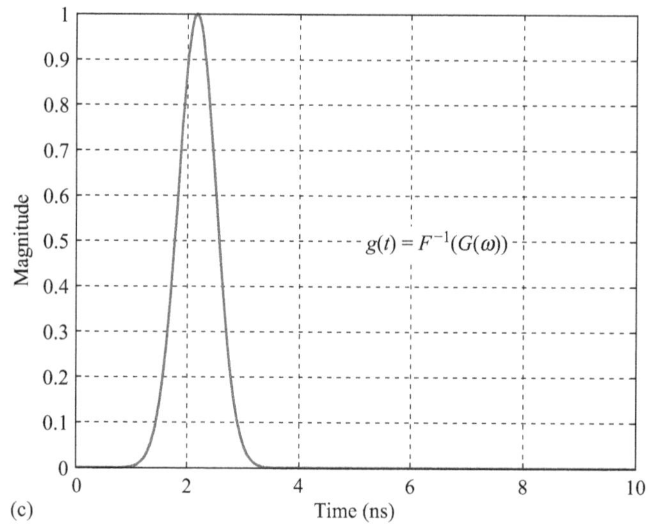

Figure 5.7 *(Continued)*

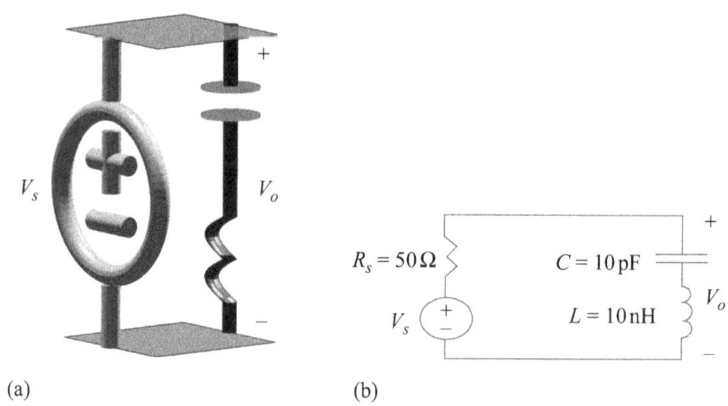

Figure 5.8 An RLC circuit: (a) circuit simulated in FDTD and (b) lumped element equivalent circuit.

5.4.2 An RLC circuit excited by a cosine-modulated Gaussian waveform

The second example is a simulation of a 10 nH inductor and a 10 pF capacitor connected in series and excited by a voltage source with 50 Ω internal resistor. The geometry of the problem is shown in Figure 5.8(a) as it is simulated by the FDTD program. The equivalent circuit representation of the problem is illustrated in Figure 5.8(b). The size of a unit cell is set to 1 mm on each side. The number of time steps to run the simulation is 2000. Two parallel PEC plates are defined 2 mm apart from each other as shown in Listing 5.11. A voltage source is placed in between these plates at one end, and an inductor and a capacitor are placed in series at the other end as shown in Listing 5.12.

Listing 5.11 define_geometry.m

```
disp('defining the problem geometry');

bricks  = [];
spheres = [];

% define a PEC plate
bricks(1).min_x = 0;
bricks(1).min_y = 0;
bricks(1).min_z = 0;
bricks(1).max_x = 1e-3;
bricks(1).max_y = 1e-3;
bricks(1).max_z = 0;
bricks(1).material_type = 2;

% define a PEC plate
bricks(2).min_x = 0;
bricks(2).min_y = 0;
bricks(2).min_z = 2e-3;
bricks(2).max_x = 1e-3;
bricks(2).max_y = 1e-3;
bricks(2).max_z = 2e-3;
bricks(2).material_type = 2;
```

Listing 5.12 define_sources_and_lumped_elements.m

```
disp('defining sources and lumped element components');

voltage_sources = [];
current_sources = [];
diodes = [];
resistors = [];
inductors = [];
capacitors = [];

% define source waveform types and parameters
waveforms.sinusoidal(1).frequency = 1e9;
waveforms.sinusoidal(2).frequency = 5e8;
waveforms.unit_step(1).start_time_step = 50;
waveforms.gaussian(1).number_of_cells_per_wavelength = 0;
waveforms.gaussian(2).number_of_cells_per_wavelength = 15;
waveforms.derivative_gaussian(1).number_of_cells_per_wavelength = 20;
waveforms.cosine_modulated_gaussian(1).bandwidth = 4e9;
waveforms.cosine_modulated_gaussian(1).modulation_frequency = 2e9;

```

```
% voltage sources
% direction: 'xp', 'xn', 'yp', 'yn', 'zp', or 'zn'
% resistance : ohms, magitude   : volts
voltage_sources(1).min_x = 0;
voltage_sources(1).min_y = 0;
voltage_sources(1).min_z = 0;
voltage_sources(1).max_x = 0;
voltage_sources(1).max_y = 1.0e-3;
voltage_sources(1).max_z = 2.0e-3;
voltage_sources(1).direction = 'zp';
voltage_sources(1).resistance = 50;
voltage_sources(1).magnitude = 1;
voltage_sources(1).waveform_type = 'cosine_modulated_gaussian';
voltage_sources(1).waveform_index = 1;

% inductors
% direction: 'x', 'y', or 'z'
% inductance : henrys
inductors(1).min_x = 1.0e-3;
inductors(1).min_y = 0.0;
inductors(1).min_z = 0.0;
inductors(1).max_x = 1.0e-3;
inductors(1).max_y = 1.0e-3;
inductors(1).max_z = 1.0e-3;
inductors(1).direction = 'z';
inductors(1).inductance = 10e-9;

% capacitors
% direction: 'x', 'y', or 'z'
% capacitance : farads
capacitors(1).min_x = 1.0e-3;
capacitors(1).min_y = 0.0;
capacitors(1).min_z = 1.0e-3;
capacitors(1).max_x = 1.0e-3;
capacitors(1).max_y = 1.0e-3;
capacitors(1).max_z = 2.0e-3;
capacitors(1).direction = 'z';
capacitors(1).capacitance = 10e-12;
```

A cosine-modulated Gaussian waveform with 2 GHz center frequency is assigned to the voltage source. A sampled voltage is defined between the parallel PEC plates at the load end as shown in Listing 5.13.

Figure 5.9 shows the time-domain results of this simulation; V_o is compared with the source waveform, V_s. The frequency-domain counterparts of these waveforms are calculated using time_to_frequency_domain and are plotted in Figure 5.10. We can find the transfer function of this circuit by normalizing the output voltage V_o to V_s. Furthermore, we can perform a circuit analysis to find the transfer function of the equivalent circuit in Figure 5.8(b), which yields

$$T(\omega) = \frac{V_o(\omega)}{V_s(\omega)} = \frac{s^2LC + 1}{s^2LC + sRC + 1}, \tag{5.19}$$

Listing 5.13 define_output_parameters.m

```
disp('defining output parameters');

sampled_electric_fields = [];
sampled_magnetic_fields = [];
sampled_voltages = [];
sampled_currents = [];

% figure refresh rate
plotting_step = 10;

% mode of operation
run_simulation = true;
show_material_mesh = true;
show_problem_space = true;

% frequency domain parameters
frequency_domain.start = 2e7;
frequency_domain.end   = 4e9;
frequency_domain.step  = 2e7;

% define sampled voltages
sampled_voltages(1).min_x = 1.0e-3;
sampled_voltages(1).min_y = 0.0;
sampled_voltages(1).min_z = 0.0;
sampled_voltages(1).max_x = 1.0e-3;
sampled_voltages(1).max_y = 1.0e-3;
sampled_voltages(1).max_z = 2.0e-3;
sampled_voltages(1).direction = 'zp';
sampled_voltages(1).display_plot = true;
```

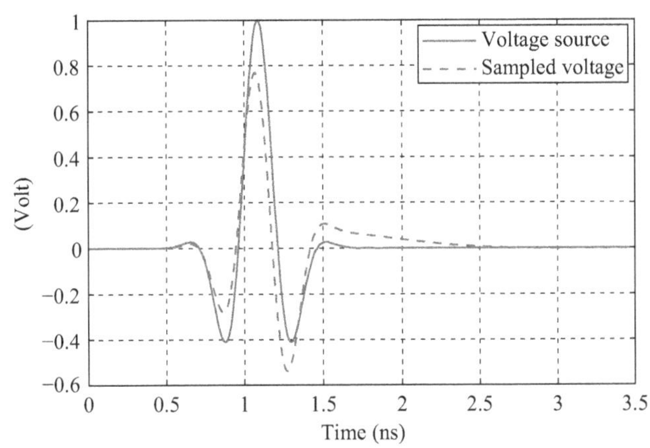

Figure 5.9 Time-domain response, V_o, compared with source waveform, V_s.

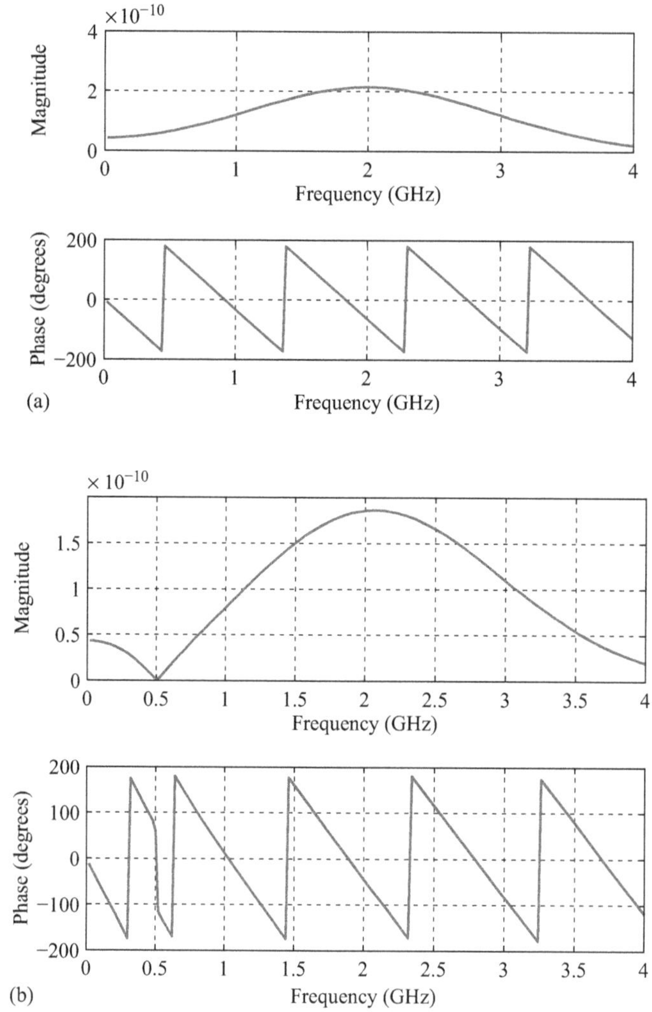

Figure 5.10 Frequency-domain response of source waveform V_s and output voltage V_o: (a) Fourier transform of V_s and (b) Fourier transform of V_o.

where $s = j\omega$. The transfer function obtained from the FDTD simulation and the exact equation (5.19) are compared as shown in Figure 5.11. These two results agree with each other very well at low frequencies; however, there is a deviation at high frequencies. One of the reasons for the difference is that the two circuits in Figure 5.8(a) and 5.8(b) are not exactly the same. For instance, the parallel plates introduce a capacitance, and the overall circuit is a loop and introduces an inductance in Figure 5.8(a). However, these additional effects are not accounted for in the lumped circuit in Figure 5.8(b). Therefore, an electromagnetic simulation including lumped element components should account for the additional effects due to the physical structure of the circuit to obtain more accurate data at higher frequencies.

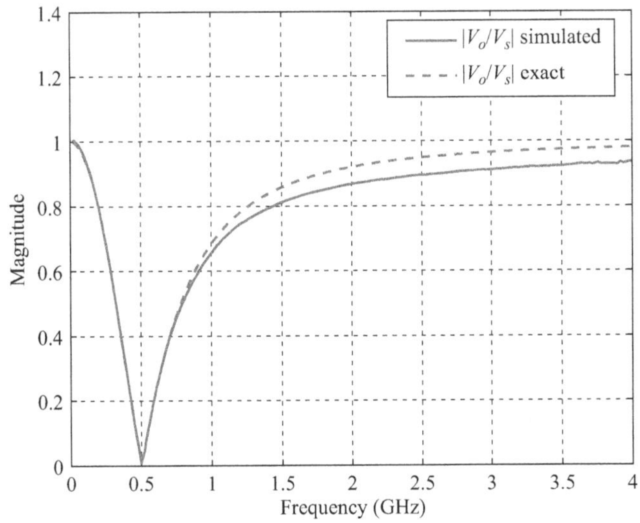

Figure 5.11 Transfer function $T(\omega) = \frac{V_o(\omega)}{V_s(\omega)}$.

5.5 Exercises

5.1 Consider the stripline structure that you constructed in Exercise 4.2. Define a Gaussian source waveform, assign it to the voltage source and run the FDTD simulation. By the time the simulation is completed, the sampled voltage and current will be captured, transformed to frequency domain, and stored in the **frequency_domain_value** field of the parameters **sampled_voltages(i)** and **sampled_currents(i)**. Input impedance of this stripline can be calculated using these sampled voltage and current arrays in the frequency domain. Since the stripline characteristic impedance is expected to be 50 Ω and the stripline is terminated by a 50 Ω resistor, the input impedance shall be 50 Ω as well. Calculate and plot the input impedance for a wide frequency band, and verify that the input impedance is close to 50 Ω.

5.2 Consider the simulation you performed in Exercise 5.1. When you examine the sampled transient voltage you will observe a Gaussian waveform generated by the voltage source followed by some reflections due to the imperfect termination of the 50 Ω resistor. If your stripline structure is long enough and the Gaussian waveform is narrow enough, the Gaussian waveform and reflections should not overlap. Determine from the transient voltage a number of time steps that would be used to run the FDTD simulations such that the simulation would end before the reflections are captured and only the Gaussian waveform would be observed as the sampled transient voltage. This way you would be simulating an infinitely long stripline structure. Therefore, when you calculate the input impedance, you would actually be calculating the characteristic impedance of the stripline. Rerun the FDTD simulation with the time step you have determined as discussed, calculate and plot the input impedance for a wide frequency band, and verify that the characteristic impedance of the stripline is 50 Ω.

5.3 Consider the circuit in Exercise 4.4. Set the waveform of the voltage source as the Gaussian waveform, and define a sampled current to capture the current flowing through the circuit. Rerun the FDTD simulation, calculate and plot the input impedance for a wide frequency band, and verify that the circuit resonates at 100 MHz. Note that you may need to run the simulation for a large number of time steps since it will take a long time for the low-frequency components in the Gaussian waveform to decay.

5.4 Consider Exercise 5.3. Now set the waveform of the voltage-source derivative of the Gaussian waveform. Rerun the FDTD simulation, calculate and plot the input impedance for a wide frequency band, and verify that the circuit resonates at 100 MHz. Note that with the derivative of the Gaussian waveform you should be able to obtain the results with a much smaller number of time steps compared with the amount of steps used in Exercise 5.3 since the derivative of the Gaussian waveform does not include zero frequency and since very low-frequency components are negligible.

CHAPTER 6

S-Parameters

Scattering parameters (S-parameters) are used to characterize the response of radiofrequency and microwave circuits, and they are more commonly used than other types of network parameters (e.g., Y-parameters, Z-parameters) because they are easier to measure and work with at high frequencies [12]. In this chapter, we discuss the methods to obtain the S-parameters from a finite-difference time-domain (FDTD) simulation of a single or multiport circuit.

6.1 Scattering parameters

S-parameters are based on the power waves concept. The incident and reflected power waves a_i and b_i associated with port i are defined as

$$a_i = \frac{V_i + Z_i \times I_i}{2\sqrt{|Re\{Z_i\}|}}, \quad b_i = \frac{V_i - Z_i^* \times I_i}{2\sqrt{|Re\{Z_i\}|}}, \tag{6.1}$$

where V_i and I_i are the voltage and the current flowing into the ith port of a junction and Z_i is the impedance looking out from the ith port as illustrated in Figure 6.1 [13]. In general, Z_i is complex; however, in most of the microwaves applications it is real and equal to 50 Ω. Then the S-parameters matrix can be expressed as

$$\begin{bmatrix} b_1 \\ b_2 \\ \vdots \\ b_N \end{bmatrix} = \begin{bmatrix} S_{11} & S_{12} & \cdots & S_{1N} \\ S_{21} & S_{22} & \cdots & S_{2N} \\ \vdots & \vdots & \ddots & \vdots \\ S_{N1} & S_{N2} & \cdots & S_{NN} \end{bmatrix} \times \begin{bmatrix} a_1 \\ a_2 \\ \vdots \\ a_N \end{bmatrix}. \tag{6.2}$$

By definition, the subscripts mn indicate output port number, m, and input port number, n, of the scattering parameter S_{mn}. If only the port n is excited while all other ports are terminated by matched loads, the output power wave at port m, b_m, and the input power wave at port n, a_n, can be used to calculate S_{mn} using

$$S_{mn} = \frac{b_m}{a_n}. \tag{6.3}$$

This technique can be applied to FDTD simulation results to obtain S-parameters for an input port n. A multiport circuit can be constructed in an FDTD problem space where all ports are terminated by matching loads and only the reference port n is excited by a source.

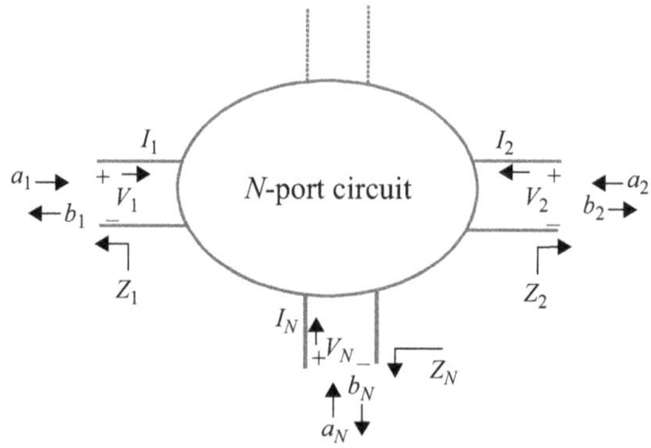

Figure 6.1 An N-port network.

Then sampled voltages and currents can be captured at all ports during the FDTD time-marching loop. S-parameters are *frequency-domain* outputs of a network. After the FDTD iterations are completed, the sampled voltages and currents can be transformed to the frequency domain using the algorithms described in Chapter 5. Then the frequency-domain sampled voltages and currents can be used in (6.1) to obtain incident and reflected power waves, a_i and b_i, from which the S-parameters can be obtained for the reference port using (6.3). One should keep in mind that in an FDTD simulation only one of the ports can be excited. Therefore, to obtain a complete set of S-parameters for all of the port excitations in the circuit, more than one FDTD run may be required depending on the type and symmetry conditions of the problem.

The S-parameters are complex quantities, as they are obtained using (6.3). Generally, S-parameters are plotted by their magnitudes in decibels such that $|S_{mn}|_{\mathrm{dB}} = 20\log_{10}(|S_{mn}|)$ and by their phases.

6.2 S-Parameter calculations

In this section, we illustrate the implementation of S-parameter calculations through an example. S-parameters are associated with ports; therefore, ports are the necessary components for the S-parameter calculations. To define a port we need a sampled voltage and a sampled current defined at the same location and associated with the port. Consider the two-port circuit shown in Figure 6.2, which has been published in [14] as an example for the application of the FDTD method to the analysis of planar microstrip circuits. The problem space is composed of cells having $\Delta x = 0.4064$ mm, $\Delta y = 0.4233$ mm, and $\Delta z = 0.265$ mm. An air gap of 5 cells is left between the circuit and the outer boundary in the *xn*, *xp*, *yn*, and *yp* directions and of 10 cells in the *zp* direction. The outer boundary is perfect electric conductor (PEC) and touches the ground of the circuit in the *zn* direction. The dimensions of the microstrips are shown in Figure 6.2(b) and are implemented in Listing 6.1. The substrate is $3 \times \Delta z$ thick and has a dielectric constant of 2.2. The microstrip filter is terminated by a voltage source with 50 Ω internal resistance on one end and by a 50 Ω resistor on the other end. The voltage source is excited by a Gaussian waveform with 20 cells per wavelength accuracy parameter. Two sampled voltages and two sampled currents are defined 10 cells away from the ends of the microstrip terminations as implemented in Listing 6.2.

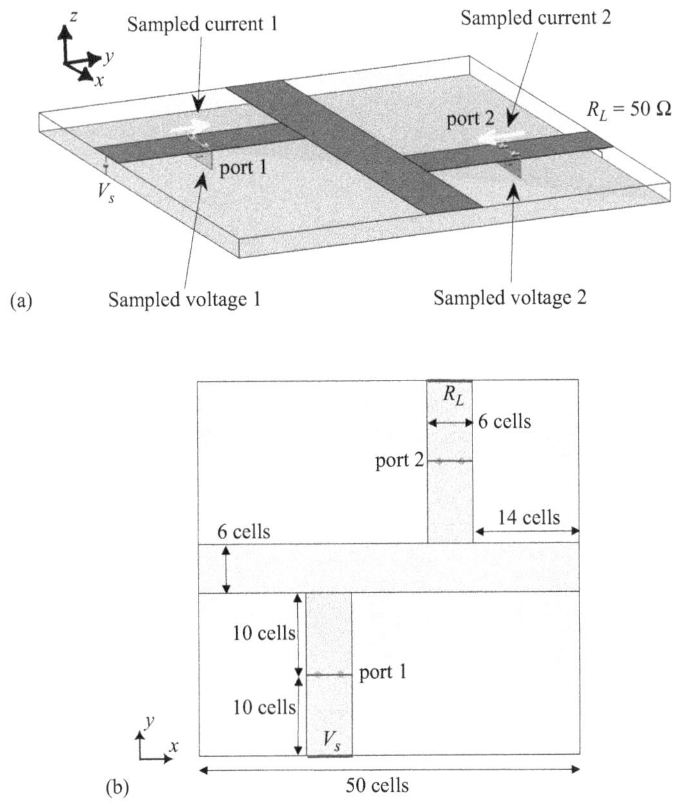

Figure 6.2 A microstrip low-pass filter terminated by a voltage source and a resistor on two ends: (a) three-dimensional view and (b) dimensions.

Listing 6.1 define_geometry.m

```
disp('defining the problem geometry');

bricks  = [];
spheres = [];

% define a substrate
bricks(1).min_x = 0;
bricks(1).min_y = 0;
bricks(1).min_z = 0;
bricks(1).max_x = 50*dx;
bricks(1).max_y = 46*dy;
bricks(1).max_z = 3*dz;
bricks(1).material_type = 4;

% define a PEC plate
bricks(2).min_x = 14*dx;
bricks(2).min_y = 0;
bricks(2).min_z = 3*dz;
```

```
bricks(2).max_x = 20*dx;
bricks(2).max_y = 20*dy;
bricks(2).max_z = 3*dz;
bricks(2).material_type = 2;

% define a PEC plate
bricks(3).min_x = 30*dx;
bricks(3).min_y = 26*dy;
bricks(3).min_z = 3*dz;
bricks(3).max_x = 36*dx;
bricks(3).max_y = 46*dy;
bricks(3).max_z = 3*dz;
bricks(3).material_type = 2;

% define a PEC plate
bricks(4).min_x = 0;
bricks(4).min_y = 20*dy;
bricks(4).min_z = 3*dz;
bricks(4).max_x = 50*dx;
bricks(4).max_y = 26*dy;
bricks(4).max_z = 3*dz;
bricks(4).material_type = 2;

% define a PEC plate as ground
bricks(5).min_x = 0;
bricks(5).min_y = 0;
bricks(5).min_z = 0;
bricks(5).max_x = 50*dx;
bricks(5).max_y = 46*dy;
bricks(5).max_z = 0;
bricks(5).material_type = 2;
```

We associate a sampled voltage and sampled current pair to a port, and thus we have two ports. In Listing 6.2 a new parameter **ports** is defined and initialized as an empty array. At the end of the listing, the sampled voltages and currents are associated to ports through their indices. For instance, the subfield **sampled_voltage_index** of **ports(2)** is assigned the value 2, implying that **sampled_voltages(2)** is the sampled voltage associated to the second port. The subfield **impedance** is assigned the impedance of the respective port, which is supposed to be equal to the microstrip-line characteristic impedance and the resistance of the voltage source and resistor terminating the microstrip line. One additional subfield of **ports** is **is_source_port**, which indicates whether the port is the source port or not. In these two ports example, the first port is the excitation port due to the voltage source; therefore, **ports(1).is_source_port** is assigned the value ***true*** whereas for the second port **ports(2).is_source_port** is ***false***.

One should notice that the direction of the second sampled current **sampled_currents(2).direction** is toward the negative x direction as ***xn***. The reference currents are defined as flowing into the circuit for the purpose of calculating the network parameters, as can be seen in Figure 6.1.

Listing 6.2 define_output_parameters.m

```
disp('defining output parameters');

sampled_electric_fields = [];
sampled_magnetic_fields = [];
sampled_voltages = [];
sampled_currents = [];
ports = [];

% figure refresh rate
plotting_step = 100;

% mode of operation
run_simulation = true;
show_material_mesh = true;
show_problem_space = true;

% frequency domain parameters
frequency_domain.start = 20e6;
frequency_domain.end   = 20e9;
frequency_domain.step  = 20e6;

% define sampled voltages
sampled_voltages(1).min_x = 14*dx;
sampled_voltages(1).min_y = 10*dy;
sampled_voltages(1).min_z = 0;
sampled_voltages(1).max_x = 20*dx;
sampled_voltages(1).max_y = 10*dy;
sampled_voltages(1).max_z = 3*dz;
sampled_voltages(1).direction = 'zp';
sampled_voltages(1).display_plot = false;

sampled_voltages(2).min_x = 30*dx;
sampled_voltages(2).min_y = 36*dy;
sampled_voltages(2).min_z = 0.0;
sampled_voltages(2).max_x = 36*dx;
sampled_voltages(2).max_y = 36*dy;
sampled_voltages(2).max_z = 3*dz;
sampled_voltages(2).direction = 'zp';
sampled_voltages(2).display_plot = false;

% define sampled currents
sampled_currents(1).min_x = 14*dx;
sampled_currents(1).min_y = 10*dy;
sampled_currents(1).min_z = 3*dz;
sampled_currents(1).max_x = 20*dx;
sampled_currents(1).max_y = 10*dy;
sampled_currents(1).max_z = 3*dz;
sampled_currents(1).direction = 'yp';
sampled_currents(1).display_plot = false;
```

```
sampled_currents(2).min_x = 30*dx;
sampled_currents(2).min_y = 36*dy;
sampled_currents(2).min_z = 3*dz;
sampled_currents(2).max_x = 36*dx;
sampled_currents(2).max_y = 36*dy;
sampled_currents(2).max_z = 3*dz;
sampled_currents(2).direction = 'yn';
sampled_currents(2).display_plot = false;

% define ports
ports(1).sampled_voltage_index = 1;
ports(1).sampled_current_index = 1;
ports(1).impedance = 50;
ports(1).is_source_port = true;

ports(2).sampled_voltage_index = 2;
ports(2).sampled_current_index = 2;
ports(2).impedance = 50;
ports(2).is_source_port = false;
```

Furthermore, one should notice that we can compute only the set of S-parameters (S_{11} and S_{21}) for which the first port is the excitation port. The other set of S-parameters (S_{12} and S_{22}) can be obtained from (S_{11} and S_{21}) due to the symmetry in the structure. If the structure is not symmetric, we should assign the voltage source to the second port location and repeat the simulation. Another approach for obtaining all S-parameters is that, instead of defining a single source, sources can be defined at every port separately, each having internal impedance equal to the port impedance. Then the FDTD simulations can be repeated in a loop a number of times equals the number of ports, where each time one of the ports is set as the excitation port. Each time, the sources associated with the excitation port are activated while other sources are deactivated. The inactive sources will not generate power and serve only as passive terminations. Each time, the S-parameter set for the excitation port as the input can be calculated and stored.

Listing 6.2 illustrates how we can define the ports. The only initialization required for the ports is the determination of the number of ports in ***initialize_output_parameters*** as illustrated in Listing 6.3. After the FDTD time-marching loop is completed, the S-parameters are calculated in ***calculate_frequency_domain_outputs*** after frequency-domain transformation of

Listing 6.3 initialize_output_parameters.m

```
disp('initializing the output parameters');

number_of_sampled_electric_fields  = size(sampled_electric_fields,2);
number_of_sampled_magnetic_fields  = size(sampled_magnetic_fields,2);
number_of_sampled_voltages  = size(sampled_voltages,2);
number_of_sampled_currents  = size(sampled_currents,2);
number_of_ports = size(ports,2);
```

sampled voltages and currents as shown in Listing 6.4. First, the incident and reflected power waves are calculated for every port using (6.1). Then the S-parameters are calculated for the active port using (6.3). The calculated S-parameters are plotted as the last step in ***display_frequency_domain_outputs***, as shown in Listing 6.5.

Listing 6.4 calculate_frequency_domain_outputs.m

```
% calculation of S-parameters
% calculate incident and reflected power waves
for ind=1:number_of_ports
    svi = ports(ind).sampled_voltage_index;
    sci = ports(ind).sampled_current_index;
    Z = ports(ind).impedance;
    V = sampled_voltages(svi).frequency_domain_value;
    I = sampled_currents(sci).frequency_domain_value;
    ports(ind).a = 0.5*(V+Z.*I)./sqrt(real(Z));
    ports(ind).b = 0.5*(V-conj(Z).*I)./sqrt(real(Z));
    ports(ind).frequencies = frequency_array;
end

% calculate the S-parameters
for ind=1:number_of_ports
     if ports(ind).is_source_port == true
         for oind=1:number_of_ports
             ports(ind).S(oind).values = ports(oind).b ./ ports(ind).a;
         end
     end
end
```

Listing 6.5 display_frequency_domain_outputs.m

```
% figures for S-parameters
for ind=1:number_of_ports
    if ports(ind).is_source_port == true
        frequencies = ports(ind).frequencies*1e-9;
        for oind=1:number_of_ports
            S = ports(ind).S(oind).values;
            Sdb = 20 * log10(abs(S));
            Sphase = angle(S)*180/pi;
            figure;
            subplot(2,1,1);
            plot(frequencies, Sdb,'b-','linewidth',1.5);
            title(['S' num2str(oind) num2str(ind)],'fontsize',12);
            xlabel('frequency␣(GHz)','fontsize',12);
            ylabel('magnitude␣(dB)','fontsize',12);
            grid on;
            subplot(2,1,2);
            plot(frequencies, Sphase,'r-','linewidth',1.5);
            xlabel('frequency␣(GHz)','fontsize',12);
            ylabel('phase␣(degrees)','fontsize',12);
            grid on;
            drawnow;
        end
    end
end
```

The sample circuit in Figure 6.1 is run for 20,000 time steps, and the S-parameters of the circuit are obtained at the reference planes 10 cells away from the terminations where the ports are defined. Although the reference planes for the S-parameters are defined away from the terminations, it is possible to define the planes at the terminations. It is sufficient to place the sampled voltages and currents on the terminations.

Figure 6.3(a) shows the calculated S_{11} up to 20 GHz, and Figure 6.3(b) shows the calculated S_{21}. It can be seen that the circuit acts as a low-pass filter where the pass band is up to 5.5 GHz. Comparing the results in Figure 6.3 with the ones published in [14] it is evident that there are differences between the sets of results: some spikes appear in Figure 6.3. This is

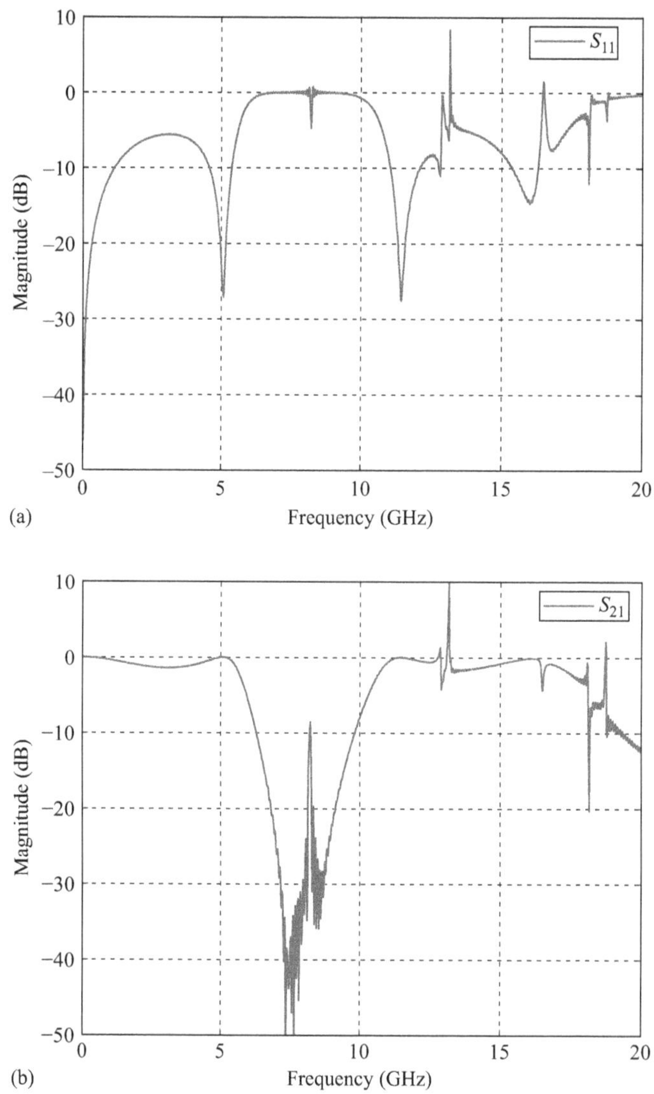

Figure 6.3 S-parameters of the microstrip low-pass filter: (a) S_{11} and (b) S_{21}.

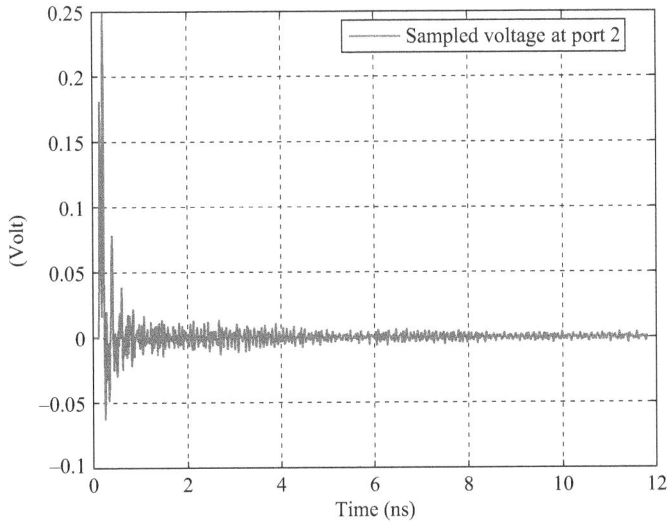

Figure 6.4 Sampled voltage at the second port of the microstrip low-pass filter.

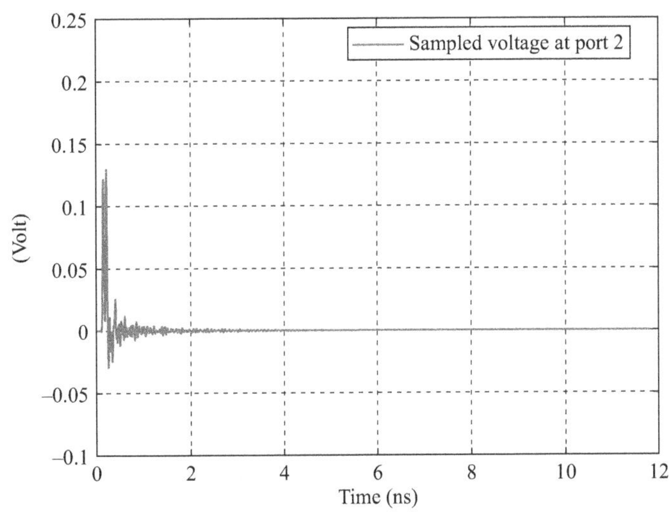

Figure 6.5 Sampled voltage at the second port of the microstrip low-pass filter with $\sigma^e = 0.2$.

because of the PEC boundaries, which make the FDTD simulation space into a cavity that resonates at certain frequencies. Figure 6.4 shows the sampled voltage captured at the second port of this microstrip filter. One can notice that, although the simulation is performed in 20,000 time steps, which is a large number of time steps for this problem, the transient response still did not damp out. This behavior exposes itself as spikes (numerical errors) in the S-parameter plots at different frequencies as shown in Figure 6.3. If the filter has slight loss, the attenuation will improve the numerical errors, as the time-domain response will be diminishing (as shown in Figure 6.5) for the filter lossy dielectric substrate. Therefore, the

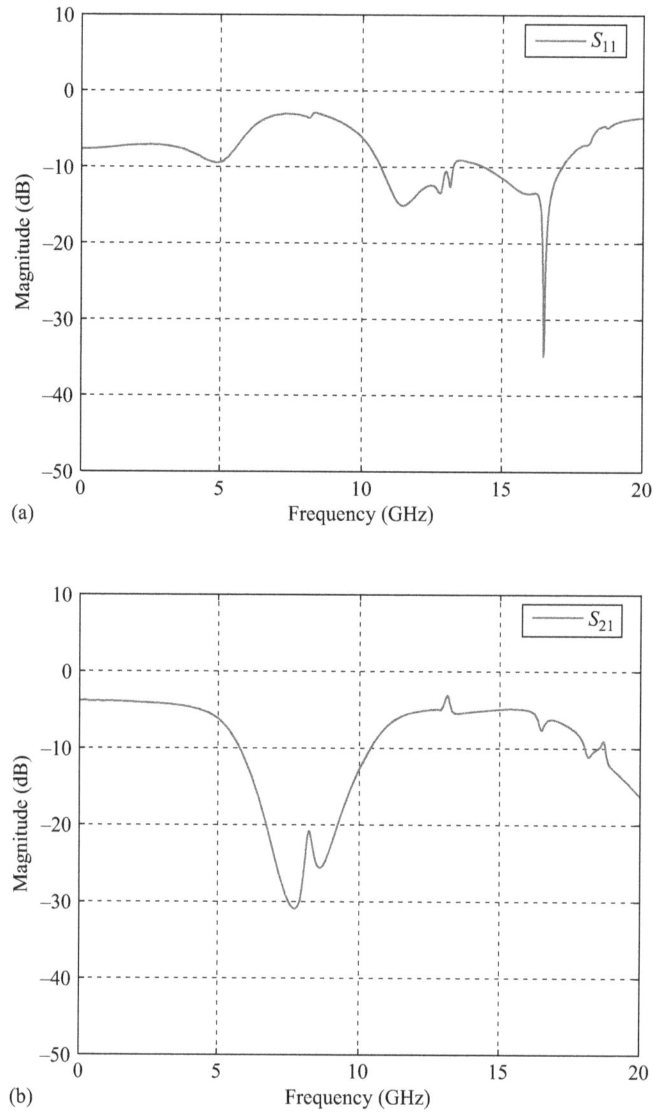

Figure 6.6 S-parameters of the microstrip low-pass filter with $\sigma^e = 0.2$: (a) S_{11} and (b) S_{21}.

truncation at a diminishing time-domain waveform will not cause significant errors even though the cavity modes of the PEC closed box still exist. This is clearly shown in Figure 6.6 (a) and 6.6(b). If the simulation had been performed such that boundaries simulate open space, the simulation time would take less and the S-parameters results would be clear of these spikes. Chapters 7 and 8 discuss algorithms that simulate open boundaries for FDTD simulations.

6.3 Simulation examples

6.3.1 Quarter-wave transformer

In this example the simulation of a microstrip quarter-wave transformer is presented. The geometry and the dimensions of the circuit are shown in Figure 6.7. The circuit is constructed on a substrate having 1 mm thickness and 4.6 dielectric constant. The index of the material type of the microstrip substrate is 4. A voltage source with internal resistance 50 Ω is connected to a 50 Ω microstrip line having 1.8 mm width and 4 mm length. This line is matched to a 100 Ω line through a 70.7 Ω line having 1 mm width and 10 mm length at 4 GHz. The 100 Ω line has 0.4 mm width and 4 mm length and is terminated by a 100 Ω resistor. The FDTD problem space is composed of cubic cells with sides each measuring 0.2 mm. The boundaries of the problem space are PEC on all sides; however, to suppress the cavity resonances another material type is defined as an absorber. The index of the material type of the absorber is 5. The relative permittivity and permeability of this absorber is 1, the electric conductivity is 1, and the magnetic conductivity is 142,130, which is the square of the intrinsic impedance of free space. The air gap between the objects and boundaries is zero on all sides. The bottom boundary serves as the ground of the microstrip circuit, while the other five sides are surrounded by the absorber material. The definition of the relevant problem space parameters and geometry are shown in Listing 6.6. The definition of the voltage source and the resistor are shown in Listing 6.7. In the previous example, the sampled voltages and sampled currents are defined on the microstrip lines away from the voltage

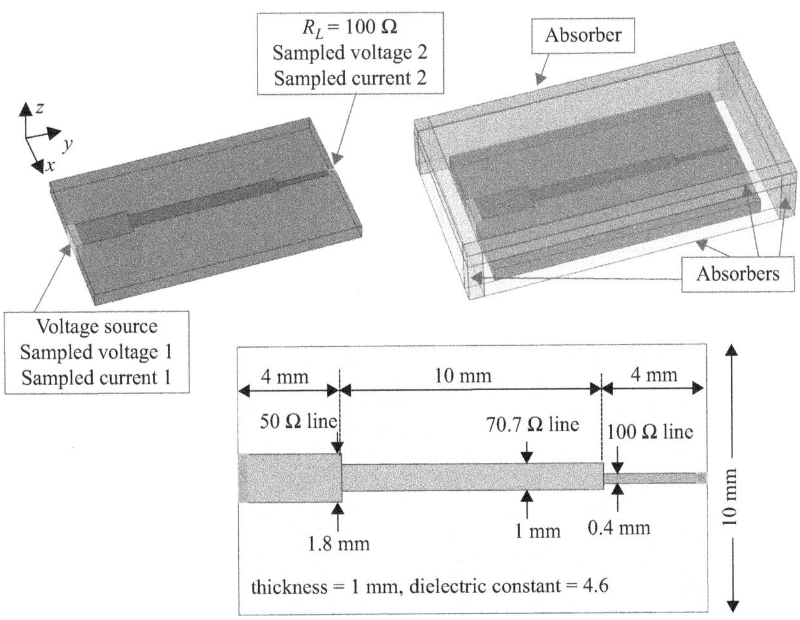

Figure 6.7 The geometry and dimensions of a microstrip quarter-wave transformer matching a 100 Ω line to 50 Ω line.

Listing 6.6 define_geometry.m

```
disp('defining the problem geometry');

bricks  = [];
spheres = [];

% define a brick with material type 4
bricks(1).min_x = 0;
bricks(1).min_y = 0;
bricks(1).min_z = 0;
bricks(1).max_x = 10e-3;
bricks(1).max_y = 18e-3;
bricks(1).max_z = 1e-3;
bricks(1).material_type = 4;

% define a brick with material type 2
bricks(2).min_x = 4e-3;
bricks(2).min_y = 0;
bricks(2).min_z = 1e-3;
bricks(2).max_x = 5.8e-3;
bricks(2).max_y = 4e-3;
bricks(2).max_z = 1e-3;
bricks(2).material_type = 2;

% define a brick with material type 2
bricks(3).min_x = 4.4e-3;
bricks(3).min_y = 4e-3;
bricks(3).min_z = 1e-3;
bricks(3).max_x = 5.4e-3;
bricks(3).max_y = 14e-3;
bricks(3).max_z = 1e-3;
bricks(3).material_type = 2;

% define a brick with material type 2
bricks(4).min_x = 4.8e-3;
bricks(4).min_y = 14e-3;
bricks(4).min_z = 1e-3;
bricks(4).max_x = 5.2e-3;
bricks(4).max_y = 18e-3;
bricks(4).max_z = 1e-3;
bricks(4).material_type = 2;

% define absorber for zp side
bricks(5).min_x = -1e-3;
bricks(5).min_y = -2e-3;
bricks(5).min_z = 3e-3;
bricks(5).max_x = 11e-3;
bricks(5).max_y = 20e-3;
bricks(5).max_z = 4e-3;
bricks(5).material_type = 5;
```

```
% define absorber for xn side
bricks(6).min_x = -1e-3;
bricks(6).min_y = -2e-3;
bricks(6).min_z = 0;
bricks(6).max_x = 0;
bricks(6).max_y = 20e-3;
bricks(6).max_z = 4e-3;
bricks(6).material_type = 5;

% define absorber for xp side
bricks(7).min_x = 10e-3;
bricks(7).min_y = -2e-3;
bricks(7).min_z = 0;
bricks(7).max_x = 11e-3;
bricks(7).max_y = 20e-3;
bricks(7).max_z = 4e-3;
bricks(7).material_type = 5;

% define absorber for yn side
bricks(8).min_x = -1e-3;
bricks(8).min_y = -2e-3;
bricks(8).min_z = 0;
bricks(8).max_x = 11e-3;
bricks(8).max_y = -1e-3;
bricks(8).max_z = 4e-3;
bricks(8).material_type = 5;
% define absorber for yp side
bricks(9).min_x = -1e-3;
bricks(9).min_y = 19e-3;
bricks(9).min_z = 0;
bricks(9).max_x = 11e-3;
bricks(9).max_y = 20e-3;
bricks(9).max_z = 4e-3;
bricks(9).material_type = 5;
```

Listing 6.7 define_sources_and_lumped_elements.m

```
voltage_sources(1).min_x = 4e-3;
voltage_sources(1).min_y = 0;
voltage_sources(1).min_z = 0;
voltage_sources(1).max_x = 5.8e-3;
voltage_sources(1).max_y = 0.4e-3;
voltage_sources(1).max_z = 1e-3;
voltage_sources(1).direction = 'zp';
voltage_sources(1).resistance = 50;
```

```
voltage_sources(1).magnitude = 1;
voltage_sources(1).waveform_type = 'gaussian';
voltage_sources(1).waveform_index = 1;

resistors(1).min_x = 4.8e-3;
resistors(1).min_y = 17.6e-3;
resistors(1).min_z = 0;
resistors(1).max_x = 5.2e-3;
resistors(1).max_y = 18e-3;
resistors(1).max_z = 1e-3;
resistors(1).direction = 'z';
resistors(1).resistance = 100;
```

source and the terminating resistor. In this example, the sampled voltages and sampled currents are defined on the voltage source and the terminating resistor. Then the sampled voltages and currents are assigned to ports. The definition of the output parameters is shown in Listing 6.8.

Listing 6.8 define_output_parameters.m

```
% frequency domain parameters
frequency_domain.start = 2e7;
frequency_domain.end   = 8e9;
frequency_domain.step  = 2e7;

% define sampled voltages
sampled_voltages(1).min_x = 4e-3;
sampled_voltages(1).min_y = 0;
sampled_voltages(1).min_z = 0;
sampled_voltages(1).max_x = 5.8e-3;
sampled_voltages(1).max_y = 0.4e-3;
sampled_voltages(1).max_z = 1e-3;
sampled_voltages(1).direction = 'zp';
sampled_voltages(1).display_plot = false;

% define sampled voltages
sampled_voltages(2).min_x = 4.8e-3;
sampled_voltages(2).min_y = 17.6e-3;
sampled_voltages(2).min_z = 0;
sampled_voltages(2).max_x = 5.2e-3;
sampled_voltages(2).max_y = 18e-3;
sampled_voltages(2).max_z = 1e-3;
sampled_voltages(2).direction = 'zp';
sampled_voltages(2).display_plot = false;

% define sampled currents
sampled_currents(1).min_x = 4e-3;
sampled_currents(1).min_y = 0;
sampled_currents(1).min_z = 0.4e-3;
sampled_currents(1).max_x = 5.8e-3;
sampled_currents(1).max_y = 0.4e-3;
sampled_currents(1).max_z = 0.6e-3;
sampled_currents(1).direction = 'zp';
sampled_currents(1).display_plot = false;
```

```
% define sampled currents
sampled_currents(2).min_x = 4.8e-3;
sampled_currents(2).min_y = 17.6e-3;
sampled_currents(2).min_z = 0.4e-3;
sampled_currents(2).max_x = 5.2e-3;
sampled_currents(2).max_y = 18e-3;
sampled_currents(2).max_z = 0.6e-3;
sampled_currents(2).direction = 'zp';
sampled_currents(2).display_plot = false;

% define ports
ports(1).sampled_voltage_index = 1;
ports(1).sampled_current_index = 1;
ports(1).impedance = 50;
ports(1).is_source_port = true;

ports(2).sampled_voltage_index = 2;
ports(2).sampled_current_index = 2;
ports(2).impedance = 100;
ports(2).is_source_port = false;
```

The FDTD simulation is run for 5,000 time steps, and the S-parameters of the simulation are plotted in Figure 6.8. The plotted S_{11} indicates that a good match has been obtained at 4 GHz. Furthermore, the results are free of spikes, indicating that the absorbers used in the simulation suppressed the cavity resonances. However, one should keep in mind that although the use of absorbers improves the FDTD simulation results, it may introduce some other undesired errors to the simulation results. In the following chapters, more advanced absorbing boundaries are discussed that minimize these errors.

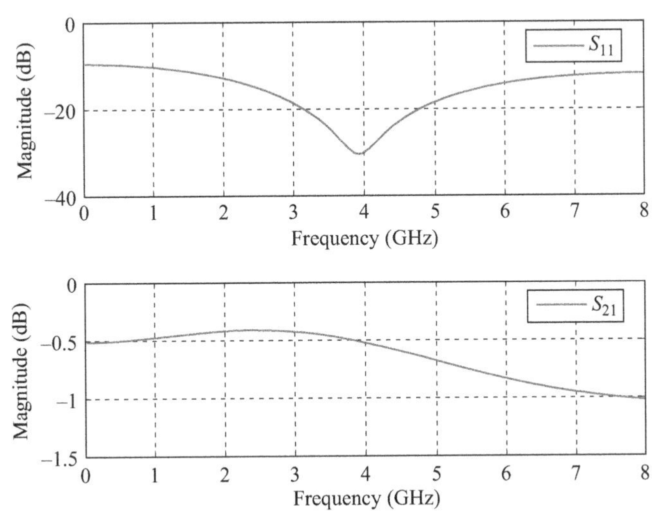

Figure 6.8 The S_{11} and S_{21} of the microstrip quarter-wave transformer circuit.

6.4 Exercises

6.1 Consider the low-pass filter circuit in the example described in Section 6.2. Use the same type of absorber as shown in the example in Section 6.3.1 to terminate the boundaries of the low-pass filter circuit. Use absorbers with 5 cells thickness, and leave at least 5 cells air gap between the circuit and the absorbers in the *xn*, *xp*, *yn*, and *yp* directions. Leave at least 10 cells air gap between the circuit and the absorbers in the *zp* direction. In the *zn* direction the PEC boundary will serve as the ground plane of the circuit. The FDTD problem space is illustrated in Figure 6.9 as a reference. Rerun the simulation for 3,000 time steps, and verify that the spikes in Figure 6.3 due to cavity resonances are suppressed.

6.2 Consider the quarter-wave transformer circuit shown in Example 6.3.1. Redefine the voltage source and the resistor such that voltage source will feed the 100 Ω line and will have 100 Ω internal resistance, while the resistor will terminate the 50 Ω line and will have 50 Ω resistance. Set port 2 as the active port, and run the simulation. Obtain the figures plotting S_{12} and S_{22}, and verify that they are similar to the S_{21} and S_{11} in Figure 6.8, respectively.

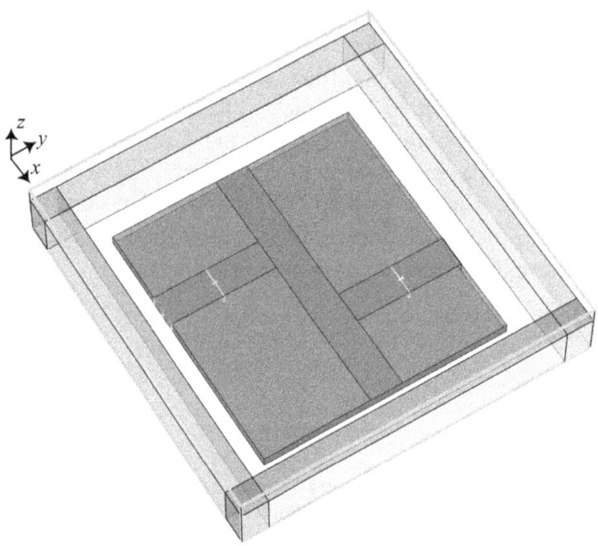

Figure 6.9 The problem space for the low-pass filter with absorbers on the *xn*, *xp*, *yn*, *yp*, and *zp* sides.

CHAPTER 7

Perfectly matched layer absorbing boundary

Because computational storage space is finite, the finite-difference time-domain (FDTD) problem space size is finite and needs to be truncated by special boundary conditions. In the previous chapters we discussed some examples for which the problem space is terminated by perfect electric conductor (PEC) boundaries. However, many applications, such as scattering and radiation problems, require the boundaries simulated as open space. The types of special boundary conditions that simulate electromagnetic waves propagating continuously beyond the computational space are called *absorbing boundary conditions* (ABCs). However, the imperfect truncation of the problem space will create numerical reflections, which will corrupt the computational results in the problem space after certain amounts of simulation time. So far, several various types of ABCs have been developed. However, the perfectly matched layer (PML) introduced by Berenger [15, 16] has been proven to be one of the most robust ABCs [17–20] in comparison with other techniques adopted in the past. PML is a finite-thickness special medium surrounding the computational space based on fictitious constitutive parameters to create a wave-impedance matching condition, which is independent of the angles and frequencies of the wave incident on this boundary. The theory and implementation of the PML boundary condition are illustrated in this chapter.

7.1 Theory of PML

In this section we demonstrate analytically the reflectionless characteristics of PML at the vacuum–PML and PML–PML interfaces [15] in detail.

7.1.1 Theory of PML at the vacuum–PML interface

We provide the analysis of reflection at a vacuum–PML interface in a two-dimensional case. Consider the TE_z polarized plane wave propagating in an arbitrary direction as shown in Figure 7.1. In the given TE_z case E_x, E_y, and H_z are the only field components that exist in

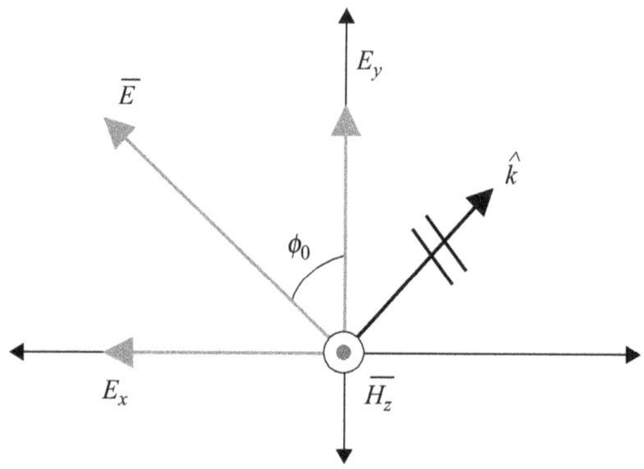

Figure 7.1 The field decomposition of a TE_z polarized plane wave.

the two-dimensional space. These field components can be expressed in the time-harmonic domain as

$$E_x = -E_0 \sin \phi_0 e^{j\omega(t-\alpha x-\beta y)}, \tag{7.1a}$$

$$E_y = E_0 \cos \phi_0 e^{j\omega(t-\alpha x-\beta y)}, \tag{7.1b}$$

$$H_z = H_0 e^{j\omega(t-\alpha x-\beta y)}. \tag{7.1c}$$

Maxwell's equations for a TE_z polarized wave are

$$\varepsilon_0 \frac{\partial E_x}{\partial t} + \sigma^e E_x = \frac{\partial H_z}{\partial y}, \tag{7.2a}$$

$$\varepsilon_0 \frac{\partial E_y}{\partial t} + \sigma^e E_y = -\frac{\partial H_z}{\partial x}, \tag{7.2b}$$

$$\mu_0 \frac{\partial H_z}{\partial t} + \sigma^m H_z = \frac{\partial E_x}{\partial y} - \frac{\partial E_y}{\partial x}. \tag{7.2c}$$

In a TE_z PML medium, H_z can be broken into two artificial components associated with the x and y directions as

$$H_{zx} = H_{zx0} e^{-j\omega\beta y} e^{j\omega(t-\alpha x)}, \tag{7.3a}$$

$$H_{zy} = H_{zy0} e^{-j\omega\alpha x} e^{j\omega(t-\beta y)}, \tag{7.3b}$$

where $H_z = H_{zx} + H_{zy}$. Therefore, a modified set of Maxwell's equations for a TE_z polarized PML medium can be expressed as

$$\varepsilon_0 \frac{\partial E_x}{\partial t} + \sigma_{pey} E_x = \frac{\partial (H_{zx} + H_{zy})}{\partial y}, \tag{7.4a}$$

$$\varepsilon_0 \frac{\partial E_y}{\partial t} + \sigma_{pex} E_y = -\frac{\partial (H_{zx} + H_{zy})}{\partial x}, \tag{7.4b}$$

$$\mu_0 \frac{\partial H_{zx}}{\partial t} + \sigma_{pmx} H_{zx} = -\frac{\partial E_y}{\partial x}, \tag{7.4c}$$

$$\mu_0 \frac{\partial H_{zy}}{\partial t} + \sigma_{pmy} H_{zy} = \frac{\partial E_x}{\partial y}, \tag{7.4d}$$

where σ_{pex}, σ_{pey}, σ_{pmx}, and σ_{pmy} are the introduced fictitious conductivities. With the given conductivities the PML medium described by (7.4) is an anisotropic medium. When $\sigma_{pmx} = \sigma_{pmy} = \sigma^m$, merging (7.4c) and (7.4d) yields (7.2c). Field components E_y and H_{zx} together can represent a wave propagating in the x direction, and field components of E_x and H_{zy} represent a wave propagating in the y direction. Substituting the field equations for the x and y propagating waves in (7.1a), (7.1b), (7.3a), and (7.3b) into the modified Maxwell's equations given, one can obtain

$$\varepsilon_0 E_0 \sin\phi_0 - j\frac{\sigma_{pey}}{\omega} E_0 \sin\phi_0 = \beta(H_{zx0} + H_{zy0}), \tag{7.5a}$$

$$\varepsilon_0 E_0 \cos\phi_0 - j\frac{\sigma_{pex}}{\omega} E_0 \cos\phi_0 = \alpha(H_{zx0} + H_{zy0}), \tag{7.5b}$$

$$\mu_0 H_{zx0} - j\frac{\sigma_{pmx}}{\omega} H_{zx0} = \alpha E_0\, cos\phi_0, \tag{7.5c}$$

$$\mu_0 H_{zy0} - j\frac{\sigma_{pmy}}{\omega} H_{zy0} = \beta E_0 \sin\phi_0. \tag{7.5d}$$

Using (7.5c) and (7.5d) to eliminate magnetic field terms from (7.5a) and (7.5b) yields

$$\varepsilon_0\mu_0\left(1 - j\frac{\sigma_{pey}}{\varepsilon_0\omega}\right)\sin\phi_0 = \beta\left[\frac{\alpha\cos\phi_0}{(1 - j(\sigma_{pmx}/\mu_0\omega))} + \frac{\beta\sin\phi_0}{(1 - j(\sigma_{pmy}/\mu_0\omega))}\right], \tag{7.6a}$$

$$\varepsilon_0\mu_0\left(1 - j\frac{\sigma_{pex}}{\varepsilon_0\omega}\right)\cos\phi_0 = \alpha\left[\frac{\alpha\cos\phi_0}{(1 - j(\sigma_{pmx}/\mu_0\omega))} + \frac{\beta\sin\phi_0}{(1 - j(\sigma_{pmy}/\mu_0\omega))}\right]. \tag{7.6b}$$

The unknown constants α and β can be obtained from (7.6a) and (7.6b) as

$$\alpha = \frac{\sqrt{\mu_0\varepsilon_0}}{G}\left(1 - j\frac{\sigma_{pex}}{\omega\varepsilon_0}\right)\cos\phi_0, \tag{7.7a}$$

$$\beta = \frac{\sqrt{\mu_0\varepsilon_0}}{G}\left(1 - j\frac{\sigma_{pey}}{\omega\varepsilon_0}\right)\sin\phi_0, \tag{7.7b}$$

where

$$G = \sqrt{w_x \cos^2\phi_0 + w_y \sin^2\phi_0}, \tag{7.8}$$

and

$$w_x = \frac{1 - j\sigma_{pex}/\omega\varepsilon_0}{1 - j\sigma_{pmx}/\omega\mu_0}, \quad w_y = \frac{1 - j\sigma_{pey}/\omega\varepsilon_0}{1 - j\sigma_{pmy}/\omega\mu_0}, \tag{7.9}$$

Therefore, the generalized field component can be expressed as

$$\psi = \psi_0 e^{j\omega\left(t - \frac{x\cos\phi_0 + y\sin\phi_0}{cG}\right)} e^{-\frac{\sigma_{pex}\cos\phi_0}{\varepsilon_0 cG}x} e^{-\frac{\sigma_{pey}\sin\phi_0}{\varepsilon_0 cG}y}, \tag{7.10}$$

where the first exponential represents the phase of a plane wave and the second and third exponentials govern the decrease in the magnitude of the wave along the x axis and y axis, respectively.

Once α and β are determined by (7.7), the split magnetic field can be determined from (7.5c) and (7.5d) as

$$H_{zx0} = E_0\sqrt{\frac{\varepsilon_0}{\mu_0}}\frac{w_x\cos^2\phi_0}{G}, \tag{7.11a}$$

$$H_{zy0} = E_0\sqrt{\frac{\varepsilon_0}{\mu_0}}\frac{w_y\sin^2\phi_0}{G}. \tag{7.11b}$$

The magnitude of the total magnetic field H_z is then given as

$$H_0 = H_{zx0} + H_{zy0} = E_0\sqrt{\frac{\varepsilon_0}{\mu_0}}G. \tag{7.12}$$

The wave impedance in a TE_z PML medium can be expressed as

$$Z = \frac{E_0}{H_0} = \sqrt{\frac{\mu_0}{\varepsilon_0}}\frac{1}{G}. \tag{7.13}$$

It is important to note that if the conductivity parameters are chosen such that

$$\frac{\sigma_{pex}}{\varepsilon_0} = \frac{\sigma_{pmx}}{\mu_0} \quad \text{and} \quad \frac{\sigma_{pey}}{\varepsilon_0} = \frac{\sigma_{pmy}}{\mu_0}, \tag{7.14}$$

then the term G becomes equal to unity as w_x and w_y becomes equal to unity. Therefore, the wave impedance of this PML medium becomes the same as that of the interior free space. In other words, when the constitutive conditions of (7.14) are satisfied, a TE_z polarized wave can propagate from free space into the PML medium without reflection for all frequencies, and all incident angles as can be concluded from (7.8). One should notice that, when the electric and magnetic losses are assigned to be zero, the field updating equation (7.4) for the PML region becomes that of a vacuum region.

7.1.2 Theory of PML at the PML–PML interface

The reflection of fields at the interface between two different PML media can be analyzed as follows. A TE_z polarized wave of arbitrary incidence traveling from PML layer "1" to PML layer "2" is depicted in Figure 7.2, where the interface is normal to the x axis. The reflection coefficient for an arbitrary incident wave between two lossy media can be expressed as

$$r_p = \frac{Z_2\cos\phi_2 - Z_1\cos\phi_1}{Z_2\cos\phi_2 + Z_1\cos\phi_1}, \tag{7.15}$$

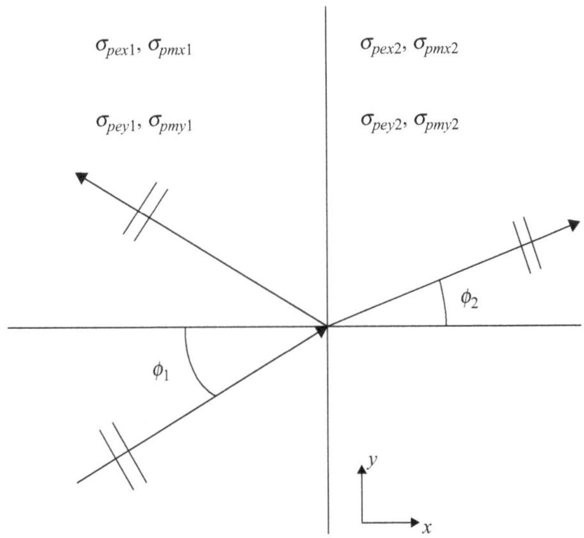

Figure 7.2 The plane wave transition at the interface between two PML media.

where Z_1 and Z_2 are the intrinsic impedances of respective mediums. Applying (7.13), the reflection coefficient r_p becomes

$$r_p = \frac{G_1 \cos\phi_2 - G_2 \cos\phi_1}{G_1 \cos\phi_2 + G_2 \cos\phi_1}. \tag{7.16}$$

The Snell–Descartes law at the interface normal to x of two lossy media can be described as

$$\left(1 - i\frac{\sigma_{y1}}{\varepsilon_0\omega}\right)\frac{\sin\phi_1}{G_1} = \left(1 - i\frac{\sigma_{y2}}{\varepsilon_0\omega}\right)\frac{\sin\phi_2}{G_2}. \tag{7.17}$$

When the two media have the same conductivities $\sigma_{pey1} = \sigma_{pey2} = \sigma_{pey}$ and $\sigma_{pmy1} = \sigma_{pmy2} = \sigma_{pmy}$, (7.17) becomes

$$\frac{\sin\phi_1}{G_1} = \frac{\sin\phi_2}{G_2}. \tag{7.18}$$

Moreover, when $(\sigma_{pex1}, \sigma_{pmx1})$, $(\sigma_{pex2}, \sigma_{pmx2})$, and $(\sigma_{pey}, \sigma_{pmy})$ satisfy the matching condition in (7.14), $G_1 = G_2 = 1$. Then (7.18) reduces to $\phi_1 = \phi_2$, and (7.16) reduces to $r_p = 0$. Therefore, theoretically when two PML media satisfy (7.14) and lie at an interface normal to the x axis with the same $(\sigma_{pey}, \sigma_{pmy})$, a wave can transmit through this interface with no reflections, at any angle of incidence and any frequency. When $(\sigma_{pex1}, \sigma_{pmx1}, \sigma_{pey1}, \sigma_{pmy1})$ are assigned to be (0, 0, 0, 0), the PML medium 1 becomes a vacuum. Therefore, when $(\sigma_{pex2}, \sigma_{pmx2})$ satisfies (7.14), the reflection coefficient at this interface is also null, which agrees with the previous vacuum–PML analysis. However, if the two media have the same $(\sigma_{pey}, \sigma_{pmy})$ but do not satisfy (7.14), then the reflection coefficient becomes

$$r_p = \frac{\sin\phi_1 \cos\phi_2 - \sin\phi_2 \cos\phi_1}{\sin\phi_1 \cos\phi_2 + \sin\phi_2 \cos\phi_1}. \tag{7.19}$$

Substituting (7.18) into (7.19), the reflection coefficient of two unmatched PML media becomes

$$r_p = \frac{\sqrt{w_{x1}} - \sqrt{w_{x2}}}{\sqrt{w_{x1}} + \sqrt{w_{x2}}}. \tag{7.20}$$

Equation (7.20) shows that the reflection coefficient for two unmatched PML media is highly dependent on frequency, regardless of the incident angle. When the two PML media follow the reflectionless condition in (7.14), $w_{x1} = w_{x2} = 1$, the reflection coefficient becomes null.

The analysis can be applied to two PML media lying at the interface normal to the y axis as well. The Snell–Descartes law related to this interface is

$$\left(1 - i\frac{\sigma_{x1}}{\varepsilon_0 \omega}\right)\frac{\sin\phi_1}{G_1} = \left(1 - i\frac{\sigma_{x2}}{\varepsilon_0 \omega}\right)\frac{\sin\phi_2}{G_2}. \tag{7.21}$$

If the two media have the same conductivities such that $\sigma_{pex1} = \sigma_{pex2} = \sigma_{pex}$ and $\sigma_{pmx1} = \sigma_{pmx2} = \sigma_{pmx}$, (7.21) reduces to (7.18). Similarly, if $(\sigma_{pey1}, \sigma_{pmy1})$, $(\sigma_{pey2}, \sigma_{pmy2})$, and $(\sigma_{pex}, \sigma_{pmx})$ satisfy the matching condition in (7.14), then $G_1 = G_2 = 1$. Equation (7.21) then reduces to $\phi_1 = \phi_2$, and the reflection coefficient at this interface is $r_p = 0$. To match the vacuum–PML interface normal to the y axis, the reflectionless condition can be achieved when $(\sigma_{pey2}, \sigma_{pmy2})$ of the PML medium satisfies (7.14).

Based on the previous discussion, if a two-dimensional FDTD problem space is attached with an adequate thickness of PML media as shown in Figure 7.3, the outgoing waves will be absorbed without any undesired numerical reflections. The PML regions must be assigned appropriate conductivity values satisfying the matching condition (7.14); the positive and

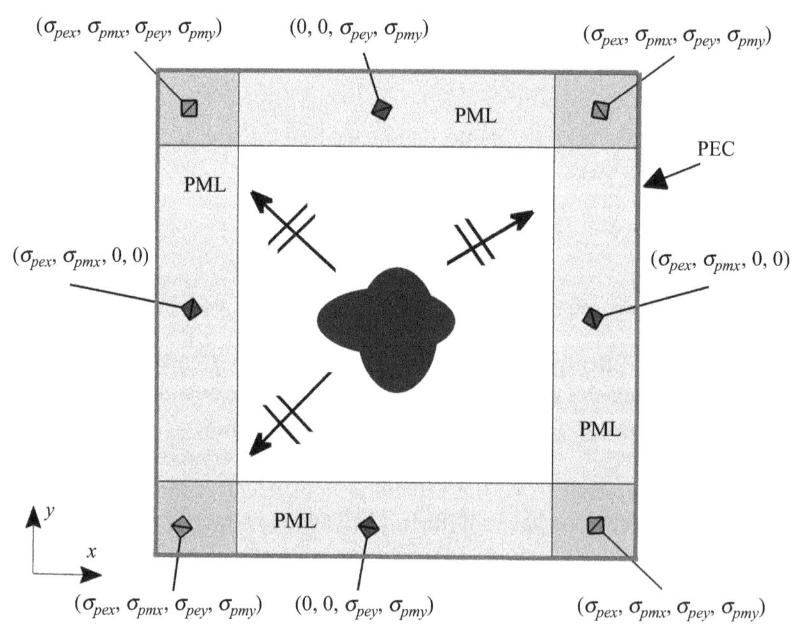

Figure 7.3 The loss distributions in two-dimensional PML regions.

negative x boundaries of the PML regions have nonzero σ_{pex}, and σ_{pmx}, whereas the positive and negative y boundaries of the PML regions have nonzero σ_{pey}, and σ_{pmy} values. The coexistence of nonzero values of σ_{pex}, σ_{pmx}, σ_{pey}, and σ_{pmy} is required at the four corner PML overlapping regions. Using a similar analysis, the conditions of (7.14) can be applied to a TM_z polarized wave to travel from free space to PML and from PML to PML without reflection [21]. Using the same impedance matching condition in (7.14), the modified Maxwell's equations for two-dimensional TM_z PML updating equations are obtained as

$$\varepsilon_0 \frac{\partial E_{zx}}{\partial t} + \sigma_{pex} E_{zx} = \frac{\partial H_y}{\partial x}, \tag{7.22a}$$

$$\varepsilon_0 \frac{\partial E_{zy}}{\partial t} + \sigma_{pey} E_{zy} = -\frac{\partial H_x}{\partial y}, \tag{7.22b}$$

$$\mu_0 \frac{\partial H_x}{\partial t} + \sigma_{pmy} H_x = -\frac{\partial (E_{zx} + E_{zy})}{\partial y}, \tag{7.22c}$$

$$\mu_0 \frac{\partial H_y}{\partial t} + \sigma_{pmx} H_y = \frac{\partial (E_{zx} + E_{zy})}{\partial x}. \tag{7.22d}$$

The finite difference approximation schemes can be applied to the modified Maxwell's equations (7.4) and (7.22) to obtain the field-updating equations for the PML regions in the two-dimensional FDTD problem space.

7.2 PML equations for three-dimensional problem space

For a three-dimensional problem space, each field component of the electric and magnetic fields is broken into two field components similar to the two-dimensional case. Therefore, the modified Maxwell's equations have 12 field components instead of the original six components. These modified split electric field equations presented in [16] are

$$\varepsilon_0 \frac{\partial E_{xy}}{\partial t} + \sigma_{pey} E_{xy} = \frac{\partial (H_{zx} + H_{zy})}{\partial y}, \tag{7.23a}$$

$$\varepsilon_0 \frac{\partial E_{xz}}{\partial t} + \sigma_{pez} E_{xz} = -\frac{\partial (H_{yx} + H_{yz})}{\partial z}, \tag{7.23b}$$

$$\varepsilon_0 \frac{\partial E_{yx}}{\partial t} + \sigma_{pex} E_{yx} = -\frac{\partial (H_{zx} + H_{zy})}{\partial x}, \tag{7.23c}$$

$$\varepsilon_0 \frac{\partial E_{yz}}{\partial t} + \sigma_{pez} E_{yz} = \frac{\partial (H_{xy} + H_{xz})}{\partial z}, \tag{7.23d}$$

$$\varepsilon_0 \frac{\partial E_{zx}}{\partial t} + \sigma_{pex} E_{zx} = \frac{\partial (H_{yx} + H_{yz})}{\partial x}, \tag{7.23e}$$

$$\varepsilon_0 \frac{\partial E_{zy}}{\partial t} + \sigma_{pey} E_{zy} = -\frac{\partial (H_{xy} + H_{xz})}{\partial y}, \tag{7.23f}$$

whereas the modified Maxwell's split magnetic field equations are

$$\mu_0 \frac{\partial H_{xy}}{\partial t} + \sigma_{pmy} H_{xy} = -\frac{\partial (E_{zx} + E_{zy})}{\partial y}, \tag{7.24a}$$

$$\mu_0 \frac{\partial H_{xz}}{\partial t} + \sigma_{pmz} H_{xz} = \frac{\partial (E_{yx} + E_{yz})}{\partial z}, \tag{7.24b}$$

$$\mu_0 \frac{\partial H_{yz}}{\partial t} + \sigma_{pmz} H_{yz} = -\frac{\partial (E_{xy} + E_{xz})}{\partial z}, \tag{7.24c}$$

$$\mu_0 \frac{\partial H_{yx}}{\partial t} + \sigma_{pmx} H_{yx} = \frac{\partial (E_{zx} + E_{zy})}{\partial x}, \tag{7.24d}$$

$$\mu_0 \frac{\partial H_{zy}}{\partial t} + \sigma_{pmy} H_{zy} = \frac{\partial (E_{xy} + E_{xz})}{\partial y}, \tag{7.24e}$$

$$\mu_0 \frac{\partial H_{zx}}{\partial t} + \sigma_{pmx} H_{zx} = -\frac{\partial (E_{yx} + E_{yz})}{\partial x}. \tag{7.24f}$$

Then the matching condition for a three-dimensional PML is given by

$$\frac{\sigma_{pex}}{\varepsilon_0} = \frac{\sigma_{pmx}}{\mu_0}, \quad \frac{\sigma_{pey}}{\varepsilon_0} = \frac{\sigma_{pey}}{\mu_0}, \quad \text{and} \quad \frac{\sigma_{pez}}{\varepsilon_0} = \frac{\sigma_{pmz}}{\mu_0}. \tag{7.25}$$

If a three-dimensional FDTD problem space is attached with adequate thickness of PML media as shown in Figure 7.4, the outgoing waves will be absorbed without any undesired numerical reflections. The PML regions must be assigned appropriate conductivity values satisfying the matching condition (7.25); the positive and negative x boundaries of PML regions have nonzero σ_{pex} and σ_{pmx}, the positive and negative y boundaries of PML regions have nonzero σ_{pey} and σ_{pmy}, and the positive and negative z boundaries of PML regions have nonzero σ_{pez} and σ_{pmz} values as illustrated in Figure 7.4. The coexistence of nonzero values of σ_{pex}, σ_{pmx}, σ_{pey}, σ_{pmy}, σ_{pez}, and σ_{pmz} is required at the PML overlapping regions.

Finally, applying the finite difference schemes to the modified Maxwell's equations (7.23) and (7.24), one can obtain the FDTD field updating equations for the three-dimensional PML regions.

7.3 PML loss functions

As discussed in the previous sections, PML regions can be formed as the boundaries of an FDTD problem space where specific conductivities are assigned such that the outgoing waves penetrate without reflection and attenuate while traveling in the PML medium. The PML medium is governed by the modified Maxwell's equations (7.4), (7.22), (7.23), and (7.24), which can be used to obtain the updating equations for the field components in the PML regions. Furthermore, the outer boundaries of the PML regions are terminated by PEC walls. When a finite-thickness PML medium backed by a PEC wall is adopted, an incident plane wave may not be totally attenuated within the PML region, and small reflections to the interior domain from the PEC back wall may occur. For a finite-width PML medium where

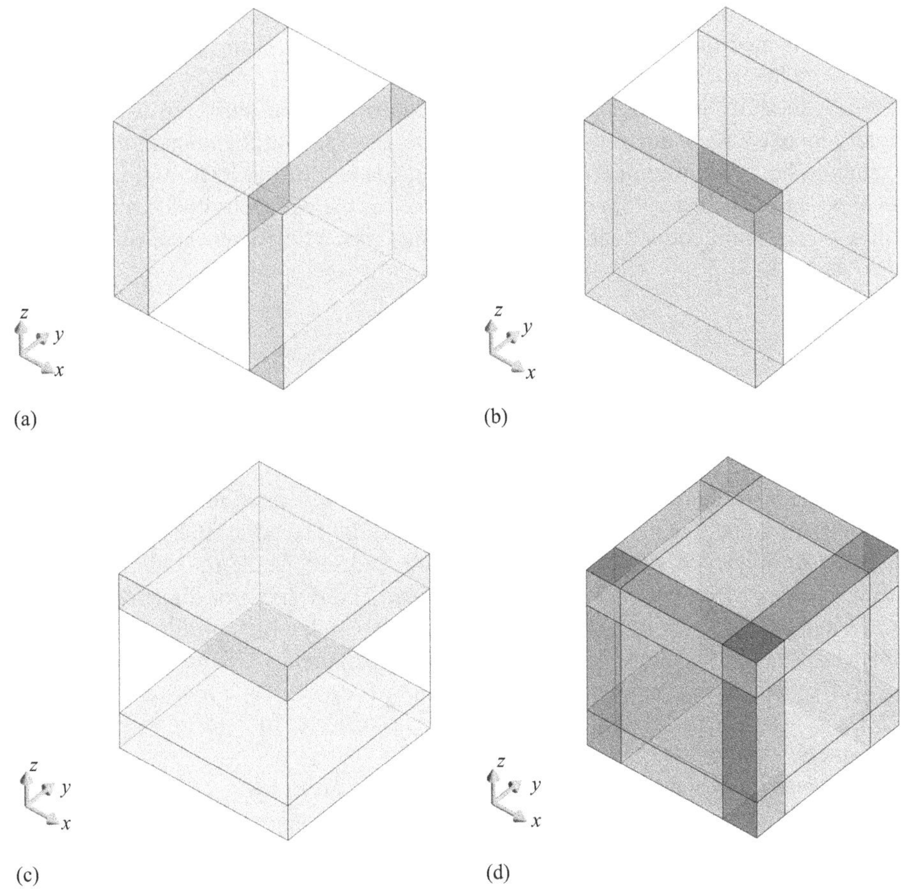

Figure 7.4 Nonzero regions of PML conductivities for a three-dimensional FDTD simulation domain: (a) nonzero σ_{pex} and σ_{pmx}; (b) nonzero σ_{pey} and σ_{pmy}; (c) nonzero σ_{pez} and σ_{pmz}; and (d) overlapping PML regions.

the conductivity distribution is uniform, there is an apparent reflection coefficient, which is expressed as

$$R(\phi_0) = e^{-2\frac{\sigma \cos \phi_0}{\varepsilon_0 c}\delta}, \tag{7.26}$$

where σ is the conductivity of the medium. Here the exponential term is the attenuation factor of the field magnitudes of the plane waves as shown in (7.10), and δ is the thickness of the PML medium. The factor 2 in the exponent is due to the travel distance, which is twice the distance between the vacuum–PML interface and the PEC backing. If ϕ_0 is 0, (7.26) is the reflection coefficient for a finite-thickness PML medium at normal incidence. If ϕ_0 is $\pi/2$, the incident plane wave is grazing to the PML medium and is attenuated by the perpendicular PML medium. From (7.26), the effectiveness of a finite-width PML is dependent on the losses within the PML medium. In addition, (7.26) can be used not only to predict the ideal performance of a finite-width PML medium but also to compute the appropriate loss distribution based on the loss profiles described in the following paragraphs.

As presented in [15], significant reflections were observed when constant uniform losses are assigned throughout the PML media, which is a result of the discrete approximation of fields and material parameters at the domain–PML interfaces and sharp variation of conductivity profiles. This mismatch problem can be tempered using a spatially gradually increasing conductivity distribution, which is zero at the domain–PML interface and tends to be a maximum conductivity $\sigma_{\max}$ at the end of the PML region. In [18] two major types of mathematical functions are proposed as the conductivity distributions or loss profiles: power and geometrically increasing functions. The power-increasing function is defined as

$$\sigma(\rho) = \sigma_{\max}\left(\frac{\rho}{\delta}\right)^{n_{pml}}, \tag{7.27a}$$

$$\sigma_{\max} = -\frac{(n_{pml}+1)\varepsilon_0 c \ln(R(0))}{2\Delta s N}, \tag{7.27b}$$

where ρ is the distance from the computational domain–PML interface to the position of the field component, and δ is the thickness of the PML cells. The parameter N is the number of PML cells, Δs is the cell size used for a PML cell, and $R(0)$ is the reflection coefficient of the finite-width PML medium at normal incidence. The distribution function is linear for $n_{pml} = 1$ and parabolic for $n_{pml} = 2$. To determine the conductivity profile using (7.27a) the parameters $R(0)$ and n_{pml} must be predefined. These parameters are used to determine $\sigma_{\max}$ using (7.27b), which is then used in the calculation of $\sigma(\rho)$. Usually n_{pml} takes a value such as 2, 3, or 4 and $R(0)$ takes a very small value such as 10^{-8} for a satisfactory PML performance.

The geometrically increasing distribution for $\sigma(\rho)$ is given by

$$\sigma(\rho) = \sigma_0 g^{\frac{\rho}{\Delta s}}, \tag{7.28a}$$

$$\sigma_0 = -\frac{\varepsilon_0 c \ln(g)}{2\Delta s g^N - 1}\ln(R(0)), \tag{7.28b}$$

where the parameter g is a real number used for a geometrically increasing function.

7.4 FDTD updating equations for PML and MATLAB® implementation

7.4.1 PML updating equations – two-dimensional TE_z case

The PML updating equations can be obtained for the two-dimensional TE_z case by applying the central difference approximation to the derivatives in the modified Maxwell's equations (7.4). After some manipulations one can obtain the two-dimensional TE_z PML updating equations based on the field positioning scheme given in Figure 1.10 as

$$E_x^{n+1}(i,j) = C_{exe}(i,j) \times E_x^n(i,j) + C_{exhz}(i,j) \times \left(H_z^{n+\frac{1}{2}}(i,j) - H_z^{n+\frac{1}{2}}(i,j-1)\right), \tag{7.29}$$

where

$$C_{exe}(i,j) = \frac{2\varepsilon_0 - \Delta t \sigma_{pey}(i,j)}{2\varepsilon_0 + \Delta t \sigma_{pey}(i,j)},$$

$$C_{exhz}(i,j) = \frac{2\Delta t}{\left(2\varepsilon_0 + \Delta t \sigma^e_{pey}(i,j)\right)\Delta y}.$$

$$E_y^{n+1}(i,j) = C_{eye}(i,j) \times E_y^n(i,j) + C_{eyhz}(i,j) \times \left(H_z^{n+\frac{1}{2}}(i,j) - H_z^{n+\frac{1}{2}}(i-1,j)\right), \tag{7.30}$$

where

$$C_{eye}(i,j) = \frac{2\varepsilon_0 - \Delta t \sigma_{pex}(i,j)}{2\varepsilon_0 + \Delta t \sigma_{pex}(i,j)},$$

$$C_{eyhz}(i,j) = -\frac{2\Delta t}{(2\varepsilon_0 + \Delta t \sigma_{pex}(i,j))\Delta x}.$$

$$H_{zx}^{n+\frac{1}{2}}(i,j) = C_{hzxh}(i,j) \times H_{zx}^{n-\frac{1}{2}}(i,j) + C_{hzxey}(i,j) \times \left(E_y^n(i+1,j) - E_y^n(i,j)\right), \tag{7.31}$$

where

$$C_{hzxh}(i,j) = \frac{2\mu_0 - \Delta t \sigma_{pmx}(i,j)}{2\mu_0 + \Delta t \sigma_{pmx}(i,j)},$$

$$C_{hzxey}(i,j) = -\frac{2\Delta t}{(2\mu_0 + \Delta t \sigma_{pmx}(i,j))\Delta x}.$$

$$H_{zy}^{n+\frac{1}{2}}(i,j) = C_{hzyh}(i,j) \times H_{zy}^{n-\frac{1}{2}}(i,j) + C_{hzyex}(i,j) \times \left(E_x^n(i,j+1) - E_x^n(i,j)\right), \tag{7.32}$$

where

$$C_{hzyh}(i,j) = \frac{2\mu_0 - \Delta t \sigma_{pmy}(i,j)}{2\mu_0 + \Delta t \sigma_{pmy}(i,j)},$$

$$C_{hzyex}(i,j) = \frac{2\Delta t}{(2\mu_0 + \Delta t \sigma_{pmy}(i,j))\Delta y}.$$

One should recall that $H_z(i,j) = H_{zx}(i,j) + H_{zy}(i,j)$. As illustrated in Figure 7.3 certain PML conductivities are defined at certain regions. Therefore, each of the equations (7.29)–(7.32) is applied in a two-dimensional problem space where its respective PML conductivity is defined. Figure 7.5 shows the PML regions where the conductivity parameters are nonzero and the respective field components that need to be updated at each region. Figure 7.5(a) shows the regions where σ_{pex} is defined. These regions are denoted as *xn* and *xp*. One can notice from (7.4b) and (7.30) that σ_{pex} appears in the equation where E_y is updated. Therefore, the components of E_y lying in the *xn* and *xp* regions are updated at every time step using (7.30). The other components of E_y that are located in the intermediate region can be updated using the regular non-PML updating equation (1.34). Similarly, σ_{pey} is nonzero in the regions that are denoted as *yn* and *yp* in Figure 7.5(b). The components of E_x lying in the *yn* and *yp*

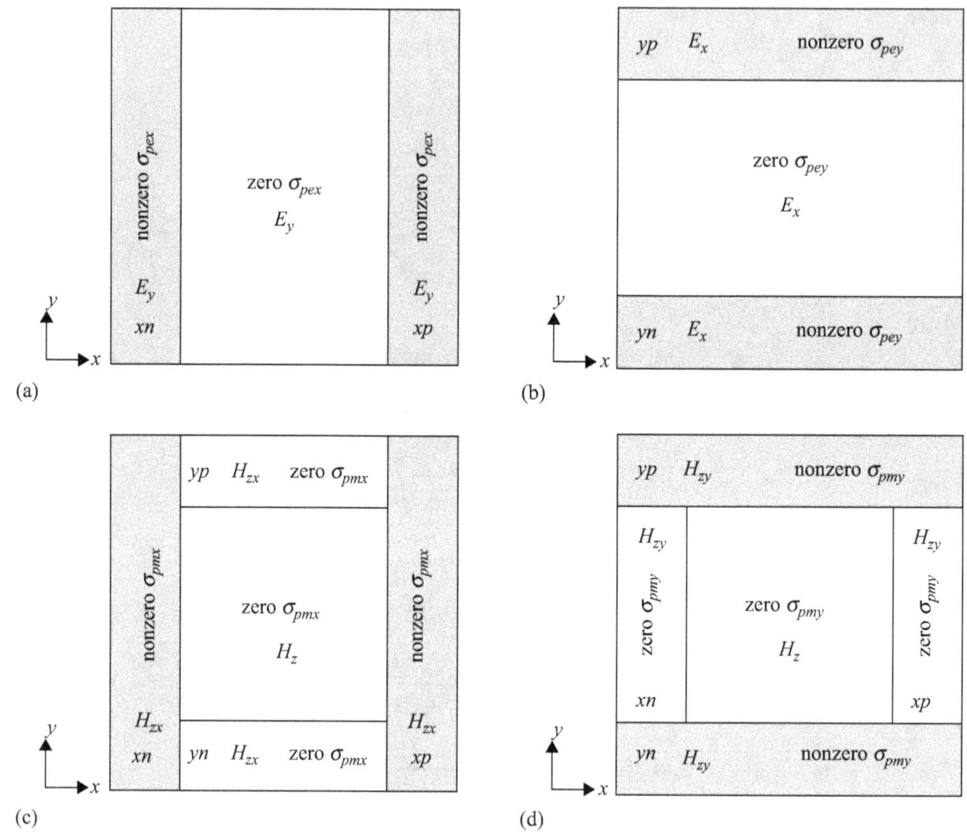

Figure 7.5 Nonzero TE_z regions of PML conductivities: (a) nonzero σ_{pex}; (b) nonzero σ_{pey}; (c) nonzero σ_{pmx}; and (d) nonzero σ_{pmy}.

regions are updated at every time step using (7.29), and the components located in the intermediate region are updated using the regular non-PML updating equation (1.33).

Since the magnetic field H_z is the sum of two split fields H_{zx} and H_{zy} in the PML regions, its update is more complicated. The PML regions and non-PML regions are shown in Figure 7.5(c) and 7.5(d), where the PML regions are denoted as *xn*, *xp*, *yn*, and *yp*. The magnetic field components of H_z lying in the non-PML region can be updated using the regular updating equation (1.35). However, the components of H_z in the PML regions are not directly calculated; the components of H_{zx} and H_{zy} are calculated in the PML regions using the appropriate updating equations, and then they are summed up to yield H_z in the PML region. The conductivity σ_{pmx} is nonzero in the *xn* and *xp* regions as shown in Figure 7.5(c). The components of H_{zx} are calculated in the same regions using (7.31). However, components of H_{zx} need to be calculated in the *yn* and *yp* regions as well. Since σ_{pmx} is zero in these regions setting σ_{pmx} zero in (7.31) will yield the required updating equation for H_{zx} in the *yn* and *yp* regions as

$$H_{zx}^{n+\frac{1}{2}}(i,j) = C_{hzxh}(i,j) \times H_{zx}^{n-\frac{1}{2}}(i,j) + C_{hzxey}(i,j) \times \left(E_y^n(i+1,j) - E_y^n(i,j)\right), \quad (7.33)$$

where

$$C_{hzxh}(i,j) = 1, \quad C_{hzxey}(i,j) = -\frac{\Delta t}{\mu_0 \Delta x}.$$

Similarly, the conductivity σ_{pmy} is nonzero in the yn and yp regions as shown in Figure 7.5(d). The components of H_{zy} are calculated in these regions using (7.32). The components of H_{zy} need to be calculated in the xn and xp regions as well. Since σ_{pmy} is zero in these regions, setting σ_{pmy} to zero in (7.32) will yield the required updating equation for H_{zy} in the xn and xp regions as

$$H_{zy}^{n+\frac{1}{2}}(i,j) = C_{hzyh}(i,j) \times H_{zy}^{n-\frac{1}{2}}(i,j) + C_{hzyex}(i,j) \times \left(E_x^n(i,j+1) - E_x^n(i,j)\right), \tag{7.34}$$

where

$$C_{hzyh}(i,j) = 1, \quad C_{hzyex}(i,j) = \frac{\Delta t}{\mu_0 \Delta y}.$$

After all the components of H_{zx} and H_{zy} are updated in the PML regions, they are added to calculate H_z.

7.4.2 PML updating equations – two-dimensional TM_z case

The PML updating equations can be obtained for the two-dimensional TM_z case by applying the central difference approximation to the derivatives in the modified Maxwell's equations (7.22). After some manipulations one can obtain the two-dimensional TM_z PML updating equations based on the field positioning scheme given in Figure 1.11 as

$$E_{zx}^{n+1}(i,j) = C_{ezxe}(i,j) \times E_{zx}^n(i,j) + C_{ezxhy}(i,j) \times \left(H_y^{n+\frac{1}{2}}(i,j) - H_y^{n+\frac{1}{2}}(i-1,j)\right), \tag{7.35}$$

where

$$C_{ezxe}(i,j) = \frac{2\varepsilon_0 - \Delta t \sigma_{pex}(i,j)}{2\varepsilon_0 + \Delta t \sigma_{pex}(i,j)},$$

$$C_{ezxhy}(i,j) = \frac{2\Delta t}{(2\varepsilon_0 + \Delta t \sigma_{pex}(i,j))\Delta x}.$$

$$E_{zy}^{n+1}(i,j) = C_{ezye}(i,j) \times E_z^n(i,j) + C_{ezyhx}(i,j) \times \left(H_x^{n+\frac{1}{2}}(i,j) - H_x^{n+\frac{1}{2}}(i,j-1)\right), \tag{7.36}$$

where

$$C_{ezye}(i,j) = \frac{2\varepsilon_0(i,j) - \Delta t \sigma_{pey}(i,j)}{2\varepsilon_0(i,j) + \Delta t \sigma_{pey}(i,j)},$$

$$C_{ezyhx}(i,j) = -\frac{2\Delta t}{(2\varepsilon_0 + \Delta t \sigma_{pey}(i,j))\Delta y}.$$

$$H_x^{n+\frac{1}{2}}(i,j) = C_{hxh}(i,j) \times H_x^{n-\frac{1}{2}}(i,j) + C_{hxez}(i,j) \times \left(E_z^n(i,j+1) - E_z^n(i,j)\right), \tag{7.37}$$

where

$$C_{hxh}(i,j) = \frac{2\mu_0 - \Delta t \sigma_{pmy}(i,j)}{2\mu_0 + \Delta t \sigma_{pmy}(i,j)},$$

$$C_{hxez}(i,j) = -\frac{2\Delta t}{(2\mu_0 + \Delta t \sigma_{pmy}(i,j))\Delta y}.$$

$$H_y^{n+\frac{1}{2}}(i,j) = C_{hyh}(i,j) \times H_y^{n-\frac{1}{2}}(i,j) + C_{hyez}(i,j) \times \left(E_z^n(i+1,j) - E_z^n(i,j)\right), \tag{7.38}$$

where

$$C_{hyh}(i,j) = \frac{2\mu_0 - \Delta t \sigma_{pmx}(i,j)}{2\mu_0 + \Delta t \sigma_{pmx}(i,j)},$$

$$C_{hyez}(i,j) = \frac{2\Delta t}{(2\mu_0 + \Delta t \sigma_{pmx}(i,j))\Delta x}.$$

One should recall that $E_z(i,j) = E_{zx}(i,j) + E_{zy}(i,j)$. Consider the nonzero PML conductivity regions given in Figure 7.6. Figure 7.6(a) shows that σ_{pmx} is defined in the regions

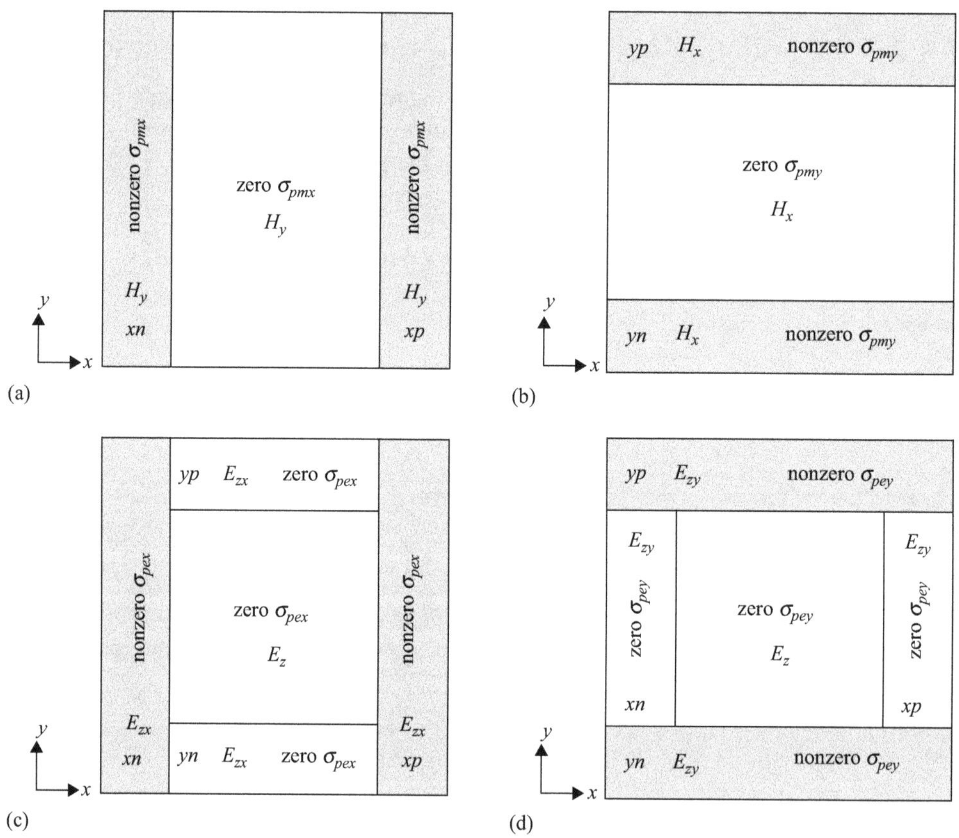

Figure 7.6 Nonzero TM_z regions of PML conductivities: (a) nonzero σ_{pmx}; (b) nonzero σ_{pmy}; (c) nonzero σ_{pex}; and (d) nonzero σ_{pey}.

denoted as xn and xp. Therefore, components of H_y lying in these regions are updated at every time step using the PML updating equation (7.38). The components of H_y lying in the intermediate region are updated using the regular updating equation (1.38). The conductivity σ_{pmy} is nonzero in the yn and yp regions shown in Figure 7.6(b); therefore, the components of H_x lying in the yn and yp regions are updated using the PML updating equation (7.37). The components of H_x lying in the intermediate region are updated using the regular updating equation (1.37).

The components of E_z are the sum of two split fields E_{zx} and E_{zy} in all of the PML regions xn, xp, yn, and yp as illustrated in Figure 7.6(c) and 7.6(d). The components of E_z in the intermediate non-PML region are updated using the regular updating equation (1.36). The conductivity σ_{pex} is nonzero in the xn and xp regions as shown in Figure 7.6(c); therefore, components of E_{zx} in these regions are updated using (7.35). Furthermore, the components of E_{zx} in the yn and yp regions need to be calculated as well. Since σ_{pex} is zero in these regions, setting σ_{pex} to zero in (7.35) yields the updating equation for E_{zx} in the yn and yp regions as

$$E_{zx}^{n+1}(i,j) = C_{ezxe}(i,j) \times E_{zx}^{n}(i,j) + C_{ezxhy}(i,j) \times \left(H_y^{n+\frac{1}{2}}(i,j) - H_y^{n+\frac{1}{2}}(i-1,j)\right), \quad (7.39)$$

where

$$C_{ezxe}(i,j) = 1, \quad C_{ezxhy}(i,j) = \frac{\Delta t}{\varepsilon_0 \Delta x}.$$

Similarly, σ_{pey} is nonzero in the yn and yp regions as shown in Figure 7.6(d); therefore, components of E_{zy} in these regions are updated using (7.36). Furthermore, the components of E_{zy} in the xn and xp regions need to be calculated as well. Since σ_{pey} is zero in these regions, setting σ_{pey} to zero in (7.36) yields the updating equation for E_{zy} in the xn and xp regions as

$$E_{zy}^{n+1}(i,j) = C_{ezye}(i,j) \times E_{z}^{n}(i,j) + C_{ezyhx}(i,j) \times \left(H_x^{n+\frac{1}{2}}(i,j) - H_x^{n+\frac{1}{2}}(i,j-1)\right), \quad (7.40)$$

where

$$C_{ezye}(i,j) = 1, \quad C_{ezyhx}(i,j) = -\frac{\Delta t}{\varepsilon_0 \Delta y}.$$

After all the components of E_{zx} and E_{zy} are updated in the PML regions, they are added to calculate E_z.

7.4.3 MATLAB® implementation of the two-dimensional FDTD method with PML

In this section we demonstrate the implementation of a two-dimensional FDTD MATLAB code including PML boundaries. The main routine of the two-dimensional program is named ***fdtd_solve_2d*** and is given in Listing 7.1. The general structure of the two-dimensional FDTD program is the same as the three-dimensional FDTD program; it is composed of problem definition, initialization, and execution sections. Many of the routines and notations of the two-dimensional FDTD program are similar to their three-dimensional FDTD counterparts; thus, the corresponding details are provided only when necessary.

Listing 7.1 fdtd_solve_2d.m

```
% initialize the matlab workspace
 clear all; close all; clc;

% define the problem
define_problem_space_parameters_2d;
define_geometry_2d;
define_sources_2d;
define_output_parameters_2d;

% initialize the problem space and parameters
initialize_fdtd_material_grid_2d;
initialize_fdtd_parameters_and_arrays_2d;
initialize_sources_2d;
initialize_updating_coefficients_2d;
initialize_boundary_conditions_2d;
initialize_output_parameters_2d;
initialize_display_parameters_2d;

% draw the objects in the problem space
draw_objects_2d;

% FDTD time marching loop
run_fdtd_time_marching_loop_2d;

% display simulation results
post_process_and_display_results_2d;
```

7.4.3.1 Definition of the two-dimensional FDTD problem

The types of boundaries surrounding the problem space are defined in the subroutine ***define_problem_space_parameters_2d***, a partial code of which is shown in Listing 7.2. Here if a boundary on one side is defined as PEC, the variable **boundary.type** takes the value "***pec***," whereas for PML it takes the value "***pml***." The parameter **air_buffer_number_of_cells** determines the distance in number of cells between the objects in the problem space and the boundaries, whether PEC or PML. The parameter **pml_number_of_cells** determines the thickness of the PML regions in number of cells. In this implementation the power increasing function (7.27a) is used for the PML conductivity distributions along the thickness of the PML regions. Two additional parameters are required for the PML, the theoretical reflection coefficient ($R(0)$) and the order of PML (n_{pml}), as discussed in Section 7.3. These two parameters are defined as **boundary_pml_R_0** and **boundary.pml_order**, respectively.

In the subroutine ***define_geometry_2d*** two-dimensional geometrical objects such as circles and rectangles can be defined by their coordinates, sizes, and material types.

The sources exciting the two-dimensional problem space are defined in the subroutine ***define_sources_2d***. Unlike the three-dimensional case, in this implementation the impressed current sources are defined as sources explicitly as shown in Listing 7.3.

Listing 7.2 define_problem_space_parameters_2d.m

```
% ==<boundary conditions>========
% Here we define the boundary conditions parameters
% 'pec' : perfect electric conductor
% 'pml' : perfectly matched layer

boundary.type_xn = 'pml';
boundary.air_buffer_number_of_cells_xn = 10;
boundary.pml_number_of_cells_xn = 5;

boundary.type_xp = 'pml';
boundary.air_buffer_number_of_cells_xp = 10;
boundary.pml_number_of_cells_xp = 5;

boundary.type_yn = 'pml';
boundary.air_buffer_number_of_cells_yn = 10;
boundary.pml_number_of_cells_yn = 5;

boundary.type_yp = 'pml';
boundary.air_buffer_number_of_cells_yp = 10;
boundary.pml_number_of_cells_yp = 5;

boundary.pml_order = 2;
boundary.pml_R_0 = 1e-8;
```

Listing 7.3 define_sources_2d.m

```
disp('defining␣sources');

impressed_J = [];
impressed_M = [];

% define source waveform types and parameters
waveforms.gaussian(1).number_of_cells_per_wavelength = 0;
waveforms.gaussian(2).number_of_cells_per_wavelength = 15;

% electric current sources
% direction: 'xp', 'xn', 'yp', 'yn', 'zp', or 'zn'
impressed_J(1).min_x = -0.1e-3;
impressed_J(1).min_y = -0.1e-3;
impressed_J(1).max_x = 0.1e-3;
impressed_J(1).max_y = 0.1e-3;
impressed_J(1).direction = 'zp';
impressed_J(1).magnitude = 1;
impressed_J(1).waveform_type = 'gaussian';
impressed_J(1).waveform_index = 1;

% % magnetic current sources
% % direction: 'xp', 'xn', 'yp', 'yn', 'zp', or 'zn'
```

```
% impressed_M(1).min_x = -0.1e-3;
% impressed_M(1).min_y = -0.1e-3;
% impressed_M(1).max_x = 0.1e-3;
% impressed_M(1).max_y = 0.1e-3;
% impressed_M(1).direction = 'zp';
% impressed_M(1).magnitude = 1;
% impressed_M(1).waveform_type = 'gaussian';
% impressed_M(1).waveform_index = 1;
```

Outputs of the program are defined in the subroutine ***define_output_parameters_2d***. In this implementation the output parameters are **sampled_electric_fields** and **sampled_magnetic_fields** captured at certain positions.

7.4.3.2 Initialization of the two-dimensional FDTD problem

The steps of the initialization process are shown in Listing 7.1. This process starts with the subroutine ***initialize_fdtd_material_grid_2d***. In this subroutine the first task is the calculation of the dimensions of the two-dimensional problem space and the number of cells nx and ny in the x and y dimensions, respectively. If some of the boundaries are defined as PML, the PML regions are also included in the problem space. Therefore, nx and ny include the PML number of cells as well. Then the material component arrays of the two-dimensional problem space are constructed using the material averaging schemes that were discussed in Section 3.2. Next, while creating the material grid, the PML regions are assigned free-space parameters. At this point the PML regions are treated as if they are free-space regions; however, later PML conductivity parameters are assigned to these regions, and special updating equations are used to update fields in these regions. Furthermore, new parameters are defined and assigned appropriate values as **n_pml_xn**, **n_pml_xp**, **n_pml_yn**, and **n_pml_yp** to hold the numbers of cells for the thickness of the PML regions. Four logical parameters, **is_pml_xn**, **is_pml_xp**, **is_pml_yn**, and **is_pml_yp**, are defined to indicate whether the respective side of the computational boundary is PML or not. Since these parameters are used frequently, shorthand notations are more appropriate to use. Figure 7.7 illustrates a problem space composed of two circles, two rectangles, an impressed current source at the center, and 5 cells thick PML boundaries. The air gap between the objects and the PML is 10 cells.

The subroutine ***initialize_fdtd_parameters_and_arrays_2d*** includes the definition of some parameters such as ε_0, μ_0, c, and Δt that are required for the FDTD calculation and definition and initialization of field arrays **Ex**, **Ey**, **Ez**, **Hx**, **Hy**, and **Hz**. The field arrays are defined for all the problem space including the PML regions.

In the subroutine ***initialize_sources_2d*** the indices indicating the positions of the impressed electric and magnetic currents are determined and stored as the subfields of the respective parameters **impressed_J** and **impressed_M**. Since the impressed currents are used as explicit sources, their source waveforms are also computed. A two-dimensional problem can have two modes of operation as TE and TM. The mode of operation is determined by the sources. For instance, an impressed electric current J_{iz} excites E_z fields in the problem space, thus giving rise to a TM_z operation. Similarly, impressed magnetic currents M_{ix} and M_{iy} also give rise to TM_z operation. The impressed currents M_{iz}, J_{ix}, and J_{iy} give rise to TE_z operation. Therefore, the mode of operation is determined by the impressed currents as shown in Listing 7.4, which includes the partial code of ***initialize_sources_2d***. Here a

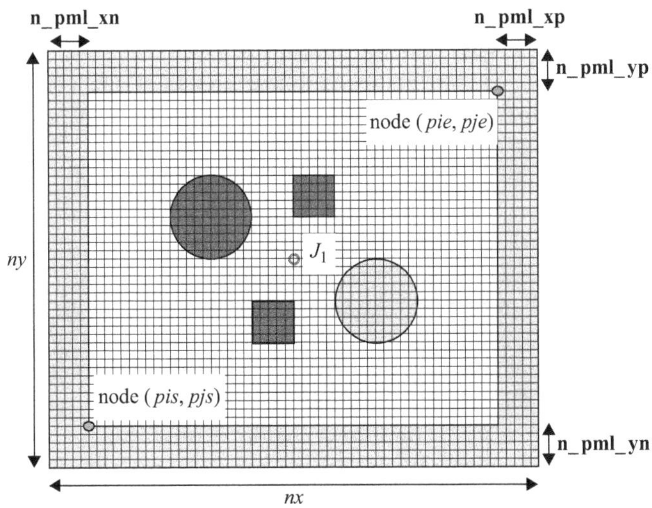

Figure 7.7 A two-dimensional FDTD problem space with PML boundaries.

Listing 7.4 initialize_sources_2d.m

```
% determine if TEz or TMz
is_TEz = false;
is_TMz = false;
for ind = 1:number_of_impressed_J
    switch impressed_J(ind).direction(1)
        case 'x'
            is_TEz = true;
        case 'y'
            is_TEz = true;
        case 'z'
            is_TMz = true;
    end
end
for ind = 1:number_of_impressed_M
    switch impressed_M(ind).direction(1)
        case 'x'
            is_TMz = true;
        case 'y'
            is_TMz = true;
        case 'z'
            is_TEz = true;
    end
end
```

logical parameter **is_TEz** is defined to indicate that the mode of operation is TE_z, whereas another logical parameter **is_TMz** is defined to indicate that the mode of operation is TM_z.

The regular updating coefficients of the two-dimensional FDTD method are calculated in the subroutine ***initialize_updating_coefficients_2d*** based on the updating equations (1.33)–(1.38), including the impressed current coefficients as well.

Some coefficients and field arrays need to be defined and initialized for the application of the PML boundary conditions. The PML initialization process is performed in the subroutine ***initialize_boundary_conditions_2d***, which is shown in Listing 7.5. The indices of the nodes determining the non-PML rectangular region as illustrated in Figure 7.7 are calculated as (*pis*, *pjs*) and (*pie*, *pje*). Then two separate subroutines dedicated to the initialization of the TE_z and TM_z cases are called based on the mode of operation.

The coefficients and fields required for the TE_z PML boundaries are initialized in the subroutine ***initialize_pml_boundary_conditions_2d_TEz***, and partial code for this case is shown in Listing 7.6. Figure 7.8 shows the field distribution for the TE_z case. The field components in the shaded region are updated by the PML updating equations. The field components on the outer boundary are not updated and are kept at zero value during the FDTD iterations since they are simulating the PEC boundaries. The magnetic field components H_{zx} and H_{zy} are defined in the four PML regions shown in Figure 7.5; therefore, the corresponding field arrays **Hzx_xn, Hzx_xp, Hzx_yn, Hzx_yp, Hzy_xn, Hzy_xp, Hzy_yn,** and **Hzy_yp** are initialized in Listing 7.6.

Then for each region the corresponding conductivity parameters and PML updating coefficients are calculated. Listing 7.6 shows the initialization for the *xn* and *yp* regions.

One should take care while calculating the conductivity arrays. The conductivities σ_{pex} and σ_{pey} are associated with the electric fields E_x and E_y, respectively, whereas σ_{pmx} and σ_{pmy} are associated with the magnetic field H_z. Therefore, the conductivity components are located in different positions. As mentioned before, the conductivity $\sigma(\rho)$ is zero at the PML interior-domain interface, and it increases to a maximum value σ_{max} at the end of the PML

Listing 7.5 initialize_boundary_conditions_2d.m

```
disp('initializing boundary conditions');

% determine the boundaries of the non-pml region
pis = n_pml_xn+1;
pie = nx-n_pml_xp+1;
pjs = n_pml_yn+1;
pje = ny-n_pml_yp+1;

if is_any_side_pml
    if is_TEz
        initialize_pml_boundary_conditions_2d_TEz;
    end
    if is_TMz
        initialize_pml_boundary_conditions_2d_TMz;
    end
end
```

Listing 7.6 initialize_pml_boundary_conditions_2d_TEz.m

```
% initializing PML boundary conditions for TEz
disp('initializing PML boundary conditions for TEz');

Hzx_xn = zeros(n_pml_xn,ny);
Hzy_xn = zeros(n_pml_xn,ny-n_pml_yn-n_pml_yp);
Hzx_xp = zeros(n_pml_xp,ny);
Hzy_xp = zeros(n_pml_xp,ny-n_pml_yn-n_pml_yp);
Hzx_yn = zeros(nx-n_pml_xn-n_pml_xp, n_pml_yn);
Hzy_yn = zeros(nx,n_pml_yn);
Hzx_yp = zeros(nx-n_pml_xn-n_pml_xp, n_pml_yp);
Hzy_yp = zeros(nx,n_pml_yp);

pml_order = boundary.pml_order;
R_0 = boundary.pml_R_0;

if is_pml_xn
    sigma_pex_xn = zeros(n_pml_xn,ny);
    sigma_pmx_xn = zeros(n_pml_xn,ny);

    sigma_max = -(pml_order+1)*eps_0*c*log(R_0)/(2*dx*n_pml_xn);
    rho_e = ([n_pml_xn:-1:1] - 0.75)/n_pml_xn;
    rho_m = ([n_pml_xn:-1:1] - 0.25)/n_pml_xn;
    for ind = 1:n_pml_xn
        sigma_pex_xn(ind,:) = sigma_max * rho_e(ind)^pml_order;
        sigma_pmx_xn(ind,:) = ...
            (mu_0/eps_0) * sigma_max * rho_m(ind)^pml_order;
    end

    % Coeffiecients updating Ey
    Ceye_xn = (2*eps_0 - dt*sigma_pex_xn)./(2*eps_0+dt*sigma_pex_xn);
    Ceyhz_xn= -(2*dt/dx)./(2*eps_0 + dt*sigma_pex_xn);

    % Coeffiecients updating Hzx
    Chzxh_xn  = (2*mu_0 - dt*sigma_pmx_xn)./(2*mu_0+dt*sigma_pmx_xn);
    Chzxey_xn = -(2*dt/dx)./(2*mu_0 + dt*sigma_pmx_xn);

    % Coeffiecients updating Hzy
    Chzyh_xn  =  1;
    Chzyex_xn =  dt/(dy*mu_0);
end
if is_pml_yp
    sigma_pey_yp = zeros(nx,n_pml_yp);
    sigma_pmy_yp = zeros(nx,n_pml_yp);

    sigma_max = -(pml_order+1)*eps_0*c*log(R_0)/(2*dy*n_pml_yp);
    rho_e = ([1:n_pml_yp] - 0.75)/n_pml_yp;
    rho_m = ([1:n_pml_yp] - 0.25)/n_pml_yp;
    for ind = 1:n_pml_yp
        sigma_pey_yp(:,ind) = sigma_max * rho_e(ind)^pml_order;
        sigma_pmy_yp(:,ind) = ...
```

```
            (mu_0/eps_0) * sigma_max * rho_m(ind)^pml_order;
    end

    % Coeffiecients updating Ex
    Cexe_yp  = (2*eps_0 - dt*sigma_pey_yp)./(2*eps_0+dt*sigma_pey_yp);
    Cexhz_yp = (2*dt/dy)./(2*eps_0 + dt*sigma_pey_yp);

    % Coeffiecients updating Hzx
    Chzxh_yp  = 1;
    Chzxey_yp = -dt/(dx*mu_0);

    % Coeffiecients updating Hzy
    Chzyh_yp  = (2*mu_0 - dt*sigma_pmy_yp)./(2*mu_0+dt*sigma_pmy_yp);
    Chzyex_yp = (2*dt/dy)./(2*mu_0 + dt*sigma_pmy_yp);
end
```

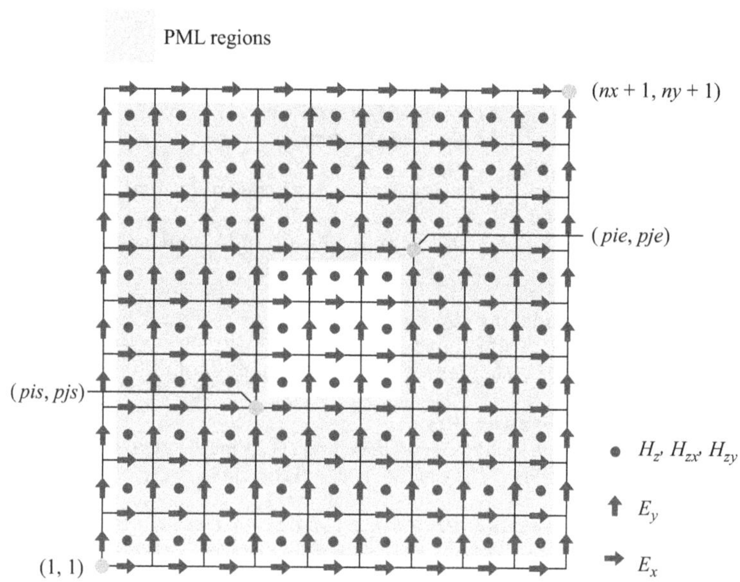

Figure 7.8 TE_z field components in the PML regions.

region. In this implementation the imaginary PML regions are shifted inward by a quarter cell size as shown in Figure 7.8. If the thickness of the PML region is N cells, this shift ensures that N electric field components and N magnetic field components are updated across the PML thickness. For instance, consider the cross-section of a two-dimensional problem space shown in Figure 7.9, which illustrates the field components updated in the *xn* and *xp* regions. The distance of the electric field components E_y from the interior boundary of the PML is denoted as ρ_e whereas for the magnetic field components H_z it is denoted as ρ_m. The distances ρ_e and ρ_m are used in (7.27a) to calculate the values of σ_{pex} and σ_{pmx}, respectively. These values are stored in arrays **sigma_pex_xn** and **sigma_pmx_xn** for the *xn* region as

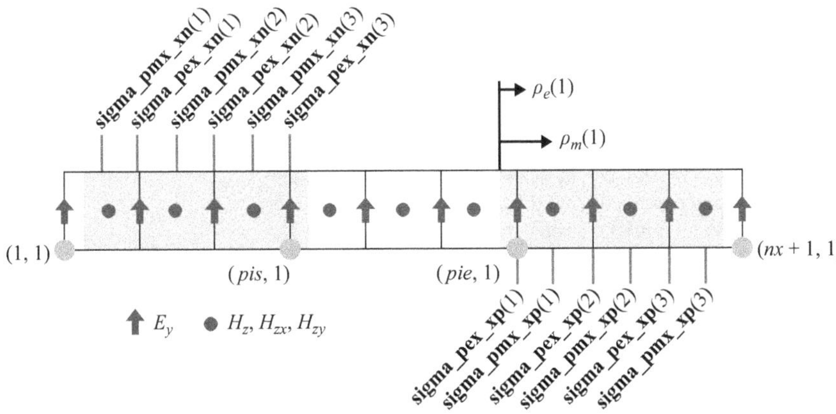

Figure 7.9 Field components updated by PML equations.

shown in Listing 7.6 and in **sigma_pex_xp** and **sigma_pmx_xp** for the *xp* region. Calculation of σ_{pey} and σ_{pmy} follows the same logic. Then these parameters are used to calculate the respective PML updating coefficients in (7.29)–(7.32).

The initialization of the auxiliary split fields and the PML updating coefficients for the two-dimensional TM_z case is performed in the subroutine ***initialize_pml_boundary_conditions_2d_Tmz***, a partial code of which is given for the *xp* and *yn* regions in Listing 7.7. The initialization of the TM_z case follows the same logic as the TE_z case. The field positioning and PML regions are shown in Figure 7.10, while the positions of the field components updated by the PML equations and conductivity positions are shown in Figure 7.11 as a reference.

7.4.3.3 Running the two-dimensional FDTD simulation: the time-marching loop

After the initialization process is completed, the subroutine ***run_fdtd_time_marching_loop_2d*** including the time-marching loop of FDTD procedure is called. The implementation of the FDTD updating loop is shown in Listing 7.8.

During the time-marching loop the first step at every iteration is the update of magnetic field components using the regular updating equations in ***update_magnetic_fields_2d*** as shown in Listing 7.9. In the TE_z case the H_z field components in the intermediate regions of Figure 7.5(c) and 7.5(d) are updated based on (1.35). In the TM_z case the H_x field components in the intermediate region of Figure 7.6(b) and H_y field components in the intermediate region of Figure 7.6(a) are updated based on (1.37) and (1.38), respectively.

Then in ***update_impressed_M*** the impressed current terms appearing in (1.35), (1.37), and (1.38) are added to their respective field terms H_z, H_x, and H_y as shown in Listing 7.10.

The subroutine ***update_magnetic_fields_for_PML_2d*** is used to update the magnetic field components needing special PML updates. As can be followed in Listing 7.11 the TE_z and TM_z cases are treated in separate subroutines.

The TM_z case is implemented in Listing 7.12, where H_x is updated in the *yn* and *yp* regions of Figure 7.6(b) using (7.37). H_y is updated in the *xn* and *xp* regions of Figure 7.6(a) using (7.38).

The TE_z case is implemented in Listing 7.13. H_{zx} is updated in the *xn* and *xp* regions of Figure 7.5(c) using (7.31). H_{zx} is updated in the *yn* and *yp* regions of Figure 7.5(c) using

Listing 7.7 initialize_pml_boundary_conditions_2d_TMz.m

```
% initializing PML boundary conditions for TMz
disp('initializing␣PML␣boundary␣conditions␣for␣TMz');

Ezx_xn = zeros(n_pml_xn,nym1);
Ezy_xn = zeros(n_pml_xn,nym1-n_pml_yn-n_pml_yp);
Ezx_xp = zeros(n_pml_xp,nym1);
Ezy_xp = zeros(n_pml_xp,nym1-n_pml_yn-n_pml_yp);
Ezx_yn = zeros(nxm1-n_pml_xn-n_pml_xp, n_pml_yn);
Ezy_yn = zeros(nxm1,n_pml_yn);
Ezx_yp = zeros(nxm1-n_pml_xn-n_pml_xp, n_pml_yp);
Ezy_yp = zeros(nxm1,n_pml_yp);

pml_order = boundary.pml_order;
R_0 = boundary.pml_R_0;

if is_pml_xp
    sigma_pex_xp = zeros(n_pml_xp,nym1);
    sigma_pmx_xp = zeros(n_pml_xp,nym1);

    sigma_max = -(pml_order+1)*eps_0*c*log(R_0)/(2*dx*n_pml_xp);
    rho_e = ([1:n_pml_xp] - 0.25)/n_pml_xp;
    rho_m = ([1:n_pml_xp] - 0.75)/n_pml_xp;
    for ind = 1:n_pml_xp
        sigma_pex_xp(ind,:) = sigma_max * rho_e(ind)^pml_order;
        sigma_pmx_xp(ind,:) = ...
            (mu_0/eps_0) * sigma_max * rho_m(ind)^pml_order;
    end

    % Coeffiecients updating Hy
    Chyh_xp  = (2*mu_0 - dt*sigma_pmx_xp)./(2*mu_0 + dt*sigma_pmx_xp);
    Chyez_xp = (2*dt/dx)./(2*mu_0 + dt*sigma_pmx_xp);

    % Coeffiecients updating Ezx
    Cezxe_xp  = (2*eps_0 - dt*sigma_pex_xp)./(2*eps_0+dt*sigma_pex_xp);
    Cezxhy_xp = (2*dt/dx)./(2*eps_0 + dt*sigma_pex_xp);

    % Coeffiecients updating Ezy
    Cezye_xp  =  1;
    Cezyhx_xp = -dt/(dy*eps_0);
end

if is_pml_yn
    sigma_pey_yn = zeros(nxm1,n_pml_yn);
    sigma_pmy_yn = zeros(nxm1,n_pml_yn);

    sigma_max = -(pml_order+1)*eps_0*c*log(R_0)/(2*dy*n_pml_yn);
    rho_e = ([n_pml_yn:-1:1] - 0.25)/n_pml_yn;
    rho_m = ([n_pml_yn:-1:1] - 0.75)/n_pml_yn;
    for ind = 1:n_pml_yp
        sigma_pey_yn(:,ind) = sigma_max * rho_e(ind)^pml_order;
```

```
        sigma_pmy_yn(:,ind) = ...
            (mu_0/eps_0) * sigma_max * rho_m(ind)^pml_order;
    end

    % Coeffiecients updating Hx
    Chxh_yn = (2*mu_0 - dt*sigma_pmy_yn)./(2*mu_0+dt*sigma_pmy_yn);
    Chxez_yn= -(2*dt/dy)./(2*mu_0 + dt*sigma_pmy_yn);

    % Coeffiecients updating Ezx
    Cezxe_yn  =  1;
    Cezxhy_yn =  dt/(dx*eps_0);

    % Coeffiecients updating Ezy
    Cezye_yn = (2*eps_0 - dt*sigma_pey_yn)./(2*eps_0+dt*sigma_pey_yn);
    Cezyhx_yn= -(2*dt/dy)./(2*eps_0 + dt*sigma_pey_yn);
end
```

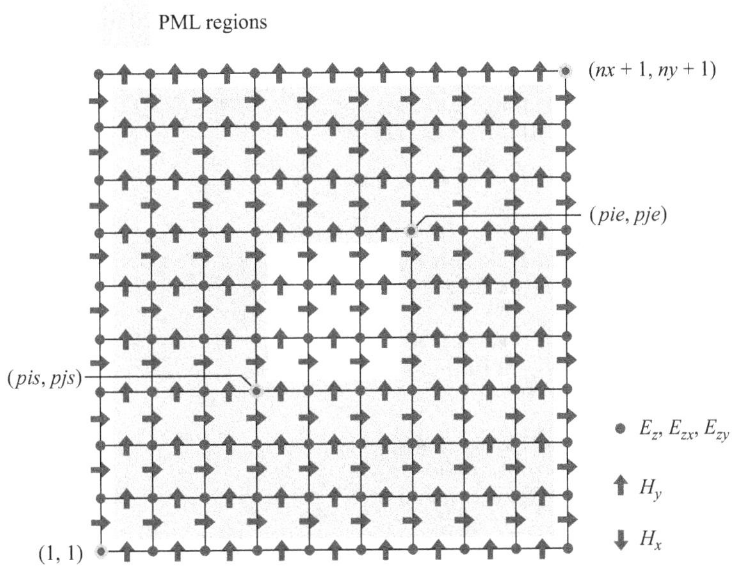

Figure 7.10 TM_z field components in the PML regions.

(7.33). H_{zy} is updated in the *yn* and *yp* regions of Figure 7.5(d) using (7.32). H_{zy} is updated in *xn* and *xp* regions of Figure 7.5(d) using (7.34). After all these updates are completed, the components of H_{zx} and H_{zy}, located at the same positions, are added to obtain H_z at the same positions.

The update of the electric field components using the regular updating equations is performed in ***update_electric_fields_2d*** as shown in Listing 7.14. In the TM_z case the E_z field components in the intermediate regions of Figure 7.6(c) and 7.6(d) are updated based on

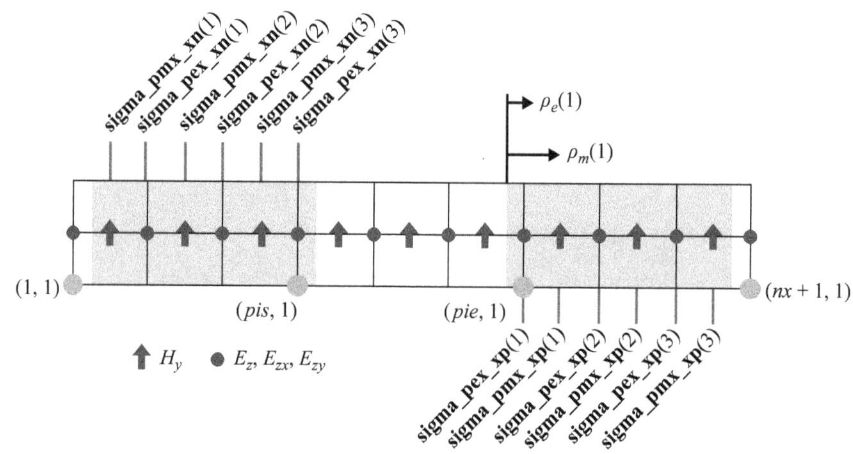

Figure 7.11 Field components updated by PML equations.

Listing 7.8 run_fdtd_time_marching_loop_2d.m

```
disp (['Starting the time marching loop']);
disp (['Total number of time steps : ' ...
    num2str(number_of_time_steps)]);

start_time = cputime;
current_time = 0;

for time_step = 1:number_of_time_steps
    update_magnetic_fields_2d;
    update_impressed_M;
    update_magnetic_fields_for_PML_2d;
    capture_sampled_magnetic_fields_2d;
    update_electric_fields_2d;
    update_impressed_J;
    update_electric_fields_for_PML_2d;
    capture_sampled_electric_fields_2d;
    display_sampled_parameters_2d;
end

end_time = cputime;
total_time_in_minutes = (end_time - start_time)/60;
disp (['Total simulation time is ' ...
    num2str(total_time_in_minutes) ' minutes.']);
```

(1.36). In the TE_z case the E_x field components in the intermediate region of Figure 7.5(b) and E_y field components in the intermediate region of Figure 7.5(a) are updated based on (1.33) and (1.34), respectively.

Then in ***update_impressed_J*** the impressed current terms appearing in (1.36), (1.33), and (1.34) are added to their respective field terms E_z, E_x, and E_y.

Listing 7.9 update_magnetic_fields_2d.m

```
% update magnetic fields

current_time  = current_time + dt/2;

% TEz
if is_TEz
Hz(pis:pie-1,pjs:pje-1) = ...
    Chzh(pis:pie-1,pjs:pje-1).* Hz(pis:pie-1,pjs:pje-1) ...
    + Chzex(pis:pie-1,pjs:pje-1) ...
    .* (Ex(pis:pie-1,pjs+1:pje)-Ex(pis:pie-1,pjs:pje-1))  ...
    + Chzey(pis:pie-1,pjs:pje-1) ...
    .*(Ey(pis+1:pie,pjs:pje-1)-Ey(pis:pie-1,pjs:pje-1));
end

% TMz
if is_TMz
    Hx(:,pjs:pje-1) = Chxh(:,pjs:pje-1) .* Hx(:,pjs:pje-1) ...
        + Chxez(:,pjs:pje-1) .* (Ez(:,pjs+1:pje)-Ez(:,pjs:pje-1));

    Hy(pis:pie-1,:) = Chyh(pis:pie-1,:) .* Hy(pis:pie-1,:) ...
        + Chyez(pis:pie-1,:) .* (Ez(pis+1:pie,:) - Ez(pis:pie-1,:));
end
```

Listing 7.10 update_impressed_M.m

```
% updating magnetic field components
% associated with the impressed magnetic currents

 for ind = 1:number_of_impressed_M
    is = impressed_M(ind).is;
    js = impressed_M(ind).js;
    ie = impressed_M(ind).ie;
    je = impressed_M(ind).je;
    switch (impressed_M(ind).direction(1))
    case 'x'
        Hx(is:ie,js:je-1) = Hx(is:ie,js:je-1) ...
        + impressed_M(ind).Chxm * impressed_M(ind).waveform(time_step);
    case 'y'
        Hy(is:ie-1,js:je) = Hy(is:ie-1,js:je) ...
        + impressed_M(ind).Chym * impressed_M(ind).waveform(time_step);
    case 'z'
        Hz(is:ie-1,js:je-1) = Hz(is:ie-1,js:je-1) ...
        + impressed_M(ind).Chzm * impressed_M(ind).waveform(time_step);
    end
end
```

Listing 7.11 update_magnetic_fields_for_PML_2d.m

```
% update magnetic fields at the PML regions
if is_any_side_pml == false
    return;
end
if is_TEz
    update_magnetic_fields_for_PML_2d_TEz;
end
if is_TMz
    update_magnetic_fields_for_PML_2d_TMz;
end
```

Listing 7.12 update_magnetic_fields_for_PML_2d_TMz.m

```
% update magnetic fields at the PML regions
% TMz
if is_pml_xn
    Hy(1:pis-1,2:ny) = Chyh_xn .* Hy(1:pis-1,2:ny) ...
        + Chyez_xn .* (Ez(2:pis,2:ny)-Ez(1:pis-1,2:ny));
end

if is_pml_xp
    Hy(pie:nx,2:ny) = Chyh_xp .* Hy(pie:nx,2:ny) ...
    + Chyez_xp .* (Ez(pie+1:nxp1,2:ny)-Ez(pie:nx,2:ny));
end

if is_pml_yn
    Hx(2:nx,1:pjs-1) = Chxh_yn .* Hx(2:nx,1:pjs-1) ...
        + Chxez_yn.*(Ez(2:nx,2:pjs)-Ez(2:nx,1:pjs-1));
end

if is_pml_yp
    Hx(2:nx,pje:ny) = Chxh_yp .* Hx(2:nx,pje:ny) ...
        + Chxez_yp.*(Ez(2:nx,pje+1:nyp1)-Ez(2:nx,pje:ny));
end
```

The subroutine ***update_electric_fields_for_PML_2d*** is used to update the electric field components needing special PML updates. As can be followed in Listing 7.15 the TE_z and TM_z cases are treated in separate subroutines.

The TE_z case is implemented in Listing 7.16, where E_x is updated in the *yn* and *yp* regions of Figure 7.5(b) using (7.29). E_y is updated in the *xn* and *xp* regions of Figure 7.5(a) using (7.30).

The TM_z case is implemented in Listing 7.17, where E_{zx} is updated in the *xn* and *xp* regions of Figure 7.6(c) using (7.35). E_{zx} is updated in the *yn* and *yp* regions of Figure 7.6(c) using (7.39). E_{zy} is updated in the *yn* and *yp* regions of Figure 7.6(d) using (7.36). E_{zy} is updated in the *xn* and *xp* regions of Figure 7.6(d) using (7.40). After all these updates are completed, the components of E_{zx} and E_{zy}, located at the same positions, are added to obtain E_z at the same positions.

Listing 7.13 update_magnetic_fields_for_PML_2d_TEz.m

```
% update magnetic fields at the PML regions
% TEz
if is_pml_xn
    Hzx_xn = Chzxh_xn .* Hzx_xn+Chzxey_xn.*(Ey(2:pis,:) - Ey(1:pis-1,:));
    Hzy_xn = Chzyh_xn .* Hzy_xn ...
        + Chzyex_xn.*(Ex(1:pis-1,pjs+1:pje)-Ex(1:pis-1,pjs:pje-1));
end
if is_pml_xp
    Hzx_xp = Chzxh_xp .* Hzx_xp ...
        + Chzxey_xp.*(Ey(pie+1:nxp1,:) - Ey(pie:nx,:));
    Hzy_xp = Chzyh_xp .* Hzy_xp ...
        + Chzyex_xp.*(Ex(pie:nx,pjs+1:pje)-Ex(pie:nx,pjs:pje-1));
end
if is_pml_yn
    Hzx_yn = Chzxh_yn .* Hzx_yn ...
        + Chzxey_yn.*(Ey(pis+1:pie,1:pjs-1)-Ey(pis:pie-1,1:pjs-1));
    Hzy_yn = Chzyh_yn .* Hzy_yn ...
        + Chzyex_yn.*(Ex(:,2:pjs)-Ex(:,1:pjs-1));
end
if is_pml_yp
    Hzx_yp = Chzxh_yp .* Hzx_yp ...
        + Chzxey_yp.*(Ey(pis+1:pie,pje:ny)-Ey(pis:pie-1,pje:ny));
    Hzy_yp = Chzyh_yp .* Hzy_yp ...
        + Chzyex_yp.*(Ex(:,pje+1:nyp1)-Ex(:,pje:ny));
end
Hz(1:pis-1,1:pjs-1) = Hzx_xn(:,1:pjs-1) + Hzy_yn(1:pis-1,:);
Hz(1:pis-1,pje:ny)  = Hzx_xn(:,pje:ny) + Hzy_yp(1:pis-1,:);
Hz(pie:nx,1:pjs-1)  = Hzx_xp(:,1:pjs-1) + Hzy_yn(pie:nx,:);
Hz(pie:nx,pje:ny)   = Hzx_xp(:,pje:ny) + Hzy_yp(pie:nx,:);
Hz(1:pis-1,pjs:pje-1) = Hzx_xn(:,pjs:pje-1) + Hzy_xn;
Hz(pie:nx,pjs:pje-1)  = Hzx_xp(:,pjs:pje-1) + Hzy_xp;
Hz(pis:pie-1,1:pjs-1) = Hzx_yn + Hzy_yn(pis:pie-1,:);
Hz(pis:pie-1,pje:ny)  = Hzx_yp + Hzy_yp(pis:pie-1,:);
```

Listing 7.14 update_electric_fields_2d.m

```
current_time  = current_time + dt/2;

if is_TEz
    Ex(:,pjs+1:pje-1) = Cexe(:,pjs+1:pje-1).*Ex(:,pjs+1:pje-1) ...
                      + Cexhz(:,pjs+1:pje-1).*...
                      (Hz(:,pjs+1:pje-1)-Hz(:,pjs:pje-2));

    Ey(pis+1:pie-1,:) = Ceye(pis+1:pie-1,:).*Ey(pis+1:pie-1,:) ...
                       + Ceyhz(pis+1:pie-1,:).* ...
                       (Hz(pis+1:pie-1,:)-Hz(pis:pie-2,:));
end
```

```
if is_TMz
    Ez(pis+1:pie-1,pjs+1:pje-1) = ...
        Ceze(pis+1:pie-1,pjs+1:pje-1).*Ez(pis+1:pie-1,pjs+1:pje-1) ...
        + Cezhy(pis+1:pie-1,pjs+1:pje-1) ...
        .* (Hy(pis+1:pie-1,pjs+1:pje-1)-Hy(pis:pie-2,pjs+1:pje-1)) ...
        + Cezhx(pis+1:pie-1,pjs+1:pje-1) ...
        .* (Hx(pis+1:pie-1,pjs+1:pje-1)-Hx(pis+1:pie-1,pjs:pje-2));
end
```

Listing 7.15 update_electric_fields_for_PML_2d.m

```
% update electric fields at the PML regions
% update magnetic fields at the PML regions
if is_any_side_pml == false
    return;
end
if is_TEz
    update_electric_fields_for_PML_2d_TEz;
end
if is_TMz
    update_electric_fields_for_PML_2d_TMz;
end
```

Listing 7.16 update_electric_fields_for_PML_2d_TEz.m

```
% update electric fields at the PML regions
% TEz
if is_pml_xn
    Ey(2:pis,:) = Ceye_xn .* Ey(2:pis,:) ...
        + Ceyhz_xn .* (Hz(2:pis,:)-Hz(1:pis-1,:));
end

if is_pml_xp
    Ey(pie:nx,:) = Ceye_xp .* Ey(pie:nx,:) ...
        + Ceyhz_xp .* (Hz(pie:nx,:)-Hz(pie-1:nx-1,:));
end

if is_pml_yn
    Ex(:,2:pjs)  = Cexe_yn .* Ex(:,2:pjs) ...
        + Cexhz_yn .* (Hz(:,2:pjs)-Hz(:,1:pjs-1));
end

if is_pml_yp
    Ex(:,pje:ny) = Cexe_yp .* Ex(:,pje:ny) ...
        + Cexhz_yp .* (Hz(:,pje:ny)-Hz(:,pje-1:ny-1));
end
```

Listing 7.17 update_electric_fields_for_PML_2d_TMz.m

```
% update electric fields at the PML regions
% TMz
if is_pml_xn
    Ezx_xn = Cezxe_xn .* Ezx_xn ...
        + Cezxhy_xn .* (Hy(2:pis,2:ny)-Hy(1:pis-1,2:ny));
    Ezy_xn = Cezye_xn .* Ezy_xn ...
        + Cezyhx_xn .* (Hx(2:pis,pjs+1:pje-1)-Hx(2:pis,pjs:pje-2));
end
if is_pml_xp
    Ezx_xp = Cezxe_xp .* Ezx_xp + Cezxhy_xp.*   ...
        (Hy(pie:nx,2:ny)-Hy(pie-1:nx-1,2:ny));
    Ezy_xp = Cezye_xp .* Ezy_xp ...
        + Cezyhx_xp .* (Hx(pie:nx,pjs+1:pje-1)-Hx(pie:nx,pjs:pje-2));
end
if is_pml_yn
    Ezx_yn = Cezxe_yn .* Ezx_yn ...
        + Cezxhy_yn .* (Hy(pis+1:pie-1,2:pjs)-Hy(pis:pie-2,2:pjs));
    Ezy_yn = Cezye_yn .* Ezy_yn ...
        + Cezyhx_yn .* (Hx(2:nx,2:pjs)-Hx(2:nx,1:pjs-1));
end
if is_pml_yp
    Ezx_yp = Cezxe_yp .* Ezx_yp ...
        + Cezxhy_yp .* (Hy(pis+1:pie-1,pje:ny)-Hy(pis:pie-2,pje:ny));
    Ezy_yp = Cezye_yp .* Ezy_yp ...
        + Cezyhx_yp .* (Hx(2:nx,pje:ny)-Hx(2:nx,pje-1:ny-1));
end
Ez(2:pis,2:pjs)    = Ezx_xn(:,1:pjs-1) + Ezy_yn(1:pis-1,:);
Ez(2:pis,pje:ny)   = Ezx_xn(:,pje-1:nym1) + Ezy_yp(1:pis-1,:);
Ez(pie:nx,pje:ny)  = Ezx_xp(:,pje-1:nym1) + Ezy_yp(pie-1:nxm1,:);
Ez(pie:nx,2:pjs)   = Ezx_xp(:,1:pjs-1) + Ezy_yn(pie-1:nxm1,:);
Ez(pis+1:pie-1,2:pjs)   = Ezx_yn + Ezy_yn(pis:pie-2,:);
Ez(pis+1:pie-1,pje:ny)  = Ezx_yp + Ezy_yp(pis:pie-2,:);
Ez(2:pis,pjs+1:pje-1)   = Ezx_xn(:,pjs:pje-2) + Ezy_xn;
Ez(pie:nx,pjs+1:pje-1)  = Ezx_xp(:,pjs:pje-2) + Ezy_xp;
```

7.5 Simulation examples

7.5.1 Validation of PML performance

In this section, we evaluate the performance of the two-dimensional PML for the TE_z case. A two-dimensional problem is constructed as shown in Figure 7.12(a). The problem space is empty (all free space) and is composed of 36 $\times$ 36 cells with cell size 1 mm on a side. There are eight cell layers of PML on the four sides of the boundaries. The order of the PML parameter n_{pml} is 2, and the theoretical reflection coefficient $R(0)$ is 10^{-8}. The problem space is excited by a z-directed impressed magnetic current as defined in Listing 7.18. The impressed magnetic current is centered at the origin and has a Gaussian waveform. A sampled magnetic field with z component is placed at the position $x = 8$ mm and $y = 8$ mm, two cells away from the upper right corner of the PML boundaries as defined in Listing 7.19. The problem is run for 1,800 time steps. The captured sampled magnetic field H_z is plotted in

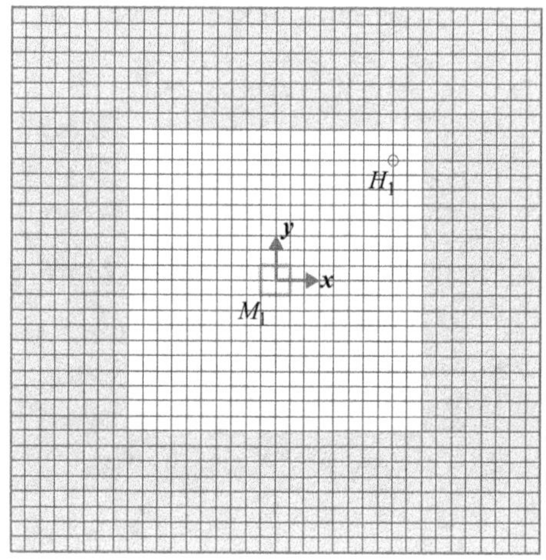

(a)

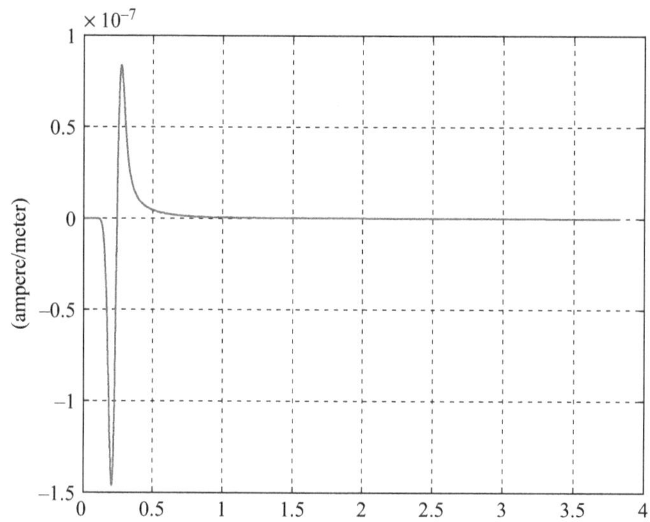

(b)

Figure 7.12 A two-dimensional TE_z FDTD problem terminated by PML boundaries and its simulation results: (a) an empty two-dimensional problem space and (b) sampled H_z in time.

Figure 7.12(b) as a function of time. The captured field includes the effect of the reflected fields from the PML boundaries as well.

To determine how well the PML boundaries simulate the open boundaries, a reference case is constructed as shown in Figure 7.13(a). The cell size, the source, and the output of this problem space are the same as the previous one, but in this case the problem space size is 600×600 cells, and it is terminated by PEC boundaries on four sides. Any fields excited by the source will propagate, will hit the PEC boundaries, and will propagate back to the center. Since the problem size is large, it will take some time until the reflected fields arrive at the

Listing 7.18 define_sources_2d.m

```
disp('defining_sources');

impressed_J = [];
impressed_M = [];

% define source waveform types and parameters
waveforms.gaussian(1).number_of_cells_per_wavelength = 0;
waveforms.gaussian(2).number_of_cells_per_wavelength = 25;

% magnetic current sources
% direction: 'xp', 'xn', 'yp', 'yn', 'zp', or 'zn'
impressed_M(1).min_x = -1e-3;
impressed_M(1).min_y = -1e-3;
impressed_M(1).max_x = 1e-3;
impressed_M(1).max_y = 1e-3;
impressed_M(1).direction = 'zp';
impressed_M(1).magnitude = 1;
impressed_M(1).waveform_type = 'gaussian';
impressed_M(1).waveform_index = 2;
```

Listing 7.19 define_output_parameters_2d.m

```
disp('defining_output_parameters');

sampled_electric_fields = [];
sampled_magnetic_fields = [];
sampled_transient_E_planes = [];
sampled_frequency_E_planes = [];

% figure refresh rate
plotting_step = 10;

% frequency domain parameters
frequency_domain.start = 20e6;
frequency_domain.end   = 20e9;
frequency_domain.step  = 20e6;

% define sampled magnetic fields
sampled_magnetic_fields(1).x = 8e-3;
sampled_magnetic_fields(1).y = 8e-3;
sampled_magnetic_fields(1).component = 'z';
```

sampling point. Therefore, the fields captured at the sampling point before any reflected fields arrive are the same as the fields that would be observed if the boundaries are open space. In the given example no reflection is observed in the 1,800 time steps of simulation. Therefore, this case can be considered as a reference case for an open space during the 1,800 time steps. The captured sampled magnetic field is shown in Figure 7.13(b) as a function of time.

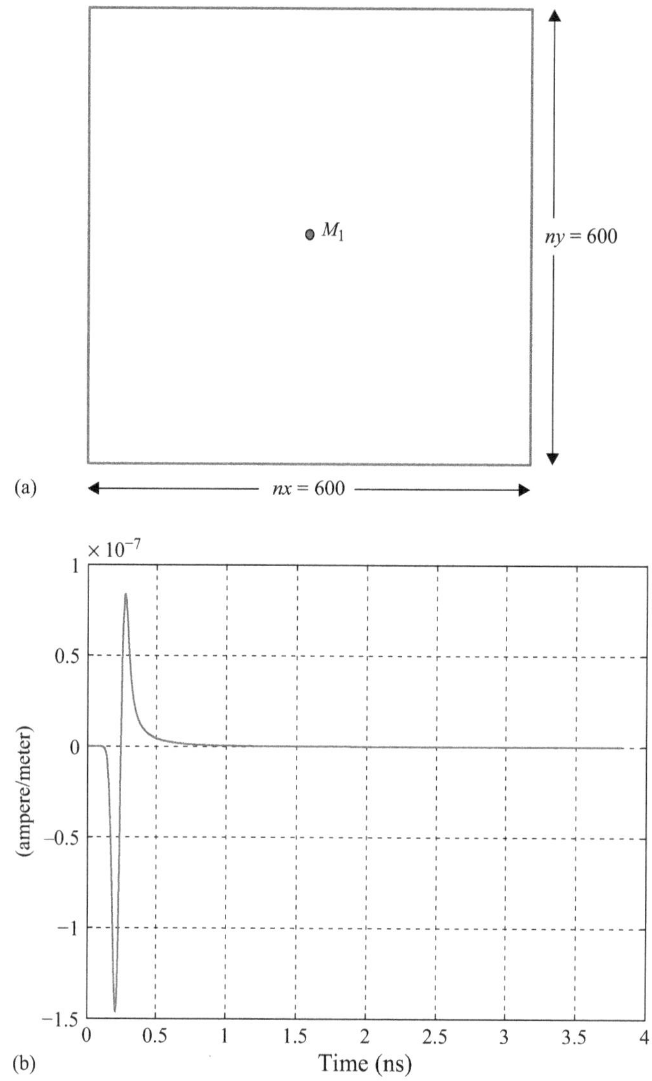

Figure 7.13 A two-dimensional TE_z FDTD problem used as open boundary reference and its simulation results: (a) an empty two-dimensional problem space and (b) sampled H_z in time.

No difference can be seen by looking at the responses of the PML case and the reference case. The difference between the two cases is a measure for the amount of reflection from the PML and can be determined numerically. The difference denoted as $error_t$ is the error as a function of time and calculated by

$$error_t = 20 \times \log_{10}\left(\frac{|H_z^{pml} - H_z^{ref}|}{max\left(|H_z^{ref}|\right)}\right), \tag{7.41}$$

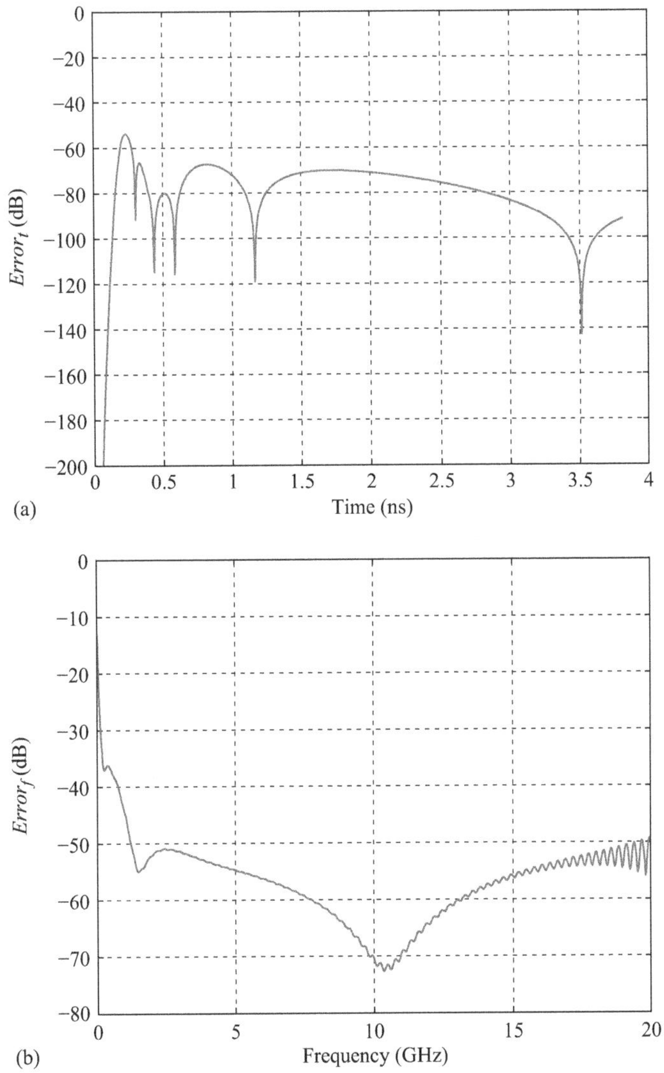

Figure 7.14 Error in time and frequency domains.

where H_z^{pml} is the sampled magnetic field in the PML problem case and H_z^{ref} is the sampled magnetic field in the reference case. The error as a function of time is plotted in Figure 7.14(a). The error in frequency domain as well is obtained as $error_f$ from the difference between the Fourier transforms of the sampled magnetic fields from the PML and reference cases by

$$error_f = 20 \times \log_{10}\left(\frac{|F(H_z^{pml}) - F(H_z^{ref})|}{F\left(|H_z^{ref}|\right)}\right), \tag{7.42}$$

where the operator $F(\cdot)$ denotes the Fourier transform. The error in frequency-domain $error_f$ is plotted in Figure 7.14(b). The errors obtained in this example can further be reduced by using a larger number of cells of PML thickness, a better choice of $R(0)$, and a higher order of PML n_{pml}. One can see in Figure 7.14(b) that the performance of the PML degrades at low frequencies.

7.5.2 Electric field distribution

Since the fields are calculated on a plane by the two-dimensional FDTD program, it is possible to capture and display the electric and magnetic field distributions as a runtime animation while the simulation is running. Furthermore, it is possible to calculate the field distribution as a response of a time-harmonic excitation at predefined frequencies. Then the time-harmonic field distribution can be compared with results obtained from simulation of the same problem using frequency-domain solvers. In this example the two-dimensional FDTD program is used to calculate the electric field distribution in a problem space including a cylinder of circular cross-section with radius 0.2 m, and dielectric constant 4, due to a current line source placed 0.2 m away from the cylinder and excited at 1 GHz frequency. Figure 7.15 illustrates the geometry of the two-dimensional problem space. The problem space is composed of square cells with 5 mm on a side and is terminated by PML boundaries with 8 cells thickness. The air gap between the cylinder and the boundaries is 30 cells in the *xn*, *yn*, and *yp* directions and 80 cells in the *xp* direction. The definition of the geometry is shown in Listing 7.20. The line source is an impressed current density with a sinusoidal waveform as shown in Listing 7.21.

In this example we define two new output types: (1) transient electric field distributions represented with a parameter named **sampled_transient_E_planes**; and (2) electric field distributions calculated at certain frequencies represented with a parameter named **sampled_frequency_E_planes**. The definition of these parameters is shown in Listing 7.22, and the initialization of these parameters are performed in the subroutine ***initialize_output_parameters_2d*** is shown in Listing 7.23.

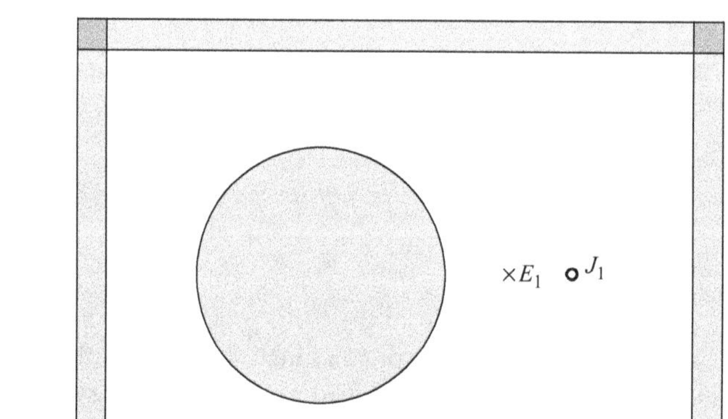

Figure 7.15 A two-dimensional problem space including a cylinder and a line source.

Listing 7.20 define_geometry_2d.m

```
% define a circle
circles(1).center_x = 0.4;
circles(1).center_y = 0.5;
circles(1).radius   = 0.2;
circles(1).material_type = 4;
```

Listing 7.21 define_sources_2d.m

```
waveforms.sinusoidal(1).frequency = 1e9;

% electric current sources
% direction: 'xp', 'xn', 'yp', 'yn', 'zp', or 'zn'
impressed_J(1).min_x = 0.8;
impressed_J(1).min_y = 0.5;
impressed_J(1).max_x = 0.8;
impressed_J(1).max_y = 0.5;
impressed_J(1).direction = 'zp';
impressed_J(1).magnitude = 1;
impressed_J(1).waveform_type = 'sinusoidal';
impressed_J(1).waveform_index = 1;
```

Listing 7.22 define_output_parameters_2d.m

```
disp('defining output parameters');

sampled_electric_fields = [];
sampled_magnetic_fields = [];
sampled_transient_E_planes = [];
sampled_frequency_E_planes = [];
% define sampled electric field distributions
% component can be 'x', 'y', 'z', or 'm' (magnitude)
% transient
sampled_transient_E_planes(1).component = 'z';

% frequency domain
sampled_frequency_E_planes(1).component = 'z';
sampled_frequency_E_planes(1).frequency = 1e9;
```

The electric fields at node positions are captured and displayed as an animation while the simulation is running in the subroutine ***display_sampled_parameters_2d*** as shown in Listing 7.24.

The electric fields at node positions are captured and calculated as the *frequency-domain* response at the given frequency in the subroutine ***capture_sampled_electric_fields_2d*** as shown in Listing 7.25. One should notice in the given code that the fields are being captured

Listing 7.23 initialize_output_parameters_2d.m

```
disp('initializing the output parameters');

number_of_sampled_electric_fields  = size(sampled_electric_fields,2);
number_of_sampled_magnetic_fields  = size(sampled_magnetic_fields,2);
number_of_sampled_transient_E_planes=size(sampled_transient_E_planes,2);
number_of_sampled_frequency_E_planes=size(sampled_frequency_E_planes,2);
% initialize sampled transient electric field
for ind=1:number_of_sampled_transient_E_planes
    sampled_transient_E_planes(ind).figure = figure;
end

% initialize sampled time harmonic electric field
for ind=1:number_of_sampled_frequency_E_planes
    sampled_frequency_E_planes(ind).sampled_field = zeros(nxp1,nyp1);
end
xcoor = linspace(fdtd_domain.min_x,fdtd_domain.max_x,nxp1);
ycoor = linspace(fdtd_domain.min_y,fdtd_domain.max_y,nyp1);
```

Listing 7.24 display_sampled_parameters_2d.m

```
% display sampled electric field distribution
for ind=1:number_of_sampled_transient_E_planes
    figure(sampled_transient_E_planes(ind).figure);
    Es = zeros(nxp1, nyp1);
    component = sampled_transient_E_planes(ind).component;
    switch (component)
        case 'x'
            Es(2:nx,:) = 0.5 * (Ex(1:nx-1,:) + Ex(2:nx,:));
        case 'y'
            Es(:,2:ny) = 0.5 * (Ey(:,1:ny-1) + Ey(:,2:ny));
        case 'z'
            Es = Ez;
        case 'm'
            Exs(2:nx,:) = 0.5 * (Ex(1:nx-1,:) + Ex(2:nx,:));
            Eys(:,2:ny) = 0.5 * (Ey(:,1:ny-1) + Ey(:,2:ny));
            Ezs = Ez;
            Es = sqrt(Exs.^2 + Eys.^2 + Ezs.^2);
    end
    imagesc(xcoor,ycoor,Es.');
    axis equal; axis xy; colorbar;
    title(['Electric field <' component '>[' num2str(ind) ']']);
    drawnow;
end
```

after 6,000 time steps. Here it is assumed that the time-domain response of the sinusoidal excitation has reached the steady state after 6,000 time steps. Then the magnitude of the steady fields is captured. Therefore, with the given code it is possible to capture the magnitude of the frequency-domain response and only at the single excitation frequency. The given code cannot

Listing 7.25 capture_sampled_electric_fields_2d.m

```
% capture sampled time harmonic electric fields on a plane
if time_step >6000
    for ind=1:number_of_sampled_frequency_E_planes
        Es = zeros(nxp1, nyp1);
        component = sampled_frequency_E_planes(ind).component;
        switch (component)
            case 'x'
                Es(2:nx,:) = 0.5 * (Ex(1:nx-1,:) + Ex(2:nx,:));
            case 'y'
                Es(:,2:ny) = 0.5 * (Ey(:,1:ny-1) + Ey(:,2:ny));
            case 'z'
                Es = Ez;
            case 'm'
                Exs(2:nx,:) = 0.5 * (Ex(1:nx-1,:) + Ex(2:nx,:));
                Eys(:,2:ny) = 0.5 * (Ey(:,1:ny-1) + Ey(:,2:ny));
                Ezs = Ez;
                Es = sqrt(Exs.^2 + Eys.^2 + Ezs.^2);
        end
        I = find(Es > sampled_frequency_E_planes(ind).sampled_field);
        sampled_frequency_E_planes(ind).sampled_field(I) = Es(I);
    end
end
```

Listing 7.26 display_frequency_domain_outputs_2d.m

```
% display sampled time harmonic electric fields on a plane
for ind=1:number_of_sampled_frequency_E_planes
    figure;
    f = sampled_frequency_E_planes(ind).frequency;
    component = sampled_frequency_E_planes(ind).component;
    Es = abs(sampled_frequency_E_planes(ind).sampled_field);
    Es = Es/max(max(Es));
    imagesc(xcoor,ycoor,Es.');
    axis equal; axis xy; colorbar;
    title(['Electric␣field␣at␣f␣=␣' ...
        num2str(f*1e-9) '␣GHz,␣<' component '>[' num2str(ind) ']']);
    drawnow;
end
```

calculate the *phase* of the response. The given algorithm can further be improved to calculate the phases as well; however, in the following example, a more efficient method based on discrete Fourier transform (DFT) is presented, which can be used to calculate both the magnitude and phase responses concurrently. After the simulation is completed, the calculated magnitude response can be plotted using the code shown in Listing 7.26.

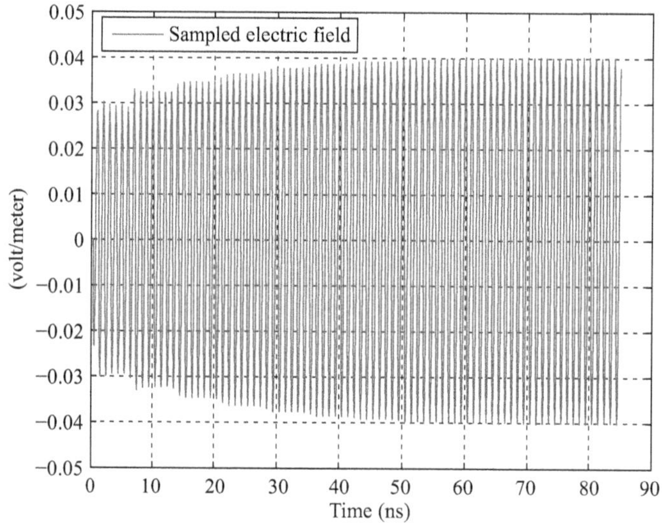

Figure 7.16 Sampled electric field at a point between the cylinder and the line source.

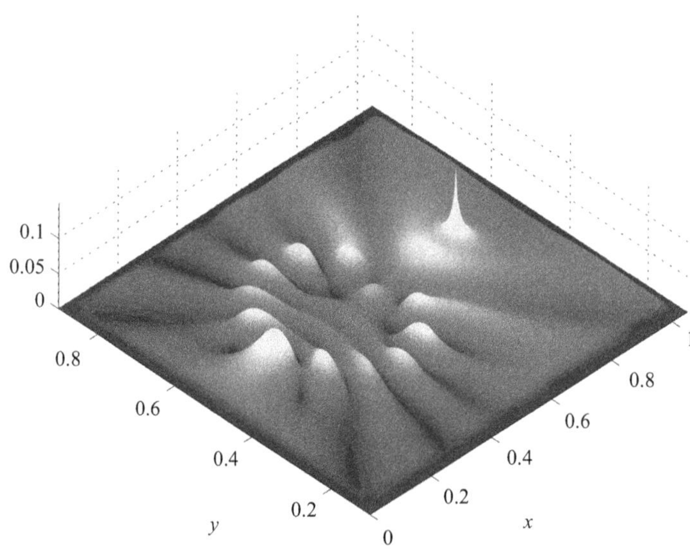

Figure 7.17 Magnitude of electric field distribution calculated by FDTD.

The two-dimensional FDTD program is excited for 8,000 time steps, and the transient electric field is sampled at a point between the cylinder and the line source as shown in Figure 7.15. The sampled electric field is plotted in Figure 7.16, which shows that the simulation has reached the steady state after 50 ns. Furthermore, the *magnitude* of the electric field distribution is captured for 1 GHz as discussed already is shown in Figure 7.17 as a surface plot. The same problem is solved using boundary value solution (BVS) [22], and

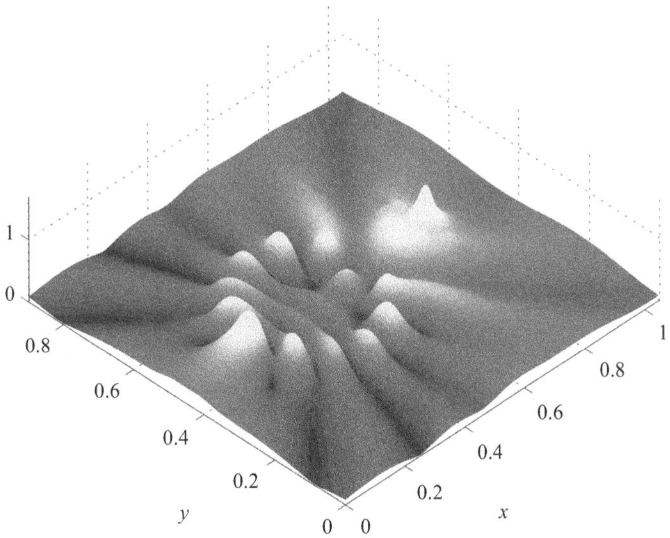

Figure 7.18 Magnitude of electric field distribution calculated by BVS.

the result is shown in Figure 7.18 for comparison. It can be seen that the results agree very well. The levels of the magnitudes are different since these figures are normalized to different values.

7.5.3 Electric field distribution using DFT

The previous example demonstrated how the magnitude of electric field distribution can be calculated as a response of a time-harmonic excitation. As discussed before, it is only possible to obtain results for a single frequency with the given technique. However, if the excitation is a waveform including a spectrum of frequencies, then it should be possible to obtain results for multiple frequencies using DFT. We modify the previous example to be able to calculate field distributions for multiple frequencies.

The output for the field distribution is defined in the subroutine ***define_output_parameters*** as shown in Listing 7.27. One can notice that it is possible to define multiple field distributions with different frequencies. The excitation waveform as well shall include the desired frequencies in its spectrum. An impressed current line source is defined as shown in Listing 7.28. To calculate field distributions for multiple frequencies we have to implement an

Listing 7.27 define_output_parameters_2d.m

```
% frequency domain
sampled_frequency_E_planes(1).component = 'z';
sampled_frequency_E_planes(1).frequency = 1e9;
sampled_frequency_E_planes(2).component = 'z';
sampled_frequency_E_planes(2).frequency = 2e9;
```

Listing 7.28 define_sources_2d.m

```
% define source waveform types and parameters
waveforms.gaussian(1).number_of_cells_per_wavelength = 0;
waveforms.gaussian(2).number_of_cells_per_wavelength = 20;
waveforms.sinusoidal(1).frequency = 1e9;

% magnetic current sources
% direction: 'xp', 'xn', 'yp', 'yn', 'zp', or 'zn'
impressed_J(1).min_x = 0.8;
impressed_J(1).min_y = 0.5;
impressed_J(1).max_x = 0.8;
impressed_J(1).max_y = 0.5;
impressed_J(1).direction = 'zp';
impressed_J(1).magnitude = 1;
impressed_J(1).waveform_type = 'gaussian';
impressed_J(1).waveform_index = 2;
```

Listing 7.29 capture_sampled_electric_fields_2d.m

```
% capture sampled time harmonic electric fields on a plane
for ind=1:number_of_sampled_frequency_E_planes
    w = 2 * pi * sampled_frequency_E_planes(ind).frequency;
    Es = zeros(nxp1, nyp1);
    component = sampled_frequency_E_planes(ind).component;
    switch (component)
        case 'x'
            Es(2:nx,:) = 0.5 * (Ex(1:nx-1,:) + Ex(2:nx,:));
        case 'y'
            Es(:,2:ny) = 0.5 * (Ey(:,1:ny-1) + Ey(:,2:ny));
        case 'z'
            Es = Ez;
        case 'm'
            Exs(2:nx,:) = 0.5 * (Ex(1:nx-1,:) + Ex(2:nx,:));
            Eys(:,2:ny) = 0.5 * (Ey(:,1:ny-1) + Ey(:,2:ny));
            Ezs = Ez;
            Es = sqrt(Exs.^2 + Eys.^2 + Ezs.^2);
    end
    sampled_frequency_E_planes(ind).sampled_field = ...
        sampled_frequency_E_planes(ind).sampled_field ...
        + dt * Es * exp(-j*w*dt*time_step);
end
```

on-the-fly DFT. Therefore, the subroutine ***capture_sampled_electric_fields_2d*** is modified as shown in Listing 7.29.

The FDTD simulation is run, and electric field distributions are calculated for 1 and 2 GHz and are plotted in Figures 7.19 and 7.20, respectively. One can notice that the result obtained at 1 GHz using the on-the-fly DFT technique is the same as the ones shown in Section 7.5.2.

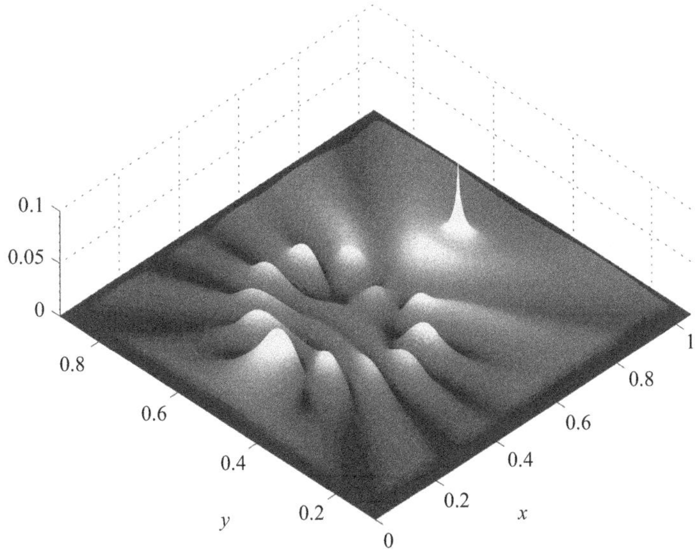

Figure 7.19 Magnitude of electric field distribution calculated by FDTD using DFT at 1 GHz.

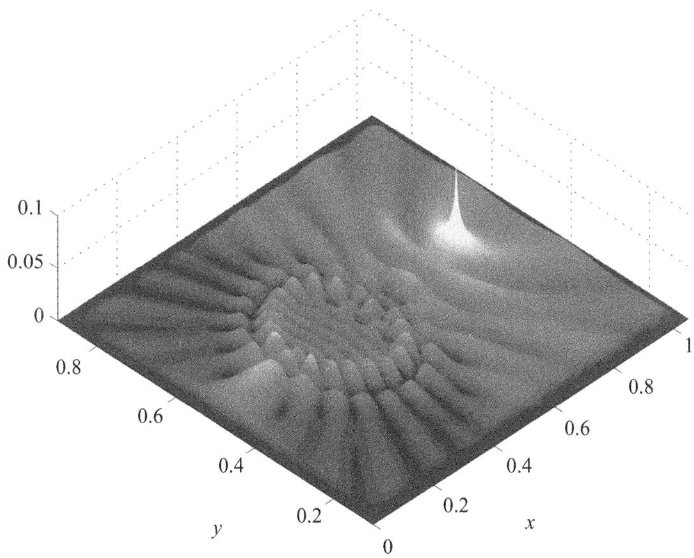

Figure 7.20 Magnitude of electric field distribution calculated by FDTD using DFT at 2 GHz.

7.6 Exercises

7.1 The performance of the two-dimensional PML for the TE_z case is evaluated in Section 7.5.1. Follow the same procedure, and evaluate the performance of the two-dimensional PML case for the TM_z case. You can use the same parameters as the given example. Notice that you need to use an electric current source to excite the TM_z mode, and you can sample E_z at a point close to the PML boundaries.

7.2 In Sections 7.5.2 and 7.5.3 new code sections are added to the two-dimensional FDTD program to add the functionality of displaying the electric field distributions. Follow the same procedure, and add new code sections to the program such that the program will display magnetic field distributions as animations while the simulation is running and it will display the magnitude of the magnetic field distribution as the response of time-harmonic excitations at a number of frequencies.

CHAPTER 8

Advanced PML formulations

In the previous chapter, we discussed the perfectly matched layers (PMLs) as a boundary to terminate the finite-difference time-domain (FDTD) lattices to simulate open boundary problems. However, the PML is shown to be ineffective for absorbing evanescent waves. As a result, the PML must be placed sufficiently far from an obstacle such that the evanescent waves have sufficiently decayed [23]. However, this increases the number of cells in an FDTD computational domain and, hence, the computational memory and time requirements. Another problem reported for PML is that it suffers from late-time reflections when terminating highly elongated lattices or when simulating fields with very long time signatures [23]. This is partly due to the weakly causal nature of the PML [24].

A strictly causal form of the PML, which is referred to as the complex frequency-shifted PML (CFS-PML) was developed in [25]. It has been shown that the CFS-PML is highly effective at absorbing evanescent waves and signals with a long time signature. Therefore, using the CFS-PML, the boundaries can be placed closer to the objects in the problem space and a time and memory saving can be achieved, thus avoiding the aforementioned weaknesses of the PML. An efficient implementation of the CFS-PML, known as the convolutional PML (CPML), is introduced in [23]. The theoretical analyses of the CPML and other forms of PML can be found in [26]. In this chapter we introduce the formulation of the CPML based on [23] and provide a MATLAB® implementation of the concept.

8.1 Formulation of CPML

In Chapter 7 we provided the PML equations for terminating a problem space surrounded by free space. However, in general a PML can be constructed to terminate a problem space surrounded by an arbitrary media. Furthermore, it can even simulate infinite length structures; these structures penetrate into the PML, and the PML absorbs the waves traveling on them.

8.1.1 PML in stretched coordinates

Without loss of generality, the PML equations for a lossy medium are posed in the stretched coordinate space [27] as

$$j\omega\varepsilon_x E_x + \sigma_x^e E_x = \frac{1}{S_{ey}}\frac{\partial H_z}{\partial y} - \frac{1}{S_{ez}}\frac{\partial H_y}{\partial z}, \tag{8.1a}$$

$$j\omega\varepsilon_y E_y + \sigma_y^e E_y = \frac{1}{S_{ez}}\frac{\partial H_x}{\partial z} - \frac{1}{S_{ex}}\frac{\partial H_z}{\partial x}, \tag{8.1b}$$

$$j\omega\varepsilon_z E_z + \sigma_z^e E_z = \frac{1}{S_{ex}}\frac{\partial H_y}{\partial x} - \frac{1}{S_{ey}}\frac{\partial H_x}{\partial y}, \tag{8.1c}$$

where S_{ex}, S_{ey}, and S_{ez} are the stretched coordinate metrics, and σ_x^e, σ_y^e, and σ_z^e are the electric conductivities of the terminating media. One can notice that (8.1) is in the frequency domain with $e^{j\omega t}$ time-harmonic convention.

Equations (8.1) reduce to Berenger's PML in the case where

$$S_{ex} = 1 + \frac{\sigma_{pex}}{j\omega\varepsilon_0}, \quad S_{ey} = 1 + \frac{\sigma_{pey}}{j\omega\varepsilon_0}, \quad \text{and} \quad S_{ez} = 1 + \frac{\sigma_{pez}}{j\omega\varepsilon_0}. \tag{8.2}$$

Here σ_{pex}, σ_{pey}, and σ_{pez} are the PML conductivities.

It should be noted that (8.1a)–(8.1c) and 8.2 follow the form given in [23], but the subscript notations have been modified to indicate the electric and magnetic parameters separately. This form of notation facilitates establishing the connection between the parameters in the formulations and their counterparts in program implementation.

The other three scalar equations that are used to construct the magnetic field update equations are

$$j\omega\mu_x H_x + \sigma_x^m H_x = -\frac{1}{S_{my}}\frac{\partial E_z}{\partial y} + \frac{1}{S_{mz}}\frac{\partial E_y}{\partial z}, \tag{8.3a}$$

$$j\omega\mu_y H_y + \sigma_y^m H_y = -\frac{1}{S_{mz}}\frac{\partial E_x}{\partial z} + \frac{1}{S_{mx}}\frac{\partial E_z}{\partial x}, \tag{8.3b}$$

$$j\omega\mu_z H_z + \sigma_z^m H_z = -\frac{1}{S_{mx}}\frac{\partial E_y}{\partial x} + \frac{1}{S_{my}}\frac{\partial E_x}{\partial y}. \tag{8.3c}$$

Equations (8.3) reduce to the PML in the case where

$$S_{mx} = 1 + \frac{\sigma_{pmx}}{j\omega\mu_0}, \quad S_{my} = 1 + \frac{\sigma_{pmy}}{j\omega\mu_0}, \quad \text{and} \quad S_{mz} = 1 + \frac{\sigma_{pmz}}{j\omega\mu_0}. \tag{8.4}$$

8.1.2 Complex stretching variables in CFS-PML

In the CPML method, the choice of the complex stretching variables follows the new definition proposed by Kuzuoglu and Mittra [25], such that

$$S_{ex} = 1 + \frac{\sigma_{pex}}{j\omega\varepsilon_0}, \quad \Rightarrow \quad S_{ex} = \kappa_{ex} + \frac{\sigma_{pex}}{\alpha_{ex} + j\omega\varepsilon_0}.$$

Then the complex stretching variables are given as

$$S_{ei} = \kappa_{ei} + \frac{\sigma_{pei}}{\alpha_{ei} + j\omega\varepsilon_0}, \quad S_{mi} = \kappa_{mi} + \frac{\sigma_{pmi}}{\alpha_{mi} + j\omega\mu_0}, \quad i = x,\ y, \text{ or } z,$$

where the parameters κ_{ei}, κ_{mi}, α_{ei}, and α_{mi} are the new parameters taking the values

$$\kappa_{ei} \geq 1, \quad \kappa_{mi} \geq 1, \quad \alpha_{ei} \geq 0, \quad \text{and} \quad \alpha_{mi} \geq 0.$$

8.1.3 The matching conditions at the PML–PML interface

For zero reflection at the PML–PML interface, one should have

$$S_{ei} = S_{mi}. \tag{8.5}$$

This condition leads to

$$\kappa_{ei} = \kappa_{mi}, \tag{8.6a}$$

$$\frac{\sigma_{pei}}{\alpha_{ei} + j\omega\varepsilon_0} = \frac{\sigma_{pmi}}{\alpha_{mi} + j\omega\mu_0}. \tag{8.6b}$$

To satisfy (8.6b), we must have

$$\frac{\sigma_{pei}}{\varepsilon_0} = \frac{\sigma_{pmi}}{\mu_0}, \quad and \quad \frac{\alpha_{ei}}{\varepsilon_0} = \frac{\alpha_{mi}}{\mu_0}. \tag{8.7}$$

8.1.4 Equations in the time domain

As mentioned before, (8.1) and (8.3) are in frequency domain. We need to have them expressed in time to obtain the field updating equations from them. For instance, (8.1a) is expressed in the time domain as

$$\varepsilon_x \frac{\partial E_x}{\partial t} + \sigma_x^e E_x = \overline{S}_{ey} * \frac{\partial H_z}{\partial y} - \overline{S}_{ez} * \frac{\partial H_y}{\partial z}, \tag{8.8}$$

where $\overline{S}_{ey}$ is a function of time that is the inverse Laplace transform of S_{ey}^{-1}, and $\overline{S}_{ez}$ is the inverse Laplace transform of S_{ez}^{-1}. One should notice that the *product* operations in the frequency-domain equations (8.1) and (8.3) are expressed as *convolution* operations in the time domain.

The terms $\overline{S}_{ei}$ and $\overline{S}_{mi}$ are then given in open form as

$$\overline{S}_{ei}(t) = \frac{\delta(t)}{\kappa_{ei}} - \frac{\sigma_{pei}}{\varepsilon_0 \kappa_{ei}^2} e^{-\left(\frac{\sigma_{pei}}{\varepsilon_0 \kappa_{ei}} + \frac{\alpha_{pei}}{\varepsilon_0}\right)t} u(t) = \frac{\delta(t)}{\kappa_{ei}} + \xi_{ei}(t), \tag{8.9a}$$

$$\overline{S}_{mi}(t) = \frac{\delta(t)}{\kappa_{mi}} - \frac{\sigma_{pmi}}{\mu_0 \kappa_{mi}^2} e^{-\left(\frac{\sigma_{pmi}}{\mu_0 \kappa_{mi}} + \frac{\alpha_{pmi}}{\mu_0}\right)t} u(t) = \frac{\delta(t)}{\kappa_{mi}} + \xi_{mi}(t), \tag{8.9b}$$

where $\delta(t)$ is the unit impulse function and $u(t)$ is the unit step function. Inserting (8.9) in (8.8) leads to

$$\varepsilon_x \frac{\partial E_x}{\partial t} + \sigma_x^e E_x = \frac{1}{\kappa_{ey}} \frac{\partial H_z}{\partial y} - \frac{1}{\kappa_{ez}} \frac{\partial H_y}{\partial z} + \xi_{ey}(t) * \frac{\partial H_z}{\partial y} - \xi_{ez}(t) * \frac{\partial H_y}{\partial z}. \tag{8.10}$$

At this point, the central difference approximation of the derivatives can be used to express (8.10) in discrete time and space and then to obtain the field updating equation for E_x^{n+1}. However, (8.10) includes two convolution terms, and these terms also need to be expressed in discrete time and space before proceeding with the construction of the updating equations.

8.1.5 Discrete convolution

The convolution terms in (8.10) can be written in open form, for instance, as

$$\xi_{ey} * \frac{\partial H_z}{\partial y} = \int_{\tau=0}^{\tau=t} \xi_{ey}(\tau) \frac{\partial H_z(t-\tau)}{\partial y} d\tau. \tag{8.11}$$

In the discrete domain, (8.11) takes the form

$$\int_{\tau=0}^{\tau=t} \xi_{ey}(\tau) \frac{\partial H_z(t-\tau)}{\partial y} d\tau \simeq \sum_{m=0}^{m=n-1} Z_{0ey}(m)\left(H_z^{n-m+\frac{1}{2}}(i,j,k) - H_z^{n-m+\frac{1}{2}}(i,j-1,k)\right), \quad (8.12)$$

where

$$\begin{aligned} Z_{0ey}(m) &= \frac{1}{\Delta y}\int_{\tau=m\Delta_t}^{\tau=(m+1)\Delta_t} \xi_{ey}(\tau)d\tau = -\frac{\sigma_{pey}}{\Delta y \varepsilon_0 \kappa_{ey}^2}\int_{\tau=m\Delta_t}^{\tau=(m+1)\Delta_t} e^{-\left(\frac{\sigma_{pey}}{\varepsilon_0\kappa_{ey}}+\frac{\alpha_{ey}}{\varepsilon_0}\right)\tau} d\tau \\ &= a_{ey} e^{-\left(\frac{\sigma_{pey}}{\kappa_{ey}}+\alpha_{ey}\right)\frac{m\,\Delta t}{\varepsilon_0}} \end{aligned} \quad (8.13)$$

and

$$a_{ey} = \frac{\sigma_{pey}}{\Delta y\left(\sigma_{pey}\kappa_{ey} + \alpha_{ey}\kappa_{ey}^2\right)}\left[e^{-\left(\frac{\sigma_{pey}}{\kappa_{ey}}+\alpha_{ey}\right)\frac{\Delta t}{\varepsilon_0}} - 1\right]. \quad (8.14)$$

Having derived the expression for $Z_{0ey}(m)$, we can express the discrete convolution term in (8.12) with a new parameter $\psi_{exy}^{n+\frac{1}{2}}(i,j,k)$, such that

$$\psi_{exy}^{n+\frac{1}{2}}(i,j,k) = \sum_{m=0}^{m=n-1} Z_{0ey}(m)\left(H_z^{n-m+\frac{1}{2}}(i,j,k) - H_z^{n-m+\frac{1}{2}}(i,j-1,k)\right). \quad (8.15)$$

Here the subscript *exy* indicates that this term is updating E_x and is associated with the derivative of the magnetic field term with respect to y. Then (8.10) can be written in discrete form as

$$\begin{aligned} &\varepsilon_x(i,j,k)\frac{E_x^{n+1}(i,j,k) - E_x^n(i,j,k)}{\Delta t} + \sigma_x^e(i,j,k)\frac{E_x^{n+1}(i,j,k) + E_x^n(i,j,k)}{2} \\ &= \frac{1}{\kappa_{ey}(i,j,k)}\frac{H_z^{n+\frac{1}{2}}(i,j,k) - H_z^{n+\frac{1}{2}}(i,j-1,k)}{\Delta y} \\ &\quad - \frac{1}{\kappa_{ez}(i,j,k)}\frac{H_y^{n+\frac{1}{2}}(i,j,k) - H_y^{n+\frac{1}{2}}(i,j,k-1)}{\Delta z} \\ &\quad + \psi_{exy}^{n+\frac{1}{2}}(i,j,k) - \psi_{exz}^{n+\frac{1}{2}}(i,j,k). \end{aligned} \quad (8.16)$$

8.1.6 The recursive convolution method

As can be followed from (8.16), the parameter ψ_{exy} has to be recalculated at every time step of the FDTD time-marching loop. However, observing expression (8.15) one can see that the values of the magnetic field component H_z calculated at all previous time steps have to be readily available to perform the discrete convolution. This implies that all the previous history of H_z must be stored in the computer memory, which is not feasible with the current computer resources. The recursive convolution technique, which is frequently used to overcome the same problem while modeling dispersive media in the FDTD method, is employed to obtain a new expression for ψ_{exy} that does not require all the previous history of H_z.

The general form of the discrete convolution (8.15) can be written as

$$\psi(n) = \sum_{m=0}^{m=n-1} Ae^{mT} B(n-m), \tag{8.17}$$

where, for instance,

$$A = a_{ey}$$
$$T = -\left(\frac{\sigma_{ey}}{\kappa_{ey}} + \alpha_{ey}\right)\frac{\Delta t}{\varepsilon_0}$$
$$B = H_z^{n-m+\frac{1}{2}}(i,j,k) - H_z^{n-m+\frac{1}{2}}(i,j-1,k).$$

Equation (8.17) can be written in open form as

$$\psi(n) = AB(n) + Ae^{T}B(n-1) + Ae^{2T}B(n-2) + \cdots + Ae^{(n-2)T}B(2) + Ae^{(n-1)T}B(1). \tag{8.18}$$

We can write the same equation in open form for one previous time step value $\psi(n-1)$ as

$$\psi(n-1) = AB(n-1) + Ae^{T}B(n-2) + Ae^{2T}B(n-3) + \cdots + Ae^{(n-2)T}B(2) + Ae^{(n-2)T}B(1). \tag{8.19}$$

Comparing the right-hand side of (8.18) with (8.19), one can realize that the right-hand side of (8.18), except the first term, is nothing but a multiplication of e^T by $\psi(n-1)$. Therefore, (8.18) can be rewritten as

$$\psi(n) = AB(n) + e^{T}\psi(n-1). \tag{8.20}$$

In this form only the previous time step value $\psi(n-1)$ is required to calculate the new value of $\psi(n)$. Hence, the need for the storage of all previous values is eliminated. Since the new value of $\psi(n)$ is calculated recursively in (8.20), this technique is known as recursive convolution.

After applying this technique to simplify (8.15), we obtain

$$\psi_{exy}^{n+\frac{1}{2}}(i,j,k) = b_{ey}\psi_{exy}^{n-\frac{1}{2}}(i,j,k) + a_{ey}\left(H_z^{n+\frac{1}{2}}(i,j,k) - H_z^{n+\frac{1}{2}}(i,j-1,k)\right), \tag{8.21}$$

where

$$a_{ey} = \frac{\sigma_{pey}}{\Delta y\left(\sigma_{pey}\kappa_{ey} + \alpha_{ey}\kappa_{ey}^2\right)}\left[b_{ey} - 1\right], \tag{8.22}$$

$$b_{ey} = e^{-\left(\frac{\sigma_{pey}}{\kappa_{ey}} + \alpha_{ey}\right)\frac{\Delta t}{\varepsilon_0}}. \tag{8.23}$$

8.2 The CPML algorithm

Examining the discrete domain equation (8.16), one can see that the form of the equation is the same as the one for a lossy medium except that two new auxiliary terms are added to the expression. This form of equation indicates that the CPML algorithm is independent of the material medium and can be extended for dispersive media, anisotropic media, or nonlinear media. In each of these cases, the left-hand side must be modified to treat the specific host medium. Yet the application of the CPML remains unchanged [23].

The FDTD flowchart can be modified to include the steps of the CPML as shown in Figure 8.1. The auxiliary parameters and coefficients required by the CPML are initialized before the time-marching loop. During the time-marching loop, at every time step, first the magnetic field components are updated in the problem space using the regular updating equation derived for the host medium. Then the CPML terms (ψ_{mxy}, ψ_{mxz}, ψ_{myx}, ψ_{myz}, ψ_{mzx}, ψ_{mzy}) are calculated using their previous time step values and the new electric field values. Afterward, these terms are added to their respective magnetic field components only at the

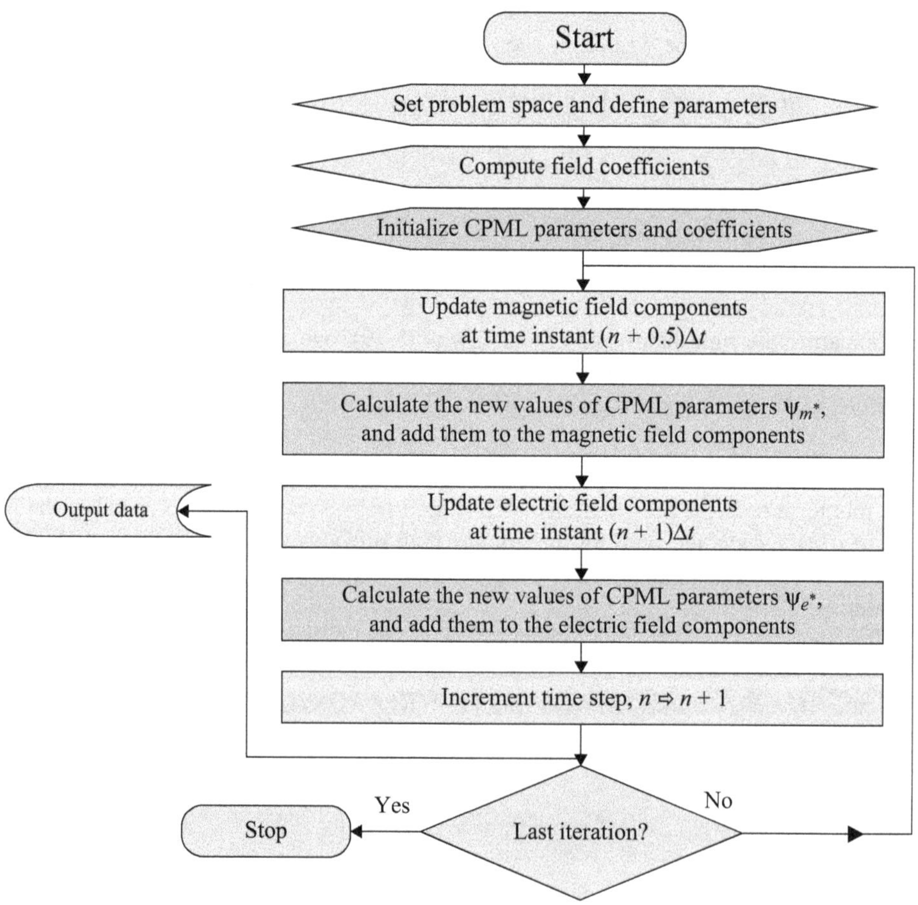

Figure 8.1 The FDTD flowchart including the steps of CPML algorithm.

CPML regions for which they were defined. The next step is updating the electric field components in the problem space using the regular updating equation derived for the host medium. Then the CPML terms (ψ_{exy}, ψ_{exz}, ψ_{eyx}, ψ_{eyz}, ψ_{ezx}, ψ_{ezy}) are calculated using their previous time step values and the new magnetic field values and are added to their respective electric field components only at the CPML regions for which they were defined. The details of the steps of this algorithm are discussed further in subsequent sections of this chapter.

8.2.1 Updating equations for CPML

The general form of the updating equation for a non-CPML lossy medium is given in Chapter 1 and is repeated here for the E_x component for convenience as

$$\begin{aligned} E_x^{n+1}(i,j,k) &= C_{exe}(i,j,k) \times E_x^n(i,j,k) \\ &+ C_{exhz}(i,j,k) \times \left(H_z^{n+\frac{1}{2}}(i,j,k) - H_z^{n+\frac{1}{2}}(i,j-1,k)\right) \\ &+ C_{exhy}(i,j,k) \times \left(H_y^{n+\frac{1}{2}}(i,j,k) - H_y^{n+\frac{1}{2}}(i,j,k-1)\right). \end{aligned} \tag{8.24}$$

Then the updating equation for the CPML region takes the form

$$\begin{aligned} E_x^{n+1}(i,j,k) &= C_{exe}(i,j,k) \times E_x^n(i,j,k) \\ &+ \left(1/\kappa_{ey}(i,j,k)\right) \times C_{exhz}(i,j,k) \times \left(H_z^{n+\frac{1}{2}}(i,j,k) - H_z^{n+\frac{1}{2}}(i,j-1,k)\right) \\ &+ \left(1/\kappa_{ez}(i,j,k)\right) \times C_{exhy}(i,j,k) \times \left(H_y^{n+\frac{1}{2}}(i,j,k) - H_y^{n+\frac{1}{2}}(i,j,k-1)\right) \\ &+ \left(\Delta y C_{exhz}(i,j,k)\right) \times \psi_{exy}^{n+\frac{1}{2}}(i,j,k) + \left(\Delta z C_{exhy}(i,j,k)\right) \times \psi_{exz}^{n+\frac{1}{2}}(i,j,k). \end{aligned} \tag{8.25}$$

Notice that the negative sign in front of the $\psi_{exz}^{n+\frac{1}{2}}(i,j,k)$ term in (8.16) is embedded in the coefficient $C_{exhy}(i,j,k)$ as can be followed from (1.26). Here two new coefficients can be defined such that

$$C_{\psi exy}(i,j,k) \Leftarrow \Delta y C_{exhz}(i,j,k). \tag{8.26a}$$

$$C_{\psi exz}(i,j,k) \Leftarrow \Delta z C_{exhy}(i,j,k). \tag{8.26b}$$

Then (8.25) reduces to

$$\begin{aligned} E_x^{n+1}(i,j,k) &= C_{exe}(i,j,k) \times E_x^n(i,j,k) \\ &+ \left(1/\kappa_{ey}(i,j,k)\right) \times C_{exhz}(i,j,k) \times \left(H_z^{n+\frac{1}{2}}(i,j,k) - H_z^{n+\frac{1}{2}}(i,j-1,k)\right) \\ &+ \left(1/\kappa_{ez}(i,j,k)\right) \times C_{exhy}(i,j,k) \times \left(H_y^{n+\frac{1}{2}}(i,j,k) - H_y^{n+\frac{1}{2}}(i,j,k-1)\right) \\ &+ C_{\psi exy}(i,j,k) \times \psi_{exy}^{n+\frac{1}{2}}(i,j,k) + C_{\psi exz}(i,j,k) \times \psi_{exz}^{n+\frac{1}{2}}(i,j,k). \end{aligned} \tag{8.27}$$

Furthermore, the terms $(1/\kappa_{ey}(i,j,k))$ and $(1/\kappa_{ez}(i,j,k))$ can be embedded into $C_{exhz}(i,j,k)$ and $C_{exhy}(i,j,k)$, respectively, such that

$$C_{exhz}(i,j,k) \Leftarrow \left(1/\kappa_{ey}(i,j,k)\right) \times C_{exhz}(i,j,k), \tag{8.28a}$$

$$C_{exhy}(i,j,k) \Leftarrow \left(1/\kappa_{ez}(i,j,k)\right) \times C_{exhy}(i,j,k). \tag{8.28b}$$

These modifications are applied to the coefficients only at the positions overlapping with the respective CPML regions and further reduce (8.27) to

$$\begin{aligned} E_x^{n+1}(i,j,k) &= C_{exe}(i,j,k) \times E_x^n(i,j,k) \\ &+ C_{exhz}(i,j,k) \times \left(H_z^{n+\frac{1}{2}}(i,j,k) - H_z^{n+\frac{1}{2}}(i,j-1,k)\right) \\ &+ C_{exhy}(i,j,k) \times \left(H_y^{n+\frac{1}{2}}(i,j,k) - H_y^{n+\frac{1}{2}}(i,j,k-1)\right) \\ &+ C_{\psi exy}(i,j,k) \times \psi_{exy}^{n+\frac{1}{2}}(i,j,k) + C_{\psi exz}(i,j,k) \times \psi_{exz}^{n+\frac{1}{2}}(i,j,k). \end{aligned} \tag{8.29}$$

With the given form of updating equation and coefficients, $E_x^{n+1}(i,j,k)$ is updated using the first three terms on the right side in the entire domain as usual. Then $\psi_{exy}^{n+\frac{1}{2}}(i,j,k)$ and $\psi_{exz}^{n+\frac{1}{2}}(i,j,k)$ are calculated (e.g., using (8.21) for $\psi_{exy}^{n+\frac{1}{2}}(i,j,k)$), and the last two terms of (8.29) are added to $E_x^{n+1}(i,j,k)$ at their respective CPML regions. It should be mentioned that the same procedure applies for updating the other electric and magnetic field components as well by using their respective updating equations and CPML parameters.

8.2.2 Addition of auxiliary CPML terms at respective regions

While discussing the PML we showed in Figure 7.4 that certain PML conductivity parameters take nonzero value at certain respective regions. In the CPML case there are three types of parameters, and the regions for which they are defined are shown in Figure 8.2, which is similar to Figure 7.4. In the CPML case the parameters σ_{pei}, σ_{pmi}, α_{ei}, and α_{mi} are nonzero, whereas the parameters κ_{ei} and κ_{mi} take values larger than one at their respective regions. These regions are named *xn*, *xp*, *yn*, *yp*, *zn*, and *zp*. The CPML auxiliary terms ψ are associated with these CPML parameters. Therefore, each term ψ is defined at the region where its associated parameters are defined. For instance, ψ_{exy} is associated with σ_{pey}, α_{ey}, and κ_{ey} as can be observed in (8.21). Then the term ψ_{exy} is defined for the *yn* and *yp* regions. Similarly, the term ψ_{exz} is defined for the *zn* and *zp* regions. One should notice in (8.27) that both the ψ_{exy} and ψ_{exz} terms are added to E_x as the requirement of the CPML algorithm; however, ψ_{exy} and ψ_{exz} are defined in different regions. This implies that the terms ψ_{exy} and ψ_{exz} should be added to E_x only in their respective regions. Therefore, one should take care while determining the regions where the CPML terms are added to the field components. For instance, ψ_{exy} should be added to E_x in the *yn* and *yp* regions, whereas ψ_{exz} should be added to E_x in the *zn* and *zp* regions. A detailed list of the CPML updating equations for all six electric and magnetic field components and the regions at which these equations are applied is given in Appendix B.

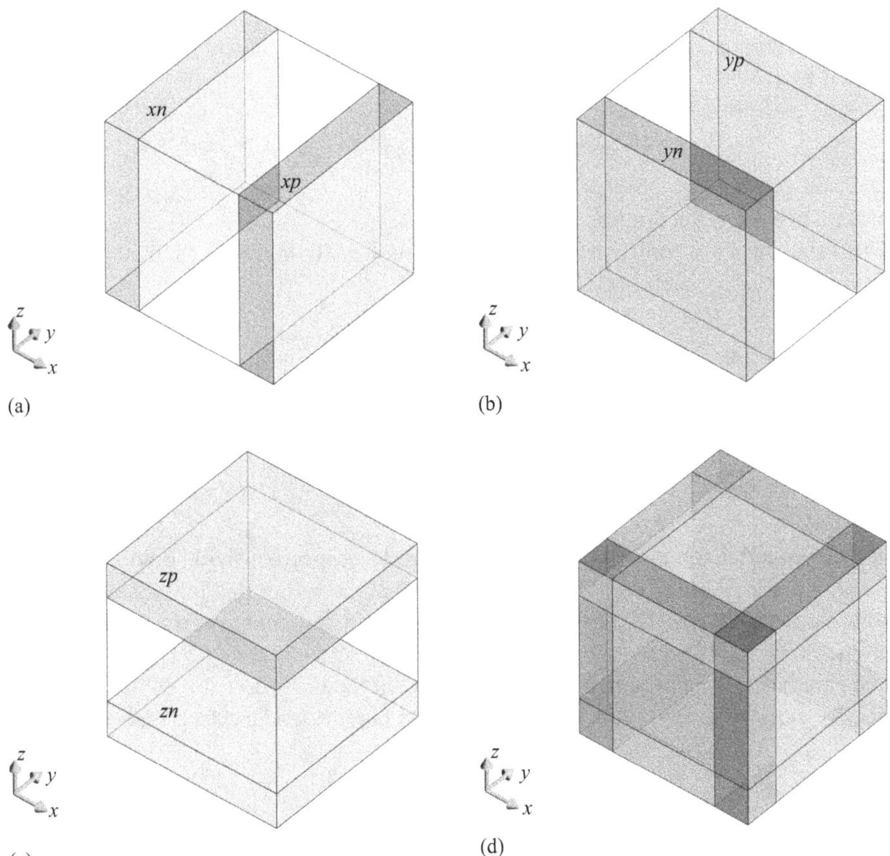

Figure 8.2 Regions where CPML parameters are defined: (a) σ_{pex}, σ_{pmx}, α_{ex}, α_{mx}, κ_{ex}, and κ_{mx}; (b) σ_{pey}, σ_{pmy}, α_{ey}, α_{my}, κ_{ey}, and κ_{my}; (c) σ_{pez}, σ_{pmz}, α_{ez}, α_{mz}, κ_{ez}, and κ_{mz}; and (d) overlapping CPML regions.

8.3 CPML parameter distribution

As described in the previous chapter, the PML conductivities are scaled along the PML region starting from a zero value at the inner domain–PML interface and increasing to a maximum value at the outer boundary. In the CPML case there are two additional types of parameter scaling profiles, which are different from the conductivity scaling profiles.

The maximum conductivity of the conductivity profile is computed in [23] using $\sigma_{\max} = \sigma_{factor} \times \sigma_{opt}$, where

$$\sigma_{opt} = \frac{n_{pml} + 1}{150\pi\sqrt{\varepsilon_r}\Delta i}. \tag{8.30}$$

Here n_{pml} is the order of the polynomial scaling, and ε_r is the relative permittivity of the background material. Then the CPML conductivity σ_{pei} can be calculated by

$$\sigma_{pei}(\rho) = \sigma_{\max}\left(\frac{\rho}{\delta}\right)^{n_{pml}}, \tag{8.31}$$

where ρ is the distance from the computational domain–PML interface to the position of the field component and δ is the thickness of the CPML layer. Similarly, σ_{pmi} can be calculated by

$$\sigma_{pmi}(\rho) = \frac{\mu_0}{\varepsilon_0}\sigma_{\max}\left(\frac{\rho}{\delta}\right)^{n_{pml}}, \tag{8.32}$$

which satisfies the reflectionless condition (8.7).

The value of κ_{ei} is unity at the inner domain–CPML interface, and it increases to a maximum value $\kappa_{\max}$ such that

$$\kappa_{ei}(\rho) = 1 + (\kappa_{\max} - 1)\left(\frac{\rho}{\delta}\right)^{n_{pml}}, \tag{8.33}$$

whereas κ_{mi} is given by

$$\kappa_{mi}(\rho) = 1 + (\kappa_{\max} - 1)\left(\frac{\rho}{\delta}\right)^{n_{pml}}. \tag{8.34}$$

In these equations, ρ indicates the distances of the respective field components from the inner domain–CPML interface. It should be noted that the values that ρ can take in (8.33) and (8.34) are different since the electric and magnetic field components are located at different positions.

The parameter α_{ei} takes a maximum value $\alpha_{\max}$ at the inner domain–CPML interface and is linearly scaled to a minimum value, $\alpha_{\min}$, at the outer boundary such that

$$\alpha_{ei}(\rho) = \alpha_{\min} + (\alpha_{\max} - \alpha_{\min})\left(1 - \frac{\rho}{\delta}\right). \tag{8.35}$$

Similarly, α_{mi} is calculated as

$$\alpha_{mi}(\rho) = \frac{\mu_0}{\varepsilon_0}\left(\alpha_{\min} + (\alpha_{\max} - \alpha_{\min})\left(1 - \frac{\rho}{\delta}\right)\right). \tag{8.36}$$

This scaling of the CPML parameter profile is chosen as before specifically to reduce the reflection error of evanescent modes; α must be nonzero at the front boundary interface. However, for the CFS-PML to absorb purely propagating modes at low frequency, α should actually decrease to zero away from the boundary interface [23].

The choice of the parameters σ_{factor}, $\kappa_{\max}$, $\alpha_{\max}$, $\alpha_{\min}$, and n_{pml} and the thickness of the CPML layer in terms of number of cells determine the performance of the CPML. In [23] parametric studies have been performed to evaluate the effects of these parameters. Usually, σ_{factor} can be taken in the range 0.7–1.5, $\kappa_{\max}$ in the range 5–11, $\alpha_{\max}$ in the range 0–0.05, thickness of CPML as 8 cells, and n_{pml} as 2, 3, or 4.

8.4 MATLAB® implementation of CPML in the three-dimensional FDTD method

In this section, we demonstrate the implementation of the CPML in the three-dimensional FDTD MATLAB code.

8.4.1 Definition of CPML

In the previous chapters, the only type of boundary available in our three-dimensional program was perfect electric conductor (PEC). The type of the boundaries were defined as PEC by assigning "***pec***" to the parameter **boundary.type**, and the air gap between the objects and the outer boundary was assigned to **boundary.air_buffer_number_of_cells** in the ***define_problem_space_parameters*** subroutine. To define the boundaries as CPML, we can update ***define_problem_space_parameters*** as shown in Listing 8.1. The type of a boundary is determined as CMPL by assigning "***cpml***" to **boundary.type**. It should be noted that the PEC boundaries can be used together with CPML boundaries; that is, some sides can be assigned PEC while the others are CPML. In Listing 8.1, the *zn* and *zp* sides are PEC while other boundaries are CPML.

Listing 8.1 define_problem_space_parameters

```
% ==<boundary conditions>========
% Here we define the boundary conditions parameters
% 'pec' : perfect electric conductor
% 'cpml' : conlvolutional PML
% if cpml_number_of_cells is less than zero
% CPML extends inside of the domain rather than outwards

boundary.type_xn = 'cpml';
boundary.air_buffer_number_of_cells_xn = 4;
boundary.cpml_number_of_cells_xn = 5;

boundary.type_xp = 'cpml';
boundary.air_buffer_number_of_cells_xp = 4;
boundary.cpml_number_of_cells_xp = 5;

boundary.type_yn = 'cpml';
boundary.air_buffer_number_of_cells_yn = 0;
boundary.cpml_number_of_cells_yn = -5;

boundary.type_yp = 'cpml';
boundary.air_buffer_number_of_cells_yp = 0;
boundary.cpml_number_of_cells_yp = -5;

boundary.type_zn = 'pec';
boundary.air_buffer_number_of_cells_zn = 0;
boundary.cpml_number_of_cells_zn = 5;

boundary.type_zp = 'pec';
boundary.air_buffer_number_of_cells_zp = 6;
boundary.cpml_number_of_cells_zp = 5;

boundary.cpml_order = 3;
boundary.cpml_sigma_factor = 1.5;
boundary.cpml_kappa_max = 7;
boundary.cpml_alpha_min = 0;
boundary.cpml_alpha_max = 0.05;
```

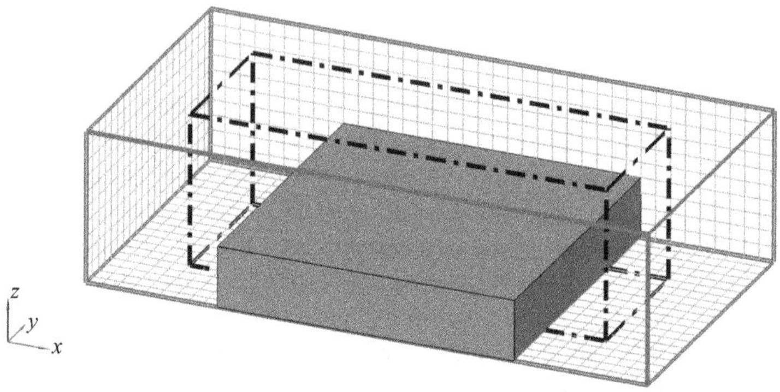

Figure 8.3 A problem space with 20 × 20 × 4 cells brick.

Another parameter required for a boundary defined as CPML is the thickness of the CPML layer in terms of number of cells. A new parameter **cpml_number_of_cells** is defined as a subfield of the parameter **boundary**, and the thicknesses of the CPML regions are assigned to this new parameter. One can notice that the thickness of CPML is 5 cells in the *xn* and *xp* regions and −5 cells in the *yn* and *yp* regions. As discussed before, objects in an FDTD problem space can be defined as penetrating into CPML; thus, these objects resemble structures extending to infinity. As a convention, if **cpml_number_of_cells** is defined as a negative number, it is assumed that the CPML layers are extending to inside of the domain rather than extending outside from the end of the air buffer region. For instance, if **air_buffer_number_of_cells** is zero, and the **cpml_number_of_cells** is a negative number at the *yn* region, then the objects on the *yn* side will be penetrating to the CPML region. Figure 8.3 illustrates a problem space including a 20 × 20 × 4 cells brick and having its boundaries defined as shown in Listing 8.1. The solid thick line shows the boundaries of the problem space. The dash-dot thick line shows the inner boundaries of the CPML regions. The brick penetrates into the CPML in *yn* and *yp* regions. The *zn* and *zp* boundaries are PEC. The other parameters described in Section 8.3 are defined in ***define_problem_space_parameters*** as well. The order of the polynomial distribution n_{pml} is defined as **boundary.cpml_order** as shown in Listing 8.1. The σ_{factor} is defined as **boundary.cpml_sigma_factor**, $\kappa_{\max}$ is defined as **boundary.cpml_kappa_max**, $\alpha_{\max}$ is defined as **boundary.cpml_alpha_max**, and similarly, $\alpha_{\min}$ is defined as **boundary.cpml_alpha_min**.

8.4.2 Initialization of CPML

To employ the CPML algorithm in the time-marching loop, several coefficient and auxiliary parameter arrays need to be defined and initialized for the respective CPML regions before the time-marching loop starts. Before defining these arrays, the size of the FDTD problem space should be determined by taking the thicknesses of the CPML regions into account. The size of the problem space and the number of cells in the domain are determined in the subroutine ***calculate_domain_size***, which was called in ***initialize_fdtd_material_grid***. The implementation of ***calculate_domain_size*** is given in Listing 3.5, where the boundaries are assumed to be PEC. As discussed in the previous section, if the thickness of the CPML is negative on a side, the CPML region extends into the problem space on this side; therefore, the position of the outer boundary of the problem space remains the same, and the calculation given for determining the position of the outer boundary in Listing 3.5 remains the same. However, if the thickness of the CPML is positive on a side, the CPML adds to the problem space on that side, and the calculation of the position of the outer boundary needs to be updated accordingly. The necessary modifications are implemented in ***calculate_domain_size***, and the modified section of the code is given in Listing 8.2. It can be noticed in the code that, for the sides where the boundary type is CPML, the problem space boundary is extended outward by the thickness of the CPML.

Listing 8.2 calculate_domain_size

```
% Determine the problem space boundaries including air buffers
fdtd_domain.min_x = fdtd_domain.min_x ...
    - dx * boundary.air_buffer_number_of_cells_xn;
fdtd_domain.min_y = fdtd_domain.min_y ...
    - dy * boundary.air_buffer_number_of_cells_yn;
fdtd_domain.min_z = fdtd_domain.min_z ...
    - dz * boundary.air_buffer_number_of_cells_zn;
fdtd_domain.max_x = fdtd_domain.max_x ...
    + dx * boundary.air_buffer_number_of_cells_xp;
fdtd_domain.max_y = fdtd_domain.max_y ...
    + dy * boundary.air_buffer_number_of_cells_yp;
fdtd_domain.max_z = fdtd_domain.max_z ...
    + dz * boundary.air_buffer_number_of_cells_zp;

% Determine the problem space boundaries including cpml layers
if strcmp(boundary.type_xn, 'cpml') && ...
        (boundary.cpml_number_of_cells_xn >0)
    fdtd_domain.min_x = fdtd_domain.min_x ...
        - dx * boundary.cpml_number_of_cells_xn;
end
if strcmp(boundary.type_xp, 'cpml') && ...
        (boundary.cpml_number_of_cells_xp >0)
    fdtd_domain.max_x = fdtd_domain.max_x ...
        + dx * boundary.cpml_number_of_cells_xp;
end
if strcmp(boundary.type_yn, 'cpml') && ...
        (boundary.cpml_number_of_cells_yn >0)
    fdtd_domain.min_y = fdtd_domain.min_y ...
        - dy * boundary.cpml_number_of_cells_yn;
end
if strcmp(boundary.type_yp, 'cpml') && ...
        (boundary.cpml_number_of_cells_yp >0)
    fdtd_domain.max_y = fdtd_domain.max_y ...
        + dy * boundary.cpml_number_of_cells_yp;
end
if strcmp(boundary.type_zn, 'cpml') && ...
        (boundary.cpml_number_of_cells_zn >0)
    fdtd_domain.min_z = fdtd_domain.min_z ...
        - dz * boundary.cpml_number_of_cells_zn;
end
if strcmp(boundary.type_zp, 'cpml') && ...
        (boundary.cpml_number_of_cells_zp >0)
    fdtd_domain.max_z = fdtd_domain.max_z ...
        + dz * boundary.cpml_number_of_cells_zp;
end
```

The initialization process of the CPML continues in ***initialize_boundary_conditions***, which is a subroutine called in the main program ***fdtd_solve*** as shown in Listing 3.1. So far we have not implemented any code in ***initialize_boundary_conditions***, since the only type of boundary we have considered is PEC; which does not require any special treatment; the tangential electric field components on the outer faces of the problem space are not updated, and their values are left zero during the time-marching loop which naturally simulates the PEC boundaries. However, the CPML requires special treatment, and the initialization of the CPML is coded in ***initialize_boundary_conditions*** accordingly as shown in Listing 8.3.

Listing 8.3 initialize_boundary_conditions

```
% initialize boundary parameters

% define logical parameters for the conditions that  will be used often
is_cpml_xn = false; is_cpml_xp = false; is_cpml_yn = false;
is_cpml_yp = false; is_cpml_zn = false; is_cpml_zp = false;
is_any_side_cpml = false;
if strcmp(boundary.type_xn, 'cpml')
    is_cpml_xn = true;
    n_cpml_xn = abs(boundary.cpml_number_of_cells_xn);
end
if strcmp(boundary.type_xp, 'cpml')
    is_cpml_xp = true;
    n_cpml_xp = abs(boundary.cpml_number_of_cells_xp);
end
if strcmp(boundary.type_yn, 'cpml')
    is_cpml_yn = true;
    n_cpml_yn = abs(boundary.cpml_number_of_cells_yn);
end
if strcmp(boundary.type_yp, 'cpml')
    is_cpml_yp = true;
    n_cpml_yp = abs(boundary.cpml_number_of_cells_yp);
end
if strcmp(boundary.type_zn, 'cpml')
    is_cpml_zn = true;
    n_cpml_zn = abs(boundary.cpml_number_of_cells_zn);
end
if strcmp(boundary.type_zp, 'cpml')
    is_cpml_zp = true;
    n_cpml_zp = abs(boundary.cpml_number_of_cells_zp);
end

if (is_cpml_xn || is_cpml_xp || is_cpml_yn ...
        || is_cpml_yp || is_cpml_zn || is_cpml_zp)
    is_any_side_cpml = true;
end

% Call CPML initialization routine if any side is CPML
if is_any_side_cpml
    initialize_CPML_ABC;
end
```

In this subroutine several frequently used parameters are defined for convenience. For instance, the parameter **is_cpml_xn** is a logical parameter that indicates whether the *xn* boundary is CPML, and **n_cpml_xn** stores the thickness of the CPML for the *xn* region. Similar parameters are defined for other sides as well. Then another subroutine, ***initialize_CPML_ABC***, is called to continue the CPML specific initialization process.

Listing 8.4 shows the contents of ***initialize_CPML_ABC*** for the *xn* and *xp* regions. The code listing for the other sides follows the same procedure. Let us consider the *xp* region, for example. In the code, distribution of the CPML parameters σ_{pex}, σ_{pmx}, κ_{ex}, κ_{mx}, α_{ex}, and α_{mx} along the CPML thickness are calculated as one-dimensional arrays first, with the respective names **sigma_pex_xp, sigma_pmx_xp, kappa_ex_xp, kappa_mx_xp, alpha_ex_xp,** and

Listing 8.4 initialize_CPML_ABC

```
% Initialize CPML boundary condition

p_order = boundary.cpml_order; % order of the polynomial distribution
sigma_ratio = boundary.cpml_sigma_factor;
kappa_max = boundary.cpml_kappa_max;
alpha_min = boundary.cpml_alpha_min;
alpha_max = boundary.cpml_alpha_max;

% Initialize cpml for xn region
if is_cpml_xn

    % define one-dimensional temporary cpml parameter arrays
    sigma_max = sigma_ratio  * (p_order+1)/(150*pi*dx);
    ncells = n_cpml_xn;
    rho_e = ([ncells:-1:1]-0.75)/ncells;
    rho_m = ([ncells:-1:1]-0.25)/ncells;
    sigma_pex_xn = sigma_max * rho_e.^p_order;
    sigma_pmx_xn = sigma_max * rho_m.^p_order;
    sigma_pmx_xn = (mu_0/eps_0) * sigma_pmx_xn;
    kappa_ex_xn = 1 + (kappa_max - 1) * rho_e.^p_order;
    kappa_mx_xn = 1 + (kappa_max - 1) * rho_m.^p_order;
    alpha_ex_xn = alpha_min + (alpha_max - alpha_min) * (1-rho_e);
    alpha_mx_xn = alpha_min + (alpha_max - alpha_min) * (1-rho_m);
    alpha_mx_xn = (mu_0/eps_0) * alpha_mx_xn;

    % define one-dimensional cpml parameter arrays
    cpml_b_ex_xn = exp((-dt/eps_0) ...
        *((sigma_pex_xn./kappa_ex_xn)+ alpha_ex_xn));
    cpml_a_ex_xn = (1/dx)*(cpml_b_ex_xn -1.0).* sigma_pex_xn ...
        ./(kappa_ex_xn.*(sigma_pex_xn+kappa_ex_xn.*alpha_ex_xn));
    cpml_b_mx_xn = exp((-dt/mu_0) ...
        *((sigma_pmx_xn./kappa_mx_xn)+ alpha_mx_xn));
    cpml_a_mx_xn = (1/dx)*(cpml_b_mx_xn -1.0) .* sigma_pmx_xn ...
        ./(kappa_mx_xn.*(sigma_pmx_xn+kappa_mx_xn.*alpha_mx_xn));

    % Create and initialize 2D cpml convolution parameters
    Psi_eyx_xn = zeros(ncells,ny,nzp1);
    Psi_ezx_xn = zeros(ncells,nyp1,nz);
    Psi_hyx_xn = zeros(ncells,nyp1,nz);
```

```
    Psi_hzx_xn = zeros(ncells,ny,nzp1);

    % Create and initialize 2D cpml convolution coefficients
    % Notice that Ey(1,:,:) and Ez(1,:,:) are not updated by cmpl
    CPsi_eyx_xn = Ceyhz(2:ncells+1,:,:)*dx;
    CPsi_ezx_xn = Cezhy(2:ncells+1,:,:)*dx;
    CPsi_hyx_xn = Chyez(1:ncells,:,:)*dx;
    CPsi_hzx_xn = Chzey(1:ncells,:,:)*dx;

    % Adjust FDTD coefficients in the CPML region
    % Notice that Ey(1,:,:) and Ez(1,:,:) are not updated by cmpl
    for i = 1: ncells
        Ceyhz(i+1,:,:) = Ceyhz(i+1,:,:)/kappa_ex_xn(i);
        Cezhy(i+1,:,:) = Cezhy(i+1,:,:)/kappa_ex_xn(i);
        Chyez(i,:,:) = Chyez(i,:,:)/kappa_mx_xn(i);
        Chzey(i,:,:) = Chzey(i,:,:)/kappa_mx_xn(i);
    end

    % Delete temporary arrays. These arrays will not be used any more.
    clear sigma_pex_xn sigma_pmx_xn;
    clear kappa_ex_xn kappa_mx_xn;
    clear alpha_ex_xn alpha_mx_xn;
end

% Initialize cpml for xp region
if is_cpml_xp

    % define one-dimensional temporary cpml parameter arrays
    sigma_max = sigma_ratio  * (p_order+1)/(150*pi*dx);
    ncells = n_cpml_xp;
    rho_e = ([1:ncells]-0.25)/ncells;
    rho_m = ([1:ncells]-0.75)/ncells;
    sigma_pex_xp = sigma_max * rho_e.^p_order;
    sigma_pmx_xp = sigma_max * rho_m.^p_order;
    sigma_pmx_xp = (mu_0/eps_0) * sigma_pmx_xp;
    kappa_ex_xp = 1 + (kappa_max - 1) * rho_e.^p_order;
    kappa_mx_xp = 1 + (kappa_max - 1) * rho_m.^p_order;
    alpha_ex_xp = alpha_min + (alpha_max - alpha_min) * (1-rho_e);
    alpha_mx_xp = alpha_min + (alpha_max - alpha_min) * (1-rho_m);
    alpha_mx_xp = (mu_0/eps_0) * alpha_mx_xp;

    % define one-dimensional cpml parameter arrays
    cpml_b_ex_xp = exp((-dt/eps_0) ...
        *((sigma_pex_xp./kappa_ex_xp)+ alpha_ex_xp));
    cpml_a_ex_xp = (1/dx)*(cpml_b_ex_xp-1.0).* sigma_pex_xp ...
        ./(kappa_ex_xp.*(sigma_pex_xp+kappa_ex_xp.*alpha_ex_xp));
    cpml_b_mx_xp = exp((-dt/mu_0) ...
        *((sigma_pmx_xp./kappa_mx_xp)+ alpha_mx_xp));
    cpml_a_mx_xp = (1/dx)*(cpml_b_mx_xp-1.0) .* sigma_pmx_xp ...
        ./(kappa_mx_xp.*(sigma_pmx_xp+kappa_mx_xp.*alpha_mx_xp));

    % Create and initialize 2D cpml convolution parameters
```

```
    Psi_eyx_xp = zeros(ncells,ny,nzp1);
    Psi_ezx_xp = zeros(ncells,nyp1,nz);
    Psi_hyx_xp = zeros(ncells,nyp1,nz);
    Psi_hzx_xp = zeros(ncells,ny,nzp1);

    % Create and initialize 2D cpml convolution coefficients
    % Notice that Ey(nxp1,:,:) and Ez(nxp1,:,:) are not updated by cmpl
    CPsi_eyx_xp = Ceyhz(nxp1-ncells:nx,:,:)*dx;
    CPsi_ezx_xp = Cezhy(nxp1-ncells:nx,:,:)*dx;
    CPsi_hyx_xp = Chyez(nxp1-ncells:nx,:,:)*dx;
    CPsi_hzx_xp = Chzey(nxp1-ncells:nx,:,:)*dx;

    % Adjust FDTD coefficients in the CPML region
    % Notice that Ey(nxp1,:,:) and Ez(nxp1,:,:) are not updated by cmpl
    for i = 1: ncells
        Ceyhz(nx-ncells+i,:,:) = Ceyhz(nx-ncells+i,:,:)/kappa_ex_xp(i);
        Cezhy(nx-ncells+i,:,:) = Cezhy(nx-ncells+i,:,:)/kappa_ex_xp(i);
        Chyez(nx-ncells+i,:,:) = Chyez(nx-ncells+i,:,:)/kappa_mx_xp(i);
        Chzey(nx-ncells+i,:,:) = Chzey(nx-ncells+i,:,:)/kappa_mx_xp(i);
    end

    % Delete temporary arrays. These arrays will not be used any more.
    clear sigma_pex_xp sigma_pmx_xp;
    clear kappa_ex_xp kappa_mx_xp;
    clear alpha_ex_xp alpha_mx_xp;
end
```

alpha_mx_xp, using the equations given in Section 8.3. The parameters σ_{pex}, κ_{ex}, and α_{ex} are used to update the E_y and E_z field components, and the distances of these field components from the interior interface of the CPML are used in (8.31), (8.33), and (8.35) as ρ_e. The positions of the field components and the respective distances of the field components from the CPML interface are illustrated in Figure 8.4. One can see that CPML interface is assumed to be offset from the first CPML cell by a quarter cell size. This is to ensure that the same number of electric and magnetic field components is updated by the CPML algorithm. Similarly, Figure 8.5 illustrates the magnetic field components of H_y and H_z being updated using the CPML in the *xp* region. The distances of these field components from the CPML interface are denoted as ρ_h and are used in (8.32), (8.34), and (8.36) to calculate σ_{pmx}, κ_{mx}, and α_{mx}.

Then Listing 8.4 continues with calculation of a_{ex}, b_{ex}, a_{mx}, and b_{mx} as **cpml_a_ex_xp**, **cpml_b_ex_xp**, **cpml_a_mx_xp**, and **cpml_b_mx_xp** using the equations in Appendix B. Then the CPML auxiliary parameters ψ_{eyx}, ψ_{ezx}, ψ_{hyx}, and ψ_{hzx} are defined as two-dimensional arrays with the names **Psi_eyx_xp**, **Psi_ezx_xp**, **Psi_hyx_xp**, and **Psi_hzx_xp** and are initialized with zero value. The coefficients $C_{\psi eyx}$, $C_{\psi ezx}$, $C_{\psi hyx}$, and $C_{\psi hzx}$ are calculated and stored as two-dimensional arrays with the respective names **CPsi_eyx_xp**, **CPsi_ezx_xp**, **CPsi_hyx_xp**, and **CPsi_hzx_xp**. Then the FDTD updating coefficients **Ceyhz**, **Cezhy**, **Chyez**, and **Chzey** are scaled with the respective κ values in the *xp* region. Finally, the parameters that will not be used anymore during the FDTD calculations are cleared from MATLAB workspace.

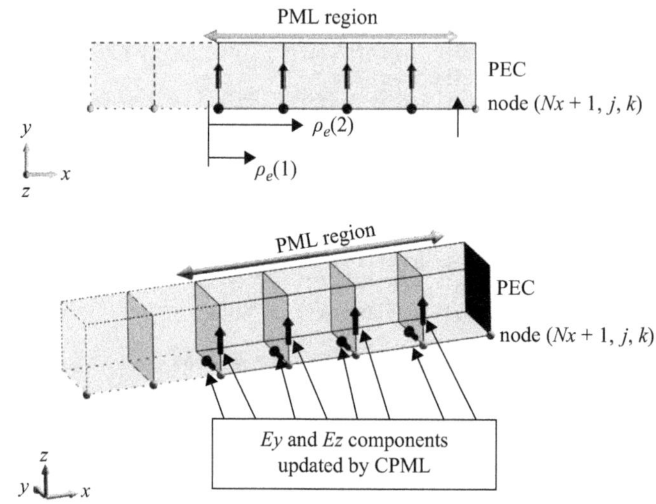

Figure 8.4 Positions of the electric field components updated by CPML in the *xp* region.

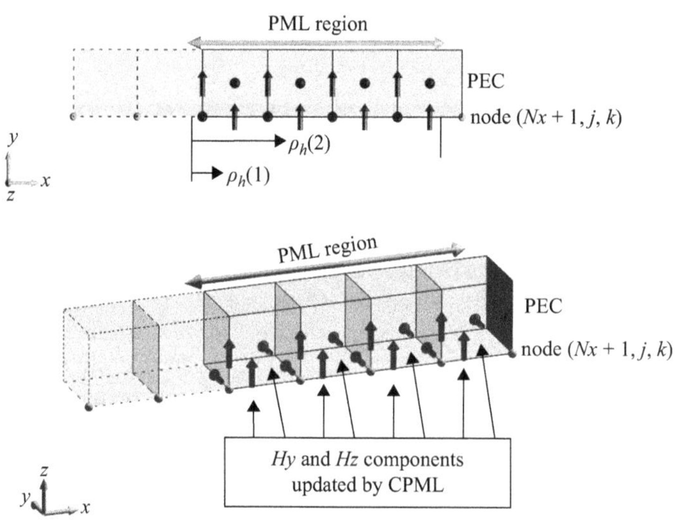

Figure 8.5 Positions of the magnetic field components updated by CPML in the *xp* region.

8.4.3 Application of CPML in the FDTD time-marching loop

The necessary CPML arrays and parameters are initialized in ***initialize_CPML_ABC***, and they are ready to use for applying the CPML in the FDTD time-marching loop. Two new subroutines are added to the subroutine ***run_fdtd_time_marching_loop*** as shown in Listing 8.5. The first one is ***update_magnetic_field_CPML_ABC***, and it follows ***update_magnetic_fields***. The second one is ***update_electric_field_CPML_ABC***, and it follows ***update_electric_fields***.

Listing 8.5 run_fdtd_time_marching_loop

```
disp (['Starting the time marching loop']);
disp(['Total number of time steps : ' ...
    num2str(number_of_time_steps)]);

start_time = cputime;
current_time = 0;

for time_step = 1:number_of_time_steps
    update_magnetic_fields;
    update_magnetic_field_CPML_ABC;
    capture_sampled_magnetic_fields;
    capture_sampled_currents;
    update_electric_fields;
    update_electric_field_CPML_ABC;
    update_voltage_sources;
    update_current_sources;
    update_inductors;
    update_diodes;
    capture_sampled_electric_fields;
    capture_sampled_voltages;
    display_sampled_parameters;
end

end_time = cputime;
total_time_in_minutes = (end_time - start_time)/60;
disp(['Total simulation time is ' ...
    num2str(total_time_in_minutes) ' minutes.']);
```

The implementation of ***update_magnetic_field_CPML_ABC*** is given in Listing 8.6. The magnetic field components are updated for the current time step using the regular updating equations in ***update_magnetic_fields***, and then in ***update_magnetic_field_CPML_ABC***, first the auxiliary ψ parameters are calculated using the current values of the electric field components. These terms are then added to the appropriate magnetic field components in their respective CPML regions.

Similarly, the implementation of ***update_electric_field_CPML_ABC*** is given in Listing 8.7. The electric field components are updated for the current time step using the regular updating equations in ***update_electric_fields***, and then in ***update_electric_field_CPML_ABC***, first the auxiliary ψ parameters are calculated using the current values of the magnetic field components. These terms are then added to the appropriate electric field components in their respective CPML regions.

One can notice in the code listings that the arrays representing ψ_e and ψ_h are two-dimensional, whereas the arrays representing a_{ei}, a_{mi}, b_{ei}, and b_{mi} are one-dimensional. Therefore, ψ_e and ψ_h are updated in a "for" loop using the one-dimensional arrays; if they were defined as two-dimensional arrays, the "for" loop would be avoided, which would speed up the calculation with the trade-off of increased memory. In the two-dimensional array case, the coefficients representing $C_{\psi e}$ and $C_{\psi m}$ can be distributed over a_{ei}, a_{mi}, b_{ei}, and b_{mi} in the initialization process ***initialize_CPML_ABC***, which further reduces the amount of calculations during the CPML updates.

Listing 8.6 update_magnetic_field_CPML_ABC

```
% apply CPML to magnetic field components
if is_cpml_xn
    for i = 1: n_cpml_xn
        Psi_hyx_xn(i,:,:) = cpml_b_mx_xn(i) * Psi_hyx_xn(i,:,:) ...
            + cpml_a_mx_xn(i)*(Ez(i+1,:,:)-Ez(i,:,:));
        Psi_hzx_xn(i,:,:) = cpml_b_mx_xn(i) * Psi_hzx_xn(i,:,:) ...
            + cpml_a_mx_xn(i)*(Ey(i+1,:,:)-Ey(i,:,:));
    end
        Hy(1:n_cpml_xn,:,:) = Hy(1:n_cpml_xn,:,:) ...
            + CPsi_hyx_xn(:,:,:) .* Psi_hyx_xn(:,:,:);
        Hz(1:n_cpml_xn,:,:) = Hz(1:n_cpml_xn,:,:) ...
            + CPsi_hzx_xn(:,:,:) .* Psi_hzx_xn(:,:,:);
end

if is_cpml_xp
    n_st = nx - n_cpml_xp;
    for i = 1:n_cpml_xp
        Psi_hyx_xp(i,:,:) = cpml_b_mx_xp(i) * Psi_hyx_xp(i,:,:) ...
            + cpml_a_mx_xp(i)*(Ez(i+n_st+1,:,:)-Ez(i+n_st,:,:));
        Psi_hzx_xp(i,:,:) = cpml_b_mx_xp(i) * Psi_hzx_xp(i,:,:) ...
            + cpml_a_mx_xp(i)*(Ey(i+n_st+1,:,:)-Ey(i+n_st,:,:));
    end

        Hy(n_st+1:nx,:,:) = Hy(n_st+1:nx,:,:) ...
            + CPsi_hyx_xp(:,:,:) .* Psi_hyx_xp(:,:,:);
        Hz(n_st+1:nx,:,:) = Hz(n_st+1:nx,:,:) ...
            + CPsi_hzx_xp(:,:,:) .* Psi_hzx_xp(:,:,:);
end
```

Listing 8.7 update_electric_field_CPML_ABC

```
% apply CPML to electric field components
if is_cpml_xn
    for i = 1:n_cpml_xn
        Psi_eyx_xn(i,:,:) = cpml_b_ex_xn(i) * Psi_eyx_xn(i,:,:) ...
            + cpml_a_ex_xn(i)*(Hz(i+1,:,:)-Hz(i,:,:));
        Psi_ezx_xn(i,:,:) = cpml_b_ex_xn(i) * Psi_ezx_xn(i,:,:) ...
            + cpml_a_ex_xn(i)*(Hy(i+1,:,:)-Hy(i,:,:));
    end
    Ey(2:n_cpml_xn+1,:,:) = Ey(2:n_cpml_xn+1,:,:) ...
            + CPsi_eyx_xn .* Psi_eyx_xn;
    Ez(2:n_cpml_xn+1,:,:) = Ez(2:n_cpml_xn+1,:,:) ...
            + CPsi_ezx_xn .* Psi_ezx_xn;
end

if is_cpml_xp
    n_st = nx - n_cpml_xp;
    for i = 1:n_cpml_xp
        Psi_eyx_xp(i,:,:) = cpml_b_ex_xp(i) * Psi_eyx_xp(i,:,:) ...
            + cpml_a_ex_xp(i)*(Hz(i+n_st,:,:) - Hz(i+n_st-1,:,:));
        Psi_ezx_xp(i,:,:) = cpml_b_ex_xp(i) * Psi_ezx_xp(i,:,:) ...
            + cpml_a_ex_xp(i)*(Hy(i+n_st,:,:) - Hy(i+n_st-1,:,:));
    end
```

```
    Ey(n_st+1:nx,:,:) = Ey(n_st+1:nx,:,:) ...
        + CPsi_eyx_xp .* Psi_eyx_xp;
    Ez(n_st+1:nx,:,:) = Ez(n_st+1:nx,:,:) ...
        + CPsi_ezx_xp .* Psi_ezx_xp;
end
```

8.5 Simulation examples

So far we have discussed the CPML algorithm and have demonstrated its implementation in the MATLAB program. In this section, we provide examples in which the boundaries are realized by the CPML.

8.5.1 Microstrip low-pass filter

In Section 6.2, we demonstrated a low-pass filter geometry, which is illustrated in Figure 6.2. We assumed that the boundaries of the problem space were PEC. In this section we repeat the same example but assume that the boundaries are CPML. The section of the code that defines the boundaries in ***define_problem_space_parameters*** is shown in Listing 8.8. As can be followed in the listing, an air gap of 5 cells thickness is surrounding the filter circuit, and the boundaries are terminated by 8 cells thickness of the CPML.

Listing 8.8 define_problem_space_parameters.m

```
% ==<boundary conditions>========
% Here we define the boundary conditions parameters
% 'pec' : perfect electric conductor
% 'cpml' : conlvolutional PML
% if cpml_number_of_cells is less than zero
% CPML extends inside of the domain rather than outwards

boundary.type_xn = 'cpml';
boundary.air_buffer_number_of_cells_xn = 5;
boundary.cpml_number_of_cells_xn = 8;

boundary.type_xp = 'cpml';
boundary.air_buffer_number_of_cells_xp = 5;
boundary.cpml_number_of_cells_xp = 8;

boundary.type_yn = 'cpml';
boundary.air_buffer_number_of_cells_yn = 5;
boundary.cpml_number_of_cells_yn = 8;

boundary.type_yp = 'cpml';
boundary.air_buffer_number_of_cells_yp = 5;
boundary.cpml_number_of_cells_yp = 8;

boundary.type_zn = 'cpml';
boundary.air_buffer_number_of_cells_zn = 5;
boundary.cpml_number_of_cells_zn = 8;
```

```
boundary.type_zp = 'cpml';
boundary.air_buffer_number_of_cells_zp = 5;
boundary.cpml_number_of_cells_zp = 8;

boundary.cpml_order = 3;
boundary.cpml_sigma_factor = 1.3;
boundary.cpml_kappa_max = 7;
boundary.cpml_alpha_min = 0;
boundary.cpml_alpha_max = 0.05;
```

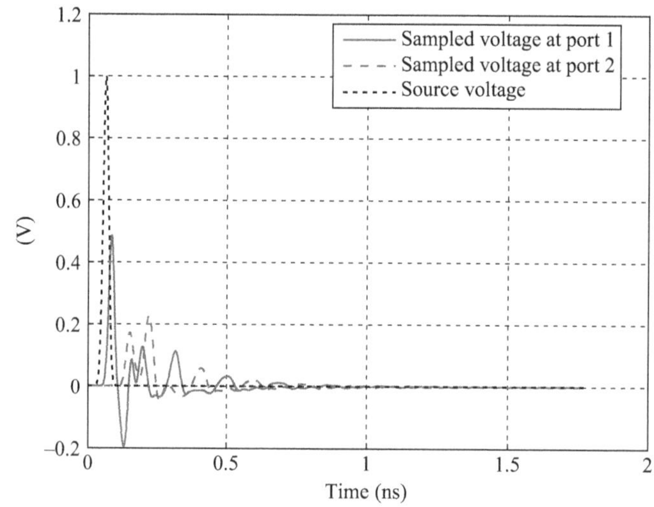

Figure 8.6 Source voltage and sampled voltages observed at the ports of the low-pass filter.

The simulation of the circuit is performed for 3,000 time steps. The results of this simulation are plotted in the Figures 8.6–8.9. Figure 8.6 shows the source voltage and sampled voltages observed at the ports of the low-pass filter, whereas Figure 8.7 shows the sampled currents observed at the ports of the low-pass filter. It can be seen that the signals are sufficiently decayed after 3,000 time steps. After the transient voltages and currents are obtained, they are used for postprocessing and scattering parameter (S-parameter) calculations. Figure 8.8 shows S_{11}, whereas Figure 8.9 shows S_{21} of the circuit. Comparing these figures with the plots in Figure 6.3 one can observe that the S-parameters calculated for the case where the boundaries are CPML are smooth and do not include glitches. This means that resonances due to the closed PEC boundaries are eliminated and the circuit is simulated as if it is surrounded by open space.

8.5.2 Microstrip branch line coupler

The second example is a microstrip branch line coupler, which was published in [14]. The circuit discussed in this section is illustrated in Figure 8.10. The cell sizes are

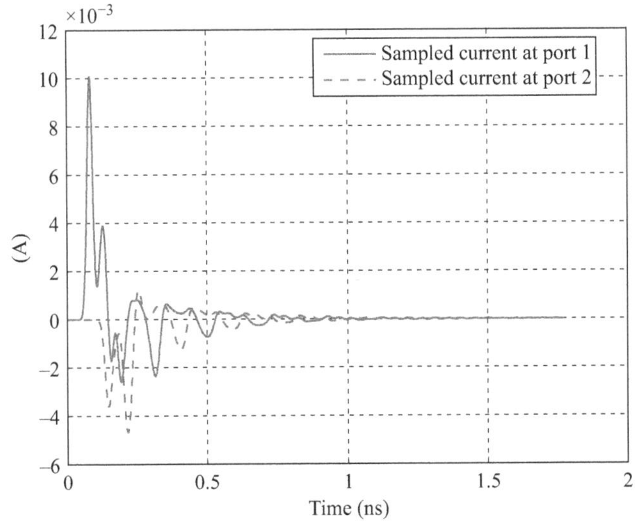

Figure 8.7 Sampled currents observed at the ports of the low-pass filter.

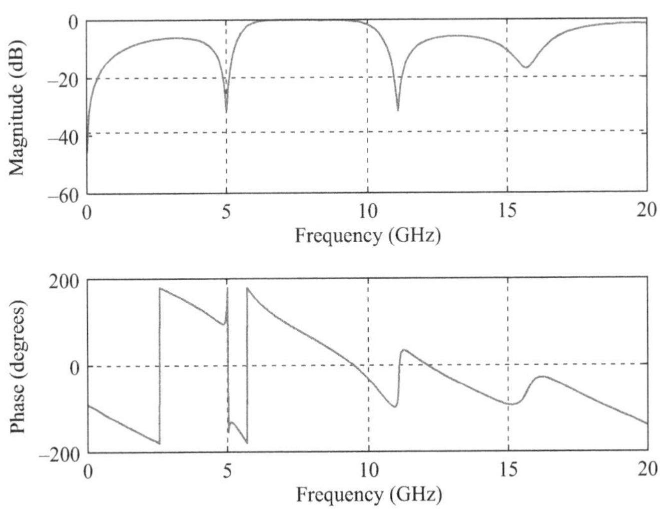

Figure 8.8 S_{11} of the low-pass filter.

$\Delta x = 0.406$ mm, $\Delta y = 0.406$ mm, and $\Delta z = 0.265$ mm. In this circuit the wide lines are 10 cells wide and the narrow lines are 6 cells wide. The center-to-center distances between the strips in the square coupler section are 24 cells. The dielectric substrate has 2.2 dielectric constant and 3 cells thickness. The definition of the geometry of the problem is shown in Listing 8.9.

The boundaries of the problem space are the same as those in Listing 8.8: that is, an 8 cells thick CPML, which terminates a 5 cells thick air gap surrounding the circuit.

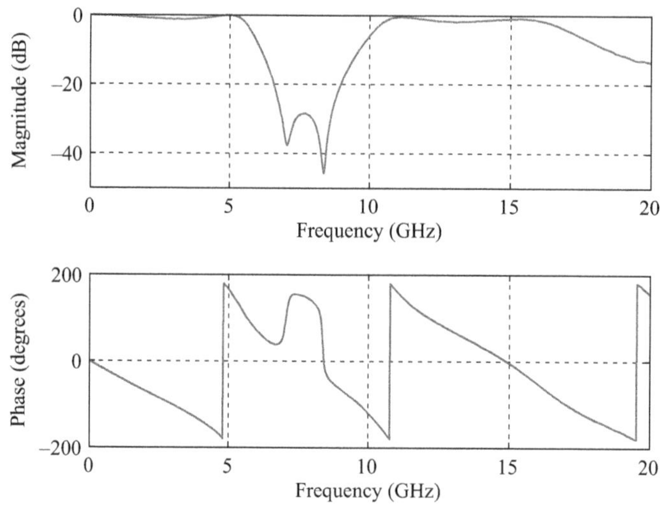

Figure 8.9 S_{21} of the low-pass filter.

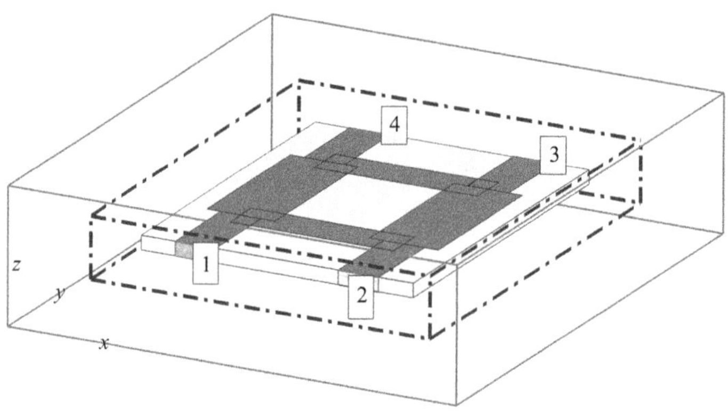

Figure 8.10 An FDTD problem space including a microstrip branch line coupler.

The voltage source with a 50 Ω internal resistance is placed at the input of the circuit indicated as port 1 as shown in Figure 8.10. The outputs of the circuit are terminated by 50 Ω resistors. The definition of the voltage source and resistors is shown in Listing 8.10.

Four sampled voltages and four sampled currents are defined as the outputs of the circuit as shown in Listing 8.11. One should notice that the sampled voltages and currents are defined directly on the voltage source and the termination resistors. Therefore, the reference planes for the S-parameter calculations are at the positions of the source and resistor terminations. Then four 50 Ω ports are defined by associating sampled voltage–current pairs with them. Then port 1 is set to be the excitation port for this FDTD simulation.

After the definition of the problem is completed, the FDTD simulation is performed for 4,000 time steps, and the S-parameters of this circuit are calculated. Figure 8.11 shows the results of this calculation, where S_{11}, S_{21}, S_{31}, and S_{41} are plotted. There is a good agreement between the plotted results and those published in [14].

Listing 8.9 define_geometry.m

```
disp('defining_the_problem_geometry');

bricks  = [];
spheres = [];

% define a substrate
bricks(1).min_x = -20*dx;
bricks(1).min_y = -25*dy;
bricks(1).min_z = 0;
bricks(1).max_x = 20*dx;
bricks(1).max_y = 25*dy;
bricks(1).max_z = 3*dz;
bricks(1).material_type = 4;

% define a PEC plate
bricks(2).min_x = -12*dx;
bricks(2).min_y = -15*dy;
bricks(2).min_z = 3*dz;
bricks(2).max_x = 12*dx;
bricks(2).max_y = -9*dy;
bricks(2).max_z = 3*dz;
bricks(2).material_type = 2;

% define a PEC plate
bricks(3).min_x = -12*dx;
bricks(3).min_y = 9*dy;
bricks(3).min_z = 3*dz;
bricks(3).max_x = 12*dx;
bricks(3).max_y = 15*dy;
bricks(3).max_z = 3*dz;
bricks(3).material_type = 2;

% define a PEC plate
bricks(4).min_x = -17*dx;
bricks(4).min_y = -12*dy;
bricks(4).min_z = 3*dz;
bricks(4).max_x = -7*dx;
bricks(4).max_y = 12*dy;
bricks(4).max_z = 3*dz;
bricks(4).material_type = 2;

% define a PEC plate
bricks(5).min_x = 7*dx;
bricks(5).min_y = -12*dy;
bricks(5).min_z = 3*dz;
bricks(5).max_x = 17*dx;
bricks(5).max_y = 12*dy;
bricks(5).max_z = 3*dz;
bricks(5).material_type = 2;

% define a PEC plate
bricks(6).min_x = -15*dx;
bricks(6).min_y = -25*dy;
bricks(6).min_z = 3*dz;
bricks(6).max_x = -9*dx;
```

```
bricks(6).max_y = -12*dy;
bricks(6).max_z = 3*dz;
bricks(6).material_type = 2;

% define a PEC plate
bricks(7).min_x = 9*dx;
bricks(7).min_y = -25*dy;
bricks(7).min_z = 3*dz;
bricks(7).max_x = 15*dx;
bricks(7).max_y = -12*dy;
bricks(7).max_z = 3*dz;
bricks(7).material_type = 2;

% define a PEC plate
bricks(8).min_x = -15*dx;
bricks(8).min_y = 12*dy;
bricks(8).min_z = 3*dz;
bricks(8).max_x = -9*dx;
bricks(8).max_y = 25*dy;
bricks(8).max_z = 3*dz;
bricks(8).material_type = 2;

% define a PEC plate
bricks(9).min_x = 9*dx;
bricks(9).min_y = 12*dy;
bricks(9).min_z = 3*dz;
bricks(9).max_x = 15*dx;
bricks(9).max_y = 25*dy;
bricks(9).max_z = 3*dz;
bricks(9).material_type = 2;

% define a PEC plate as ground
bricks(10).min_x = -20*dx;
bricks(10).min_y = -25*dy;
bricks(10).min_z = 0;
bricks(10).max_x = 20*dx;
bricks(10).max_y = 25*dy;
bricks(10).max_z = 0;
bricks(10).material_type = 2;
```

Listing 8.10 define_sources_and_lumped_elements.m

```
disp('defining sources and lumped element components');

voltage_sources = [];
current_sources = [];
diodes = [];
resistors = [];
inductors = [];
capacitors = [];

% define source waveform types and parameters
waveforms.gaussian(1).number_of_cells_per_wavelength = 0;
waveforms.gaussian(2).number_of_cells_per_wavelength = 15;
```

```

% voltage sources
% direction: 'xp', 'xn', 'yp', 'yn', 'zp', or 'zn'
% resistance : ohms, magitude   : volts
voltage_sources(1).min_x = -15*dx;
voltage_sources(1).min_y = -25*dy;
voltage_sources(1).min_z = 0;
voltage_sources(1).max_x = -9*dx;
voltage_sources(1).max_y = -25*dy;
voltage_sources(1).max_z = 3*dz;
voltage_sources(1).direction = 'zp';
voltage_sources(1).resistance = 50;
voltage_sources(1).magnitude = 1;
voltage_sources(1).waveform_type = 'gaussian';
voltage_sources(1).waveform_index = 1;

% resistors
% direction: 'x', 'y', or 'z'
% resistance : ohms
resistors(1).min_x = 9*dx;
resistors(1).min_y = -25*dy;
resistors(1).min_z = 0;
resistors(1).max_x = 15*dx;
resistors(1).max_y = -25*dy;
resistors(1).max_z = 3*dz;
resistors(1).direction = 'z';
resistors(1).resistance = 50;

% resistors
% direction: 'x', 'y', or 'z'
% resistance : ohms
resistors(2).min_x = 9*dx;
resistors(2).min_y = 25*dy;
resistors(2).min_z = 0;
resistors(2).max_x = 15*dx;
resistors(2).max_y = 25*dy;
resistors(2).max_z = 3*dz;
resistors(2).direction = 'z';
resistors(2).resistance = 50;

% resistors
% direction: 'x', 'y', or 'z'
% resistance : ohms
resistors(3).min_x = -15*dx;
resistors(3).min_y = 25*dy;
resistors(3).min_z = 0;
resistors(3).max_x = -9*dx;
resistors(3).max_y = 25*dy;
resistors(3).max_z = 3*dz;
resistors(3).direction = 'z';
resistors(3).resistance = 50;
```

Listing 8.11 define_output_parameters.m

```
disp('defining output parameters');

sampled_electric_fields = [];
sampled_magnetic_fields = [];
sampled_voltages = [];
sampled_currents = [];
ports = [];

% figure refresh rate
plotting_step = 5;

% mode of operation
run_simulation = true;
show_material_mesh = true;
show_problem_space = true;

% frequency domain parameters
frequency_domain.start = 20e6;
frequency_domain.end   = 10e9;
frequency_domain.step  = 20e6;

% define sampled voltages
sampled_voltages(1).min_x = -15*dx;
sampled_voltages(1).min_y = -25*dy;
sampled_voltages(1).min_z = 0;
sampled_voltages(1).max_x = -9*dx;
sampled_voltages(1).max_y = -25*dy;
sampled_voltages(1).max_z = 3*dz;
sampled_voltages(1).direction = 'zp';
sampled_voltages(1).display_plot = false;

% define sampled voltages
sampled_voltages(2).min_x = 9*dx;
sampled_voltages(2).min_y = -25*dy;
sampled_voltages(2).min_z = 0;
sampled_voltages(2).max_x = 15*dx;
sampled_voltages(2).max_y = -25*dy;
sampled_voltages(2).max_z = 3*dz;
sampled_voltages(2).direction = 'zp';
sampled_voltages(2).display_plot = false;

% define sampled voltages
sampled_voltages(3).min_x = 9*dx;
sampled_voltages(3).min_y = 25*dy;
sampled_voltages(3).min_z = 0;
sampled_voltages(3).max_x = 15*dx;
sampled_voltages(3).max_y = 25*dy;
sampled_voltages(3).max_z = 3*dz;
sampled_voltages(3).direction = 'zp';
sampled_voltages(3).display_plot = false;
% define sampled voltages
sampled_voltages(4).min_x = -15*dx;
```

```
sampled_voltages(4).min_y = 25*dy;
sampled_voltages(4).min_z = 0;
sampled_voltages(4).max_x = -9*dx;
sampled_voltages(4).max_y = 25*dy;
sampled_voltages(4).max_z = 3*dz;
sampled_voltages(4).direction = 'zp';
sampled_voltages(4).display_plot = false;

% define sampled currents
sampled_currents(1).min_x = -15*dx;
sampled_currents(1).min_y = -25*dy;
sampled_currents(1).min_z = 2*dz;
sampled_currents(1).max_x = -9*dx;
sampled_currents(1).max_y = -25*dy;
sampled_currents(1).max_z = 2*dz;
sampled_currents(1).direction = 'zp';
sampled_currents(1).display_plot = false;

% define sampled currents
sampled_currents(2).min_x = 9*dx;
sampled_currents(2).min_y = -25*dy;
sampled_currents(2).min_z = 2*dz;
sampled_currents(2).max_x = 15*dx;
sampled_currents(2).max_y = -25*dy;
sampled_currents(2).max_z = 2*dz;
sampled_currents(2).direction = 'zp';
sampled_currents(2).display_plot = false;

% define sampled currents
sampled_currents(3).min_x = 9*dx;
sampled_currents(3).min_y = 25*dy;
sampled_currents(3).min_z = 2*dz;
sampled_currents(3).max_x = 15*dx;
sampled_currents(3).max_y = 25*dy;
sampled_currents(3).max_z = 2*dz;
sampled_currents(3).direction = 'zp';
sampled_currents(3).display_plot = false;

% define sampled currents
sampled_currents(4).min_x = -15*dx;
sampled_currents(4).min_y = 25*dy;
sampled_currents(4).min_z = 2*dz;
sampled_currents(4).max_x = -9*dx;
sampled_currents(4).max_y = 25*dy;
sampled_currents(4).max_z = 2*dz;
sampled_currents(4).direction = 'zp';
sampled_currents(4).display_plot = false;

% define ports
ports(1).sampled_voltage_index = 1;
ports(1).sampled_current_index = 1;
ports(1).impedance = 50;
ports(1).is_source_port = true;
```

```
ports(2).sampled_voltage_index = 2;
ports(2).sampled_current_index = 2;
ports(2).impedance = 50;
ports(2).is_source_port = false;

ports(3).sampled_voltage_index = 3;
ports(3).sampled_current_index = 3;
ports(3).impedance = 50;
ports(3).is_source_port = false;

ports(4).sampled_voltage_index = 4;
ports(4).sampled_current_index = 4;
ports(4).impedance = 50;
ports(4).is_source_port = false;

% define animation
% field_type shall be 'e' or 'h'
% plane cut shall be 'xy', yz, or zx
% component shall be 'x', 'y', 'z', or 'm;
animation(1).field_type = 'e';
animation(1).component = 'm';
animation(1).plane_cut(1).type = 'xy';
animation(1).plane_cut(1).position = 2*dz;
animation(1).enable = true;
animation(1).display_grid = false;
animation(1).display_objects = true;

% display problem space parameters
problem_space_display.labels = false;
problem_space_display.axis_at_origin = false;
problem_space_display.axis_outside_domain = false;
problem_space_display.grid_xn = false;
problem_space_display.grid_xp = false;
problem_space_display.grid_yn = false;
problem_space_display.grid_yp = false;
problem_space_display.grid_zn = false;
problem_space_display.grid_zp = false;
problem_space_display.outer_boundaries = false;
problem_space_display.cpml_boundaries = false;
```

8.5.3 Characteristic impedance of a microstrip line

We have demonstrated the use of CPML for two microstrip circuits in the previous examples. These circuits are fed using microstrip lines, the characteristic impedance of which is assumed to be 50 Ω. In this section we provide an example showing the calculation of characteristic impedance of a microstrip line.

We use the same microstrip feeding line as the one used in previous example; the line is 2.4 mm wide, and the dielectric substrate has 2.2 dielectric constant and 0.795 mm thickness. A single line microstrip circuit is constructed in the FDTD method as illustrated in Figure 8.12. The microstrip line is fed with a voltage source. There is a 5 cells air gap on the *yn* and *zp* sides

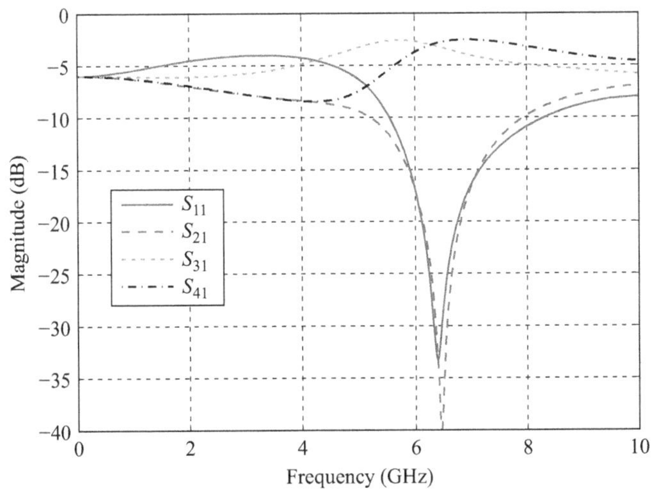

Figure 8.11 S-parameters of the branch line coupler.

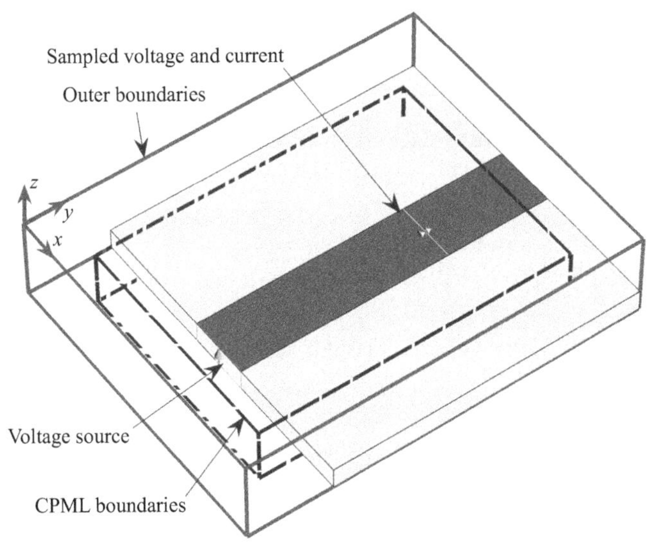

Figure 8.12 An FDTD problem space including a microstrip line.

of the substrate. On the *zn* side of the substrate the boundary is PEC, which serves as the ground plane for the microstrip line. The other sides of the geometry are terminated by CPML. One should notice that the substrate and microstrip line penetrate into the CPML regions on the *yp*, *xn*, and *xp* sides, thus simulating structures extending to infinity in these directions. The definition of the problem space including the cell sizes and boundaries is shown in Listing 8.12. The definition of the geometry is given in Listing 8.13. The microstrip line is fed with voltage source as defined in Listing 8.14. A sampled voltage and a sampled current are defined 10 cells away from the source, and they are tied to form a port as shown in Listing 8.15.

Listing 8.12 define_problem_space_parameters.m

```
disp('defining the problem space parameters');

% maximum number of time steps to run FDTD simulation
number_of_time_steps = 1000;

% A factor that determines duration of a time step
% wrt CFL limit
courant_factor = 0.9;

% A factor determining the accuracy limit of FDTD results
number_of_cells_per_wavelength = 20;

% Dimensions of a unit cell in x, y, and z directions (meters)
dx = 2.030e-4;
dy = 2.030e-4;
dz = 1.325e-4;

% ==<boundary conditions>========
% Here we define the boundary conditions parameters
% 'pec' : perfect electric conductor
% 'cpml' : conlvolutional PML
% if cpml_number_of_cells is less than zero
% CPML extends inside of the domain rather than outwards

boundary.type_xn = 'cpml';
boundary.air_buffer_number_of_cells_xn = 0;
boundary.cpml_number_of_cells_xn = -8;

boundary.type_xp = 'cpml';
boundary.air_buffer_number_of_cells_xp = 0;
boundary.cpml_number_of_cells_xp = -8;

boundary.type_yn = 'cpml';
boundary.air_buffer_number_of_cells_yn = 8;
boundary.cpml_number_of_cells_yn = 8;

boundary.type_yp = 'cpml';
boundary.air_buffer_number_of_cells_yp = 0;
boundary.cpml_number_of_cells_yp = -8;

boundary.type_zn = 'pec';
boundary.air_buffer_number_of_cells_zn = 0;
boundary.cpml_number_of_cells_zn = 8;

boundary.type_zp = 'cpml';
boundary.air_buffer_number_of_cells_zp = 10;
boundary.cpml_number_of_cells_zp = 8;

boundary.cpml_order = 3;
boundary.cpml_sigma_factor = 1;
boundary.cpml_kappa_max = 10;
boundary.cpml_alpha_min = 0;
boundary.cpml_alpha_max = 0.01;
```

Listing 8.13 define_geometry.m

```
disp('defining the problem geometry');

bricks  = [];
spheres = [];

% define a substrate
bricks(1).min_x = 0;
bricks(1).min_y = 0;
bricks(1).min_z = 0;
bricks(1).max_x = 60*dx;
bricks(1).max_y = 60*dy;
bricks(1).max_z = 6*dz;
bricks(1).material_type = 4;

% define a PEC plate
bricks(2).min_x = 24*dx;
bricks(2).min_y = 0;
bricks(2).min_z = 6*dz;
bricks(2).max_x = 36*dx;
bricks(2).max_y = 60*dy;
bricks(2).max_z = 6.02*dz;
bricks(2).material_type = 2;
```

Listing 8.14 define_sources_and_lumped_elements.m

```
disp('defining sources and lumped element components');

voltage_sources = [];
current_sources = [];
diodes = [];
resistors = [];
inductors = [];
capacitors = [];

% define source waveform types and parameters
waveforms.gaussian(1).number_of_cells_per_wavelength = 0;
waveforms.gaussian(2).number_of_cells_per_wavelength = 15;

% voltage sources
% direction: 'xp', 'xn', 'yp', 'yn', 'zp', or 'zn'
% resistance : ohms, magitude   : volts
voltage_sources(1).min_x = 24*dx;
voltage_sources(1).min_y = 0;
voltage_sources(1).min_z = 0;
voltage_sources(1).max_x = 36*dx;
voltage_sources(1).max_y = 0;
voltage_sources(1).max_z = 6*dz;
voltage_sources(1).direction = 'zp';
voltage_sources(1).resistance = 50;
voltage_sources(1).magnitude = 1;
voltage_sources(1).waveform_type = 'gaussian';
voltage_sources(1).waveform_index = 1;
```

Listing 8.15 define_output_parameters.m

```
disp('defining_output_parameters');

sampled_electric_fields = [];
sampled_magnetic_fields = [];
sampled_voltages = [];
sampled_currents = [];
ports = [];

% figure refresh rate
plotting_step = 20;

% mode of operation
run_simulation = true;
show_material_mesh = true;
show_problem_space = true;

% frequency domain parameters
frequency_domain.start = 20e6;
frequency_domain.end   = 10e9;
frequency_domain.step  = 20e6;

% define sampled voltages
sampled_voltages(1).min_x = 24*dx;
sampled_voltages(1).min_y = 40*dy;
sampled_voltages(1).min_z = 0;
sampled_voltages(1).max_x = 36*dx;
sampled_voltages(1).max_y = 40*dy;
sampled_voltages(1).max_z = 6*dz;
sampled_voltages(1).direction = 'zp';
sampled_voltages(1).display_plot = false;

% define sampled currents
sampled_currents(1).min_x = 24*dx;
sampled_currents(1).min_y = 40*dy;
sampled_currents(1).min_z = 6*dz;
sampled_currents(1).max_x = 36*dx;
sampled_currents(1).max_y = 40*dy;
sampled_currents(1).max_z = 6*dz;
sampled_currents(1).direction = 'yp';
sampled_currents(1).display_plot = false;

% define ports
ports(1).sampled_voltage_index = 1;
ports(1).sampled_current_index = 1;
ports(1).impedance = 50;
ports(1).is_source_port = true;
```

An FDTD simulation of the microstrip line is performed for 1,000 time steps. Figure 8.13 shows the source voltage and sampled voltage captured at the port position, whereas Figure 8.14 shows the sampled current. The sampled voltages and currents are transformed to the frequency domain, and S_{11} of the circuit is calculated using these frequency-domain voltage and

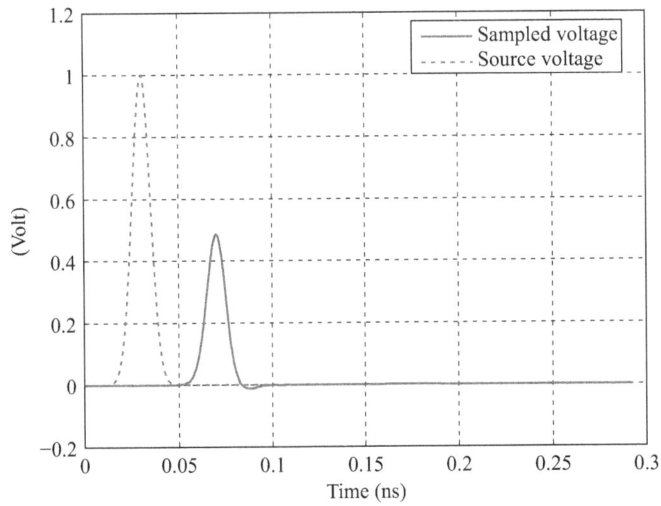

Figure 8.13 Source voltage and sampled voltage of the microstrip line.

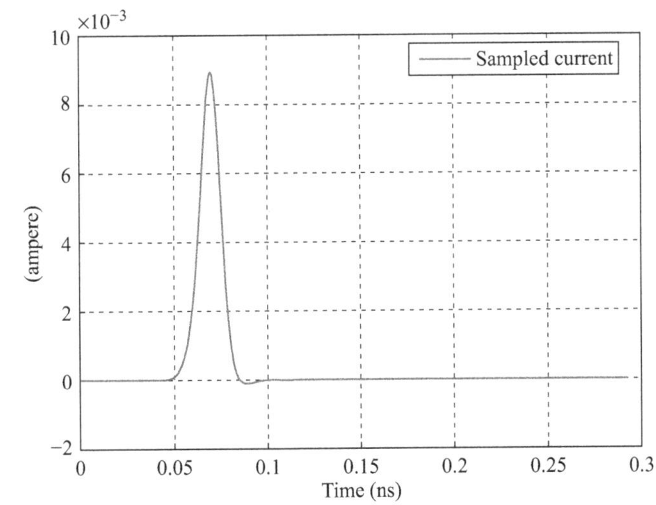

Figure 8.14 Current on the microstrip line.

current values. The S_{11} of the circuit is plotted in Figure 8.15 and indicates that there is better than −35 dB matching of the line to 50 Ω. Furthermore, these frequency-domain values are used to calculate the input impedance of the microstrip line. Since there is no visible reflection observed from the CPML terminated end of the microstrip line, the input impedance of the line is equivalent to the characteristic impedance of the line. The characteristic impedance is plotted in Figure 8.16, which again indicates a good matching to 50 Ω.

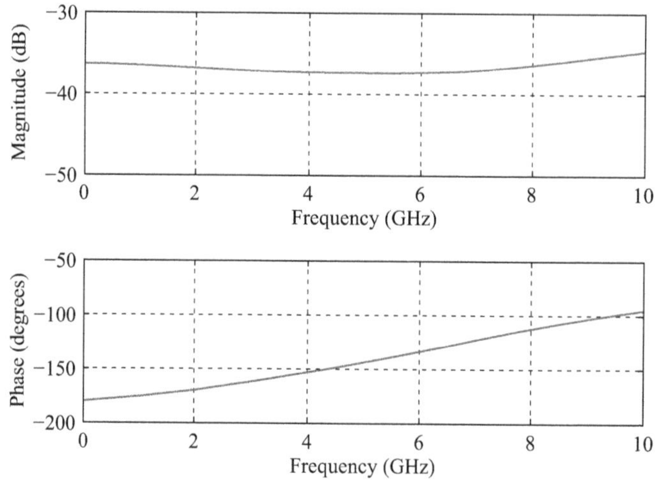

Figure 8.15 S_{11} of the microstrip line.

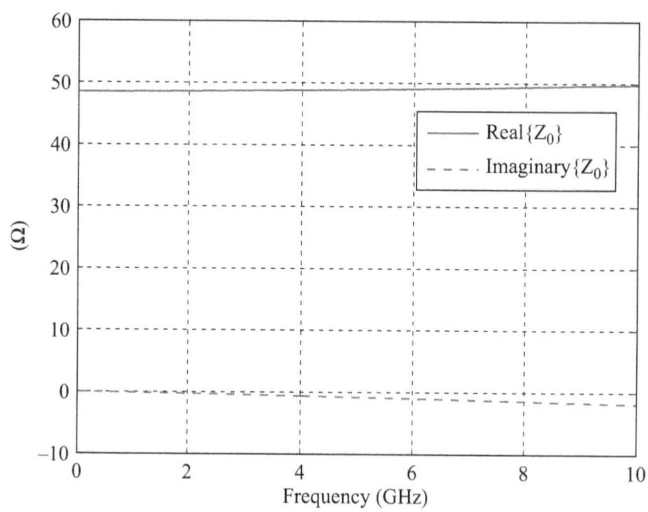

Figure 8.16 Characteristic impedance of the microstrip line.

8.6 CPML in the two-dimensional FDTD method

In Chapter 7, we presented the two-dimensional FDTD method along with the absorbing boundaries based on the PML technique. Since we already have developed the CPML formulations for the three-dimensional FDTD in this chapter, it is straightforward to reduce the CPML formulations to the two-dimensional case. As illustrated in Chapter 7, there is no field and geometric variation in one of the dimensions in a two-dimensional space; any partial derivative with respect to the constant dimension becomes zero. Thus, one needs to start with three-dimensional equations, and just eliminate the respective terms, which become zero, in these equations. For instance, if z is the constant dimension, (8.1) and (8.3) reduce to

$$j\omega\varepsilon_x E_x + \sigma_x^e E_x = \frac{1}{S_{ey}}\frac{\partial H_z}{\partial y}, \tag{8.37a}$$

$$j\omega\varepsilon_y E_y + \sigma_y^e E_y = -\frac{1}{S_{ex}}\frac{\partial H_z}{\partial x}, \tag{8.37b}$$

$$j\omega\varepsilon_z E_z + \sigma_z^e E_z = \frac{1}{S_{ex}}\frac{\partial H_y}{\partial x} - \frac{1}{S_{ey}}\frac{\partial H_x}{\partial y}, \tag{8.37c}$$

and

$$j\omega\mu_x H_x + \sigma_x^m H_x = -\frac{1}{S_{my}}\frac{\partial E_z}{\partial y}, \tag{8.38a}$$

$$j\omega\mu_y H_y + \sigma_y^m H_y = \frac{1}{S_{mx}}\frac{\partial E_z}{\partial x}, \tag{8.38b}$$

$$j\omega\mu_z H_z + \sigma_z^m H_z = -\frac{1}{S_{mx}}\frac{\partial E_y}{\partial x} + \frac{1}{S_{my}}\frac{\partial E_x}{\partial y}, \tag{8.38c}$$

where (8.37a), (8.37b), and (8.38c) constitute the two-dimensional TE_z case, whereas (8.37c), (8.38a), and (8.38b) constitute the TM_z case. One can start with (8.37) and (8.38), follow the steps discussed in Sections 8.1 and 8.2, and obtain the updating equations sought for CPML, however, as a short cut, it is sufficient to start with the three-dimensional CPML updating equations, and just eliminate the terms associated with the partial derivatives of z. For instance, (8.29) can be modified to obtain

$$\begin{aligned} E_x^{n+1}(i,j,k) &= C_{exe}(i,j,k) \times E_x^n(i,j,k) \\ &+ C_{exhz}(i,j,k) \times \left(H_z^{n+\frac{1}{2}}(i,j,k) - H_z^{n+\frac{1}{2}}(i,j-1,k)\right) \\ &+ C_{\psi exy}(i,j,k) \times \psi_{exy}^{n+\frac{1}{2}}(i,j,k). \end{aligned} \tag{8.39}$$

The remaining updating equations can be obtained in a similar fashion.

In Chapter 7 it has been discussed that the two-dimensional updating equations update the fields in partial regions of a problem space because of the split field PML algorithm, and one should be very careful determining which equation updates which partial region. Unlike the two-dimensional PML case, in the CPML algorithm, the main terms of an updating equation is applied to the entire problem space, and the additional CPML parameters, such as $C_{\psi exy}$ and ψ_{exy}, are defined only at their respective CPML regions, and are used to update fields at these regions only. The updating coefficients, such as C_{exhz}, are also modified as in (8.28) at the respective regions. Figures 8.17 and 8.18 illustrate these CPML regions for the TE_z and TM_z cases, respectively. For instance, as shown in Figure 8.17, H_z is first updated in the entire problem space using normal two-dimensional equations, and then additional components using ψ_{hzx} in the xn and xp regions and using ψ_{hzy} in the yn and yp regions are added. E_x and E_y are first updated in the entire problem space, and then E_y is modified using the term

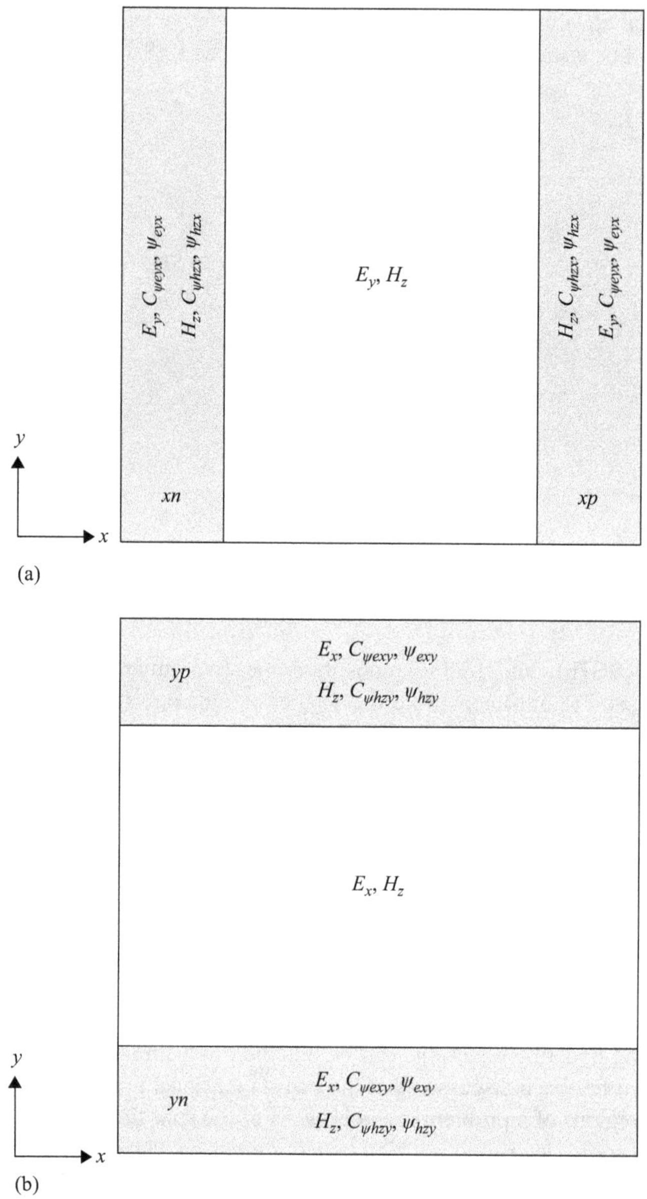

Figure 8.17 Updated fields and CPML parameters in TE_z case: (a) *xn* and *xp* CPML regions and (b) *yn* and *yp* CPML regions.

containing ψ_{eyx} in the *xn* and *xp* regions while E_x is modified using the term containing ψ_{exy} in the *yn* and *yp* regions.

The other CPML parameters (σ_{pei}, σ_{pmi}, κ_{ei}, κ_{mi}, α_{ei}, σ_{mi}), which have specific distribution profiles along the CPML regions, can be calculated using (8.31)–(8.36) as discussed in Section 8.3.

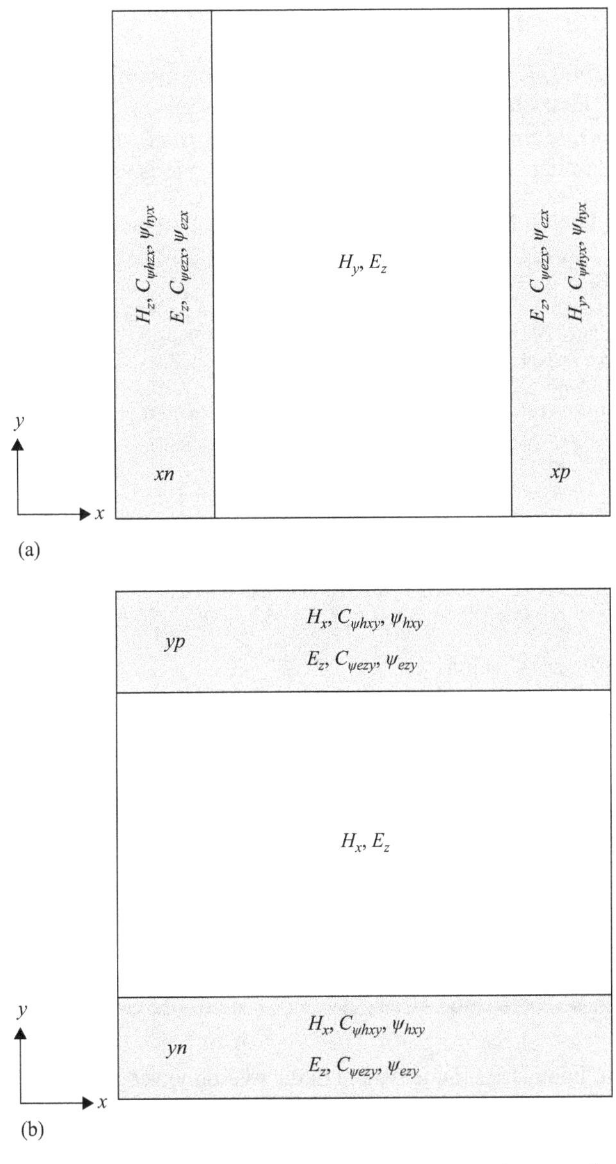

Figure 8.18 Updated fields and CPML parameters in TM_z case: (a) *xn* and *xp* CPML regions and (b) *yn* and *yp* CPML regions.

8.7 MATLAB® implementation of CPML in the two-dimensional FDTD method

We have presented a two-dimensional FDTD code in Chapter 7, where PML is used as the absorbing boundary. Since we have already implemented CPML in the three-dimensional FDTD program in Section 8.4, it is easy to modify the two-dimensional FDTD code to replace PML by CPML by using the code sections from the three-dimensional program.

8.7.1 Definition of CPML

In Listing 7.2, ***define_problem_space_parameters_2d*** is modified to accommodate CPML instead of PML. First, **boundary.type** parameter is set as "*cpml*," then **cpml_order**, **cpml_sigma_factor**, **cpml_kappa_max**, **cpml_alpha_max**, and **cpml_alpha_min** parameters are added similar to Listing 8.1. The modified code is shown in Listing 8.16.

Listing 8.16 define_problem_space_parameters_2d.m

```
% ==<boundary conditions>========
% Here we define the boundary conditions parameters
% 'pec' : perfect electric conductor
% 'cpml' : convolutional PML
% if cpml_number_of_cells is less than zero
% CPML extends inside of the domain rather than outwards

boundary.type_xn = 'cpml';
boundary.air_buffer_number_of_cells_xn = 10;
boundary.cpml_number_of_cells_xn = 8;

boundary.type_xp = 'cpml';
boundary.air_buffer_number_of_cells_xp = 10;
boundary.cpml_number_of_cells_xp = 8;

boundary.type_yn = 'cpml';
boundary.air_buffer_number_of_cells_yn = 10;
boundary.cpml_number_of_cells_yn = 8;

boundary.type_yp = 'cpml';
boundary.air_buffer_number_of_cells_yp = 10;
boundary.cpml_number_of_cells_yp = 8;

boundary.cpml_order = 3;
boundary.cpml_sigma_factor = 1.5;
boundary.cpml_kappa_max = 7;
boundary.cpml_alpha_min = 0;
boundary.cpml_alpha_max = 0.05;
```

Unlike the PML boundaries implemented in the two-dimensional FDTD program presented in Chapter 7, it is allowed to have objects in the problem space to penetrate into the CPML regions to simulate structures extending to infinity in the current CPML implementation. Relevant code for the three-dimensional case can be copied from Listing 8.2 into ***calculate_domain_size_2d*** for the two-dimensional domain size calculations in x and y directions.

8.7.2 Initialization of CPML

Next is the initialization of boundary conditions. Some Boolean parameters that will be used throughout the program execution are initialized in ***initialize_boundary_conditions_2d*** as shown in Listing 8.17.

At the end of Listing 8.17 two separate subroutines are used to initialize the CPML parameters: one for TE_z case and one for TM_z case. Code sections from Listing 8.4 can

Listing 8.17 initialize_boundary_conditions_2d.m

```
% define logical parameters for the conditions that will be used often
is_cpml_xn = false; is_cpml_xp = false;
is_cpml_yn = false; is_cpml_yp = false;
is_any_side_cpml = false;

if strcmp(boundary.type_xn, 'cpml')
    is_cpml_xn = true;
    n_cpml_xn = abs(boundary.cpml_number_of_cells_xn);
end
if strcmp(boundary.type_xp, 'cpml')
    is_cpml_xp = true;
    n_cpml_xp = abs(boundary.cpml_number_of_cells_xp);
end
if strcmp(boundary.type_yn, 'cpml')
    is_cpml_yn = true;
    n_cpml_yn = abs(boundary.cpml_number_of_cells_yn);
end
if strcmp(boundary.type_yp, 'cpml')
    is_cpml_yp = true;
    n_cpml_yp = abs(boundary.cpml_number_of_cells_yp);
end

if (is_cpml_xn || is_cpml_xp || is_cpml_yn || is_cpml_yp)
    is_any_side_cpml = true;
end

% Call CPML initialization routine if any side is CPML
if is_any_side_cpml
    if is_TEz
        initialize_cpml_boundary_conditions_2d_TEz;
    end
    if is_TMz
        initialize_cpml_boundary_conditions_2d_TMz;
    end
end
```

be copied into these subroutines and adopted for the respective two-dimensional case. Listing 8.18 shows a section of such a code modified for the TE_z case. One should notice that parameters such as **Psi_eyx_xn** and **Psi_hzx_xn** are two-dimensional arrays in Listing 8.18 while they are three-dimensional arrays in Listing 8.4.

8.7.3 Application of CPML in the FDTD time-marching loop

Once the initializations are complete, the field updating and boundary updating routines in the FDTD time-marching loop can be modified as follows: the fields need to be updated in the entire problem space. Listing 8.19 illustrates the subroutine that updates the magnetic fields. Magnetic field updates are followed by boundary updates. A section of the code that recalculates CPML parameters **Psi_hzx_xn** and **Psi_hzx_xn** for the current time step and adds their contribution to **Hz** in *xn* and *xp* CPML regions is shown in Listing 8.20.

Listing 8.18 initialize_cpml_boundary_conditions_2d_TEz.m

```
% Initialize CPML boundary condition

p_order = boundary.cpml_order; % order of the polynomial distribution
sigma_ratio = boundary.cpml_sigma_factor;
kappa_max = boundary.cpml_kappa_max;
alpha_min = boundary.cpml_alpha_min;
alpha_max = boundary.cpml_alpha_max;

% Initialize cpml for xn region
if is_cpml_xn

    % define one-dimensional temporary cpml parameter arrays
    sigma_max = sigma_ratio  * (p_order+1)/(150*pi*dx);
    ncells = n_cpml_xn;
    rho_e = ([ncells:-1:1]-0.75)/ncells;
    rho_m = ([ncells:-1:1]-0.25)/ncells;
    sigma_pex_xn = sigma_max * rho_e.^p_order;
    sigma_pmx_xn = sigma_max * rho_m.^p_order;
    sigma_pmx_xn = (mu_0/eps_0) * sigma_pmx_xn;
    kappa_ex_xn = 1 + (kappa_max - 1) * rho_e.^p_order;
    kappa_mx_xn = 1 + (kappa_max - 1) * rho_m.^p_order;
    alpha_ex_xn = alpha_min + (alpha_max - alpha_min) * (1-rho_e);
    alpha_mx_xn = alpha_min + (alpha_max - alpha_min) * (1-rho_m);
    alpha_mx_xn = (mu_0/eps_0) * alpha_mx_xn;

    % define one-dimensional cpml parameter arrays
    cpml_b_ex_xn = exp((-dt/eps_0) ...
        *((sigma_pex_xn./kappa_ex_xn)+ alpha_ex_xn));
    cpml_a_ex_xn = (1/dx)*(cpml_b_ex_xn-1.0).* sigma_pex_xn ...
        ./(kappa_ex_xn.*(sigma_pex_xn+kappa_ex_xn.*alpha_ex_xn));
    cpml_b_mx_xn = exp((-dt/mu_0) ...
        *((sigma_pmx_xn./kappa_mx_xn)+ alpha_mx_xn));
    cpml_a_mx_xn = (1/dx)*(cpml_b_mx_xn-1.0) .* sigma_pmx_xn ...
        ./(kappa_mx_xn.*(sigma_pmx_xn+kappa_mx_xn.*alpha_mx_xn));

    % Create and initialize cpml convolution parameters
    Psi_eyx_xn = zeros(ncells,ny);
    Psi_hzx_xn = zeros(ncells,ny);

    % Create and initialize cpml convolution coefficients
    CPsi_eyx_xn = Ceyhz(2:ncells+1,:)*dx;
    CPsi_hzx_xn = Chzey(1:ncells,:)*dx;

    % Adjust FDTD coefficients in the CPML region
    for i = 1: ncells
        Ceyhz(i+1,:) = Ceyhz(i+1,:)/kappa_ex_xn(i);
        Chzey(i,:) = Chzey(i,:)/kappa_mx_xn(i);
    end

    % Delete temporary arrays. These arrays will not be used any more.
    clear sigma_pex_xn sigma_pmx_xn;
    clear kappa_ex_xn kappa_mx_xn;
    clear alpha_ex_xn alpha_mx_xn;
end
```

Listing 8.19 update_magnetic_fields_2d.m

```
% update magnetic fields
current_time  = current_time + dt/2;

if is_TEz
Hz(1:nx,1:ny) = Chzh(1:nx,1:ny).* Hz(1:nx,1:ny) ...
    + Chzex(1:nx,1:ny) .* (Ex(1:nx,2:nyp1)-Ex(1:nx,1:ny))  ...
    + Chzey(1:nx,1:ny) .*(Ey(2:nxp1,1:ny)-Ey(1:nx,1:ny));
end

if is_TMz
    Hx(1:nxp1,1:ny) = Chxh(1:nxp1,1:ny) .* Hx(1:nxp1,1:ny) ...
        + Chxez(1:nxp1,1:ny) .* (Ez(1:nxp1,2:nyp1)-Ez(1:nxp1,1:ny));

    Hy(1:nx,1:nyp1) = Chyh(1:nx,1:nyp1) .* Hy(1:nx,1:nyp1) ...
        + Chyez(1:nx,1:nyp1) .* (Ez(2:nxp1,1:nyp1)-Ez(1:nx,1:nyp1));
end
```

Listing 8.20 update_magnetic_fields_for_CPML_2d_TEz.m

```
% TEz: apply CPML to magnetic field components
if is_cpml_xn
    for i = 1: n_cpml_xn
        Psi_hzx_xn(i,:,:) = cpml_b_mx_xn(i) * Psi_hzx_xn(i,:,:) ...
            + cpml_a_mx_xn(i)*(Ey(i+1,:,:)-Ey(i,:,:));
    end
        Hz(1:n_cpml_xn,:,:) = Hz(1:n_cpml_xn,:,:) ...
            + CPsi_hzx_xn(:,:,:) .* Psi_hzx_xn(:,:,:);
end

if is_cpml_xp
    n_st = nx - n_cpml_xp;
    for i = 1:n_cpml_xp
        Psi_hzx_xp(i,:,:) = cpml_b_mx_xp(i) * Psi_hzx_xp(i,:,:) ...
            + cpml_a_mx_xp(i)*(Ey(i+n_st+1,:,:)-Ey(i+n_st,:,:));
    end

        Hz(n_st+1:nx,:,:) = Hz(n_st+1:nx,:,:) ...
            + CPsi_hzx_xp(:,:,:) .* Psi_hzx_xp(:,:,:);
end
```

Similarly, electric fields as well are updated in the entire problem space and respective CPML parameters and boundary updates are added to them.

8.7.4 Validation of CPML performance

In this section, we evaluate the performance of the two-dimensional CPML for the TE_z case. The test case that is illustrated in Section 7.5.1 is repeated here except that the PML boundaries are replaced by CPML boundaries. The CPML specific parameters n_{pml}, σ_{factor},

κ_{max}, α_{min}, and α_{max} are set as 3, 1.5, 7, 0, and 0.05, respectively. The problem space is excited by a z-directed impressed magnetic current and a sampled H_z field component is captured at the position $x = 8$ mm and $y = 8$ mm while running the simulation for 1,800 time steps. The captured H_z is then compared with the corresponding results from the case discussed in Section 7.5.1 and errors are obtained in time and frequency domains using (7.41) and (7.42), respectively. These errors are plotted in Figure 8.19(a) and (b) where they are compared with PML errors shown in Figure 7.14. The figures validate the performance

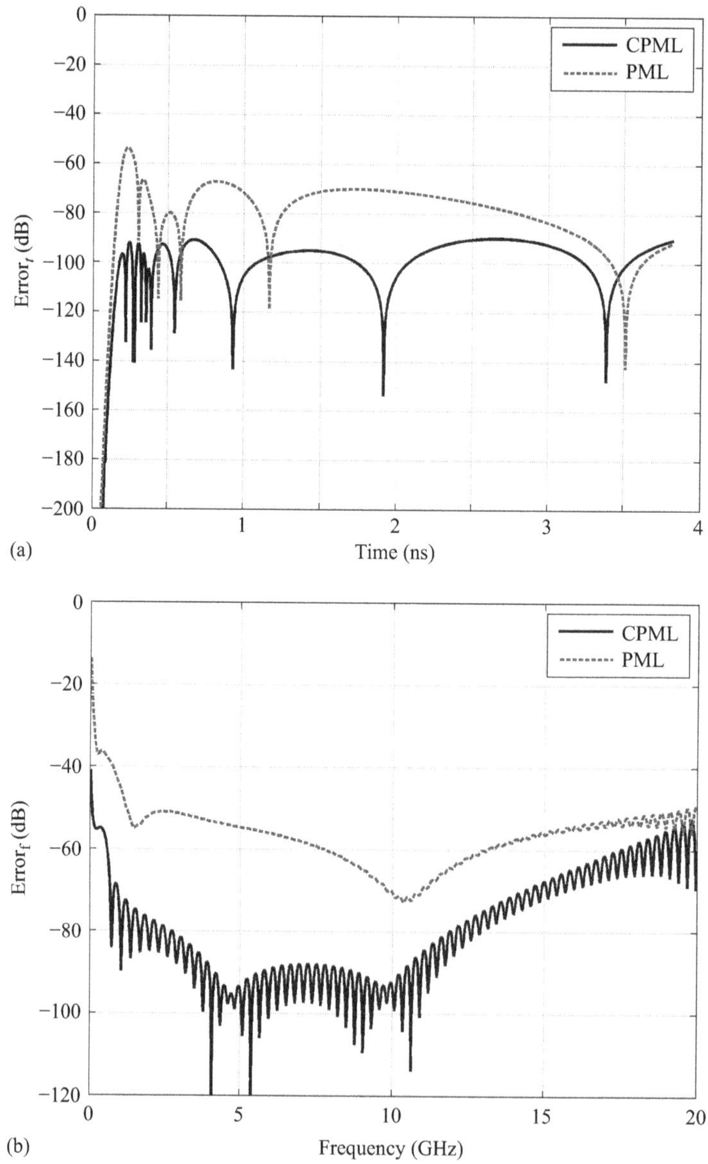

Figure 8.19 CPML and PML errors of sampled H_z in time and frequency domains: (a) errors in time domain and (b) errors in frequency domain.

of the implemented CPML and show the significant improvement CPML has provided over the split field PML in terms of accuracy and ease of implementation.

8.8 Auxiliary differential equation PML

The CPML, discussed above, was introduced by Roden and Gedney [23] as an efficient implementation of the CFS-PML, and has been the most popular implementation used since then. However, derivation of the updating equations for the CPML is very complicated as illustrated in Section 8.1. Recently, another formulation for the CFS-PML has been presented by Gedney and Zhao in [28], which uses an auxiliary differential equation (ADE), thus referred to as ADE-PML. As will be shown in the following section, the final updating equations of the ADE-PML are the same as those of the CPML. Moreover, numerical simulations presented in [28] shows that the ADE-CPML has comparable reflection error as the CPML for the same constitutive parameters. Besides these similarities, the updating equations of the ADE-PML are easier to derive and this approach is presented here to show that alternate formulation can be developed to solve the same problem with the same simulation accuracy.

8.8.1 Derivation of the ADE-PML formulation

Derivation of CPML formulation starts with the PML equations for a lossy medium posed in the stretched coordinate space as shown in (8.1) and (8.3). For instance, repeated here for convenience, (8.1a) is

$$j\omega\varepsilon_x E_x + \sigma_x^e E_x = \frac{1}{S_{ey}}\frac{\partial H_z}{\partial y} - \frac{1}{S_{ez}}\frac{\partial H_y}{\partial z}, \tag{8.40}$$

where S_{ey} is expressed for CFS-PML as

$$S_{ey} = \kappa_{ey} + \frac{\sigma_{pey}}{\alpha_{ey} + j\omega\varepsilon_0}. \tag{8.41}$$

Therefore,

$$\frac{1}{S_{ey}} = \frac{1}{\kappa_{ey}} + \frac{\xi_{ey}}{j\omega + \zeta_{ey}}, \tag{8.42}$$

where

$$\xi_{ey} = \frac{-\sigma_{pey}}{\varepsilon_0\kappa_{ey}^2}, \tag{8.43a}$$

$$\zeta_{ey} = \frac{\sigma_{pey}\kappa_{ey} + \alpha_{ey}\kappa_{ey}^2}{\varepsilon_0\kappa_{ey}^2}. \tag{8.43b}$$

Using (8.42) in (8.40) yields

$$j\omega\varepsilon_x E_x + \sigma_x^e E_x = \frac{1}{\kappa_{ey}}\frac{\partial H_z}{\partial y} + \frac{\xi_{ey}}{j\omega + \zeta_{ey}}\frac{\partial H_z}{\partial y} - \frac{1}{\kappa_{ez}}\frac{\partial H_y}{\partial z} - \frac{\xi_{ez}}{j\omega + \zeta_{ez}}\frac{\partial H_y}{\partial z}. \tag{8.44}$$

When (8.44) is transformed to time domain, the following convolution is obtained for the second term on the right-hand side:

$$\frac{\xi_{ey}}{j\omega + \zeta_{ey}} \frac{\partial H_z}{\partial y} \Rightarrow \xi_{ey} e^{-\zeta_{ey} t} * \frac{\partial H_z}{\partial y} \tag{8.45}$$

Direct convolution is costly if programmed as is, however recursive convolution makes the convolution computationally convenient. Thus, the discrete recursive convolution is proposed to handle this convolution, hence the CPML is used as shown in the previous sections.

An alternate approach can be developed as follows. One can notice that the term in (8.45) is an additional term in (8.44). We can denote this additional term as

$$\psi_{exy} = \frac{\xi_{ey}}{j\omega + \zeta_{ey}} \frac{\partial H_z}{\partial y}, \tag{8.46}$$

the same way as shown in (8.10) or (8.16). In discrete time, this additional term ψ_{exy} has to be calculated at every time step and added to the E_x^{n+1} as in (8.16).

Equation (8.46) can be arranged as

$$j\omega\psi_{exy} + \zeta_{ey}\psi_{exy} = \xi_{ey} \frac{\partial H_z}{\partial y}, \tag{8.47}$$

and transformed to time domain as

$$\frac{\partial}{\partial t}\psi_{exy} + \zeta_{ey}\psi_{exy} = \xi_{ey} \frac{\partial H_z}{\partial y}. \tag{8.48}$$

One should notice that when representing (8.44) in discrete time domain the time derivative of E_x will be expressed in terms of finite difference about the time instant $(n + 0.5)\Delta t$, therefore the value of $\psi_{exy}^{n+0.5}$ is needed to be added to the future value of E_x^{n+1} in the final updating equation. It is also important to decide on the finite difference approximation for the partial derivative of time in (8.48) when transforming to discrete time domain. One choice is to use backward difference with $n + 0.5$ being the reference, as done in [28], and obtain

$$\frac{\psi_{exy}^{n+0.5} - \psi_{exy}^{n-0.5}}{\Delta t} + \zeta_{ey}\psi_{exy}^{n+0.5} = \frac{\xi_{ey}}{\Delta y}\left(H_z^{n+0.5}(i, j, k) - H_z^{n+0.5}(i, j-1, k)\right), \tag{8.49}$$

which can be rearranged such that

$$\psi_{exy}^{n+0.5}(i, j, k) = b_{ey}\psi_{exy}^{n-0.5}(i, j, k) + a_{ey}\left(H_z^{n+0.5}(i, j, k) - H_z^{n+0.5}(i, j-1, k)\right), \tag{8.50}$$

where

$$b_{ey} = \frac{1}{\left(1 + \Delta t\zeta_{ey}\right)}, \tag{8.51a}$$

$$a_{ey} = \frac{\Delta t\xi_{ey}}{\Delta y} \frac{1}{\left(1 + \Delta t\zeta_{ey}\right)}. \tag{8.51b}$$

The terms b_{ey} and a_{ey} can be written in open form as

$$b_{ey} = \frac{\varepsilon_0 \kappa_{ey}}{\varepsilon_0 \kappa_{ey} + \Delta t\left(\sigma_{pey} + \alpha_{ey}\kappa_{ey}\right)}, \tag{8.52a}$$

$$a_{ey} = \frac{-\sigma_{pey}\Delta t}{\kappa_{ey}\Delta y\left(\varepsilon_0 \kappa_{ey} + \Delta t\left(\sigma_{pey} + \alpha_{ey}\kappa_{ey}\right)\right)}. \tag{8.52b}$$

The form of the equation to update $\psi_{exy}^{n+0.5}$ at every time step of the FDTD time-marching loop is derived through the ADE approach as (8.50), which is the same as (8.21) that is derived through the discrete convolution approach; the difference is in the terms b_{ey} and a_{ey}.

8.8.2 MATLAB® implementation of the ADE-PML formulation

As illustrated, the only difference between the CPML and the ADE-PML is the terms b_{ei}, a_{ei}, b_{mi} and a_{mi}, and besides that everything is the same. Since the MATLAB implementation of CPML is already available, as discussed in Section 8.4, it only requires to redefine these terms following (8.52) to convert a CPML code to ADE-PML code. For instance, the associated one-dimensional arrays are defined as **cpml_b_ex_xn**, **cpml_a_ex xn**, **cpml_b_mx_xn**, and **cpml_a_mx_xn**, respectively, on lines 26–34 in Listing 8.4 for the *xn* boundary of the problem space. These lines can be modified as shown in Listing 8.21 to obtain the ADE-PML code. Here the one-dimensional arrays are renamed to reflect the ADE-PML as **ade_pml_b_ex_xn**, **ade_pml_a_ex xn**, **ade_pml_b_mx_xn**, and **ade_pml_a_mx_xn**.

Listing 8.21 initialize_ADE_PML_ABC.m

```
% define one-dimensional ade_pml parameter arrays
ade_pml_b_ex_xn = (eps_0*kappa_ex_xn) ...
   ./(eps_0*kappa_ex_xn+dt*(sigma_pex_xn+alpha_ex_xn.*kappa_ex_xn));
ade_pml_a_ex_xn = (-dt*sigma_pex_xn)./(dy*kappa_ex_xn ...
                 .* (eps_0*kappa_ex_xn+dt*(sigma_pex_xn+alpha_ex_xn.*kappa_ex_xn)));
ade_pml_b_mx_xn = (mu_0*kappa_mx_xn) ...
   ./(mu_0*kappa_mx_xn+dt*(sigma_pmx_xn+alpha_mx_xn.*kappa_mx_xn));
ade_pml_a_mx_xn = (-dt*sigma_pmx_xn)./(dy*kappa_mx_xn ...
                 .* (mu_0*kappa_mx_xn+dt*(sigma_pmx_xn+alpha_mx_xn.*kappa_mx_xn)));
```

8.9 Exercises

8.1 Consider the quarter-wave transformer circuit in Section 6.3.1. Remove the absorbers from the circuit, and use 8 cells thick CPML boundaries instead. Leave 5 cells air gap between the substrate and the CPML on the *xn*, *xp*, *yn*, and *yp* sides. Leave 10 cells air gap on the *zp* side. The geometry of the problem is illustrated in Figure 8.20 as a reference. Run the simulation, obtain the S-parameters of the circuit, and compare the results with those shown in Section 6.3.1.

8.2 Consider the quarter-wave transformer circuit in Exercise 8.1. Redefine the CPML absorbing boundary condition (ABC) such that the substrate penetrates into the CPML on the *xn* and *xp* sides. The geometry of the problem is illustrated in Figure 8.21. Run the simulation, obtain the S-parameters of the circuit, and compare the results with those you obtained for Exercise 8.1.

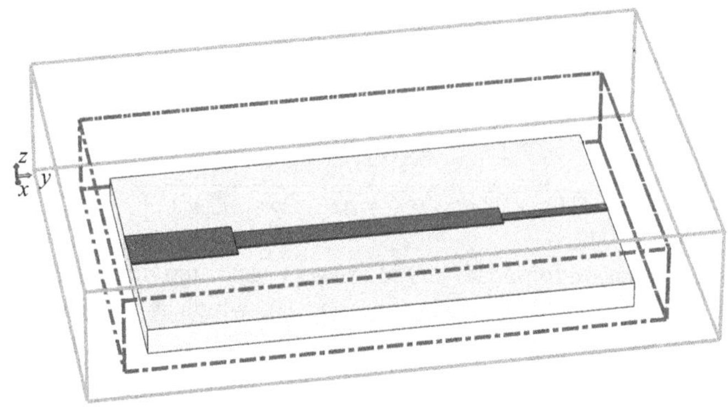

Figure 8.20 The problem space for the quarter-wave transformer with CPML boundaries on the *xn*, *xp*, *yn*, *yp*, and *zp* sides.

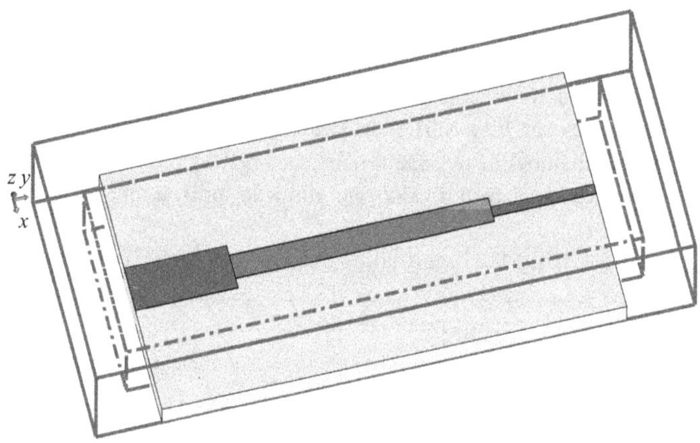

Figure 8.21 The problem space for the quarter-wave transformer with the substrate penetrating into the CPML boundaries on the *xn* and *xp* sides.

8.3 Consider the microstrip line circuit in Section 8.5.3. In this example, the performance of the CPML is not examined numerically. We need a reference case to examine the CPML performance. Extend the length of the circuit twice on the *yp* side, and change the boundary type on that side to PEC such that the microstrip line is touching the PEC boundary. The geometry of the problem is illustrated in Figure 8.22. Run the simulation a number of time steps such that the reflected pulse from the PEC termination will not be observed at the sampled voltage. This simulation result serves as the reference case since it does not include any reflected signals as if the microstrip line is infinitely long. Compare the sampled voltage that you obtained from the circuit in Section 8.5.3 with the reference case. The difference between the sampled voltage magnitudes is the reflection. Find the maximum value of the reflection.

8.4 In Exercise 8.3 you constructed a reference case for examining the CPML performance. Change the thickness of the CPML to 5 cells on the *yp* side in the microstrip

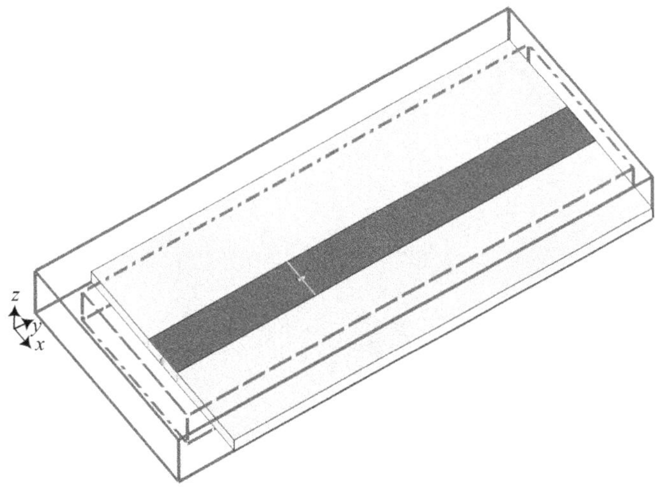

Figure 8.22 The microstrip line reference case.

line circuit in Section 8.5.3. Run the simulation, and then calculate the reflected voltage as described in Exercise 8.3. Examine whether the reflection increases or decreases. Then repeat the same exercise by increasing the CPML thickness to 10 cells. Examine whether the reflection increases or decreases.

8.5 The CPML parameters σ_{factor}, $\kappa_{\max}$, $\alpha_{\max}$, $\alpha_{\min}$, and n_{pml} used in the example in Section 8.5.3 are not necessarily the optimum parameters for the CPML performance. Change the values of some of these parameters, and examine whether the reflections from the 8 cell thick CPML boundary increases or decreases. For instance, try $\sigma_{factory} = 1.5$, $\kappa_{\max} = 11$, $\alpha_{\min} = 0$, $\alpha_{\max} = 0.02$, and $n_{pml} = 3$.

CHAPTER 9

Near-field to far-field transformation

In previous chapters, the finite-difference time-domain (FDTD) method is used to compute electric and magnetic fields within a finite space around an electromagnetic object (i.e., the near-zone electromagnetic fields). In many applications, such as antennas and radar cross-section, it is necessary to find the radiation or scattered fields in the region that is far away from an antenna or scatterer. With the FDTD technique, the direct evaluation of the far field calls for an excessively large computational domain, which is not practical in applications. Instead, the far-zone electromagnetic fields are computed from the near-field FDTD data through a near-field to far-field (NF–FF) transformation technique [1].

A simple condition for the far field is defined as follows:

$$kR \gg 1 \Rightarrow \frac{2\pi R}{\lambda} \gg 1, \tag{9.1}$$

where R is the distance from radiator to the observation point, k is the wavenumber in free space, and λ is the wavelength. For an electrically large antenna such as a parabolic reflector, the aperture size D is often used to determine the far-field condition [29]:

$$r > \frac{2D^2}{\lambda}, \tag{9.2}$$

where r is the distance from the center of the antenna aperture to the observation point. In the far-field region, the electromagnetic field at an observation point (r, θ, ϕ) can be expressed as

$$\vec{E}(r, \theta, \phi) = \frac{e^{-jkr}}{4\pi r}\vec{F}(\theta, \phi), \tag{9.3a}$$

$$\vec{H} = \hat{r} \times \frac{\vec{E}}{\eta_0}, \tag{9.3b}$$

where η_0 is the wave impedance of free space, and $\vec{F}(\theta, \phi)$ is a term determining the angular variations of the far-field pattern of the electric field. Thus, the radiation pattern of the antenna is only a function of the angular position (θ, ϕ) and is independent of the distance r.

In general, the near-field to far-field transformation technique is implemented in a two-step procedure. First, an imaginary surface is selected to enclose the antenna, as shown in Figure 9.1. The currents $\vec{J}$ and $\vec{M}$ on the surface are determined by the computed $\vec{E}$ and $\vec{H}$ fields inside the computational domain. According to the equivalence theorem, the radiation field from the currents is equivalent to the radiation field from the antenna. Next, the vector potentials $\vec{A}$ and $\vec{F}$ are used to compute the radiation fields from the equivalent currents $\vec{J}$ and $\vec{M}$. The far-field conditions are used in the derivations to obtain the appropriate analytical formulas.

Compared with the direct FDTD simulation that requires a mesh extending many wavelengths from the object, a much smaller FDTD mesh is needed to evaluate the equivalent currents $\vec{J}$ and $\vec{M}$. Thus, it is much more computationally efficient to use this near-field to far-field transformation technique.

According to different computation objectives, the transformation technique can be applied in both the time and frequency domains, as shown in Figure 9.2. When transient or broadband frequency-domain results are required at a limited number of observation angles, the left path in Figure 9.2 is adopted. For these situations, the time-domain transformation is used and the transient far-zone fields at each angle of interest are stored while updating the field components [30,31].

In contrast, when the far fields at all observation angles are required for a limited number of frequencies, the right path in Figure 9.2 is adopted. For each frequency of interest a running discrete Fourier transform (DFT) of the tangential fields (surface currents) on a closed surface is updated at each time step. The complex frequency-domain currents obtained from the DFT are then used to compute the far-zone fields at all observation angles through the frequency-domain transformation.

This chapter is organized as follows. First, the surface equivalence theorem is discussed, and the equivalent currents $\vec{J}$ and $\vec{M}$ are obtained from the near-field FDTD data. Then the important radiation formulas are presented to calculate the far fields from the equivalent currents. The implementation procedure is illustrated with full details. Finally, two antenna examples are provided to demonstrate the validity of the near- to far-field technique.

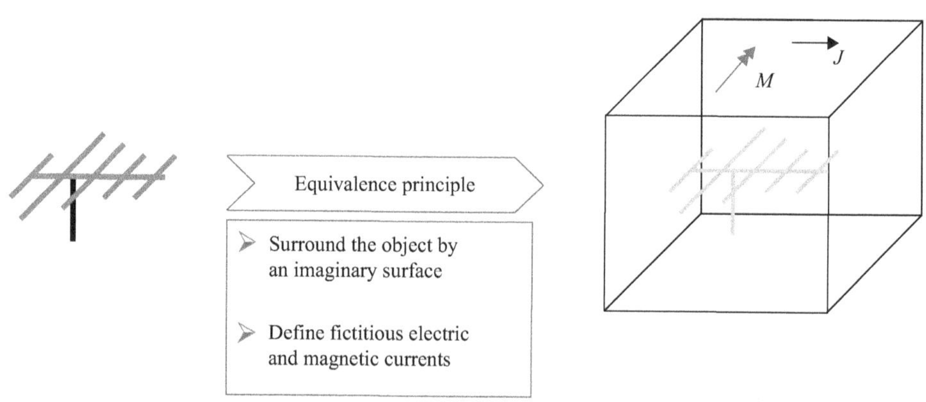

Figure 9.1 Near-field to far-field transformation technique: equivalent currents on an imaginary surface.

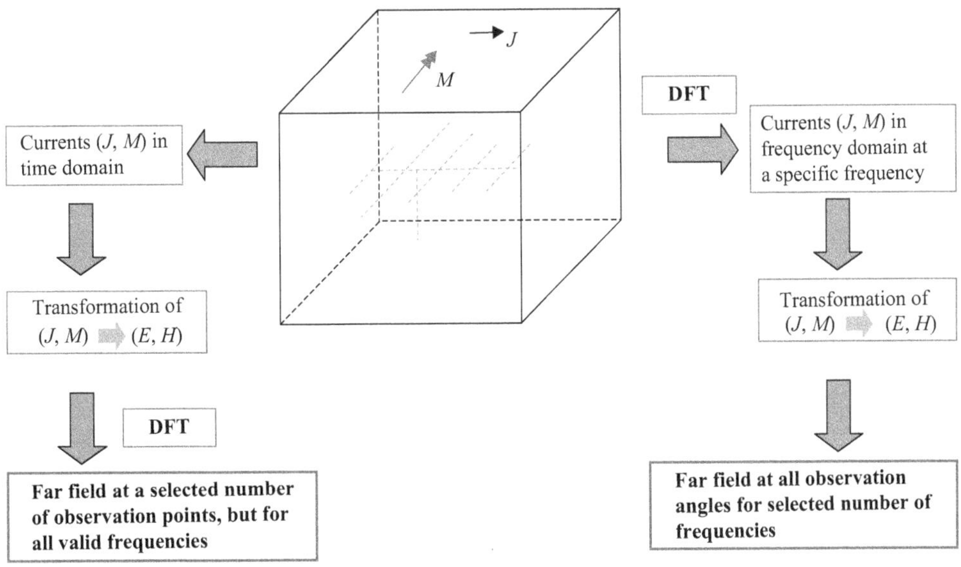

Figure 9.2 Two paths of the near-field to far-field transformation technique are implemented to achieve different computation objectives.

9.1 Implementation of the surface equivalence theorem

9.1.1 Surface equivalence theorem

The surface equivalence theorem was introduced in 1936 by Sckelkunoff [32] and is now widely used in electromagnetic and antenna problems [33]. The basic idea is to replace the actual sources such as antennas or scatterers with fictitious surface currents on a surrounding closed surface.

Within a specific region, the fields generated by the fictitious currents represent the original fields.

Figure 9.3 illustrates a typical implementation of the surface equivalent theorem. Assume that fields generated by an arbitrary source are $(\vec{E}, \vec{H})$. An imaginary surface S is selected to enclose *all* the sources and scattering objects, as shown in Figure 9.3(a). Outside the surface S is only free space. An equivalent problem is set up in Figure 9.3(b), where the fields outside the surface S remain the same but inside the surface S are set to zero. It is obvious that this setup is feasible because the fields satisfy Maxwell's equations both inside and outside the surface S. To comply with the boundary conditions on the surface, equivalent surface currents must be introduced on S:

$$\vec{J}_S = \hat{n} \times \left(\vec{H}^{out} - \vec{H}^{in} \right) = \hat{n} \times \vec{H}, \tag{9.4a}$$

$$\vec{M}_S = -\hat{n} \times \left(\vec{E}^{out} - \vec{E}^{in} \right) = -\hat{n} \times \vec{E}. \tag{9.4b}$$

It is worthwhile to emphasize that Figures 9.3(a) and 9.3(b) have the same fields outside the surface S but different fields inside the surface S.

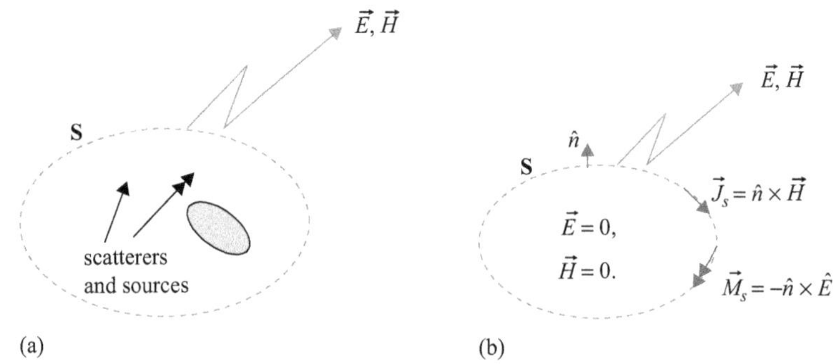

Figure 9.3 Surface equivalence theorem: (a) original problem and (b) equivalent problem for region outside *S*.

If the field values on the surface *S* in the original problem Figure 9.3(a) can be accurately obtained by some means, the surface currents in Figure 9.3(b) can be determined from (9.4). Then the fields at any arbitrary far observation point in Figure 9.3(b) can be readily calculated from the vector potential approach. Based on the uniqueness theorem, the computed fields are the only solution of the problem in Figure 9.3(b). According to the relations between Figures 9.3(a) and 9.3(b), the computed fields outside the surface *S* are also the solution for the original problem.

The implementation of the surface equivalence theorem simplifies the far-field calculation. In the original Figure 9.3(a) problem, materials with different permittivities and permeabilities may exist inside the surface *S*. Thus, a complex Green's function needs to be derived to calculate the radiating field. In the problem in Figure 9.3(b), the fields inside the surface are zero, and the permittivity and permeability can be set the same as the outside free space. Hence, the simple free space Green's function is used to compute the radiating field.

9.1.2 Equivalent surface currents in FDTD simulation

From the previous discussion, it is clear that the key point of the equivalence theorem implementation is to accurately obtain the equivalent currents on the imaginary surface *S*. In the FDTD simulation, the surface currents can be readily computed from the following procedure.

First, a closed surface is selected around the antennas or scatterers, as shown in Figure 9.4. The selected surface is usually a rectangular box that fits the FDTD grid. It is set between the analyzed objects and the outside absorbing boundary. The location of the box can be defined by two corners: lowest coordinate (*li*, *lj*, *lk*) corner and upper coordinate (*ui*, *uj*, *uk*) corner. It is critical that all the antennas or scatterers must be enclosed by this rectangular box so that the equivalent theorem can be implemented. It is also important to have this box in the air buffer area between all objects and the first interface of the absorbing boundary.

Once the imaginary closed surface is selected, the equivalent surface currents are computed next. There are six surfaces of the rectangular box, and each surface has four

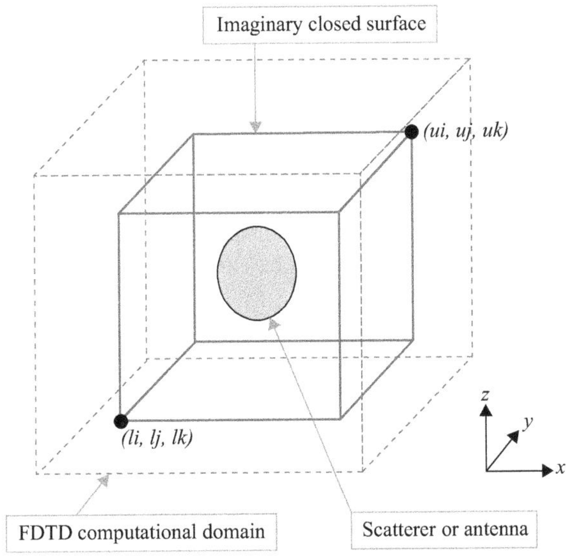

Figure 9.4 An imaginary surface is selected to enclose the antennas or scatterers.

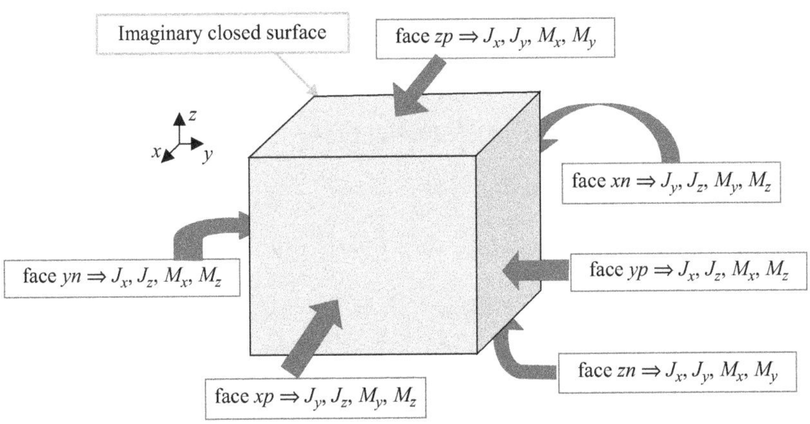

Figure 9.5 Equivalent surface currents on the imaginary closed surface.

scalar electric and magnetic currents, as shown in Figure 9.5. For the top surface, the normal direction is $\hat{z}$. From (9.4), the equivalent surface currents are calculated as

$$\vec{J}_S = \hat{z} \times \vec{H} = \hat{z} \times (\hat{x}H_x + \hat{y}H_y + \hat{z}H_z) = -\hat{x}H_y + \hat{y}H_x, \tag{9.5a}$$

$$\vec{M}_S = -\hat{z} \times \vec{E} = -\hat{z} \times (\hat{x}E_x + \hat{y}E_y + \hat{z}E_z) = \hat{x}E_y - \hat{y}E_x. \tag{9.5b}$$

Thus, the scalar surface currents can be obtained:

$$\vec{J}_S = \hat{x}J_x + \hat{y}J_y \Rightarrow J_x = -H_y, \quad J_y = H_x, \tag{9.6a}$$

$$\vec{M}_S = \hat{x}M_x + \hat{y}M_y \Rightarrow M_x = E_y, \quad M_y = -E_x. \tag{9.6b}$$

Note that the E and H fields used in (9.6) are computed from the FDTD simulation. For a time-domain far-field calculation, the time-domain data are used directly. For a frequency-domain far-field calculation, a DFT needs to be carried out to obtain the desired frequency components of the fields.

Similar methodology is used to obtain the surface currents on the other five surfaces. On the bottom surface,

$$\vec{J}_S = \hat{x}J_x + \hat{y}J_y \Rightarrow J_x = H_y, \quad J_y = -H_x, \tag{9.7a}$$

$$\vec{M}_S = \hat{x}M_x + \hat{y}M_y \Rightarrow M_x = -E_y, \quad M_y = E_x. \tag{9.7b}$$

On the left surface,

$$\vec{J}_S = \hat{x}J_x + \hat{z}J_z \Rightarrow J_x = -H_z, \quad J_z = H_x, \tag{9.8a}$$

$$\vec{M}_S = \hat{x}M_x + \hat{z}M_z \Rightarrow M_x = E_z, \quad M_z = -E_x. \tag{9.8b}$$

On the right surface,

$$\vec{J}_S = \hat{x}J_x + \hat{z}J_z \Rightarrow J_x = H_z, \quad J_z = -H_x, \tag{9.9a}$$

$$\vec{M}_S = \hat{x}M_x + \hat{z}M_z \Rightarrow M_x = -E_z, \quad M_z = E_x. \tag{9.9b}$$

On the front surface,

$$\vec{J}_S = \hat{y}J_y + \hat{z}J_z \Rightarrow J_y = -H_z, \quad J_z = H_y, \tag{9.10a}$$

$$\vec{M}_S = \hat{y}M_y + \hat{z}M_z \Rightarrow M_y = E_z, \quad M_z = -E_y. \tag{9.10b}$$

On the back surface,

$$\vec{J}_S = \hat{y}J_y + \hat{z}J_z \Rightarrow J_y = H_z, \quad J_z = -H_y, \tag{9.11a}$$

$$\vec{M}_S = \hat{y}M_y + \hat{z}M_z \Rightarrow M_y = -E_z, \quad M_z = E_y. \tag{9.11b}$$

To obtain complete source currents for the far-field calculation, (9.6)–(9.11) must be calculated at every FDTD cell on the equivalent closed surface. It is preferable that the magnetic and electric currents should be located at the same position, namely, the center of each Yee's cell surface that touches the equivalent surface S. Because of the spatial offset between the components of the E and H field locations on Yee's cell, averaging of the field components may be performed to obtain the value of the current components at the center location. The obtained surface currents are then used to compute the far-field pattern in the next section.

9.1.3 Antenna on infinite ground plane

It was mentioned earlier that the imaginary surface must surround all the scatterers or antennas so that the equivalent currents radiate in a homogeneous medium, usually free space. For many antenna applications, the radiator is mounted on a large or infinite ground plane. In this situation, it is impractical to select a large surface to enclose the ground plane. Instead, the selected surface that encloses the radiator lies on top of the ground plane, and the ground plane effect is considered using image theory, as shown in Figure 9.6. The image currents have the same direction for horizontal magnetic current and vertical electric current. In contrast, the image currents flow along the opposite direction for horizontal electric current and vertical magnetic currents. This arrangement is valid for ground planes simulating infinite perfect electric conductor (PEC).

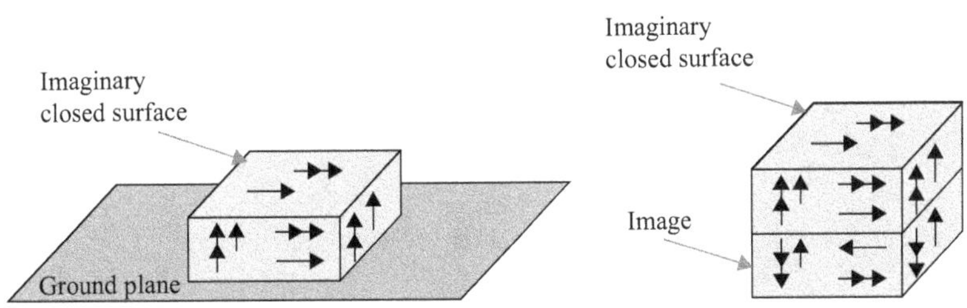

Figure 9.6 An imaginary closed surface on an infinite PEC ground plane using the image theory.

9.2 Frequency domain near-field to far-field transformation

In this section, the obtained equivalent surface currents are used to calculate the far-field radiation patterns. Antenna polarization and radiation efficiency are also discussed.

9.2.1 Time-domain to frequency-domain transformation

This section focuses on the frequency-domain far-field calculation. For the time-domain analysis, readers are suggested to study [30, 31]. The first thing in the frequency-domain calculation is to convert the time-domain FDTD data into frequency-domain data using the DFT. For example, the surface current J_y in (9.6) can be calculated as follows:

$$J_y(u, v, w; f_1) = H_x(u, v, w; f_1) = \sum_{n=1}^{N_{steps}} H_x(u, v, w; n)e^{-j2\pi f_1 n\Delta t}\Delta t. \tag{9.12}$$

Here (u, v, w) is the index for the space location, and n is the time step index. N_{steps} is the maximum number of the time steps used in the time-domain simulation. Similar formulas can be applied in calculating other surface currents in (9.6)–(9.11). Therefore, the frequency-domain patterns are calculated after all the time steps of the FDTD computation is finished. For a cubic imaginary box with $N \times N$ cells on each surface, the total required storage size for the surface currents is $4 \times 6 \times N^2$. Note that the frequency-domain data in (9.12) have complex values.

If radiation patterns at multiple frequencies are required, a pulse excitation is used in the FDTD simulation. For each frequency of interest, the DFT in (9.12) is performed with the corresponding frequency value. One round of FDTD simulation is capable to provide surface currents and radiation patterns at multiple frequencies.

9.2.2 Vector potential approach

For radiation problems, a vector potential approach is well developed to compute the unknown far fields from the known electric and magnetic currents [33]. A pair of vector potential functions is defined as

$$\vec{A} = \frac{\mu_0 e^{-jkR}}{4\pi R}\vec{N}, \tag{9.13a}$$

$$\vec{F} = \frac{\varepsilon_0 e^{-jkR}}{4\pi R}\vec{L}, \tag{9.13b}$$

where

$$\vec{N} = \int_S \vec{J}_S e^{-jkr'\cos(\psi)} dS', \tag{9.14a}$$

$$\vec{L} = \int_S \vec{M}_S e^{-jkr'\cos(\psi)} dS'. \tag{9.14b}$$

As illustrated in Figure 9.7, the vector $\vec{r} = r\hat{r}$ denotes the position of the observation point (x, y, z), whereas the vector $\vec{r}' = r'\hat{r}'$ denotes the position of source point (x', y', z') on the surface S. The vector $\vec{R} = R\hat{R}$ is between the source point and the observation point, and

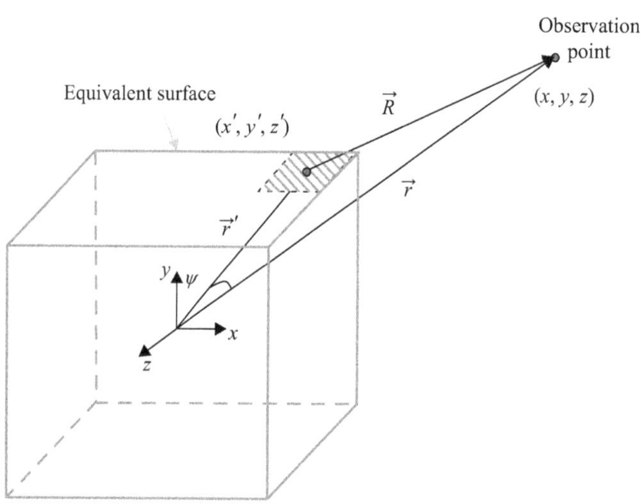

Figure 9.7 The equivalent surface current source and far field.

the angle ψ represents the angle between $\vec{r}$ and $\vec{r}'$. In the far-field calculation, the distance R is approximated by

$$R = \sqrt{r^2 + (r')^2 - 2rr'\cos(\psi)} = \begin{cases} r - r'\cos(\psi) & \text{for the phase term} \\ r & \text{for the amplitude term} \end{cases} \tag{9.15}$$

The computation of the components of E and H in the far fields can then be obtained using the vector potentials, which are expressed as

$$E_r = 0, \tag{9.16a}$$

$$E_\theta = -\frac{jke^{-jkr}}{4\pi r}\left(L_\phi + \eta_0 N_\theta\right), \tag{9.16b}$$

$$E_\phi = +\frac{jke^{-jkr}}{4\pi r}\left(L_\theta - \eta_0 N_\phi\right), \tag{9.16c}$$

$$H_r = 0, \tag{9.16d}$$

$$H_\theta = +\frac{jke^{-jkr}}{4\pi r}\left(N_\phi - \frac{L_\theta}{\eta_0}\right), \tag{9.16e}$$

$$H_\phi = -\frac{jke^{-jkr}}{4\pi r}\left(N_\theta + \frac{L_\phi}{\eta_0}\right). \tag{9.16f}$$

When the closed surface S is chosen as in Figure 9.4, the equivalent surface currents are computed based on (9.6)–(9.11), and the DFT in (9.12) is performed to obtain frequency-domain components; the auxiliary functions N and L are calculated as

$$N_\theta = \int_S \left(J_x\cos(\theta)\cos(\phi) + J_y\cos(\theta)\sin(\phi) - J_z\sin(\theta)\right)e^{jkr'\cos(\psi)}dS', \tag{9.17a}$$

$$N_\phi = \int_S \left(-J_x\sin(\phi) + J_y\cos(\phi)\right)e^{jkr'\cos(\psi)}dS', \tag{9.17b}$$

$$L_\theta = \int_S \left(M_x\cos(\theta)\cos(\phi) + M_y\cos(\theta)\sin(\phi) - M_z\sin(\theta)\right)e^{jkr'\cos(\psi)}dS', \tag{9.17c}$$

$$L_\phi = \int_S \left(-M_x\sin(\phi) + M_y\cos(\phi)\right)e^{jkr'\cos(\psi)}dS'. \tag{9.17d}$$

Substituting (9.17) into (9.16), the far-field pattern can be obtained at any observation point (r, θ, ϕ).

9.2.3 Polarization of radiation field

The E and H fields calculated in (9.16) are linearly polarized (LP) components. In some antenna applications such as satellite communications, it is desired to obtain circularly polarized (CP) field components. This can be done through unit vector transformations between LP and CP components, such that

$$\hat{\theta} = \hat{\theta} - j\frac{\hat{\phi}}{2} + \hat{\theta} + j\frac{\hat{\phi}}{2} = \frac{\hat{E}_R}{\sqrt{2}} + \frac{\hat{E}_L}{\sqrt{2}}, \tag{9.18a}$$

$$\hat{\phi} = \hat{\theta} + j\frac{\hat{\phi}}{2j} - \hat{\theta} - j\frac{\hat{\phi}}{2j} = \frac{\hat{E}_L}{j\sqrt{2}} - \frac{\hat{E}_R}{j\sqrt{2}}, \tag{9.18b}$$

where $\hat{E}_R$ and $\hat{E}_L$ are unit vectors for the right-hand circularly polarized (RHCP) field and left-hand circularly polarized (LHCP) field. Substituting (9.18) into (9.16), we obtain

$$\begin{aligned}\vec{E} = \hat{\theta}E_\theta + \hat{\phi}E_\phi &= \left(\frac{\hat{E}_R}{\sqrt{2}} + \frac{\hat{E}_L}{\sqrt{2}}\right)E_\theta + \left(\frac{\hat{E}_L}{j\sqrt{2}} - \frac{\hat{E}_R}{j\sqrt{2}}\right)E_\phi \\ &= \hat{E}_R\left(\frac{E_\theta}{\sqrt{2}} - \frac{E_\phi}{j\sqrt{2}}\right) + \hat{E}_L\left(\frac{E_\theta}{\sqrt{2}} + \frac{E_\phi}{j\sqrt{2}}\right) = \hat{E}_R E_R + \hat{E}_L E_L,\end{aligned}$$

$$E_R = \frac{E_\theta}{\sqrt{2}} - \frac{E_\phi}{j\sqrt{2}}, \tag{9.19}$$

$$E_L = \frac{E_\theta}{\sqrt{2}} + \frac{E_\phi}{j\sqrt{2}}. \tag{9.20}$$

The magnitudes of the RHCP component (E_R) and the LHCP component (E_L) are then obtained. The axial ratio is defined to describe the polarization purity of the propagating waves and is calculated as follows [33]:

$$AR = -\frac{|E_R| + |E_L|}{|E_R| - |E_L|}. \tag{9.21}$$

For an LP wave, AR goes to infinity. For an RHCP wave $AR = -1$ and for an LHCP wave $AR = 1$. For a general elliptically polarized wave, $1 \le |AR| \le \infty$. Other expressions for direct computation of the AR from the far-field components E_θ and E_ϕ are given in [34]:

$$\mathrm{AR} = 20\log_{10}\left(\frac{\left[\frac{1}{2}\left(E_\phi^2 + E_\theta^2 + \left[E_\theta^4 + E_\phi^4 + 2E_\theta^2E_\phi^2\cos(2\delta)\right]^{\frac{1}{2}}\right)\right]^{\frac{1}{2}}}{\left[\frac{1}{2}\left(E_\phi^2 + E_\theta^2 + \left[E_\theta^4 + E_\phi^4 - 2E_\theta^2E_\phi^2\cos(2\delta)\right]^{\frac{1}{2}}\right)\right]^{\frac{1}{2}}}\right), \tag{9.22}$$

and in [35]

$$\mathrm{AR} = 20\log_{10}\left(\frac{|E_\phi|^2\sin^2(\tau) + |E_\theta|^2\cos^2(\tau) + |E_\phi||E_\theta|cos(\delta)\sin(2\tau)}{|E_\phi|^2\sin^2(\tau) + |E_\theta|^2\cos^2(\tau) - |E_\phi||E_\theta|cos(\delta)\sin(2\tau)}\right), \tag{9.23}$$

where

$$2\tau = \tan^{-1}\left(\frac{2|E_\phi||E_\theta|\cos(\delta)}{|E_\theta|^2 - |E_\phi|^2}\right),$$

and δ is the phase difference between E_θ and E_ϕ.

9.2.4 Radiation efficiency

The radiation efficiency is a very important indication for the effectiveness of an antenna, which can also be obtained using the FDTD technique. First of all, the radiation power of an antenna is obtained by applying the surface equivalence theorem to obtain

$$P_{rad} = \frac{1}{2}Re\left\{\int_S \vec{E} \times \vec{H}^* \cdot \hat{n}dS'\right\} = \frac{1}{2}Re\left\{\int_S \vec{J}* \times \vec{M} \cdot \hat{n}dS'\right\}. \tag{9.24}$$

The delivered power to an antenna is determined by the product of the voltage and current provided from the voltage source and can be expressed as

$$P_{del} = \frac{1}{2}Re\{V_s(\omega)I_s^*(\omega)\}, \tag{9.25}$$

where $V_s(\omega)$ and $I_s(\omega)$ represent the Fourier transformed values of the source voltage and current. The antenna's radiation efficiency η_a is then defined as [36]:

$$\eta_a = \frac{P_{rad}}{P_{del}}. \tag{9.26}$$

9.3 MATLAB® implementation of near-field to far-field transformation

In this section we demonstrate the implementation of near-field to far-field transformation in the three-dimensional FDTD MATLAB code.

9.3.1 Definition of NF–FF parameters

The implementation of the CPML boundary in the three-dimensional code is demonstrated in Chapter 8. The availability of CPML makes it possible to simulate open boundary problems; the electric and magnetic fields in a problem space can be calculated as if the boundaries are free space extending to infinity. Moreover, the near-field to far-field transformation technique described in the previous section can be used to compute the far-field patterns for a number of frequencies. The desired frequencies as well as some other parameters for far-field radiation are defined in the definition section of the FDTD program. The definition of certain NF–FF transformation parameters is implemented in the subroutine ***define_output_parameters,*** partial implementation of which is shown in Listing 9.1.

In Listing 9.1, a structure named **farfield** is defined with a field **frequencies**, which is initialized as an empty array. Then the frequencies for which the far-field patterns are sought are assigned to the array **farfield.frequencies**. Another variable for **farfield** that needs to be initialized is **number_of_cells_from_outer_boundary**, which indicates the position of the imaginary NF–FF transformation surface enclosing the radiators or scatterers existing in the problem space. This imaginary surface must not coincide with any objects in the problem space or with the CPML boundaries. Therefore, **number_of_cells_from_outer_boundary** should be chosen such that it is larger than the thickness of the CPML medium and less than

Listing 9.1 define_output_parameters

```
disp('defining output parameters');

sampled_electric_fields = [];
sampled_magnetic_fields = [];
sampled_voltages = [];
sampled_currents = [];
ports = [];
farfield.frequencies = [];

% figure refresh rate
plotting_step = 10;

% mode of operation
run_simulation = true;
show_material_mesh = true;
show_problem_space = true;

% far field calculation parameters
farfield.frequencies(1) = 3e9;
farfield.frequencies(2) = 5e9;
farfield.frequencies(3) = 6e9;
farfield.frequencies(4) = 9e9;
farfield.number_of_cells_from_outer_boundary = 10;
```

the sum of the thicknesses of the CPML medium and the air gap surrounding the objects as illustrated in Figure 9.4. This parameter is used to determine the nodes (*li*, *lj*, *lk*) and (*ui*, *uj*, *uk*) indicating the boundaries of the imaginary surface in Figure 9.4.

9.3.2 Initialization of NF–FF parameters

A new subroutine, **initialize_farfield_arrays,** is added to the main FDTD program **fdtd_solve** as shown in Listing 9.2. Implementation of **initialize_farfield_arrays** is given in Listing 9.3.

The first step in the NF–FF initialization process is to determine the nodes indicating the boundaries of the imaginary NF–FF surface: (*li*, *lj*, *lk*) and (*ui*, *uj*, *uk*). Once the start and end nodes are determined they are used to construct the NF–FF auxiliary arrays. Two sets of arrays are needed for the NF–FF transformation: The first set is used to capture and store the fictitious electric and magnetic currents on the faces of the imaginary surface at every time step of the FDTD time-marching loop, and the second set is used to store the on-the-fly DFTs of these currents. The naming conventions of these arrays are elaborated as follows.

There are six faces on the imaginary surface, and the fictitious electric and magnetic currents are calculated for each face. Therefore, 12 two-dimensional arrays are needed. The names of these arrays start with the letter "t," which is followed by a letter "j" or "m." Here "j" indicates the electric current whereas "m" indicates the magnetic current being stored in the array.

Listing 9.2 fdtd_solve

```
% initialize the matlab workspace
clear all; close all; clc;

% define the problem
define_problem_space_parameters;
define_geometry;
define_sources_and_lumped_elements;
define_output_parameters;

% initialize the problem space and parameters
initialize_fdtd_material_grid;
display_problem_space;
display_material_mesh;
if run_simulation
    initialize_fdtd_parameters_and_arrays;
    initialize_sources_and_lumped_elements;
    initialize_updating_coefficients;
    initialize_boundary_conditions;
    initialize_output_parameters;
    initialize_farfield_arrays;
    initialize_display_parameters;

    % FDTD time marching loop
    run_fdtd_time_marching_loop;

    % display simulation results
    post_process_and_display_results;
end
```

Listing 9.3 initialize_farfield_arrays

```
% initialize farfield arrays
number_of_farfield_frequencies = size(farfield.frequencies,2);

if number_of_farfield_frequencies == 0
    return;
end
nc_farbuffer = farfield.number_of_cells_from_outer_boundary;
li = nc_farbuffer + 1;
lj = nc_farbuffer + 1;
lk = nc_farbuffer + 1;
ui = nx - nc_farbuffer+1;
uj = ny - nc_farbuffer+1;
uk = nz - nc_farbuffer+1;

farfield_w = 2*pi*farfield.frequencies;
```

```
tjxyp = zeros(1,ui-li,1,uk-lk);
tjxzp = zeros(1,ui-li,uj-lj,1);
tjyxp = zeros(1,1,uj-lj,uk-lk);
tjyzp = zeros(1,ui-li,uj-lj,1);
tjzxp = zeros(1,1,uj-lj,uk-lk);
tjzyp = zeros(1,ui-li,1,uk-lk);
tjxyn = zeros(1,ui-li,1,uk-lk);
tjxzn = zeros(1,ui-li,uj-lj,1);
tjyxn = zeros(1,1,uj-lj,uk-lk);
tjyzn = zeros(1,ui-li,uj-lj,1);
tjzxn = zeros(1,1,uj-lj,uk-lk);
tjzyn = zeros(1,ui-li,1,uk-lk);
tmxyp = zeros(1,ui-li,1,uk-lk);
tmxzp = zeros(1,ui-li,uj-lj,1);
tmyxp = zeros(1,1,uj-lj,uk-lk);
tmyzp = zeros(1,ui-li,uj-lj,1);
tmzxp = zeros(1,1,uj-lj,uk-lk);
tmzyp = zeros(1,ui-li,1,uk-lk);
tmxyn = zeros(1,ui-li,1,uk-lk);
tmxzn = zeros(1,ui-li,uj-lj,1);
tmyxn = zeros(1,1,uj-lj,uk-lk);
tmyzn = zeros(1,ui-li,uj-lj,1);
tmzxn = zeros(1,1,uj-lj,uk-lk);
tmzyn = zeros(1,ui-li,1,uk-lk);

cjxyp = zeros(number_of_farfield_frequencies,ui-li,1,uk-lk);
cjxzp = zeros(number_of_farfield_frequencies,ui-li,uj-lj,1);
cjyxp = zeros(number_of_farfield_frequencies,1,uj-lj,uk-lk);
cjyzp = zeros(number_of_farfield_frequencies,ui-li,uj-lj,1);
cjzxp = zeros(number_of_farfield_frequencies,1,uj-lj,uk-lk);
cjzyp = zeros(number_of_farfield_frequencies,ui-li,1,uk-lk);
cjxyn = zeros(number_of_farfield_frequencies,ui-li,1,uk-lk);
cjxzn = zeros(number_of_farfield_frequencies,ui-li,uj-lj,1);
cjyxn = zeros(number_of_farfield_frequencies,1,uj-lj,uk-lk);
cjyzn = zeros(number_of_farfield_frequencies,ui-li,uj-lj,1);
cjzxn = zeros(number_of_farfield_frequencies,1,uj-lj,uk-lk);
cjzyn = zeros(number_of_farfield_frequencies,ui-li,1,uk-lk);
cmxyp = zeros(number_of_farfield_frequencies,ui-li,1,uk-lk);
cmxzp = zeros(number_of_farfield_frequencies,ui-li,uj-lj,1);
cmyxp = zeros(number_of_farfield_frequencies,1,uj-lj,uk-lk);
cmyzp = zeros(number_of_farfield_frequencies,ui-li,uj-lj,1);
cmzxp = zeros(number_of_farfield_frequencies,1,uj-lj,uk-lk);
cmzyp = zeros(number_of_farfield_frequencies,ui-li,1,uk-lk);
cmxyn = zeros(number_of_farfield_frequencies,ui-li,1,uk-lk);
cmxzn = zeros(number_of_farfield_frequencies,ui-li,uj-lj,1);
cmyxn = zeros(number_of_farfield_frequencies,1,uj-lj,uk-lk);
cmyzn = zeros(number_of_farfield_frequencies,ui-li,uj-lj,1);
cmzxn = zeros(number_of_farfield_frequencies,1,uj-lj,uk-lk);
cmzyn = zeros(number_of_farfield_frequencies,ui-li,1,uk-lk);
```

The third character in the array name is a letter "x," "y," or "z," which indicates the direction of the current in consideration. The last two characters are "xn," "yn," "zn," "xp," "yp," or "zp" indicating the face of the imaginary surface with which the array is associated. These arrays are four-dimensional, though they are effectively two-dimensional. The size of the first dimension is 1; this dimension is added to facilitate the on-the-fly DFT calculations, as is illustrated later. The fields are calculated in three-dimensional space and are stored in three-dimensional arrays. However, it is required to capture the fields coinciding with the faces of the imaginary surface. Therefore, the sizes of the other three dimensions of the fictitious current arrays are determined by the number of field components coinciding with imaginary surface faces.

The fictitious current arrays are used to calculate DFTs of these currents. Then the currents obtained in the frequency domain are stored in their respective arrays. Thus, each fictitious current array is associated with another array in the frequency domain. Therefore, the naming conventions of the frequency-domain arrays are the same as for the ones in time domain except that they are preceded by a letter "c." Furthermore, the sizes of the first dimension of the frequency domain arrays are equal to the number of frequencies for which the far-field calculation is sought.

One additional parameter defined in Listing 9.3 is **farfield_w**. Since the far-field frequencies are used as angular frequencies, with the unit *radians/second*, during the calculations they are calculated once and stored in **farfield_w** for future use.

9.3.3 NF–FF DFT during time-marching loop

While the FDTD time-marching loop is running, the electric and magnetic fields are captured and used to calculate fictitious magnetic and electric currents using (9.6)–(9.11) on the NF–FF imaginary surface. A new subroutine called ***calculate_JandM*** is added to ***run_fdtd_time_marching_loop*** as shown in Listing 9.4. The implementation of ***calculate_JandM*** is given in Listing 9.5.

One can notice in Listing 9.5 that an average of two electric field components is used to calculate the value of a fictitious magnetic current component. Here the purpose is to obtain the values of electric field components at the centers of the faces of the cells coinciding with the NF–FF surface. For instance, Figure 9.8 illustrates the electric field components E_x on the *yn* face of the imaginary NF–FF surface. The E_x components are located on the edges of the faces of cells coinciding with the NF–FF surface. To obtain an equivalent E_x value at the centers of the faces of the cells, the average of two E_x components are calculated for each cell. Thus, the fictitious magnetic current components M_z generated by the averaged E_x components using (9.10b) are located at the centers of the faces of the cells.

Similarly, every fictitious electric current component is calculated at the centers of the faces of the cells by averaging four magnetic field components. For example, Figure 9.9 illustrates the magnetic field components H_x around the *yn* face of the imaginary NF–FF surface. The H_x components are located at the centers of the faces of cells, which are not coinciding with the NF–FF surface. To obtain equivalent H_x values at the centers of the cell faces, the average of four H_x components are calculated for each cell. Thus, the fictitious electric current components J_z generated by the averaged H_x components using (9.10a) are located at the centers of the cell faces.

Listing 9.4 run_fdtd_time_marching_loop

```
disp (['Starting the time marching loop']);
disp (['Total number of time steps : ' ...
    num2str(number_of_time_steps)]);

start_time = cputime;
current_time = 0;

for time_step = 1:number_of_time_steps
    update_magnetic_fields;
    update_magnetic_field_CPML_ABC;
    capture_sampled_magnetic_fields;
    capture_sampled_currents;
    update_electric_fields;
    update_electric_field_CPML_ABC;
    update_voltage_sources;
    update_current_sources;
    update_inductors;
    update_diodes;
    capture_sampled_electric_fields;
    capture_sampled_voltages;
    calculate_JandM;
    display_sampled_parameters;
end

end_time = cputime;
total_time_in_minutes = (end_time - start_time)/60;
disp (['Total simulation time is ' ...
    num2str(total_time_in_minutes) ' minutes.']);
```

Listing 9.5 calculate_JandM

```
% Calculate J and M on the imaginary farfiled surface
if number_of_farfield_frequencies == 0
    return;
end
j = sqrt(-1);

tmyxp(1,1,:,:) =  0.5*(Ez(ui,lj:uj-1,lk:uk-1)+Ez(ui,lj+1:uj,lk:uk-1));
tmzxp(1,1,:,:) = -0.5*(Ey(ui,lj:uj-1,lk:uk-1)+Ey(ui,lj:uj-1,lk+1:uk));
tmxyp(1,:,1,:) = -0.5*(Ez(li:ui-1,uj,lk:uk-1)+Ez(li+1:ui,uj,lk:uk-1));
tmzyp(1,:,1,:) =  0.5*(Ex(li:ui-1,uj,lk:uk-1)+Ex(li:ui-1,uj,lk+1:uk));
tmxzp(1,:,:,1) =  0.5*(Ey(li:ui-1,lj:uj-1,uk)+Ey(li+1:ui,lj:uj-1,uk));
tmyzp(1,:,:,1) = -0.5*(Ex(li:ui-1,lj:uj-1,uk)+Ex(li:ui-1,lj+1:uj,uk));

tjyxp(1,1,:,:) =-0.25*(Hz(ui,lj:uj-1,lk:uk-1)+Hz(ui,lj:uj-1,lk+1:uk) ...
 + Hz (ui-1,lj:uj-1,lk:uk-1) + Hz (ui-1,lj:uj-1,lk+1:uk));

tjzxp(1,1,:,:) =0.25*(Hy(ui,lj:uj-1,lk:uk-1)+Hy(ui,lj+1:uj,lk:uk-1) ...
 + Hy (ui-1,lj:uj-1,lk:uk-1) + Hy (ui-1,lj+1:uj,lk:uk-1));

tjzyp(1,:,1,:) =-0.25*(Hx(li:ui-1,uj,lk:uk-1)+Hx(li+1:ui,uj,lk:uk-1) ...
```

```
+ Hx (li:ui-1,uj-1,lk:uk-1) + Hx (li+1:ui,uj-1,lk:uk-1));

tjxyp(1,:,1,:) = 0.25*(Hz(li:ui-1,uj,lk:uk-1)+Hz(li:ui-1,uj,lk+1:uk) ...
+ Hz (li:ui-1,uj-1,lk:uk-1) + Hz (li:ui-1,uj-1,lk+1:uk));

tjyzp(1,:,:,1) = 0.25*(Hx(li:ui-1,lj:uj-1,uk)+Hx(li+1:ui,lj:uj-1,uk) ...
+ Hx (li:ui-1,lj:uj-1,uk-1) + Hx (li+1:ui,lj:uj-1,uk-1));

tjxzp(1,:,:,1) =-0.25*(Hy(li:ui-1,lj:uj-1,uk)+Hy(li:ui-1,lj+1:uj,uk) ...
+ Hy (li:ui-1,lj:uj-1,uk-1) + Hy (li:ui-1,lj+1:uj,uk-1));

tmyxn(1,1,:,:) = -0.5 * (Ez(li,lj:uj-1,lk:uk-1)+Ez(li,lj+1:uj,lk:uk-1));
tmzxn(1,1,:,:) =  0.5 * (Ey(li,lj:uj-1,lk:uk-1)+Ey(li,lj:uj-1,lk+1:uk));

tmxyn(1,:,1,:) =  0.5 * (Ez(li:ui-1,lj,lk:uk-1)+Ez(li+1:ui,lj,lk:uk-1));
tmzyn(1,:,1,:) = -0.5 * (Ex(li:ui-1,lj,lk:uk-1)+Ex(li:ui-1,lj,lk+1:uk));

tmxzn(1,:,:,1) = -0.5 * (Ey(li:ui-1,lj:uj-1,lk)+Ey(li+1:ui,lj:uj-1,lk));
tmyzn(1,:,:,1) =  0.5 * (Ex(li:ui-1,lj:uj-1,lk)+Ex(li:ui-1,lj+1:uj,lk));

tjyxn(1,1,:,:) = 0.25*(Hz(li,lj:uj-1,lk:uk-1)+Hz(li,lj:uj-1,lk+1:uk) ...
+ Hz (li-1,lj:uj-1,lk:uk-1) + Hz (li-1,lj:uj-1,lk+1:uk));

tjzxn(1,1,:,:) =-0.25*(Hy(li,lj:uj-1,lk:uk-1)+Hy(li,lj+1:uj,lk:uk-1) ...
+ Hy (li-1,lj:uj-1,lk:uk-1) + Hy (li-1,lj+1:uj,lk:uk-1));

tjzyn(1,:,1,:) = 0.25*(Hx(li:ui-1,lj,lk:uk-1)+Hx(li+1:ui,lj,lk:uk-1) ...
+ Hx (li:ui-1,lj-1,lk:uk-1) + Hx (li+1:ui,lj-1,lk:uk-1));

tjxyn(1,:,1,:) =-0.25*(Hz(li:ui-1,lj,lk:uk-1)+Hz(li:ui-1,lj,lk+1:uk) ...
+ Hz (li:ui-1,lj-1,lk:uk-1) + Hz (li:ui-1,lj-1,lk+1:uk));

tjyzn(1,:,:,1) =-0.25*(Hx(li:ui-1,lj:uj-1,lk)+Hx(li+1:ui,lj:uj-1,lk) ...
+ Hx (li:ui-1,lj:uj-1,lk-1)+Hx(li+1:ui,lj:uj-1,lk-1));

tjxzn(1,:,:,1) = 0.25*(Hy(li:ui-1,lj:uj-1,lk)+Hy(li:ui-1,lj+1:uj,lk) ...
+ Hy (li:ui-1,lj:uj-1,lk-1) + Hy (li:ui-1,lj+1:uj,lk-1));

% fourier transform
for mi=1:number_of_farfield_frequencies
    exp_h = dt * exp(-j*farfield_w(mi)*(time_step-0.5)*dt);
    cjxyp(mi,:,:,:) = cjxyp(mi,:,:,:) + exp_h * tjxyp(1,:,:,:);
    cjxzp(mi,:,:,:) = cjxzp(mi,:,:,:) + exp_h * tjxzp(1,:,:,:);
    cjyxp(mi,:,:,:) = cjyxp(mi,:,:,:) + exp_h * tjyxp(1,:,:,:);
    cjyzp(mi,:,:,:) = cjyzp(mi,:,:,:) + exp_h * tjyzp(1,:,:,:);
    cjzxp(mi,:,:,:) = cjzxp(mi,:,:,:) + exp_h * tjzxp(1,:,:,:);
    cjzyp(mi,:,:,:) = cjzyp(mi,:,:,:) + exp_h * tjzyp(1,:,:,:);

    cjxyn(mi,:,:,:) = cjxyn(mi,:,:,:) + exp_h * tjxyn(1,:,:,:);
    cjxzn(mi,:,:,:) = cjxzn(mi,:,:,:) + exp_h * tjxzn(1,:,:,:);
    cjyxn(mi,:,:,:) = cjyxn(mi,:,:,:) + exp_h * tjyxn(1,:,:,:);
    cjyzn(mi,:,:,:) = cjyzn(mi,:,:,:) + exp_h * tjyzn(1,:,:,:);
```

```
        cjzxn(mi,:,:,:) = cjzxn(mi,:,:,:) + exp_h * tjzxn(1,:,:,:);
        cjzyn(mi,:,:,:) = cjzyn(mi,:,:,:) + exp_h * tjzyn(1,:,:,:);

        exp_e = dt * exp(-j*farfield_w(mi)*time_step*dt);

        cmxyp(mi,:,:,:) = cmxyp(mi,:,:,:) + exp_e * tmxyp(1,:,:,:);
        cmxzp(mi,:,:,:) = cmxzp(mi,:,:,:) + exp_e * tmxzp(1,:,:,:);
        cmyxp(mi,:,:,:) = cmyxp(mi,:,:,:) + exp_e * tmyxp(1,:,:,:);
        cmyzp(mi,:,:,:) = cmyzp(mi,:,:,:) + exp_e * tmyzp(1,:,:,:);
        cmzxp(mi,:,:,:) = cmzxp(mi,:,:,:) + exp_e * tmzxp(1,:,:,:);
        cmzyp(mi,:,:,:) = cmzyp(mi,:,:,:) + exp_e * tmzyp(1,:,:,:);

        cmxyn(mi,:,:,:) = cmxyn(mi,:,:,:) + exp_e * tmxyn(1,:,:,:);
        cmxzn(mi,:,:,:) = cmxzn(mi,:,:,:) + exp_e * tmxzn(1,:,:,:);
        cmyxn(mi,:,:,:) = cmyxn(mi,:,:,:) + exp_e * tmyxn(1,:,:,:);
        cmyzn(mi,:,:,:) = cmyzn(mi,:,:,:) + exp_e * tmyzn(1,:,:,:);
        cmzxn(mi,:,:,:) = cmzxn(mi,:,:,:) + exp_e * tmzxn(1,:,:,:);
        cmzyn(mi,:,:,:) = cmzyn(mi,:,:,:) + exp_e * tmzyn(1,:,:,:);
end
```

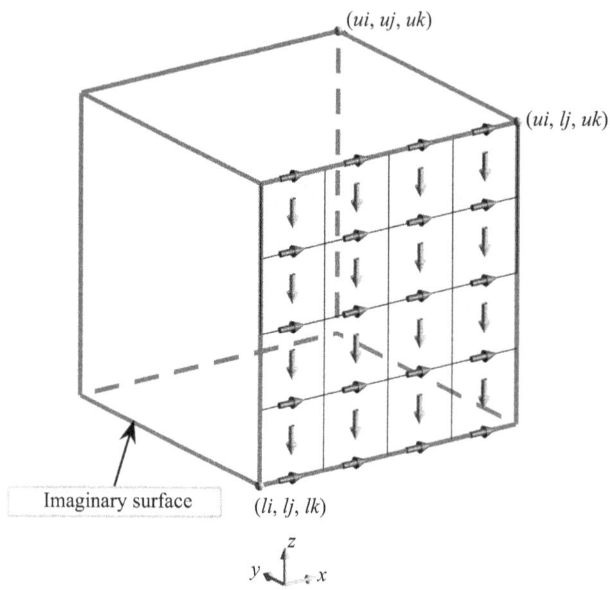

Figure 9.8 E_x field components on the yn face of the NF–FF imaginary surface and the magnetic currents generated by them.

After the fictitious electric and magnetic currents are sampled on the NF–FF surface at the current time step, they are used to update the on-the-fly DFTs for the far-field frequencies defined in the subroutine ***define_output_parameters***. For instance, one iteration of (9.12) is performed at each time step for calculating J_y in frequency domain. Therefore, when the FDTD time-marching loop is completed, the DFT of J_y is completed together.

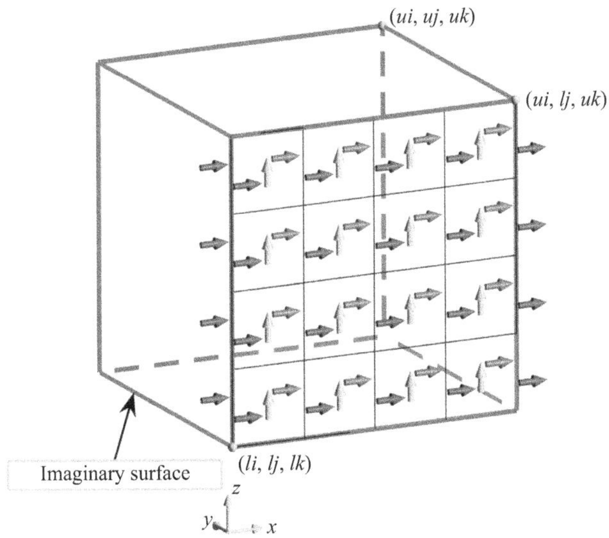

Figure 9.9 H_x field components around the *yn* face of the NF–FF imaginary surface and the electric currents generated by them.

9.3.4 Postprocessing for far-field calculation

As the FDTD time-marching loop is completed, the frequency-domain values of fictitious currents for the far-field frequencies over the NF–FF surface are available. Then they can be used at the postprocessing stage of the FDTD simulation to calculate the far-field auxiliary fields based on (9.17). Then these fields can be used to calculate the desired far-field patterns and to plot them.

A new subroutine named ***calculate_and_display_farfields*** is added to the postprocessing routine ***post_process_and_display_results*** as shown in Listing 9.6, which performs the tasks of far-field calculation and display.

Implementation of ***calculate_and_display_farfields*** is given in Listing 9.7. This subroutine initializes necessary arrays and parameters for calculation and display of far-field patterns at three principal planes: namely, the *xy*, *xz*, and *yz* planes. For instance, to calculate

Listing 9.6 post_process_and_display_results

```
disp('postprocessing and displaying simulation results');

display_transient_parameters;
calculate_frequency_domain_outputs;
display_frequency_domain_outputs;
calculate_and_display_farfields;
```

Listing 9.7 calculate_and_display_farfields

```
% This file calls the routines necessary for calculating
% farfield patterns in xy, xz, and yz plane cuts, and displays them.
% The display can be modified as desired.
% You will find the instructions the formats for
% radiation pattern plots can be set by the user.

if number_of_farfield_frequencies == 0
    return;
end

calculate_radiated_power;

j = sqrt(-1);
number_of_angles = 360;

% parameters used by polar plotting functions
step_size = 10;        % increment between the rings in the polar grid
Nrings = 4;            % number of rings in the polar grid
line_style1 = 'b-';    % line style for theta component
line_style2 = 'r--';   % line style for phi component
scale_type = 'dB';     % linear or dB
plot_type = 'D';       % the calculated data is directivity

% xy plane
% ===============================================
farfield_theta = zeros(number_of_angles, 1);
farfield_phi   = zeros(number_of_angles, 1);
farfield_theta = farfield_theta + pi/2;
farfield_phi = (pi/180)*[-180:1:179].';
const_theta = 90; % used for plot

% calculate farfields
calculate_farfields_per_plane;

% plotting the farfield data
for mi=1:number_of_farfield_frequencies
 f = figure;
 pat1 = farfield_dataTheta(mi,:).';
 pat2 = farfield_dataPhi(mi,:).';

 % if scale_type is db use these, otherwise comment these two lines
 pat1 = 10*log10(pat1);
 pat2 = 10*log10(pat2);

 max_val = max(max([pat1 pat2]));
 max_val = step_size * ceil(max_val/step_size);

 legend_str1 = [plot_type '_{\theta}, f=' ...
     num2str(farfield.frequencies(mi)*1e-9) ' GHz'];
 legend_str2 = [plot_type '_{\phi}, f=' ...
```

```
        num2str(farfield.frequencies(mi)*1e-9) '␣GHz'];

    polar_plot_constant_theta(farfield_phi,pat1,pat2,max_val, ...
            step_size, Nrings,line_style1,line_style2,const_theta, ...
            legend_str1,legend_str2,scale_type);
end

% xz plane
% ================================================
farfield_theta = zeros(number_of_angles, 1);
farfield_phi   = zeros(number_of_angles, 1);
farfield_theta = (pi/180)*[-180:1:179].';
const_phi = 0; % used for plot

% calculate farfields
calculate_farfields_per_plane;

% plotting the farfield data
for mi=1:number_of_farfield_frequencies
  f = figure;
  pat1 = farfield_dataTheta(mi,:).';
  pat2 = farfield_dataPhi(mi,:).';

% if scale_type is db use these, otherwise comment these two lines
  pat1 = 10*log10(pat1);
  pat2 = 10*log10(pat2);

  max_val = max(max([pat1 pat2]));
  max_val = step_size * ceil(max_val/step_size);

  legend_str1 = ...
  [plot_type '_{\theta},␣f=' ...
  num2str(farfield.frequencies(mi)*1e-9) '␣GHz'];
  legend_str2 = ...
  [plot_type '_{\phi},␣f=' ...
  num2str(farfield.frequencies(mi)*1e-9) '␣GHz'];

  polar_plot_constant_phi(farfield_theta,pat1,pat2,max_val, ...
          step_size, Nrings,line_style1,line_style2,const_phi, ...
          legend_str1,legend_str2,scale_type);
end

% yz plane
% ================================================
farfield_theta = zeros(number_of_angles, 1);
farfield_phi   = zeros(number_of_angles, 1);
farfield_phi = farfield_phi + pi/2;
farfield_theta = (pi/180)*[-180:1:179].';
const_phi = 90; % used for plot

% calculate farfields
calculate_farfields_per_plane;
```

```
% plotting the farfield data
for mi=1:number_of_farfield_frequencies
 f = figure;
 pat1 = farfield_dataTheta(mi,:).';
 pat2 = farfield_dataPhi(mi,:).';

% if scale_type is db use these, otherwise comment these two lines
 pat1 = 10*log10(pat1);
 pat2 = 10*log10(pat2);

 max_val = max(max([pat1 pat2]));
 max_val = step_size * ceil(max_val/step_size);

 legend_str1 = ...
 [plot_type '_{\theta},␣f=' ...
 num2str(farfield.frequencies(mi)*1e-9) '␣GHz'];
 legend_str2 = ...
 [plot_type '_{\phi},␣f=' ...
 num2str(farfield.frequencies(mi)*1e-9) '␣GHz'];

 polar_plot_constant_phi(farfield_theta,pat1,pat2,max_val, ...
         step_size, Nrings,line_style1,line_style2,const_phi, ...
         legend_str1,legend_str2,scale_type);
end
```

the far-fields in the xy plane, two arrays, **farfield_theta** and **farfield_phi**, are constructed to store the angles that represent the xy plane. For the xy plane E is 90° and ϕ sweeps from −180° to 180°. Then a function ***calculate_farfields_per_plane*** is called to calculate far-field directivity data at the given angles. These data are plotted using another function ***polar_plot_constant_theta***. For the xy plane E is a constant value of $\pi/2$ radians. For the xz and yz planes ϕ takes constant values of 0 and $\pi/2$ radians, respectively, hence another function, which is called for plotting patterns in the xz and yz planes is ***polar_plot_constant_phi***. The implementations of ***polar_plot_constant_theta*** and ***polar_plot_constant_phi*** are given in Appendix C.

The near-field to far-field calculations are performed mainly in the function ***calculate_farfields_per_plane***, implementation of which is given in Listing 9.9.

One of the initial tasks in ***calculate_farfields_per_plane*** is the calculation of total radiated power. A subroutine named ***calculate_radiated_power*** is called to perform this task. The total radiated power is stored in the parameter **radiated_power**, as can be seen in Listing 9.8. Calculation of radiated power is based on the discrete summation representation of (9.24).

Before calculating any far-field data, the auxiliary fields N_θ, N_ϕ, L_E, and L_ϕ need to be calculated based on (9.17). In (9.17) the parameters E and ϕ are angles indicating the position vector of the observation point $\vec{r}$ as illustrated in Figure 9.7. The parameters J_x, J_y, J_z, M_x, M_y, and M_z are the fictitious currents, which have already been calculated at the center

Listing 9.8 calculate_radiated_power

```
% Calculate total radiated power
radiated_power = zeros(number_of_farfield_frequencies,1);

for mi=1:number_of_farfield_frequencies
    powr = 0;
    powr = dx*dy* sum(sum(sum(cmyzp(mi,:,:,:).* ...
        conj(cjxzp(mi,:,:,:)) - cmxzp(mi,:,:,:) ...
        .* conj(cjyzp(mi,:,:,:)))));
    powr = powr - dx*dy* sum(sum(sum(cmyzn(mi,:,:,:) ...
        .* conj(cjxzn(mi,:,:,:)) - cmxzn(mi,:,:,:) ...
        .* conj(cjyzn(mi,:,:,:)))));
    powr = powr + dx*dz* sum(sum(sum(cmxyp(mi,:,:,:) ...
        .* conj(cjzyp(mi,:,:,:)) - cmzyp(mi,:,:,:) ...
        .* conj(cjxyp(mi,:,:,:)))));
    powr = powr - dx*dz* sum(sum(sum(cmxyn(mi,:,:,:) ...
        .* conj(cjzyn(mi,:,:,:)) - cmzyn(mi,:,:,:) ...
        .* conj(cjxyn(mi,:,:,:)))));
    powr = powr + dy*dz* sum(sum(sum(cmzxp(mi,:,:,:) ...
        .* conj(cjyxp(mi,:,:,:)) - cmyxp(mi,:,:,:) ...
        .* conj(cjzxp(mi,:,:,:)))));
    powr = powr - dy*dz* sum(sum(sum(cmzxn(mi,:,:,:) ...
        .* conj(cjyxn(mi,:,:,:)) - cmyxn(mi,:,:,:) ...
        .* conj(cjzxn(mi,:,:,:)))));
    radiated_power(mi) = 0.5 * real(powr);
end
```

points of the faces of the cells coinciding with the NF–FF surface for the given far-field frequencies. The parameter k is the wavenumber expressed by

$$k = 2\pi f \sqrt{\mu_0 \varepsilon_0}, \tag{9.27}$$

where f is the far-field frequency under consideration. Yet another term in (9.17) is $r' \cos(\psi)$, which needs to be further defined explicitly. As shown in Figure 9.7, $\bar{r}$ is the position vector indicating the observation point while $\bar{r}'$ is the position vector indicating the source point. The term $r' \cos(\psi)$ is actually obtained from $\bar{r}' \cdot \hat{r}$ where $\hat{r}$ is a unit vector in the direction of $\bar{r}$. The unit vector $\hat{r}$ can be expressed in Cartesian coordinates as

$$\hat{r} = \sin(\theta)\cos(\phi)\hat{x} + \sin(\theta)\sin(\phi)\hat{y} + \cos(\theta)\hat{z}. \tag{9.28}$$

As mentioned already, the vector $\bar{r}'$ is the position vector indicating the source point and is expressed in terms of source positions. The sources are located at the centers of the faces, and the source position vectors can be expressed as follows. A position vector for a source point located on the zn face of the NF–FF surface can be expressed as

$$\bar{r}' = (mi + 0.5 - ci)\Delta x\hat{x} + (mj + 0.5 - cj)\Delta y\hat{y} + (lk - ck)\Delta z\hat{z}, \tag{9.29}$$

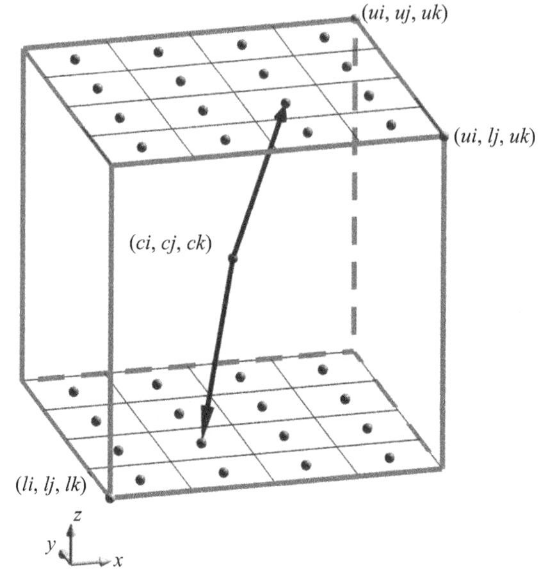

Figure 9.10 Position vectors for sources on the *zn* and *zp* faces.

as can be observed in Figure 9.10. Here (ci, cj, ck) is the center node of the problem space, which is assumed to be the origin for the far-field calculations. The parameters mi and mj are the indices of the nodes of the faces, which include the sources under consideration. Then the term $r' \cos(\psi)$ in (9.17) can be expressed explicitly using (9.28) and (9.29) as

$$r'\cos(\psi) = \bar{r}' \cdot \hat{r} = (mi + 0.5 - ci)\Delta x \sin(\theta)\cos(\phi) \\ + (mj + 0.5 - cj)\Delta y \sin(\theta)\sin(\phi) + (lk - ck)\Delta z \cos(\theta). \tag{9.30}$$

Similarly, $r' \cos(\psi)$ can be obtained for the *zp* face as

$$r'\cos(\psi) = \bar{r}' \cdot \hat{r} = (mi + 0.5 - ci)\Delta x \sin(\theta)\cos(\phi) \\ + (mj + 0.5 - cj)\Delta y \sin(\theta)\sin(\phi) + (uk - ck)\Delta z \cos(\theta). \tag{9.31}$$

The equations in (9.17) are integrations of continuous current distributions over the imaginary surface. However, we have obtained the currents at discrete points. Therefore, these integrations can be represented by discrete summations. For instance, for the *zn* and *zp* faces (9.17) can be rewritten as

$$N_\theta = \sum_{mi=li}^{ui-1} \sum_{mj=lj}^{uj-1} \Delta x \Delta y \left(J_x \cos(\theta)\cos(\phi) + J_y \cos(\theta)\sin(\phi)\right) e^{jkr'\cos(\psi)}, \tag{9.32a}$$

$$N_\phi = \sum_{mi=li}^{ui-1} \sum_{mj=lj}^{uj-1} \Delta x \Delta y \left(-J_x \sin(\phi) + J_y \cos(\phi)\right) e^{jkr'\cos(\psi)}, \tag{9.32b}$$

$$L_\theta = \sum_{mi=li}^{ui-1} \sum_{mj=lj}^{uj-1} \Delta x \Delta y \left(M_x \cos(\theta) \cos(\phi) + M_y \cos(\theta) \sin(\phi) \right) e^{jkr' \cos(\psi)}, \tag{9.32c}$$

$$L_\phi = \sum_{mi=li}^{ui-1} \sum_{mj=lj}^{uj-1} \Delta x \Delta y \left(-M_x \sin(\phi) + M_y \cos(\phi) \right) e^{jkr' \cos(\psi)}. \tag{9.32d}$$

Referring to Figure 9.11 one can obtain $r' \cos(\psi)$ for the *xn* and *xp* faces, respectively, as

$$r' \cos(\psi) = \bar{r}' \cdot \hat{r} = (li - ci) \Delta x \sin(\theta) \cos(\phi) \\ + (mj + 0.5 - cj) \Delta y \sin(\theta) \sin(\phi) + (mk + 0.5 - ck) \Delta z \cos(\theta), \tag{9.33}$$

$$r' \cos(\psi) = \bar{r}' \cdot \hat{r} = (ui - ci) \Delta x \sin(\theta) \cos(\phi) \\ + (mj + 0.5 - cj) \Delta y \sin(\theta) \sin(\phi) + (mk + 0.5 - ck) \Delta z \cos(\theta), \tag{9.34}$$

where *mj* and *mk* are the indices of the nodes of the faces that include the sources. Then the summations in 9.32 are expressed in terms of these indices *mj* and *mk* as

$$N_\theta = \sum_{mj=lj}^{uj-1} \sum_{mk=lk}^{uk-1} \Delta y \Delta z \left(J_y \cos(\theta) \sin(\phi) - J_z \sin(\theta) \right) e^{jkr' \cos(\psi)}, \tag{9.35a}$$

$$N_\phi = \sum_{mj=lj}^{uj-1} \sum_{mk=lk}^{uk-1} \Delta y \Delta z \left(J_y \cos(\phi) \right) e^{jkr' \cos(\psi)}, \tag{9.35b}$$

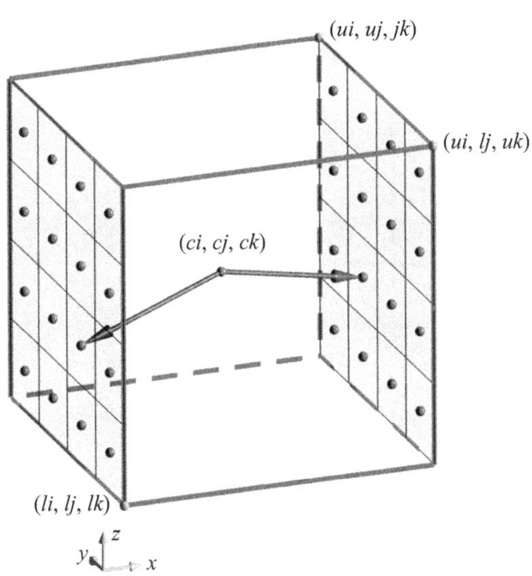

Figure 9.11 Position vectors for sources on the *xn* and *xp* faces.

$$L_\theta = \sum_{mj=lj}^{uj-1} \sum_{mk=lk}^{uk-1} \Delta y \Delta z \big(M_y \cos(\theta) \sin(\phi) - M_z \sin(\theta) \big) e^{jkr' \cos(\psi)}, \tag{9.35c}$$

$$L_\phi = \sum_{mj=lj}^{uj-1} \sum_{mk=lk}^{uk-1} \Delta y \Delta z \big(M_y \cos(\phi) \big) e^{jkr' \cos(\psi)}. \tag{9.35d}$$

Similarly, referring to Figure 9.12 $r' \cos(\psi)$ can be obtained for the *yn* and *yp* faces, respectively, as

$$\begin{aligned} r'\cos(\psi) &= (mi + 0.5 - ci)\overline{r}' \cdot \hat{r} = \Delta x \sin(\theta) \cos(\phi) \\ &\quad + (lj - cj) \Delta y \sin(\theta) \sin(\phi) + (mk + 0.5 - ck) \Delta z \cos(\theta), \end{aligned} \tag{9.36}$$

$$\begin{aligned} r'\cos(\psi) &= (mi + 0.5 - ci)\overline{r}' \cdot \hat{r} = \Delta x \sin(\theta) \cos(\phi) \\ &\quad + (uj - cj) \Delta y \sin(\theta) \sin(\phi) + (mk + 0.5 - ck) \Delta z \cos(\theta), \end{aligned} \tag{9.37}$$

Then the summations in (9.32) are expressed in terms of these indices *mi* and *mk* as

$$N_\theta = \sum_{mi=li}^{ui-1} \sum_{mk=lk}^{uk-1} \Delta x \Delta z (J_x \cos(\theta) \cos(\phi) - J_z \sin(\theta)) e^{jkr' \cos(\psi)}, \tag{9.38a}$$

$$N_\phi = \sum_{mi=li}^{ui-1} \sum_{mk=lk}^{uk-1} \Delta x \Delta z (-J_x \sin(\phi)) e^{jkr' \cos(\psi)}, \tag{9.38b}$$

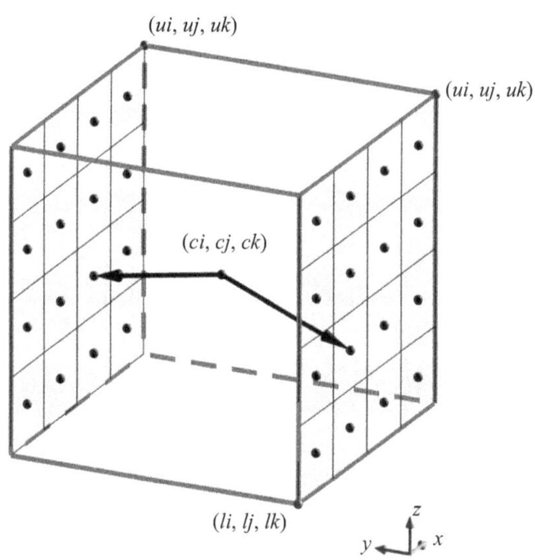

Figure 9.12 Position vectors for sources on the *yn* and *yp* faces.

$$L_\theta = \sum_{mi=li}^{ui-1} \sum_{mk=lk}^{uk-1} \Delta x \Delta z (M_x \cos(\theta)\cos(\phi) - M_z \sin(\theta)) e^{jkr'\cos(\psi)}, \tag{9.38c}$$

$$L_\phi = \sum_{mi=li}^{ui-1} \sum_{mk=lk}^{uk-1} \Delta x \Delta z (-M_x \sin(\phi)) e^{jkr'\cos(\psi)}. \tag{9.38d}$$

The implementation of the discrete summation just described can be followed in Listing 9.9.

Finally, having obtained the auxiliary fields N_θ, N_ϕ, L_E, and L_ϕ and the total radiated power, the far-field directivity data are calculated based on [1] as

$$D_\theta = \frac{k^2}{8\pi\eta_0 P_{rad}} |L_\phi + \eta_0 N_\theta|^2, \tag{9.39a}$$

$$D_\phi = \frac{k^2}{8\pi\eta_0 P_{rad}} |L_\theta - \eta_0 N_\phi|^2. \tag{9.39b}$$

Listing 9.9 calculate_farfields_per_plane

```
if number_of_farfield_frequencies == 0
    return;
end
c          = 2.99792458e+8;        % speed of light in free space
mu_0   = 4 * pi * 1e-7;        % permeability of free space
eps_0  = 1.0/(c*c*mu_0);           % permittivity of free space
eta_0  = sqrt(mu_0/eps_0);     % intrinsic impedance of free space

exp_jk_rpr = zeros(number_of_angles,1);
dx_sinth_cosphi = zeros(number_of_angles,1);
dy_sinth_sinphi = zeros(number_of_angles,1);
dz_costh = zeros(number_of_angles,1);
dy_dz_costh_sinphi = zeros(number_of_angles,1);
dy_dz_sinth = zeros(number_of_angles,1);
dy_dz_cosphi = zeros(number_of_angles,1);
dx_dz_costh_cosphi = zeros(number_of_angles,1);
dx_dz_sinth = zeros(number_of_angles,1);
dx_dz_sinphi = zeros(number_of_angles,1);
dx_dy_costh_cosphi = zeros(number_of_angles,1);
dx_dy_costh_sinphi = zeros(number_of_angles,1);
dx_dy_sinphi = zeros(number_of_angles,1);
dx_dy_cosphi = zeros(number_of_angles,1);
farfield_dataTheta = ...
    zeros(number_of_farfield_frequencies,number_of_angles);
farfield_dataPhi = ...
    zeros(number_of_farfield_frequencies,number_of_angles);

dx_sinth_cosphi = dx*sin(farfield_theta).*cos(farfield_phi);
dy_sinth_sinphi = dy*sin(farfield_theta).*sin(farfield_phi);
dz_costh = dz*cos(farfield_theta);
```

```
dy_dz_costh_sinphi = dy*dz*cos(farfield_theta).*sin(farfield_phi);
dy_dz_sinth = dy*dz*sin(farfield_theta);
dy_dz_cosphi = dy*dz*cos(farfield_phi);
dx_dz_costh_cosphi = dx*dz*cos(farfield_theta).*cos(farfield_phi);
dx_dz_sinth = dx*dz*sin(farfield_theta);
dx_dz_sinphi = dx*dz*sin(farfield_phi);
dx_dy_costh_cosphi = dx*dy*cos(farfield_theta).*cos(farfield_phi);
dx_dy_costh_sinphi = dx*dy*cos(farfield_theta).*sin(farfield_phi);
dx_dy_sinphi = dx*dy*sin(farfield_phi);
dx_dy_cosphi = dx*dy*cos(farfield_phi);
ci = 0.5*(ui+li);
cj = 0.5*(uj+lj);
ck = 0.5*(uk+lk);

% calculate directivity
for mi=1:number_of_farfield_frequencies
    disp(['Calculating directivity for ', ...
        num2str(farfield.frequencies(mi)) ' Hz']);
    k = 2*pi*farfield.frequencies(mi)*(mu_0*eps_0)^0.5;

    Ntheta = zeros(number_of_angles,1);
    Ltheta = zeros(number_of_angles,1);
    Nphi = zeros(number_of_angles,1);
    Lphi = zeros(number_of_angles,1);
    rpr = zeros(number_of_angles,1);

    for nj = lj:uj-1
        for nk =lk:uk-1
        % for +ax direction

        rpr = (ui - ci)*dx_sinth_cosphi ...
            + (nj-cj+0.5)*dy_sinth_sinphi ...
            + (nk-ck+0.5)*dz_costh;
        exp_jk_rpr = exp(j*k*rpr);
        Ntheta = Ntheta ...
            + (cjyxp(mi,1,nj-lj+1,nk-lk+1).*dy_dz_costh_sinphi ...
            - cjzxp(mi,1,nj-lj+1,nk-lk+1).*dy_dz_sinth).*exp_jk_rpr;
        Ltheta = Ltheta ...
            + (cmyxp(mi,1,nj-lj+1,nk-lk+1).*dy_dz_costh_sinphi ...
            - cmzxp(mi,1,nj-lj+1,nk-lk+1).*dy_dz_sinth).*exp_jk_rpr;
        Nphi = Nphi ...
            + (cjyxp(mi,1,nj-lj+1,nk-lk+1).*dy_dz_cosphi).*exp_jk_rpr;
        Lphi = Lphi ...
            + (cmyxp(mi,1,nj-lj+1,nk-lk+1).*dy_dz_cosphi).*exp_jk_rpr;

        % for -ax direction
        rpr = (li - ci)*dx_sinth_cosphi ...
            + (nj-cj+0.5)*dy_sinth_sinphi ...
            + (nk-ck+0.5)*dz_costh;
        exp_jk_rpr = exp(j*k*rpr);
        Ntheta = Ntheta ...
            + (cjyxn(mi,1,nj-lj+1,nk-lk+1).*dy_dz_costh_sinphi ...
```

```
            - cjzxn(mi,1,nj-lj+1,nk-lk+1).*dy_dz_sinth).*exp_jk_rpr;
        Ltheta = Ltheta ...
            + (cmyxn(mi,1,nj-lj+1,nk-lk+1).*dy_dz_costh_sinphi ...
            - cmzxn(mi,1,nj-lj+1,nk-lk+1).*dy_dz_sinth).*exp_jk_rpr;
        Nphi = Nphi ...
            + (cjyxn(mi,1,nj-lj+1,nk-lk+1).*dy_dz_cosphi).*exp_jk_rpr;
        Lphi = Lphi ...
            + (cmyxn(mi,1,nj-lj+1,nk-lk+1).*dy_dz_cosphi).*exp_jk_rpr;
        end
    end
    for ni =li:ui-1
        for nk =lk:uk-1
        % for +ay direction

        rpr = (ni - ci + 0.5)*dx_sinth_cosphi ...
            + (uj-cj)*dy_sinth_sinphi ...
            + (nk-ck+0.5)*dz_costh;
        exp_jk_rpr = exp(j*k*rpr);

        Ntheta = Ntheta ...
            + (cjxyp(mi,ni-li+1,1,nk-lk+1).*dx_dz_costh_cosphi ...
            - cjzyp(mi,ni-li+1,1,nk-lk+1).*dx_dz_sinth).*exp_jk_rpr;
        Ltheta = Ltheta ...
            + (cmxyp(mi,ni-li+1,1,nk-lk+1).*dx_dz_costh_cosphi ...
            - cmzyp(mi,ni-li+1,1,nk-lk+1).*dx_dz_sinth).*exp_jk_rpr;
        Nphi = Nphi ...
            + (-cjxyp(mi,ni-li+1,1,nk-lk+1).*dx_dz_sinphi).*exp_jk_rpr;
        Lphi = Lphi ...
            + (-cmxyp(mi,ni-li+1,1,nk-lk+1).*dx_dz_sinphi).*exp_jk_rpr;

        % for -ay direction
        rpr = (ni - ci + 0.5)*dx_sinth_cosphi ...
            + (lj-cj)*dy_sinth_sinphi ...
            + (nk-ck+0.5)*dz_costh;
        exp_jk_rpr = exp(j*k*rpr);

        Ntheta = Ntheta ...
            + (cjxyn(mi,ni-li+1,1,nk-lk+1).*dx_dz_costh_cosphi ...
            - cjzyn(mi,ni-li+1,1,nk-lk+1).*dx_dz_sinth).*exp_jk_rpr;
        Ltheta = Ltheta ...
            + (cmxyn(mi,ni-li+1,1,nk-lk+1).*dx_dz_costh_cosphi ...
            - cmzyn(mi,ni-li+1,1,nk-lk+1).*dx_dz_sinth).*exp_jk_rpr;
        Nphi = Nphi ...
            + (-cjxyn(mi,ni-li+1,1,nk-lk+1).*dx_dz_sinphi).*exp_jk_rpr;
        Lphi = Lphi ...
            + (-cmxyn(mi,ni-li+1,1,nk-lk+1).*dx_dz_sinphi).*exp_jk_rpr;
        end
    end

    for ni =li:ui-1
        for nj =lj:uj-1
        % for +az direction
```

```
            rpr = (ni-ci+0.5)*dx_sinth_cosphi ...
                + (nj - cj + 0.5)*dy_sinth_sinphi ...
                + (uk-ck)*dz_costh;
            exp_jk_rpr = exp(j*k*rpr);

            Ntheta = Ntheta ...
                + (cjxzp(mi,ni-li+1,nj-lj+1,1).*dx_dy_costh_cosphi ...
                + cjyzp(mi,ni-li+1,nj-lj+1,1).*dx_dy_costh_sinphi) ...
                .*exp_jk_rpr;
            Ltheta = Ltheta ...
                + (cmxzp(mi,ni-li+1,nj-lj+1,1).*dx_dy_costh_cosphi ...
                + cmyzp(mi,ni-li+1,nj-lj+1,1).*dx_dy_costh_sinphi) ...
                .*exp_jk_rpr;
            Nphi = Nphi + (-cjxzp(mi,ni-li+1,nj-lj+1,1) ...
                .*dx_dy_sinphi+cjyzp(mi,ni-li+1,nj-lj+1,1).*dx_dy_cosphi)...
                .*exp_jk_rpr;
            Lphi = Lphi + (-cmxzp(mi,ni-li+1,nj-lj+1,1) ...
                .*dx_dy_sinphi+cmyzp(mi,ni-li+1,nj-lj+1,1).*dx_dy_cosphi)...
                .*exp_jk_rpr;

            % for -az direction

            rpr = (ni-ci+0.5)*dx_sinth_cosphi ...
                + (nj - cj + 0.5)*dy_sinth_sinphi ...
                + (lk-ck)*dz_costh;
            exp_jk_rpr = exp(j*k*rpr);

            Ntheta = Ntheta ...
                + (cjxzn(mi,ni-li+1,nj-lj+1,1).*dx_dy_costh_cosphi ...
                + cjyzn(mi,ni-li+1,nj-lj+1,1).*dx_dy_costh_sinphi)...
                .*exp_jk_rpr;
            Ltheta = Ltheta ...
                + (cmxzn(mi,ni-li+1,nj-lj+1,1).*dx_dy_costh_cosphi ...
                + cmyzn(mi,ni-li+1,nj-lj+1,1).*dx_dy_costh_sinphi) ...
                .*exp_jk_rpr;
            Nphi = Nphi + (-cjxzn(mi,ni-li+1,nj-lj+1,1) ...
                .*dx_dy_sinphi+cjyzn(mi,ni-li+1,nj-lj+1,1).*dx_dy_cosphi)...
                .*exp_jk_rpr;
            Lphi = Lphi + (-cmxzn(mi,ni-li+1,nj-lj+1,1) ...
                .*dx_dy_sinphi+cmyzn(mi,ni-li+1,nj-lj+1,1).*dx_dy_cosphi)...
                .*exp_jk_rpr;
            end
        end
    % calculate directivity
    farfield_dataTheta(mi,:)=(k^2./(8*pi*eta_0*radiated_power(mi))) ...
        .* (abs(Lphi+eta_0*Ntheta).^2);
    farfield_dataPhi(mi,:)  =(k^2./(8*pi*eta_0*radiated_power(mi))) ...
        .* (abs(Ltheta-eta_0*Nphi).^2);
end
```

9.4 Simulation examples

In the previous sections we discussed the NF–FF transformation algorithm and demonstrated its implementation in MATLAB programs. In this section we provide examples of antennas and their simulation results including the radiation patterns.

9.4.1 Inverted-F antenna

In this section an inverted-F antenna is discussed, and its FDTD simulation results are presented and compared with the published results in [37]. The antenna layout and dimensions are shown in Figure 9.13; the dimensions are slightly modified compared with the ones in [37] to have the conductor patches snapped to a 0.4 mm × 0.4 mm grid in the *yz* plane. The antenna substrate is 0.787 mm thick and has 2.2 dielectric constant.

The FDTD problem space is composed of cells with $\Delta x = 0.262$ mm, $\Delta y = 0.4$ mm, and $\Delta z = 0.4$ mm. The boundaries are terminated by an 8 cells thickness convolutional perfectly matched layer (CPML) and a 10 cells air gap is left between the objects in the problem space and the CPML boundaries. The definition of the cell sizes, material types, and CPML parameters are illustrated in Listing 9.10, and the definition of the geometry is given in Listing 9.11. The antenna is excited using a voltage source, and a sampled voltage and a sampled current are defined on and around the voltage source to form an input port. The definitions of the source as well as the outputs are shown in Listings 9.12 and 9.13, respectively. One can notice in Listing 9.13 that the definition of the NF–FF transformation is part of the ***define_outputs_parameters*** routine. In this example the far-field frequencies are 2.4 and 5.8 GHz. One additional parameter for the NF–FF transform is the **number_of_cells_from_outer_boundary**. The value of this parameter is 13, which means that the imaginary NF–FF surface is 13

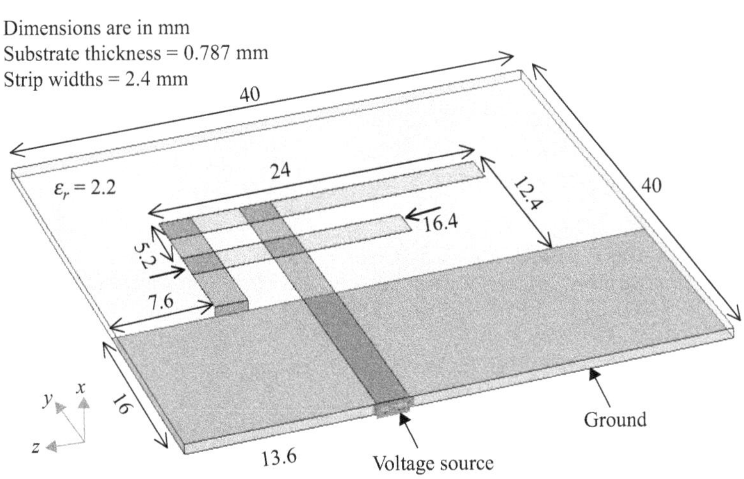

Figure 9.13 An inverted-F antenna.

Listing 9.10 define_problem_space_parameters.m

```
disp('defining the problem space parameters');

% maximum number of time steps to run FDTD simulation
number_of_time_steps = 7000;

% A factor that determines duration of a time step
% wrt CFL limit
courant_factor = 0.9;

% A factor determining the accuracy limit of FDTD results
number_of_cells_per_wavelength = 20;

% Dimensions of a unit cell in x, y, and z directions (meters)
dx = 0.262e-3;
dy = 0.4e-3;
dz = 0.4e-3;

% ==<boundary conditions>========
% Here we define the boundary conditions parameters
% 'pec' : perfect electric conductor
% 'cpml' : conlvolutional PML
% if cpml_number_of_cells is less than zero
% CPML extends inside of the domain rather than outwards

boundary.type_xn = 'cpml';
boundary.air_buffer_number_of_cells_xn = 10;
boundary.cpml_number_of_cells_xn = 8;

boundary.type_xp = 'cpml';
boundary.air_buffer_number_of_cells_xp = 10;
boundary.cpml_number_of_cells_xp = 8;

boundary.type_yn = 'cpml';
boundary.air_buffer_number_of_cells_yn = 10;
boundary.cpml_number_of_cells_yn = 8;

boundary.type_yp = 'cpml';
boundary.air_buffer_number_of_cells_yp = 10;
boundary.cpml_number_of_cells_yp = 8;

boundary.type_zn = 'cpml';
boundary.air_buffer_number_of_cells_zn = 10;
boundary.cpml_number_of_cells_zn = 8;

boundary.type_zp = 'cpml';
boundary.air_buffer_number_of_cells_zp = 10;
boundary.cpml_number_of_cells_zp = 8;

boundary.cpml_order = 3;
boundary.cpml_sigma_factor = 1.3;
boundary.cpml_kappa_max = 7;
boundary.cpml_alpha_min = 0;
boundary.cpml_alpha_max = 0.05;
```

```
% PEC : perfect electric conductor
material_types(2).eps_r   = 1;
material_types(2).mu_r    = 1;
material_types(2).sigma_e = 1e10;
material_types(2).sigma_m = 0;
material_types(2).color   = [1 0 0];

% substrate
material_types(4).eps_r   = 2.2;
material_types(4).mu_r    = 1;
material_types(4).sigma_e = 0;
material_types(4).sigma_m = 0;
material_types(4).color   = [0 0 1];
```

Listing 9.11 define_geometry.m

```
disp('defining the problem geometry');

bricks  = [];
spheres = [];

% define substrate
bricks(1).min_x = -0.787e-3;
bricks(1).min_y = 0;
bricks(1).min_z = 0;
bricks(1).max_x = 0;
bricks(1).max_y = 40e-3;
bricks(1).max_z = 40e-3;
bricks(1).material_type = 4;

bricks(2).min_x = 0;
bricks(2).min_y = 0;
bricks(2).min_z = 24e-3;
bricks(2).max_x = 0;
bricks(2).max_y = 28.4e-3;
bricks(2).max_z = 26.4e-3;
bricks(2).material_type = 2;

bricks(3).min_x = 0;
bricks(3).min_y = 16e-3;
bricks(3).min_z = 30e-3;
bricks(3).max_x = 0;
bricks(3).max_y = 28.4e-3;
bricks(3).max_z = 32.4e-3;
bricks(3).material_type = 2;

bricks(4).min_x = 0;
bricks(4).min_y = 26e-3;
bricks(4).min_z = 8.4e-3;
bricks(4).max_x = 0;
bricks(4).max_y = 28.4e-3;
bricks(4).max_z = 32.4e-3;
bricks(4).material_type = 2;
```

```

bricks(5).min_x = 0;
bricks(5).min_y = 20.8e-3;
bricks(5).min_z = 16e-3;
bricks(5).max_x = 0;
bricks(5).max_y = 23.2e-3;
bricks(5).max_z = 32.4e-3;
bricks(5).material_type = 2;

bricks(6).min_x = -0.787e-3;
bricks(6).min_y = 16e-3;
bricks(6).min_z = 30e-3;
bricks(6).max_x = 0;
bricks(6).max_y = 16e-3;
bricks(6).max_z = 32.4e-3;
bricks(6).material_type = 2;

bricks(7).min_x = -0.787e-3;
bricks(7).min_y = 0;
bricks(7).min_z = 0;
bricks(7).max_x = -0.787e-3;
bricks(7).max_y = 16e-3;
bricks(7).max_z = 40e-3;
bricks(7).material_type = 2;
```

Listing 9.12 define_sources_and_lumped_elements.m

```
disp('defining sources and lumped element components');

voltage_sources = [];
current_sources = [];
diodes = [];
resistors = [];
inductors = [];
capacitors = [];

% define source waveform types and parameters
waveforms.gaussian(1).number_of_cells_per_wavelength = 0;
waveforms.gaussian(2).number_of_cells_per_wavelength = 15;

% voltage sources
% direction: 'xp', 'xn', 'yp', 'yn', 'zp', or 'zn'
% resistance : ohms, magitude  : volts
voltage_sources(1).min_x = -0.787e-3;
voltage_sources(1).min_y = 0;
voltage_sources(1).min_z =  24e-3;
voltage_sources(1).max_x = 0;
voltage_sources(1).max_y = 0;
voltage_sources(1).max_z =  26.4e-3;
voltage_sources(1).direction = 'xp';
voltage_sources(1).resistance = 50;
voltage_sources(1).magnitude = 1;
voltage_sources(1).waveform_type = 'gaussian';
voltage_sources(1).waveform_index = 1;
```

Listing 9.13 define_output_parameters.m

```
disp('defining output parameters');

sampled_electric_fields = [];
sampled_magnetic_fields = [];
sampled_voltages = [];
sampled_currents = [];
ports = [];
farfield.frequencies = [];

% figure refresh rate
plotting_step = 100;

% mode of operation
run_simulation = true;
show_material_mesh = true;
show_problem_space = true;

% far field calculation parameters
farfield.frequencies(1) = 2.4e9;
farfield.frequencies(2) = 5.8e9;
farfield.number_of_cells_from_outer_boundary = 13;

% frequency domain parameters
frequency_domain.start = 20e6;
frequency_domain.end   = 10e9;
frequency_domain.step  = 20e6;

% define sampled voltages
sampled_voltages(1).min_x = -0.787e-3;
sampled_voltages(1).min_y = 0;
sampled_voltages(1).min_z = 24.4e-3;
sampled_voltages(1).max_x = 0;
sampled_voltages(1).max_y = 0;
sampled_voltages(1).max_z = 26.4e-3;
sampled_voltages(1).direction = 'xp';
sampled_voltages(1).display_plot = false;

% define sampled currents
sampled_currents(1).min_x = -0.39e-3;
sampled_currents(1).min_y = 0;
sampled_currents(1).min_z = 24e-3;
sampled_currents(1).max_x = -0.39e-3;
sampled_currents(1).max_y = 0;
sampled_currents(1).max_z = 26.4e-3;
sampled_currents(1).direction = 'xp';
sampled_currents(1).display_plot = false;

% define ports
ports(1).sampled_voltage_index = 1;
ports(1).sampled_current_index = 1;
ports(1).impedance = 50;
ports(1).is_source_port = true;
```

cells away from the outer boundaries, thus 5 cells away from the CPML interface and residing in the air gap region.

The FDTD simulation of the inverted-F antenna is performed with 7,000 time steps, and S_{11} of the input port is calculated. The calculated input reflection coefficient is plotted in Figure 9.14 and shows a good agreement with the published measurement data in [37].

Furthermore, the directivity patterns are calculated in the *xy*, *xz*, and *yz* plane cuts at 2.4 and 5.8 GHz. These patterns are plotted in Figures 9.15–9.17, respectively. The simulated FDTD radiation patterns show very good agreement with the published measurement data [37].

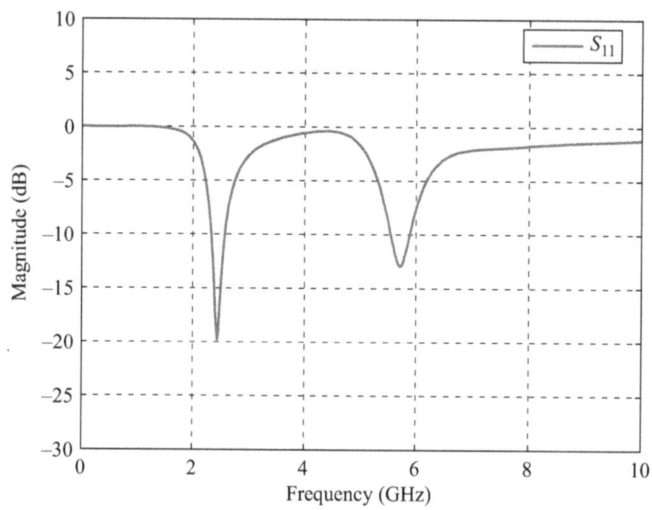

Figure 9.14 Input reflection coefficient of the inverted-F antenna.

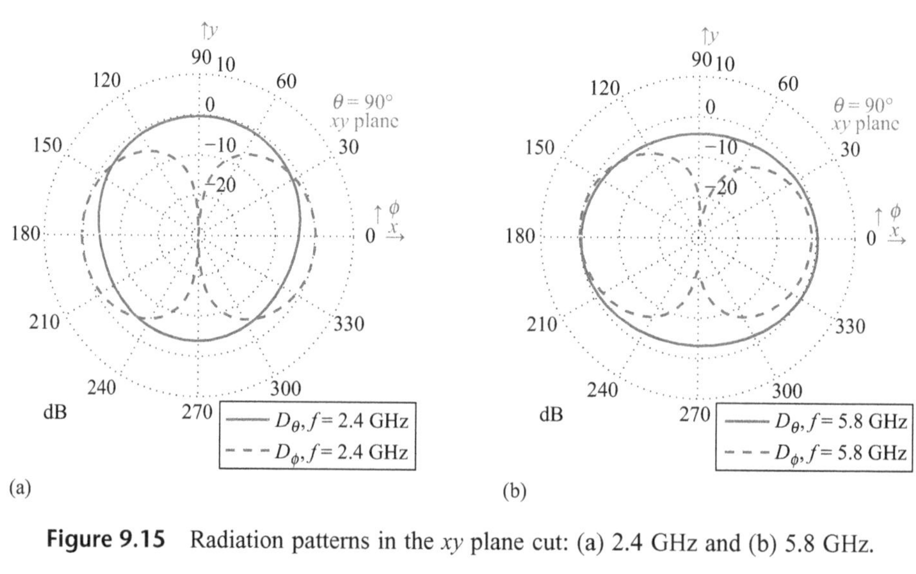

Figure 9.15 Radiation patterns in the *xy* plane cut: (a) 2.4 GHz and (b) 5.8 GHz.

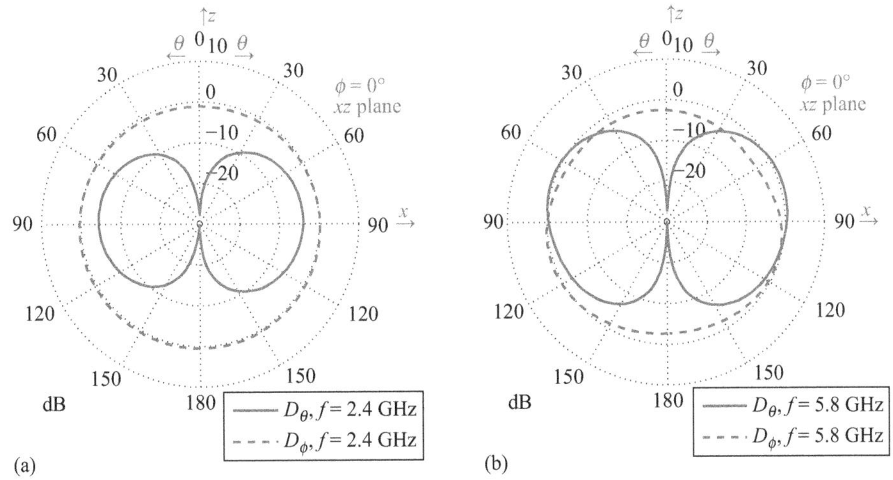

Figure 9.16 Radiation patterns in the *xz* plane cut: (a) 2.4 GHz and (b) 5.8 GHz.

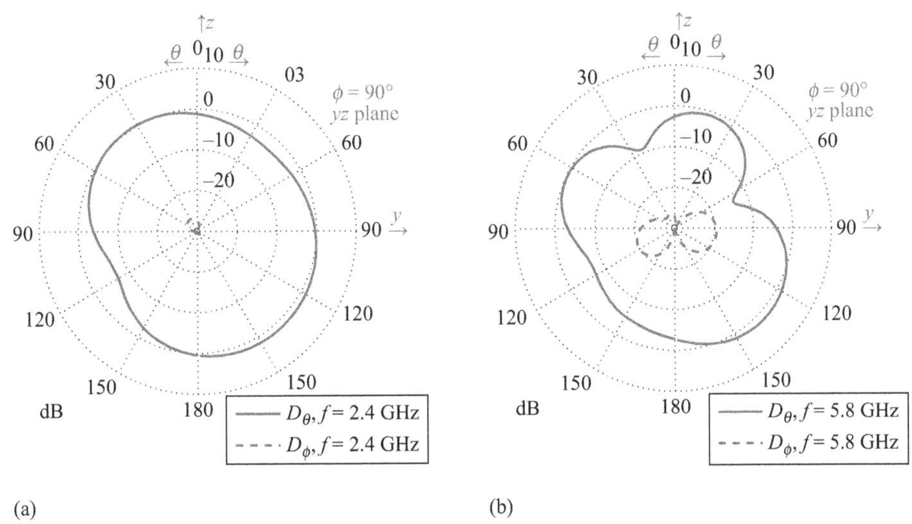

Figure 9.17 Radiation patterns in the *yz* plane cut: (a) 2.4 GHz and (b) 5.8 GHz.

9.4.2 Strip-fed rectangular dielectric resonator antenna

In this section, the FDTD simulation of a strip-fed dielectric resonator antenna (DRA) is discussed [38]. The example antenna is illustrated in Figure 9.18 together with its dimensions. The rectangular dielectric resonator has dimensions of 14.3 mm, 25.4 mm, and 26.1 mm

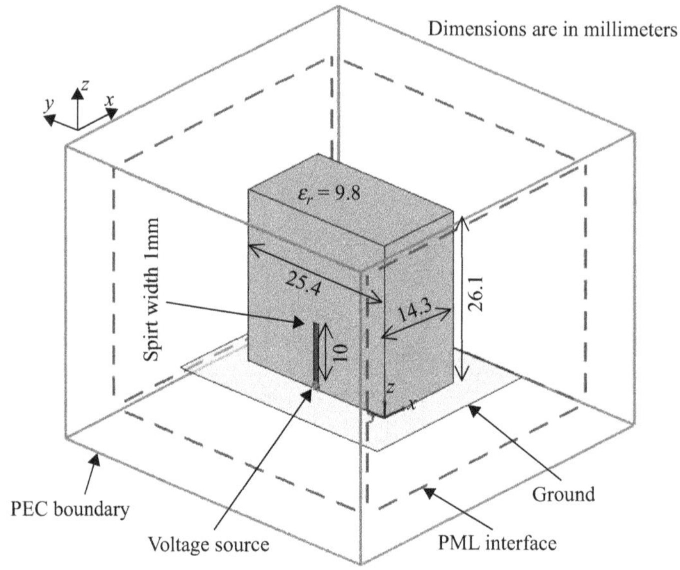

Figure 9.18 A strip-fed rectangular DRA.

in x, y, and z directions, respectively, and dielectric constant of 9.8. The resonator stands on a ground plane and is fed by a strip having 1 mm width and 10 mm height. The antenna is simulated by a probe feeding. To simulate the probe feeding, a 1 mm gap is left between the ground plane and the strip, and a voltage source is placed in this gap. Voltage and current are sampled across and around the voltage source, and they are tied together to form a port. The boundaries are terminated by an 8 cells thickness CPML, and a 10 cells air gap is left between the objects in the problem space and the CPML boundaries. The dimensions of a unit cell are $\Delta x = 0.715$ mm, $\Delta y = 0.508$ mm, and $\Delta z = 0.5$ mm. The definition of the geometry is described in Listing 9.14, and the definition of the voltage source is shown in Listing 9.15.

Listing 9.14 define_geometry.m

```
disp('defining the problem geometry');

bricks  = [];
spheres = [];

% define dielectric
bricks(1).min_x = 0;
bricks(1).min_y = 0;
bricks(1).min_z = 0;
bricks(1).max_x = 14.3e-3;
bricks(1).max_y = 25.4e-3;
bricks(1).max_z = 26.1e-3;
bricks(1).material_type = 4;
```

```
bricks(2).min_x = -8e-3;
bricks(2).min_y = -6e-3;
bricks(2).min_z = 0;
bricks(2).max_x = 22e-3;
bricks(2).max_y = 31e-3;
bricks(2).max_z = 0;
bricks(2).material_type = 2;

bricks(3).min_x = 0;
bricks(3).min_y = 12.2e-3;
bricks(3).min_z = 1e-3;
bricks(3).max_x = 0;
bricks(3).max_y = 13.2e-3;
bricks(3).max_z = 10e-3;
bricks(3).material_type = 2;
```

Listing 9.15 define_sources_and_lumped_elements.m

```
disp('defining sources and lumped element components');

voltage_sources = [];
current_sources = [];
diodes = [];
resistors = [];
inductors = [];
capacitors = [];

% define source waveform types and parameters
waveforms.gaussian(1).number_of_cells_per_wavelength = 0;
waveforms.gaussian(2).number_of_cells_per_wavelength = 15;

% voltage sources
% direction: 'xp', 'xn', 'yp', 'yn', 'zp', or 'zn'
% resistance : ohms, magitude   : volts
voltage_sources(1).min_x = 0;
voltage_sources(1).min_y = 12.2e-3;
voltage_sources(1).min_z = 0;
voltage_sources(1).max_x = 0;
voltage_sources(1).max_y = 13.2e-3;
voltage_sources(1).max_z = 1e-3;
voltage_sources(1).direction = 'zp';
voltage_sources(1).resistance = 50;
voltage_sources(1).magnitude = 1;
voltage_sources(1).waveform_type = 'gaussian';
voltage_sources(1).waveform_index = 1;
```

The definition of the output parameters is performed in ***define_output_parameters***, the contents of which are shown in Listing 9.16. The output parameters to be defined are a sampled voltage, a sampled current, and a port. Furthermore, far-field frequencies are defined for the NF–FF transformation. The far-field frequencies are 3.5 and 4.3 GHz.

Listing 9.16 define_output_parameters.m

```
disp('defining_output_parameters');

sampled_electric_fields = [];
sampled_magnetic_fields = [];
sampled_voltages = [];
sampled_currents = [];
ports = [];
farfield.frequencies = [];

% figure refresh rate
plotting_step = 20;

% mode of operation
run_simulation = true;
show_material_mesh = true;
show_problem_space = true;

% far field calculation parameters
farfield.frequencies(1) = 3.5e9;
farfield.frequencies(2) = 4.3e9;
farfield.number_of_cells_from_outer_boundary = 13;

% frequency domain parameters
frequency_domain.start = 2e9;
frequency_domain.end   = 6e9;
frequency_domain.step  = 20e6;

% define sampled voltages
sampled_voltages(1).min_x = 0;
sampled_voltages(1).min_y = 12.2e-3;
sampled_voltages(1).min_z = 0;
sampled_voltages(1).max_x = 0;
sampled_voltages(1).max_y = 13.2e-3;
sampled_voltages(1).max_z = 1e-3;
sampled_voltages(1).direction = 'zp';
sampled_voltages(1).display_plot = false;

% define sampled currents
sampled_currents(1).min_x = 0;
sampled_currents(1).min_y = 12.2e-3;
sampled_currents(1).min_z = 0.5e-3;
sampled_currents(1).max_x = 0;
sampled_currents(1).max_y = 13.2e-3;
sampled_currents(1).max_z = 0.5e-3;
sampled_currents(1).direction = 'zp';
sampled_currents(1).display_plot = false;

% define ports
ports(1).sampled_voltage_index = 1;
ports(1).sampled_current_index = 1;
ports(1).impedance = 50;
ports(1).is_source_port = true;
```

Similar to the previous example, the value of **number_of_cells_from_outer_boundary** is 13, which means that the imaginary NF–FF surface is 13 cells away from the outer boundaries.

The FDTD simulation of the DRA is performed with 10,000 time steps, and S_{11} of the input port is calculated. The calculated input reflection coefficient is plotted in Figure 9.19, and it shows good agreement with the published simulation and measurement data in [38].

Furthermore, the directivity patterns are calculated in the *xy*, *xz*, and *yz* plane cuts at 3.5 and 4.3 GHz. These patterns are plotted in Figures 9.20, 9.21, and 9.22, respectively. The simulated FDTD radiation patterns show good agreement with the published data as well [38].

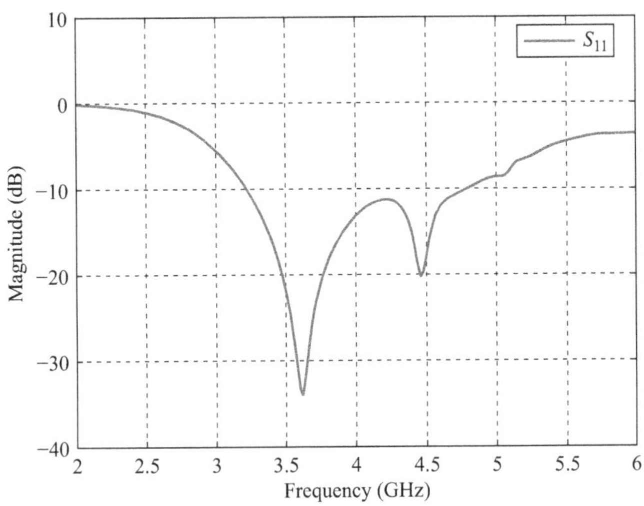

Figure 9.19 Input reflection coefficient of the strip-fed rectangular DRA.

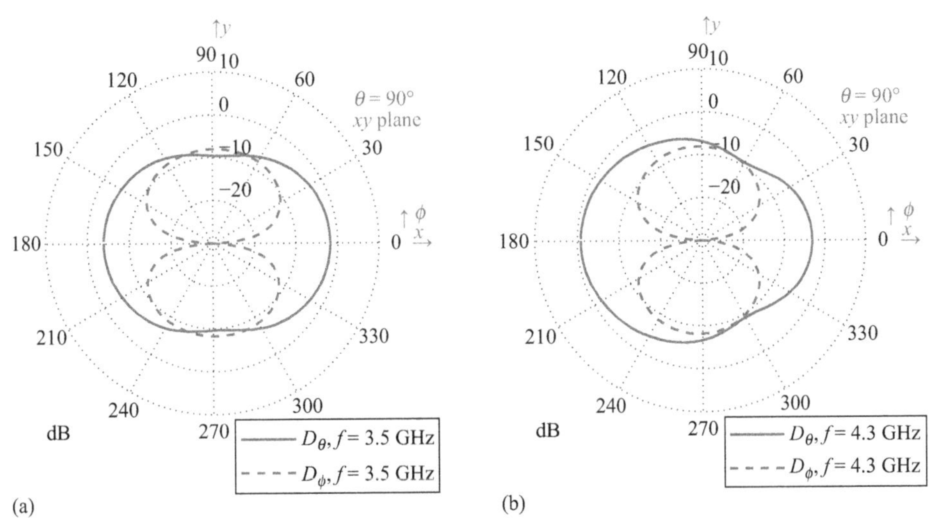

Figure 9.20 Radiation patterns in the *xy* plane cut: (a) 3.5 GHz and (b) 4.3 GHz.

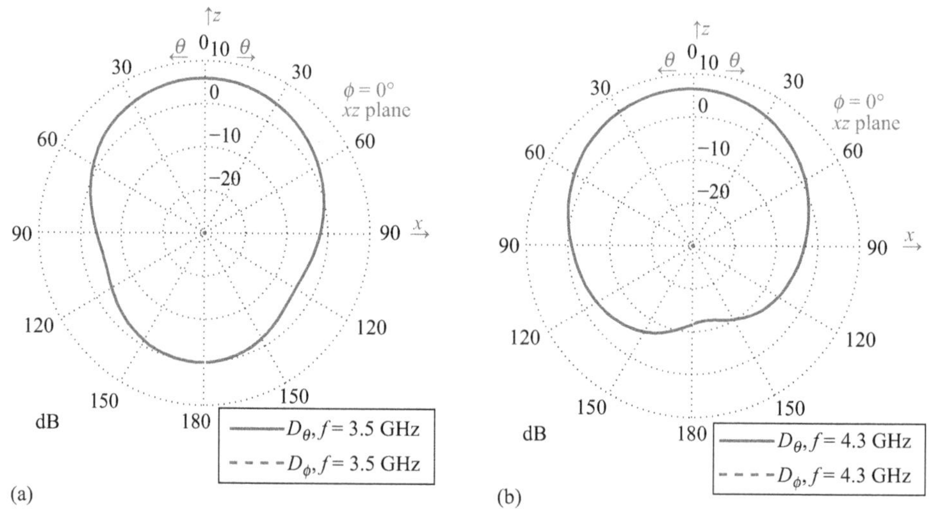

Figure 9.21 Radiation patterns in the *xz* plane cut: (a) 3.5 GHz and (b) 4.3 GHz.

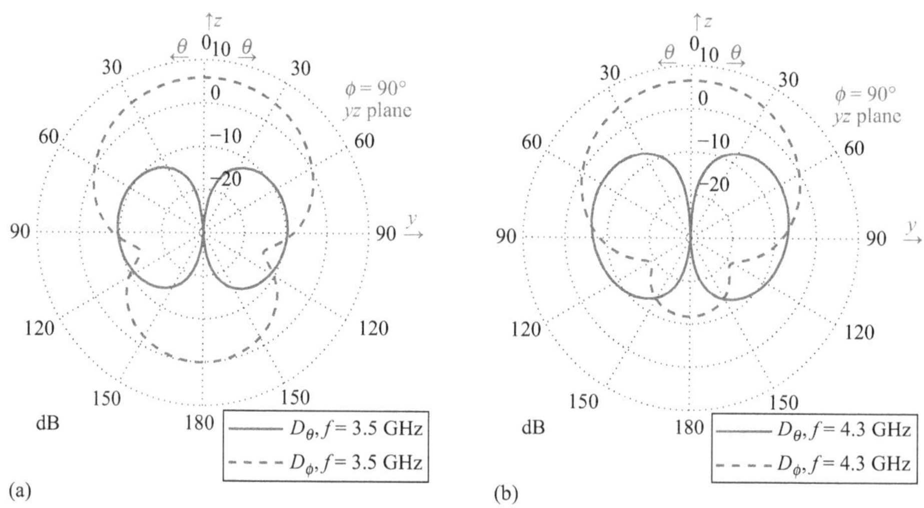

Figure 9.22 Radiation patterns in the *yz* plane cut: (a) 3.5 GHz and (b) 4.3 GHz.

9.5 Exercises

9.1 In this exercise we construct and study the characteristics of a half-wave dipole antenna. Create a problem space composed of a cubic Yee cell with 0.5 mm size on a side. Set the boundaries as CPML, and leave 10 cells air gap between the antenna and the CPML boundaries on all sides. Place a brick of square cross-section with 1 mm width and 14.5 mm height 0.5 mm above the origin. Place another brick with the same dimensions 0.5 mm below the origin. Thus, the dipole will be oriented in the *z* direction with 1 mm gap between its poles. Place a voltage source with 50 Ω internal impedance

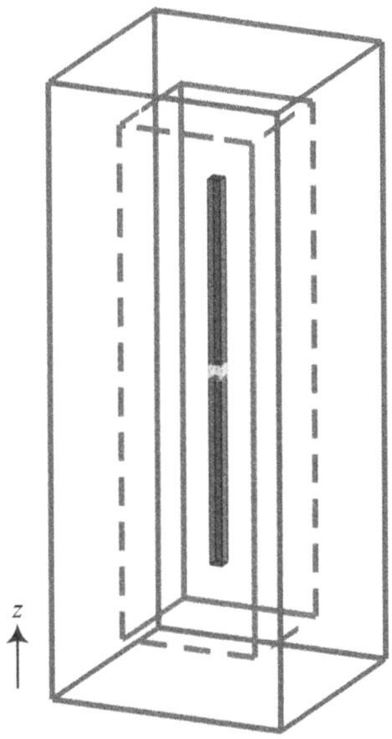

Figure 9.23 A dipole antenna.

between the poles. Then define a sampled voltage and a sampled current across and around the voltage source and associate them to a port. The geometry of the problem is illustrated in Figure 9.23. Run the simulation, and calculate the input impedance of the antenna using the frequency-domain simulation results. Identify the resonance frequency of the antenna from the input impedance results. Then rerun the simulation to calculate the radiation pattern at the resonance frequency.

9.2 Consider the dipole antenna that you constructed in Exercise 9.1. Examine the input impedance that you calculated, and record the real part of the input impedance at resonance frequency. Then change the resistance of the voltage source and the port to the resistance you recorded. Rerun the simulation, and obtain the scattering parameter (S-parameter) and radiation pattern results. Examine whether the S-parameters and the radiation patterns have changed.

9.3 In this exercise we try to construct a microstrip rectangular microstrip rectangular patch antenna. Define a problem space with grid size $\Delta x = 2$ mm, $\Delta y = 2$ mm, and $\Delta z = 0.95$ mm. Place a rectangular brick in the problem space as the substrate of the antenna with dimensions 60 mm $\times$ 40 mm $\times$ 1.9 mm and 2.2 dielectric constant. Place a PEC plate as the ground of the antenna on the bottom side of the substrate covering the entire surface area. Then place a PEC patch on the top surface of the substrate with 56 mm width and 20 mm length in the x and y directions, respectively. Make sure that the top patch is centered on the top surface of the substrate. The feeding point to the top patch will be at the center of one of the long edges of the patch. Place a voltage source with 50 Ω internal resistance between the ground plane and the feeding point, define a sampled voltage and a sampled current on the voltage source, and associate them to a port. The geometry of the problem is illustrated in Figure 9.24. Run the simulation, and obtain the input reflection coefficient (S_{11}) of the antenna. Verify that the antenna operates around 4.35 GHz. Then rerun the simulation to obtain the radiation patterns at the operation frequency.

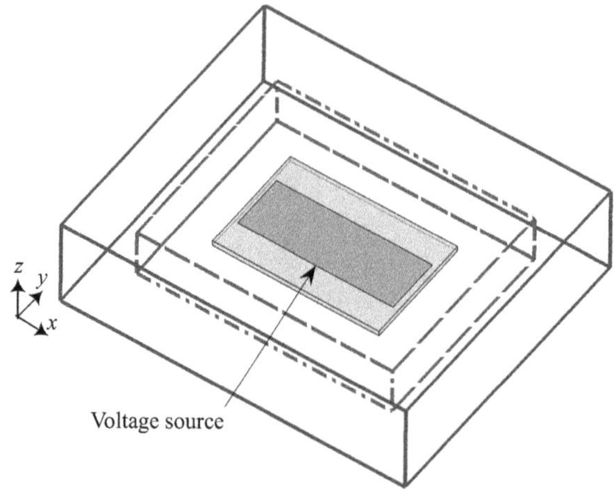

Figure 9.24 A microstrip patch antenna.

9.4 Consider the microstrip rectangular square patch antenna you constructed in Exercise 9.3. In this configuration the patch feeding is placed right at the center of the edge of the antenna. In this example the antenna will be fed through a microstrip line. Place a microstrip line with 6 mm width between the antenna feeding point and the edge of the substrate. This microstrip line with 6 mm width will simulate a characteristic impedance of 50 Ω. Notice that you will need to place the microstrip line off center by 1 mm to have it conforming to the FDTD grid. Move the voltage source, the sampled voltage, and the sampled current to the edge of the substrate where the microstrip line is extended. The geometry of the problem is illustrated in Figure 9.25. Run the simulation, and obtain the input reflection coefficient and the radiation pattern at the corresponding operating frequency. Does the existence of a feeding line change the radiation pattern?

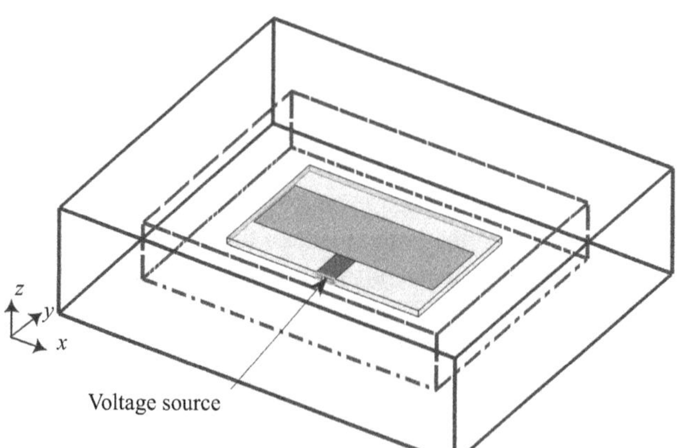

Figure 9.25 A microstrip patch antenna with a microstrip line feeding.

CHAPTER 10

Thin-wire modeling

So far we have assumed that all of the objects in the finite-difference time-domain (FDTD) problem space conform to the FDTD grid. Some subcell modeling techniques are developed to model geometries that do not conform to the grid or have dimensions smaller than cell dimensions. One of the most common such geometries is a thin wire that has a radius less than a cell size. In this chapter we discuss one of the subcell modeling techniques that is proposed to model thin wires in FDTD simulations.

There are various techniques proposed for modeling thin wires. However, we discuss the one proposed in [39], which is based on Faraday's law contour-path formulation. This model is easy to understand and to implement in the FDTD program.

10.1 Thin-wire formulation

Figure 10.1 illustrates a thin wire of radius a with its axis coinciding with a field component $E_z(i, j, k)$ on the FDTD grid. In the given example the radius a is smaller than the x and y cell dimensions. There are four magnetic field components circulating around $E_z(i, j, k)$: namely, $H_y(i, j, k)$, $H_x(i, j, k)$, $H_y(i-1, j, k)$, and $H_x(i, j-1, k)$ as shown in Figure 10.2. The thin wire model proposes special updating equations for these magnetic field components. We now demonstrate the derivation of the updating equation for the component $H_y(i, j, k)$; the derivation of updating equations for other field components follows the same procedure.

Figure 10.1 shows the magnetic field component $H_y(i, j, k)$ and four electric field components circulating around $H_y(i, j, k)$. Faraday's law given in integral form as

$$-\mu \int_S \frac{\partial \vec{H}}{\partial t} \cdot d\vec{s} = \oint_L \vec{E} \cdot d\vec{l} \tag{10.1}$$

can be applied on the enclosed surface shown in Figure 10.1 to establish the relation between $H_y(i, j, k)$ and the electric field components located on the boundaries of the enclosed surface. Before utilizing (10.1) it should be noted that the variation of the fields around the thin wire is assumed to be a function of $1/r$, where r represents the distance to the field position from the thin-wire axis [39]. In more explicit form H_y can be expressed as

$$H_y(r) = \frac{H_{y0}}{r}, \tag{10.2}$$

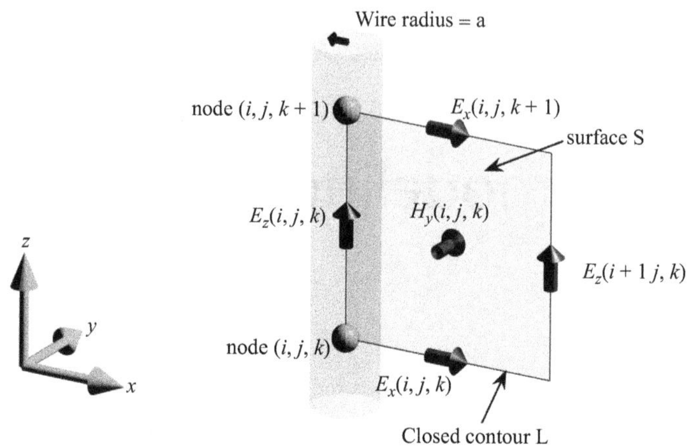

Figure 10.1 A thin wire with its axis coinciding with $E_z(i, j, k)$ and field components surrounding $H_y(i, j, k)$.

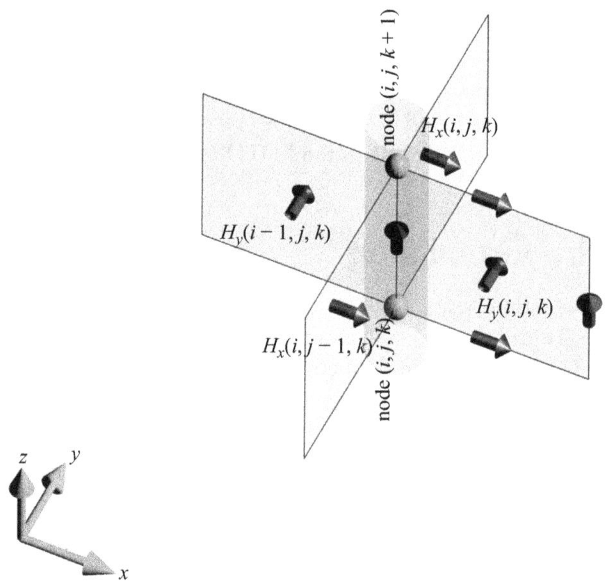

Figure 10.2 Magnetic field components surrounding the thin wire.

and, similarly, E_x can be expressed as

$$E_x(r) = \frac{E_{x0}}{r}, \tag{10.3}$$

where H_{y0} and E_{x0} are constants. In Figure 10.1 one can observe that the field components $H_y(i, j, k)$, $E_x(i, j, k)$, and $E_x(i, j, k + 1)$ are located at $r = \Delta x/2$. For instance,

$$H_y\left(\frac{\Delta x}{2}\right) = \frac{2H_{y0}}{\Delta x} = H_y(i, j, k). \tag{10.4}$$

Then

$$H_{y0} = \frac{H_y(i,j,k)\Delta x}{2}, \tag{10.5}$$

hence,

$$H_y(r) = \frac{H_y(i,j,k)\Delta x}{2r}. \tag{10.6}$$

Similarly,

$$E_x(r)|_{j,k} = \frac{E_x(i,j,k)\Delta x}{2r}, \tag{10.7}$$

and

$$E_x(r)|_{j,k+1} = \frac{E_x(i,j,k+1)\Delta x}{2r}. \tag{10.8}$$

Using (10.6), (10.7), and (10.8) in (10.1) one can obtain

$$\begin{aligned} -\mu \int_{z=0}^{z=\Delta z} \int_{r=a}^{r=\Delta x} \frac{\partial}{\partial t} \frac{H_y(i,j,k)\Delta x}{2r} drdz \\ = \int_{z=0}^{z=\Delta z} E_z(i,j,k)dz + \int_{z=\Delta z}^{z=0} E_z(i+1,j,k)dz \\ + \int_{r=a}^{r=\Delta x} \frac{E_x(i,j,k+1)\Delta x}{2r} dr + \int_{r=\Delta x}^{r=a} \frac{E_x(i,j,k)\Delta x}{2r} dr. \end{aligned} \tag{10.9}$$

Notice that the electric field should vanish inside the wire; therefore, the integration limits are from a to Δx. Due to the same reasoning, $E_z(i, j, k)$ is zero as well. Then evaluation of (10.9) yields

$$\begin{aligned} \frac{-\mu\Delta z\Delta x}{2} \ln\left(\frac{\Delta x}{a}\right) \frac{\partial H_y(i,j,k)}{\partial t} = -\Delta z E_z(i+1,j,k) \\ + \ln\left(\frac{\Delta x}{a}\right) \frac{\Delta x}{2} E_x(i,j,k+1) - ln\left(\frac{\Delta x}{a}\right) \frac{\Delta x}{2} E_x(i,j,k). \end{aligned} \tag{10.10}$$

After applying the central difference approximation to the time derivative of magnetic field component and rearranging the terms, one can obtain the new value of $H_y(i, j, k)$ in terms of other components as

$$\begin{aligned} H_y^{n+1/2}(i,j,k) = H_y^{n-1/2}(i,j,k) + \frac{2\Delta t}{\mu\Delta x \ln\left(\frac{\Delta x}{a}\right)} E_z^n(i+1,j,k) \\ - \frac{\Delta t}{\mu\Delta z}\left(E_x^n(i,j,k+1) - E_x^n(i,j,k)\right). \end{aligned} \tag{10.11}$$

Equation (10.12) can be written in the same form as the general updating equation for H_y (1.30), such that

$$H_y^{n+\frac{1}{2}}(i,j,k) = C_{hyh}(i,j,k) \times H_y^{n-\frac{1}{2}}(i,j,k) + C_{hyez}(i,j,k) \times \left(E_z^n(i+1,j,k) - E_z^n(i,j,k)\right) + C_{hyex}(i,j,k) \times \left(E_x^n(i,j,k+1) - E_x^n(i,j,k)\right), \tag{10.12}$$

where

$$C_{hyh}(i,j,k) = 1, C_{hyez}(i,j,k) = \frac{2\Delta t}{\mu_y(i,j,k)\Delta x ln\left(\frac{\Delta x}{a}\right)},$$

$$C_{hyex}(i,j,k) = -\frac{\Delta t}{\mu_y(i,j,k)\Delta z}.$$

Then one only need to modify the updating coefficients before the FDTD time-marching loop according to (10.12) to have $H_y^{n+\frac{1}{2}}(i,j,k)$ updated due to a thin wire between nodes (i, j, k) and $(i, j, k + 1)$. Furthermore, we have enforced $E_z(i, j, k)$ to be zero. This can be accomplished similarly by setting the updating coefficients for $E_z^n(i, j, k)$ in (1.28) as zero. That means $C_{eze}(i, j, k)$, $C_{ezby}(i, j, k)$, and $C_{ezby}(i, j, k)$ should be assigned zeros before the FDTD time-marching procedure begins.

As mentioned before, there are four magnetic field components circulating around $E_z^n(i, j, k)$ as illustrated in Figure 10.2; therefore, all these magnetic field components need to be updated based on thin-wire modeling. The updating equation for $H_y(i, j, k)$ is given in (10.11). Similarly, updating equations can be obtained for the other three components as

$$H_y^{n+\frac{1}{2}}(i-1,j,k) = C_{hyh}(i-1,j,k) \times H_y^{n-\frac{1}{2}}(i-1,j,k) + C_{hyez}(i-1,j,k) \times \left(E_z^n(i,j,k) - E_z^n(i-1,j,k)\right) + C_{hyex}(i-1,j,k) \times \left(E_x^n(i-1,j,k+1) - E_x^n(i-1,j,k)\right), \tag{10.13}$$

where

$$C_{hyh}(i-1,j,k) = 1, C_{hyez}(i-1,j,k) = \frac{2\Delta t}{\mu_y(i-1,j,k)\Delta x ln\left(\frac{\Delta x}{a}\right)},$$

$$C_{hyex}(i-1,j,k) = -\frac{\Delta t}{\mu_y(i-1,j,k)\Delta z}.$$

$$H_x^{n+\frac{1}{2}}(i,j,k) = C_{hxh}(i,j,k) \times H_x^{n-\frac{1}{2}}(i,j,k) + C_{hxey}(i,j,k) \times \left(E_y^n(i,j,k+1) - E_y^n(i,j,k)\right) + C_{hxez}(i,j,k) \times \left(E_z^n(i,j+1,k) - E_z^n(i,j,k)\right), \tag{10.14}$$

where

$$C_{hxh}(i,j,k) = 1, \quad C_{hyey}(i,j,k) = \frac{\Delta t}{\mu_x(i,j,k)\Delta z},$$

$$C_{hxez}(i,j,k) = -\frac{2\Delta t}{\mu_x(i,j,k)\Delta y \ln\left(\frac{\Delta y}{a}\right)}.$$

and

$$\begin{aligned} H_x^{n+\frac{1}{2}}(i,j-1,k) &= C_{hxh}(i,j-1,k) \times H_x^{n-\frac{1}{2}}(i,j-1,k) \\ &\quad + C_{hxey}(i,j-1,k) \times \left(E_y^n(i,j-1,k+1) - E_y^n(i,j-1,k)\right) \\ &\quad + C_{hxez}(i,j-1,k) \times \left(E_z^n(i,j,k) - E_z^n(i,j-1,k)\right), \end{aligned} \tag{10.15}$$

where

$$C_{hxh}(i,j-1,k) = 1, \quad C_{hxey}(i,j-1,k) = \frac{\Delta t}{\mu_x(i,j-1,k)\Delta z},$$
$$C_{hxez}(i,j-1,k) = -\frac{2\Delta t}{\mu_x(i,j-1,k)\Delta y \ln\left(\frac{\Delta y}{a}\right)}.$$

10.2 MATLAB® implementation of the thin-wire formulation

Having derived the updating equations for modeling thin wires, these equations can be implemented in the FDTD program. The first step is the definition of the thin-wire parameters. The thin wires are defined as a part of the geometry, and their definition implementation is provided in the subroutine ***define_geometry***, as shown in Listing 10.1.

Listing 10.1 define_geometry.m

```
disp('defining the problem geometry');

bricks  = [];
spheres = [];
thin_wires = [];

% define a thin wire
thin_wires(1).min_x = 0;
thin_wires(1).min_y = 0;
thin_wires(1).min_z = 1e-3;
thin_wires(1).max_x = 0;
thin_wires(1).max_y = 0;
thin_wires(1).max_z = 10e-3;
thin_wires(1).radius = 0.25e-3;
thin_wires(1).direction = 'z';

% define a thin wire
thin_wires(2).min_x = 0;
thin_wires(2).min_y = 0;
thin_wires(2).min_z = -10e-3;
thin_wires(2).max_x = 0;
thin_wires(2).max_y = 0;
thin_wires(2).max_z = -1e-3;
thin_wires(2).radius = 0.25e-3;
thin_wires(2).direction = 'z';
```

A new parameter called **thin_wires** is defined and initialized as an empty structure array. This parameter is used to hold the respective thin-wire parameter values. It is assumed that thin wires are aligned parallel to either the x, y, or z axis; therefore, the **direction** field of **thin_wires** is set accordingly. Then other parameters **min_x**, **min_y**, **min_z**, **max_x**, **max_y**, and **max_z** are assigned values to indicate the position of the thin wire in three-dimensional space. Since a thin wire is like a one-dimensional object, its length is determined by the given position parameters, while its thickness is determined by its radius. Therefore, **radius** is another field of **thin_wires** that holds the value of the radius. Although there are six position parameters, two of these are actually redundant in the current implementation. For instance, if the thin wire is aligned in the x direction, the parameters **min_x**, **min_y**, **min_z**, and **max_x** are sufficient to describe the location of the thin wire. The other two parameters **max_y** and **max_z** are assigned the same values as **min_y** and **min_z** as shown in Listing 10.1.

After the definition process, the initialization process is performed. Since the thin wire is a new type of geometry that would determine the size of the problem space, necessary codes must be implemented in the subroutine ***calculate_domain_size***. The code for this task is shown in Listing 10.2.

The next step in the thin-wire implementation is the initialization of the thin-wire parameters through the updating coefficients. A new subroutine named ***initialize_thin_wire_updating_coefficients*** is implemented for initialization of the thin-wire updating coefficients. This subroutine is named ***initialize_updating_coefficients*** and is called after the initialization subroutines for the updating coefficients of other types of objects are called. The implementation of ***initialize_thin_wire_updating_coefficients*** is shown in Listing 10.3. In this subroutine, the electric and magnetic field coefficients associated with thin wires are updated based on the equations derived in Section 10.1.

As the updating coefficients associated with the thin wires are updated appropriately, the implementation of the thin-wire formulation is completed. The thin-wire implementation only requires an extra preprocessing step. Then during the FDTD time-marching loop the updating coefficients will manipulate the fields for modeling the thin-wire behavior.

Listing 10.2 calculate_domain_size.m

```
disp('calculating the number of cells in the problem space');

number_of_spheres = size(spheres,2);
number_of_bricks  = size(bricks,2);
number_of_thin_wires  = size(thin_wires,2);
for i=1:number_of_thin_wires
    min_x(number_of_objects) = thin_wires(i).min_x;
    min_y(number_of_objects) = thin_wires(i).min_y;
    min_z(number_of_objects) = thin_wires(i).min_z;
    max_x(number_of_objects) = thin_wires(i).max_x;
    max_y(number_of_objects) = thin_wires(i).max_y;
    max_z(number_of_objects) = thin_wires(i).max_z;
    number_of_objects = number_of_objects + 1;
end
```

Listing 10.3 initialize_thin_wire_updating_coefficients.m

```
disp('initializing thin wire updating coefficients');

dtm = dt/mu_0;

for ind = 1:number_of_thin_wires
    is = round((thin_wires(ind).min_x - fdtd_domain.min_x)/dx)+1;
    js = round((thin_wires(ind).min_y - fdtd_domain.min_y)/dy)+1;
    ks = round((thin_wires(ind).min_z - fdtd_domain.min_z)/dz)+1;
    ie = round((thin_wires(ind).max_x - fdtd_domain.min_x)/dx)+1;
    je = round((thin_wires(ind).max_y - fdtd_domain.min_y)/dy)+1;
    ke = round((thin_wires(ind).max_z - fdtd_domain.min_z)/dz)+1;
    r_o = thin_wires(ind).radius;

    switch (thin_wires(ind).direction(1))
        case 'x'
            Cexe (is:ie-1,js,ks) = 0;
            Cexhy(is:ie-1,js,ks) = 0;
            Cexhz(is:ie-1,js,ks) = 0;
            Chyh (is:ie-1,js,ks-1:ks) = 1;
            Chyez(is:ie-1,js,ks-1:ks) = dtm ...
                ./ (mu_r_y(is:ie-1,js,ks-1:ks) * dx);
            Chyex(is:ie-1,js,ks-1:ks) = -2 * dtm ...
                ./ (mu_r_y(is:ie-1,js,ks-1:ks) * dz * log(dz/r_o));
            Chzh( is:ie-1,js-1:js,ks) = 1;
            Chzex(is:ie-1,js-1:js,ks) = 2 * dtm ...
                ./ (mu_r_z(is:ie-1,js-1:js,ks) * dy * log(dy/r_o));
            Chzey(is:ie-1,js-1:js,ks) = -dtm ...
                ./ (mu_r_z(is:ie-1,js-1:js,ks) * dx);
        case 'y'
            Ceye (is,js:je-1,ks) = 0;
            Ceyhx(is,js:je-1,ks) = 0;
            Ceyhz(is,js:je-1,ks) = 0;
            Chzh (is-1:is,js:je-1,ks) = 1;
            Chzex(is-1:is,js:je-1,ks) = dtm ...
                ./ (mu_r_z(is-1:is,js:je-1,ks) * dy);
            Chzey(is-1:is,js:je-1,ks) = -2 * dtm ...
                ./ (mu_r_z(is-1:is,js:je-1,ks) * dx * log(dx/r_o));
            Chxh (is,js:je-1,ks-1:ks) = 1;
            Chxey(is,js:je-1,ks-1:ks) = 2 * dtm ...
                ./ (mu_r_x(is,js:je-1,ks-1:ks) * dz * log(dz/r_o));
            Chxez(is,js:je-1,ks-1:ks) = -dtm ...
                ./ (mu_r_x(is,js:je-1,ks-1:ks) * dy);

        case 'z'
            Ceze (is,js,ks:ke-1) = 0;
            Cezhx(is,js,ks:ke-1) = 0;
            Cezhy(is,js,ks:ke-1) = 0;
            Chxh (is,js-1:js,ks:ke-1) = 1;
            Chxey(is,js-1:js,ks:ke-1) = dtm ...
                ./ (mu_r_x(is,js-1:js,ks:ke-1) * dz);
            Chxez(is,js-1:js,ks:ke-1) = -2 * dtm ...
                ./ (mu_r_x(is,js-1:js,ks:ke-1) * dy * log(dy/r_o));
            Chyh (is-1:is,js,ks:ke-1) = 1;
            Chyez(is-1:is,js,ks:ke-1) = 2 * dtm ...
                ./ (mu_r_y(is-1:is,js,ks:ke-1) * dx * log(dx/r_o));
            Chyex(is-1:is,js,ks:ke-1) = -dtm ...
                ./ (mu_r_y(is-1:is,js,ks:ke-1) * dz);
    end
end
```

10.3 Simulation examples

10.3.1 Thin-wire dipole antenna

Simulation of a thin-wire dipole antenna is presented in this section. The problem space is composed of cells with $\Delta x = 0.25$ mm, $\Delta y = 0.25$ mm, and $\Delta z = 0.25$ mm. The boundaries are CPML with 8 cells thickness and a 10 cells air gap on all sides. The dipole is composed of two thin wires having 0.05 mm radius and 9.75 mm length. There is a 0.5 mm gap between the wires where a voltage source is placed. Figure 10.3 illustrates the problem geometry. The definition of the geometry is illustrated in Listing 10.4. The definition of the voltage source is given in Listing 10.5, and the definitions of the output parameters are given in Listing 10.6. One can notice in Listing 10.6 that a far-field frequency is defined as 7 GHz to obtain the far-field radiation patterns at this frequency.

The simulation of this problem is performed for 4,000 time steps. The same problem is simulated using WIPL-D [40], which is a three-dimensional EM simulation software. Figure 10.4 shows the S_{11} of the thin-wire antenna calculated by FDTD and WIPL-D. The magnitude and phase curves show a good agreement. The input impedance of a dipole antenna is very much affected by the thickness of the dipole wires. The input impedances obtained from the two simulations are compared in Figure 10.5, and they show very good agreement over the simulated frequency band from zero to 20 GHz.

Since this is a radiation problem the far-field patterns are also calculated, as indicated before, at 7 GHz. The directivity patterns are plotted in Figures 10.6, 10.7, and 10.8 for *xy*, *xz*, and *yz* planes, respectively.

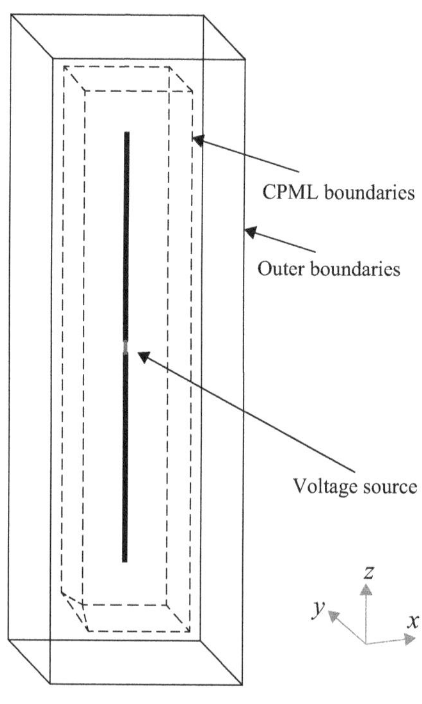

Figure 10.3 A thin-wire dipole antenna.

Listing 10.4 define_geometry.m

```
disp('defining the problem geometry');

bricks  = [];
spheres = [];
thin_wires = [];

% define a thin wire
thin_wires(1).min_x = 0;
thin_wires(1).min_y = 0;
thin_wires(1).min_z = 0.25e-3;
thin_wires(1).max_x = 0;
thin_wires(1).max_y = 0;
thin_wires(1).max_z = 10e-3;
thin_wires(1).radius = 0.05e-3;
thin_wires(1).direction = 'z';

% define a thin wire
thin_wires(2).min_x = 0;
thin_wires(2).min_y = 0;
thin_wires(2).min_z = -10e-3;
thin_wires(2).max_x = 0;
thin_wires(2).max_y = 0;
thin_wires(2).max_z = -0.25e-3;
thin_wires(2).radius = 0.05e-3;
thin_wires(2).direction = 'z';
```

Listing 10.5 define_secources_and_lumped_elements.m

```
disp('defining sources and lumped element components');

voltage_sources = [];
current_sources = [];
diodes = [];
resistors = [];
inductors = [];
capacitors = [];

% define source waveform types and parameters
waveforms.gaussian(1).number_of_cells_per_wavelength = 0;
waveforms.gaussian(2).number_of_cells_per_wavelength = 15;

% voltage sources
% direction: 'xp', 'xn', 'yp', 'yn', 'zp', or 'zn'
% resistance : ohms, magitude   : volts
voltage_sources(1).min_x = 0;
voltage_sources(1).min_y = 0;
voltage_sources(1).min_z = -0.25e-3;
voltage_sources(1).max_x = 0;
voltage_sources(1).max_y = 0;
voltage_sources(1).max_z = 0.25e-3;
voltage_sources(1).direction = 'zp';
voltage_sources(1).resistance = 50;
voltage_sources(1).magnitude = 1;
voltage_sources(1).waveform_type = 'gaussian';
voltage_sources(1).waveform_index = 1;
```

Listing 10.6 define_output_parameters.m

```
disp('defining output parameters');

sampled_electric_fields = [];
sampled_magnetic_fields = [];
sampled_voltages = [];
sampled_currents = [];
ports = [];
farfield.frequencies = [];

% figure refresh rate
plotting_step = 10;

% mode of operation
run_simulation = true;
show_material_mesh = true;
show_problem_space = true;

% far field calculation parameters
farfield.frequencies(1) = 7.0e9;
farfield.number_of_cells_from_outer_boundary = 13;

% frequency domain parameters
frequency_domain.start = 20e6;
frequency_domain.end   = 20e9;
frequency_domain.step  = 20e6;

% define sampled voltages
sampled_voltages(1).min_x = 0;
sampled_voltages(1).min_y = 0;
sampled_voltages(1).min_z = -0.25e-3;
sampled_voltages(1).max_x = 0;
sampled_voltages(1).max_y = 0;
sampled_voltages(1).max_z = 0.25e-3;
sampled_voltages(1).direction = 'zp';
sampled_voltages(1).display_plot = false;

% define sampled currents
sampled_currents(1).min_x = 0;
sampled_currents(1).min_y = 0;
sampled_currents(1).min_z = 0;
sampled_currents(1).max_x = 0;
sampled_currents(1).max_y = 0;
sampled_currents(1).max_z = 0;
sampled_currents(1).direction = 'zp';
sampled_currents(1).display_plot = false;

% define ports
ports(1).sampled_voltage_index = 1;
ports(1).sampled_current_index = 1;
ports(1).impedance = 50;
ports(1).is_source_port = true;
```

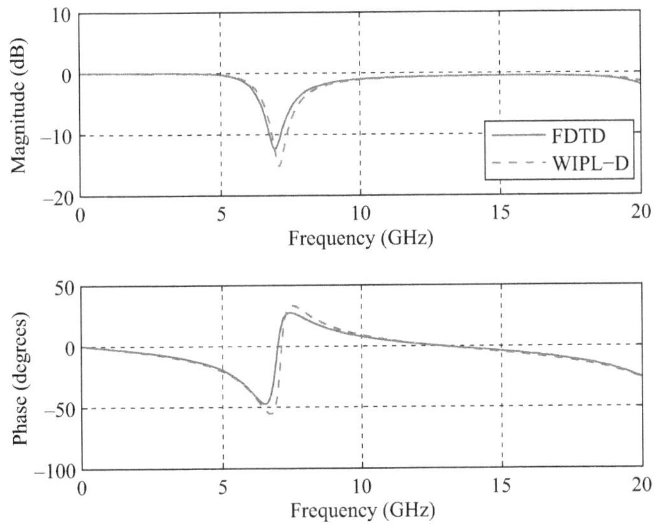

Figure 10.4 S_{11} of the thin-wire dipole antenna.

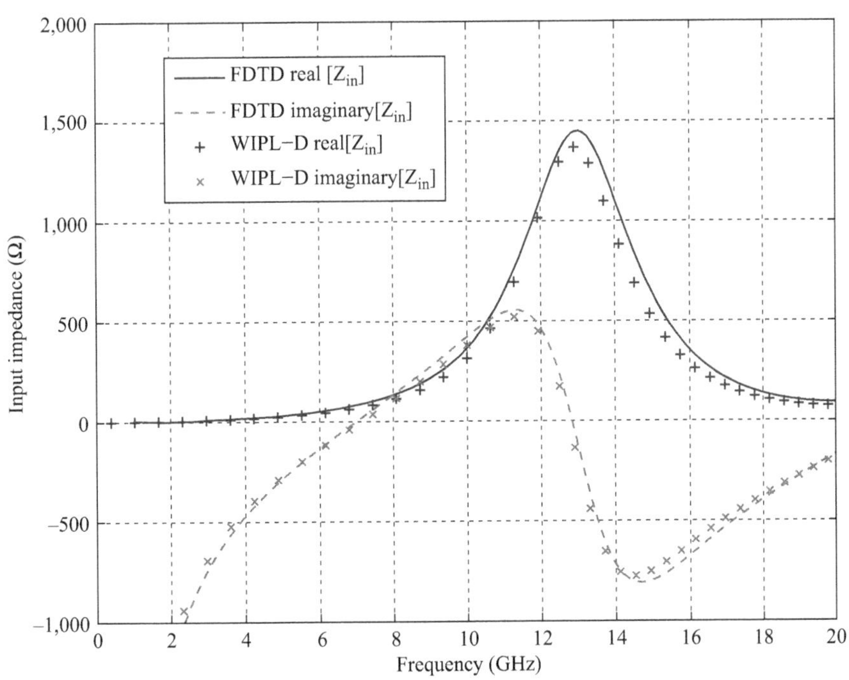

Figure 10.5 Input impedance of the thin-wire dipole antenna.

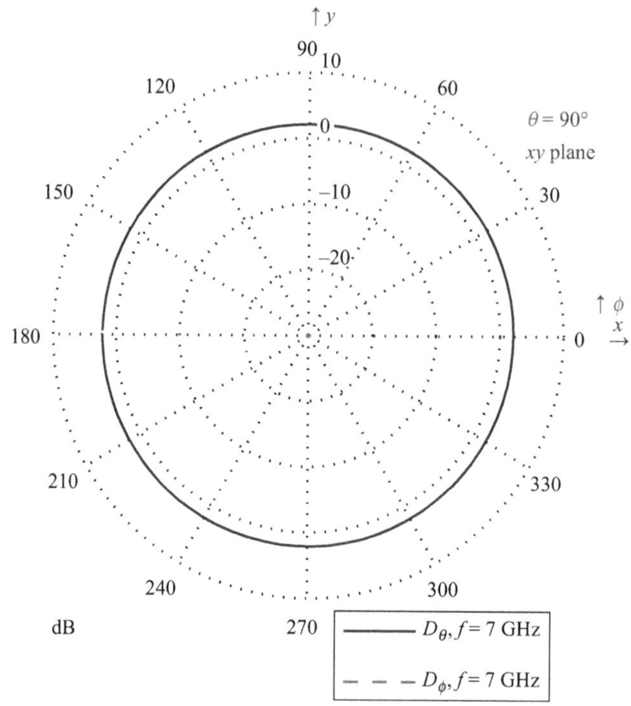

Figure 10.6 Radiation pattern in the *xy* plane cut.

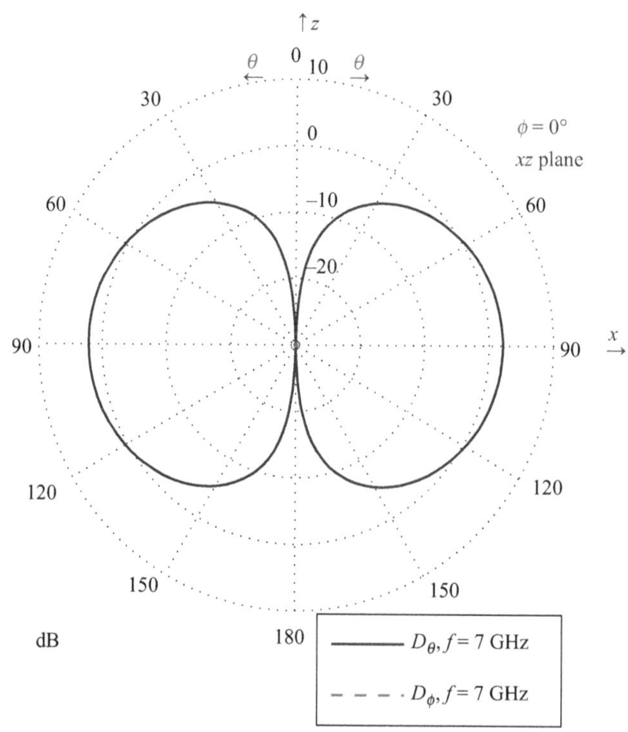

Figure 10.7 Radiation pattern in the *xz* plane cut.

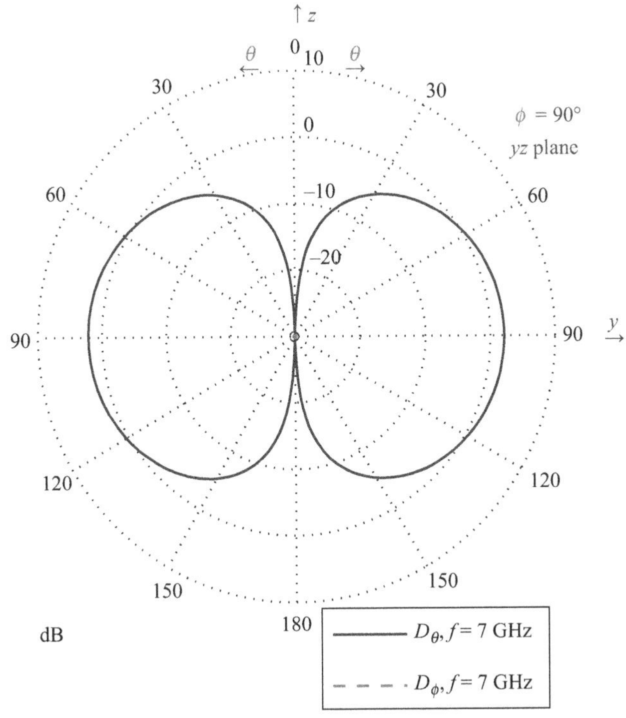

Figure 10.8 Radiation pattern in the *yz* plane cut.

10.4 An improved thin-wire model

In this section, we will discuss an improvement on the thin-wire model of [39], discussed in the previous sections, as presented in [41]. As in [39], the formulation that updates the magnetic field components circulating around a thin-wire is modified to account for the thin-wire in [41], albeit with a correction. In this section the formulation by [39] is referred to as the original thin-wire formulation while the one by [41] is referred to as the improved thin-wire formulation.

In Section 10.1, the field variation around a thin-wire is assumed to be a function of $1/r$ as in (10.2) and (10.3), which follows a static field distribution around a perfect electric conductor wire. The magnetic field component $H_y(i,j,k)$ in Figure 10.1, as calculated using (10.12), is the point-wise value of the field at the center of the surface S. It is also a constant value of magnetic field on a circular arc illustrated in Figure 10.9 due to the $1/r$ dependence assumption. When $H_y(i,j,k)$ is used in the update of an electric field as in (1.26) or (1.28), it is assumed to be an average of H_y along the edge of a rectangular path surrounding the electric field components, as illustrated in Figure 10.10 for E_z. This is due to the fact that Ampere's law is being imposed on a Cartesian grid in FDTD. Since magnetic field is constant on the circular loop, thus variable on the Cartesian cell edge due to the $1/r$ dependence, $H_y(i,j,k)$ is not an accurate average value of the field along the edge. As a solution to this contradiction, the looping magnetic field is projected on the respective cell edge and an average value of H_y that can be used to update the neighboring electric field components is obtained.

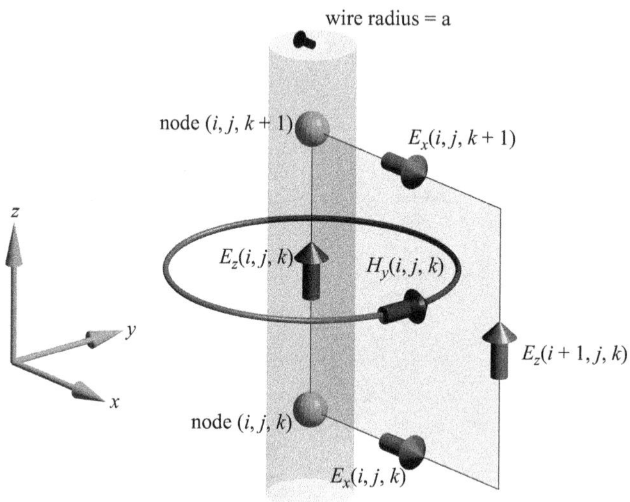

Figure 10.9 Constant magnetic field on a loop circulating around a thin wire.

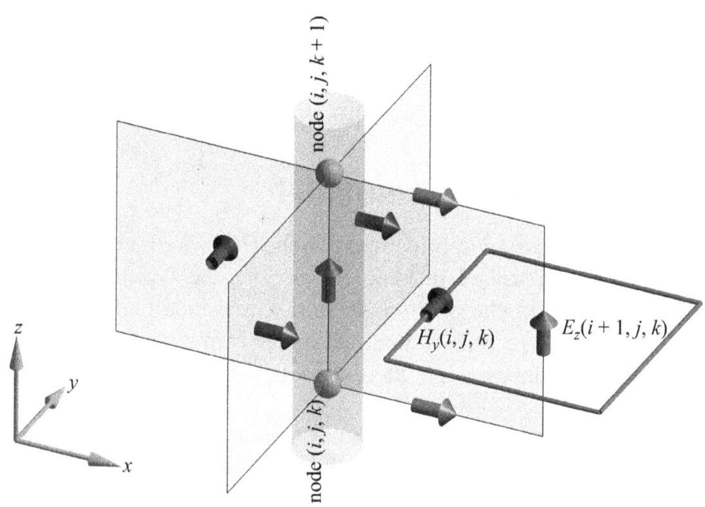

Figure 10.10 A magnetic field component on the edge of a rectangular loop.

Consider the circular contour on which $H_y(i,j,k)$ is constant and the cross-section of a cell where $\Delta x \neq \Delta y$ in Figure 10.11(a). Figure 10.11(b) shows the arc between the angles 0 and ϕ' radians zoomed in. Notice that the constant value of $H_y(i,j,k)$ is H_ϕ on the arc while the magnetic field on the right edge is a function of y as $H_y(y)$. The distance between $H_y(y)$ and the cell center is $r = \sqrt{(\Delta x/2)^2 + y^2}$. To accurately compute $H_y(y)$, the H_ϕ is to be

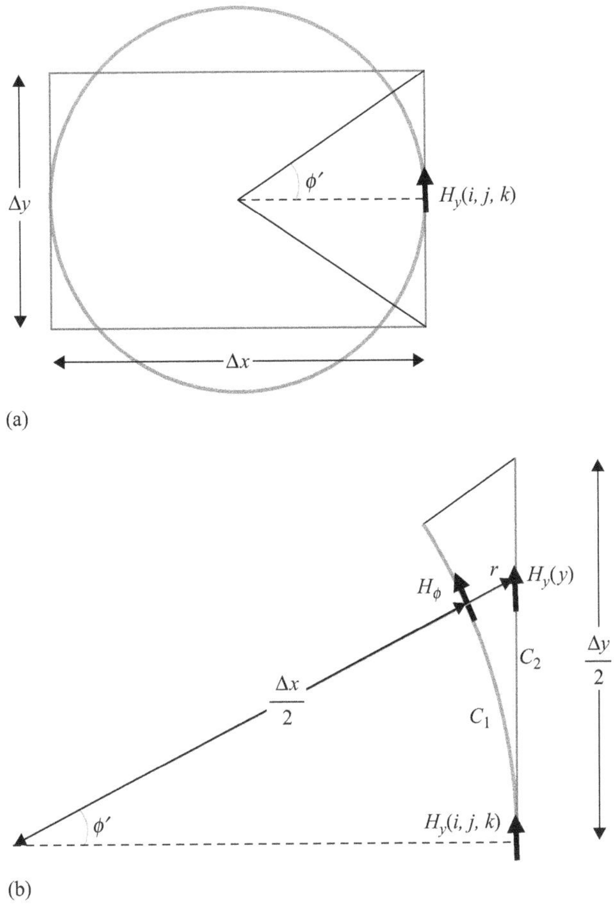

Figure 10.11 Projection of magnetic field from a circular loop to a rectangular edge: (a) large view and (b) details of the right edge.

scaled by the factor $\frac{\Delta x}{2r} = \frac{\Delta x/2}{\sqrt{(\Delta x/2)^2+y^2}}$ and then projected on the y axis using $\hat{\phi} \cdot \hat{y} = \cos\phi$ which then leads to

$$H_y(y) = \cos\phi \frac{\Delta x/2}{\sqrt{(\Delta x/2)^2 + y^2}} H_\phi = \frac{(\Delta x/2)^2}{(\Delta x/2)^2 + y^2} H_\phi. \tag{10.16}$$

To find an average value for $H_y(y)$ on the right edge, we can integrate $H_y(y)$ along the right edge and divide by Δy as

$$H_{y,avg} = \frac{1}{\Delta y} \int_{y=-\Delta y/2}^{\Delta y/2} H_y(y) dy = \frac{H_\phi}{\Delta y} \int_{y=-\Delta y/2}^{\Delta y/2} \frac{(\Delta x/2)^2}{(\Delta x/2)^2 + y^2} dy. \tag{10.17}$$

The solution of the integral (10.17) yields

$$H_{y,avg} = k_{Hy} H_\phi = k_{Hy} H_y(i,j,k), \tag{10.18}$$

where

$$k_{Hy} = \frac{\Delta x}{\Delta y}\tan^{-1}\left(\frac{\Delta y}{\Delta x}\right). \tag{10.19}$$

Notice that k_{Hy} is just a scaling factor. Thus, to scale $H_y(i,j,k)$ by k_{Hy} one can simply scale the coefficients $C_{hyez}(i,j,k)$ and $C_{hyex}(i,j,k)$ in (10.12) by k_{Hy} as

$$C_{hyez}(i,j,k) = k_{Hy}\frac{2\Delta t}{\mu_y(i,j,k)\Delta x \ln\left(\frac{\Delta x}{a}\right)}, \tag{10.20}$$

$$C_{hyex}(i,j,k) = -k_{Hy}\frac{\Delta t}{\mu_y(i,j,k)\Delta z}. \tag{10.21}$$

Note that $H_y(i-1,j,k)$ in Figure 10.2 as well need to be scaled by k_{Hy}, whereas $H_x(i,j,k)$ and $H_x(i,j-1,k)$ need to be scaled by a factor of $k_{Hx} = \frac{\Delta y}{\Delta x}\tan^{-1}\left(\frac{\Delta x}{\Delta y}\right)$.

The presented modification to the original thin-wire formulation in terms of the calculation of magnetic field components led to more accurate update values of electric field components adjacent to the thin wire. The expressions of the electric field components updates remain unchanged as the standard FDTD updating equations in the original thin-wire formulation.

In the standard FDTD updating equation for an electric field component it is assumed that the field component is an average value of the field distribution on the face of a cell face through which it flows. For instance, the radial electric field component $E_x(i,j,k)$, illustrated in Figure 10.12, is an average value of E_x on the surface S1. However, when developing the thin-wire updating equation for $H_y(i,j,k)$, the radial component of electric field as well is assumed to vary with $1/r$. Thus, when using $E_x(i,j,k)$ and $E_x(i,j,k+1)$ in (10.12), their $1/r$ dependent values at half-cell distance from the thin-wire axis need to be used instead of their

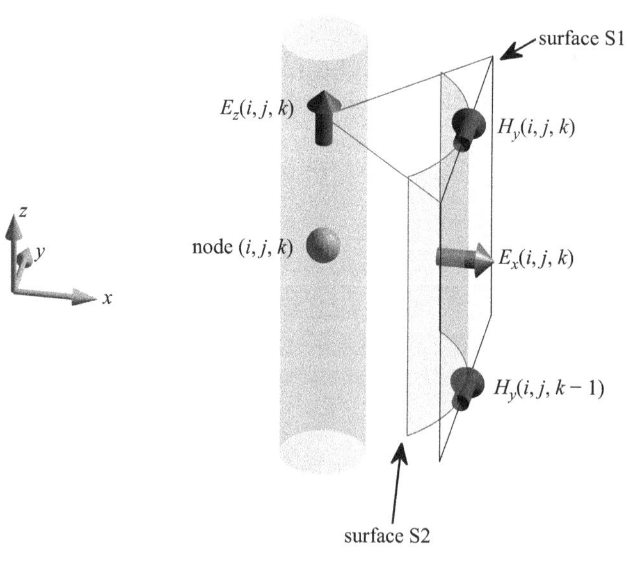

Figure 10.12 Electric field component E_x at the center of a cell face.

average values on respective cell faces. The radial electric field component E_r is constant on the cylindrical surface S2 in Figure 10.12, thus its projection on the surface S1 is a function of y. Note that $E_x(i,j,k)$ is the average of the projection of E_r on surface S1, while the value of E_r at the position of $E_x(i,j,k)$ is needed in (10.12). To obtain E_r one needs to project $E_x(i,j,k)$ from surface S1 to surface S2. Since the fields are constant in the z-direction, the required projection for $E_x(i,j,k)$ is the reverse of the projection done with $H_y(i,j,k)$ illustrated in Figure 10.11. Thus, utilizing (10.19), a scaling factor can be defined as

$$k_{Ex} = \left(\frac{\Delta x}{\Delta y}\tan^{-1}\left(\frac{\Delta y}{\Delta x}\right)\right)^{-1}, \tag{10.22}$$

which can be used to scale $E_x(i,j,k)$ and $E_x(i,j,k+1)$ before they are used to update $H_y(i,j,k)$. This scaling can be performed simply by multiplying the coefficient $C_{hyex}(i,j,k)$ given in (10.21) by k_{Ex}. Since k_{Hy} and k_{Ex} are inverse, scaling $C_{hyex}(i,j,k)$ by k_{Ex} returns it to its original value before it was scaled by k_{Hy} as

$$C_{hyex}(i,j,k) = -\frac{\Delta t}{\mu_y(i,j,k)\Delta z}. \tag{10.23}$$

Thus, the only modification needed for the improved thin-wire formulation over the original thin-wire formulation in updating $H_y(i,j,k)$ is an adjustment to the updating coefficient as in (10.20). Similar adjustments are needed for the respective updating coefficients that update $H_y(i-1,j,k)$, $H_x(i,j,k)$, and $H_x(i,j-1,k)$.

It should be noted that the presented improvement of thin-wire formulation is not the only contribution presented in [41]. Also presented in [41] is an end-cap model that includes the effect of charge accumulation at wire end caps which further improves the accuracy of modeling unconnected thin wires. Moreover, Mâkinen et al. employed the precepts of the improved thin-wire formulation to develop a stabilized resistive voltage source (RVS) model in [42] to feed thin wires that further improves the stability and accuracy of thin-wire simulations.

10.5 MATLAB® implementation of the improved thin-wire formulation

The only modification needed for the improved thin-wire formulation is the scaling of the coefficients used to update the magnetic field components circulating around the thin wire due to the electric field components parallel to the thin-wire axis. Listing 10.3 shows the initialization of thin wire updating coefficients based on the original thin-wire formulation as in (10.20). This code is modified to accommodate the scaling factors introduced by the improved thin-wire formulation as shown in Listing 10.7.

10.6 Simulation example

Simulation of a thin-wire dipole antenna is presented in Section 10.3.1. The same example is repeated here with the simulation performed using the improved thin-wire formulation. The

Listing 10.7 initialize_thin_wire_updating_coefficients.m

```
disp('initializing thin wire updating coefficients');

dtm = dt/mu_0;

for ind = 1:number_of_thin_wires
    is  = round((thin_wires(ind).min_x - fdtd_domain.min_x)/dx)+1;
    js  = round((thin_wires(ind).min_y - fdtd_domain.min_y)/dy)+1;
    ks  = round((thin_wires(ind).min_z - fdtd_domain.min_z)/dz)+1;
    ie  = round((thin_wires(ind).max_x - fdtd_domain.min_x)/dx)+1;
    je  = round((thin_wires(ind).max_y - fdtd_domain.min_y)/dy)+1;
    ke  = round((thin_wires(ind).max_z - fdtd_domain.min_z)/dz)+1;
    r_o = thin_wires(ind).radius;

   switch (thin_wires(ind).direction(1))
      case 'x'
         khy = (dz/dy)*atan(dy/dz);
         khz = (dy/dz)*atan(dz/dy);
         Cexe (is:ie-1,js,ks)  = 0;
         Cexhy(is:ie-1,js,ks)  = 0;
         Cexhz(is:ie-1,js,ks)  = 0;
         Chyh (is:ie-1,js,ks-1:ks) = 1;
         Chyex(is:ie-1,js,ks-1:ks) = -2 * dtm * khy...
            ./ (mu_r_y(is:ie-1,js,ks-1:ks) * dz * log(dz/r_o));
         Chzh( is:ie-1,js-1:js,ks) = 1;
         Chzex(is:ie-1,js-1:js,ks) = 2 * dtm * khz...
            ./ (mu_r_z(is:ie-1,js-1:js,ks) * dy * log(dy/r_o));
      case 'y'
         khz = (dx/dz)*atan(dz/dx);
         khx = (dz/dx)*atan(dx/dz);
         Ceye (is,js:je-1,ks)  = 0;
         Ceyhx(is,js:je-1,ks)  = 0;
         Ceyhz(is,js:je-1,ks)  = 0;
         Chzh (is-1:is,js:je-1,ks) = 1;
         Chzey(is-1:is,js:je-1,ks) = -2 * dtm * khz...
            ./ (mu_r_z(is-1:is,js:je-1,ks) * dx * log(dx/r_o));
         Chxh (is,js:je-1,ks-1:ks) = 1;
         Chxey(is,js:je-1,ks-1:ks) = 2 * dtm *khx...
            ./ (mu_r_x(is,js:je-1,ks-1:ks) * dz * log(dz/r_o));
      case 'z'
         khx = (dy/dx)*atan(dx/dy);
         khy = (dx/dy)*atan(dy/dx);
         Ceze (is,js,ks:ke-1)  = 0;
         Cezhx(is,js,ks:ke-1)  = 0;
         Cezhy(is,js,ks:ke-1)  = 0;
         Chxh (is,js-1:js,ks:ke-1) = 1;
         Chxez(is,js-1:js,ks:ke-1) = -2 * dtm * khx...
            ./ (mu_r_x(is,js-1:js,ks:ke-1) * dy * log(dy/r_o));
         Chyh (is-1:is,js,ks:ke-1) = 1;
         Chyez(is-1:is,js,ks:ke-1) = 2 * dtm * khy...
            ./ (mu_r_y(is-1:is,js,ks:ke-1) * dx * log(dx/r_o));
   end
end
```

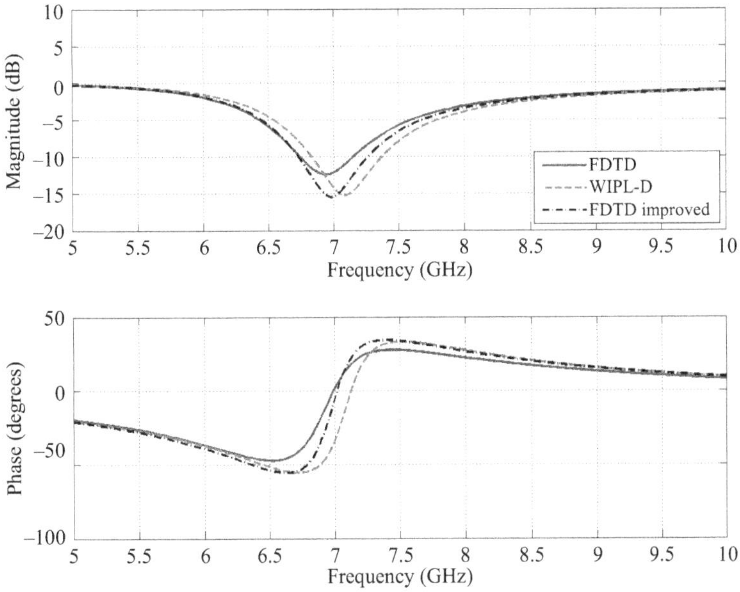

Figure 10.13 Power reflection coefficient of the thin-wire dipole antenna.

calculated power reflection coefficient of the dipole antenna is plotted in Figure 10.13 along with the results previously presented in Figure 10.4. One can notice the improvement of agreement between the FDTD and WIPL-D results when the improved thin-wire formulation is used. Moreover, the input impedance as well calculated and compared with the results in Figure 10.5. The comparison is shown in Figure 10.14.

As mentioned in Section 10.5, accuracy of thin-wire simulations can further be improved by including the end-cap model [41] and the stabilized RVS model [42]. Although these models are not presented here, they are implemented in an FDTD code, although not presented here, and the presented thin-wire dipole antenna is resimulated. Figure 10.15 shows the power reflection coefficient of the thin-wire dipole antenna including the effects of the end-caps and the stabilized RVS model, and illustrates the further improvement in accuracy.

10.7 Exercises

10.1 Construct a problem space composed of cells with size 1 mm on a side. Define a dipole antenna using thin wires similar to the dipole antenna discussed in Section 10.3.1. Set the length of each wire to 10 mm, and leave a 1 mm gap between the wires. Place a voltage source, a sampled voltage, and a sampled current in the gap between the wires, and associate them to a port. Set the radius of the wire as 0.1 mm, run the simulation, and obtain the input reflection coefficient of the antenna. Then increase the radius of the wire by 0.2 mm, repeat the simulation, and observe the change in the input reflection coefficient with incremented increase of wire radius by 0.2 mm until the increased wire radius causes instability in the simulation. Can you determine the maximum wire radius for accurate results in this configuration?

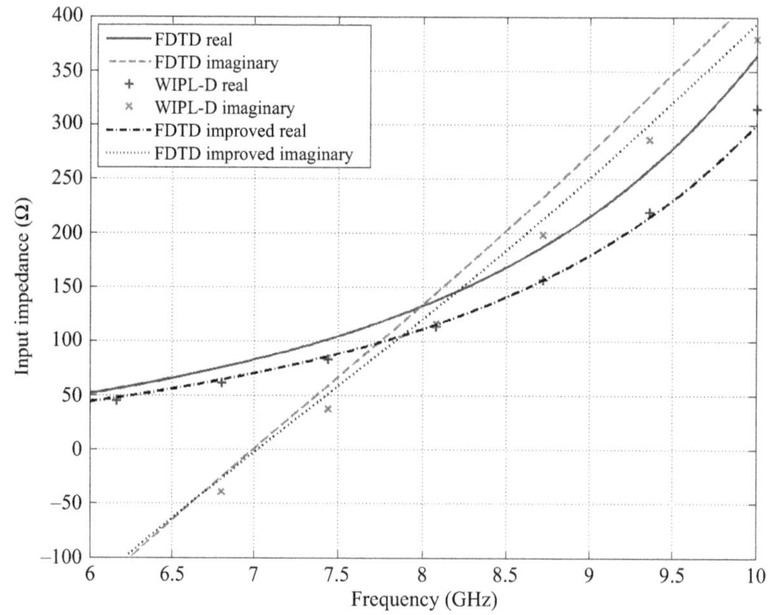

Figure 10.14 Input impedance of the thin-wire dipole antenna.

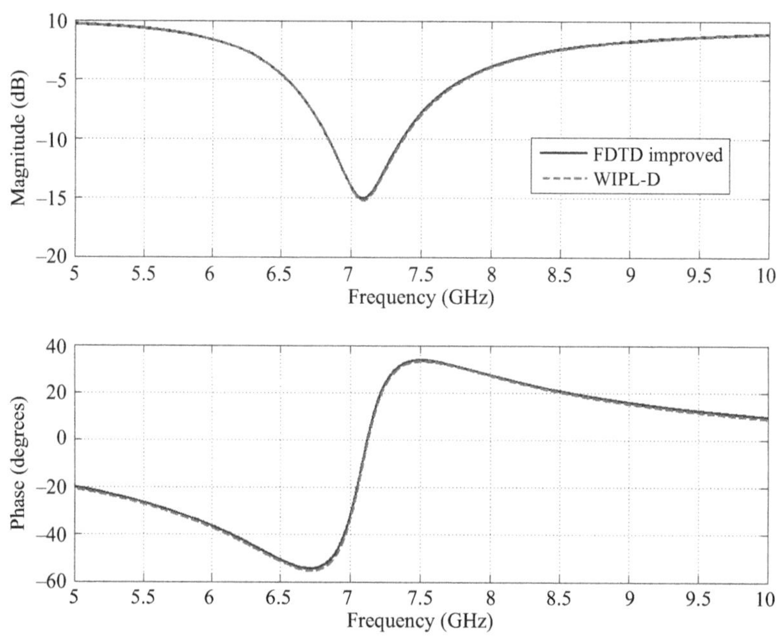

Figure 10.15 Power reflection coefficient of the thin-wire dipole antenna: FDTD implementation includes the end-caps model [41] and the stabilized RVS model [42].

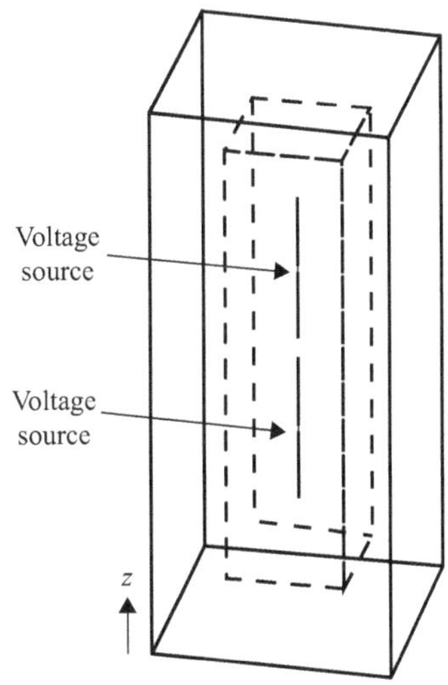

Figure 10.16 An antenna array composed of two thin-wire dipole antennas.

10.2 In this example we construct an antenna array composed of two thin-wire dipole antennas. Consider the dipole antenna that you constructed in Exercise 10.1. Run the dipole simulation using the 0.1 mm wire radius, and record the operation frequency. Rerun the simulation, and obtain the radiation pattern at the frequency of operation. Then define another thin-wire dipole antenna with the same dimensions as the current one, and place it 2 mm above the current one, such that the center to center distance between antennas will be 23 mm. Place a voltage source at the feeding gap of the second antenna that has the same properties of the voltage source of the first antenna. Do not define any ports, and if there are already defined ports, remove them. The geometry of the problem is illustrated in Figure 10.16. Run the simulation, and calculate the radiation patterns at the frequency of operation. Examine the directivity patterns of the single- and two-dipole configurations.

10.3 Consider the dipole antenna array that you constructed in Exercise 10.2. Invert the polarity of one of the voltage sources. For instance, if the direction of the voltage source is "***zp***," set the direction as "***zn***." This will let the antennas be excited by 180° phase difference. Run the simulation, and obtain the radiation pattern at the frequency of operation.

10.4 In this exercise we construct a square loop inductor. Construct a problem space composed of cells with size 1 mm on a side. Place three thin wires, each of which is 10 mm long, such that they will form the three sides of a square loop. Then place two thin wires, each of which is 4 mm long, on the fourth side, leaving a 2 mm gap in between at the center of the fourth side. Place a voltage source, a sampled voltage, and a sampled current in this gap. Set the radius of the wires as 0.1 mm. The geometry of the problem is illustrated in Figure 10.17. Run the simulation, and calculate the input impedance. At low frequencies the imaginary part of the input impedance is positive and linear, which implies that the inductance is constant at low frequencies. Calculate

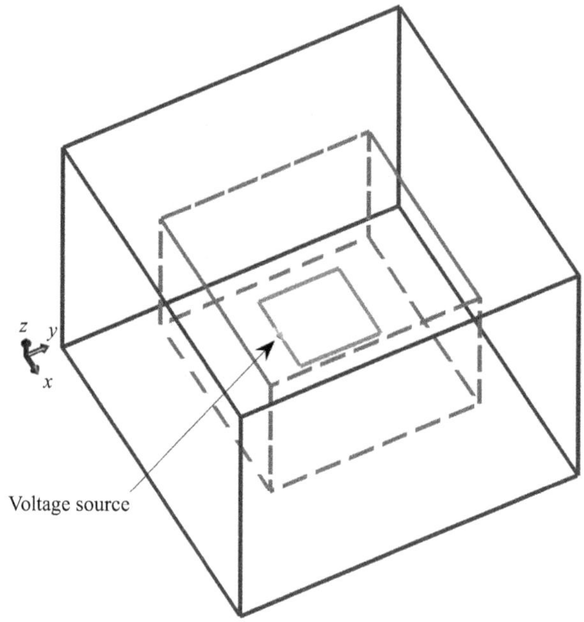

Figure 10.17 A thin-wire loop.

the inductance of the loop. The inductance of such a square loop can be calculated as 30.7 nH using the formula in [43].

10.5 Consider the square loop that you constructed in Exercise 10.4. Now reduce the cell size to 0.5 mm on a side, and reduce the length of the source gap to 1 mm. Run the simulation, and then calculate the inductance of the loop. Examine if the calculated inductance got closer to the expected value.

CHAPTER 11

Scattered field formulation

As discussed before, sources are the necessary components of a finite-difference time-domain (FDTD) simulation, and their types vary depending on the type of the problem under consideration. Usually sources are of two types: (1) *near-zone sources*, such as the voltage and current sources described in Chapter 4; and (2) *far-zone sources*, such as the incident fields in scattering problems. In the previous chapters we demonstrated several examples utilizing near-zone sources. In this chapter we present *scattered field formulation*, one of the techniques that integrates far-zone sources into the FDTD method.

11.1 Scattered field basic equations

Far-zone sources are the fields generated somewhere outside the FDTD problem space that illuminate the objects in the problem space. Therefore, they are the *incident fields* exciting the FDTD problem space. The incident field exists in a problem space in which there are no scatterers. Therefore, they are the fields that can be described by analytical expressions and would satisfy Maxwell's curl equations where the problem space medium is free space, such that

$$\nabla \times \vec{H}_{inc} = \varepsilon_0 \frac{\partial \vec{E}_{inc}}{\partial t}, \tag{11.1a}$$

$$\nabla \times \vec{E}_{inc} = -\mu_0 \frac{\partial \vec{H}_{inc}}{\partial t}. \tag{11.1b}$$

The most common type of incident field is the plane wave. The expressions for the field components generated by plane waves are discussed in a subsequent section.

When the incident field illuminates the objects in an FDTD problem space, *scattered fields* are generated and thus need to be calculated. The scattered field formulation is one of the simplest techniques that can be used for calculating scattered fields.

The fields in a problem space are referred to in general as *total fields*. The total fields satisfy Maxwell's curl equations for a general medium, such that

$$\nabla \times \vec{H}_{tot} = \varepsilon \frac{\partial \vec{E}_{tot}}{\partial t} + \sigma^e \vec{E}_{tot}, \tag{11.2a}$$

$$\nabla \times \vec{E}_{tot} = -\mu \frac{\partial \vec{H}_{tot}}{\partial t} - \sigma^m \vec{H}_{tot}. \tag{11.2b}$$

Then the scattered fields are defined as the difference between the total fields and incident fields. Therefore, one can write

$$E_{tot} = E_{inc} + E_{scat}, \quad \text{and} \quad H_{tot} = H_{inc} + H_{scat}, \tag{11.3}$$

where E_{scat} and H_{scat} are the scattered electric and magnetic fields. In light of (11.3), (11.2) can be rewritten in terms of incident and scattered field terms as

$$\nabla \times \vec{H}_{scat} + \nabla \times \vec{H}_{inc} = \varepsilon \frac{\partial \vec{E}_{scat}}{\partial t} + \varepsilon \frac{\partial \vec{E}_{inc}}{\partial t} + \sigma^e \vec{E}_{scat} + \sigma^e \vec{E}_{inc}, \tag{11.4a}$$

$$\nabla \times \vec{E}_{scat} + \nabla \times \vec{E}_{inc} = -\mu \frac{\partial \vec{H}_{scat}}{\partial t} - \mu \frac{\partial \vec{H}_{inc}}{\partial t} - \sigma^m \vec{H}_{scat} - \sigma^m \vec{H}_{inc}. \tag{11.4b}$$

The curls of the incident fields in (11.4) can be replaced by the time-derivative terms from (11.1), which yields

$$\nabla \times \vec{H}_{scat} + \varepsilon_0 \frac{\partial \vec{E}_{inc}}{\partial t} = \varepsilon \frac{\partial \vec{E}_{scat}}{\partial t} + \varepsilon \frac{\partial \vec{E}_{inc}}{\partial t} + \sigma^e \vec{E}_{scat} + \sigma^e \vec{E}_{inc}, \tag{11.5a}$$

$$\nabla \times \vec{E}_{scat} - \mu_0 \frac{\partial \vec{H}_{inc}}{\partial t} = -\mu \frac{\partial \vec{H}_{scat}}{\partial t} - \mu \frac{\partial \vec{H}_{inc}}{\partial t} - \sigma^m \vec{H}_{scat} - \sigma^m \vec{H}_{inc}. \tag{11.5b}$$

After rearranging the terms in (11.5) one can obtain

$$\varepsilon \frac{\partial \vec{E}_{scat}}{\partial t} + \sigma^e \vec{E}_{scat} = \nabla \times \vec{H}_{scat} + (\varepsilon_0 - \varepsilon) \frac{\partial \vec{E}_{inc}}{\partial t} - \sigma^e \vec{E}_{inc}, \tag{11.6a}$$

$$\mu \frac{\partial \vec{H}_{scat}}{\partial t} + \sigma^m \vec{H}_{scat} = -\nabla \times \vec{E}_{scat} + (\mu_0 - \mu) \frac{\partial \vec{H}_{inc}}{\partial t} - \sigma^m \vec{H}_{inc}. \tag{11.6b}$$

At this point the derivatives in (11.6) can be represented by central finite difference approximations, and updating equations can be obtained for the scattered field formulation.

11.2 The scattered field updating equations

After applying the central finite difference approximation, (11.6a) can be expressed for the x component as

$$\begin{aligned}
&\varepsilon_x(i,j,k) \frac{E^{n+1}_{scat,x}(i,j,k) - E^{n}_{scat,x}(i,j,k)}{\Delta t} + \sigma^e_x(i,j,k) \frac{E^{n+1}_{scat,x}(i,j,k) + E^{n}_{scat,x}(i,j,k)}{2} \\
&= \frac{H^{n+\frac{1}{2}}_{scat,z}(i,j,k) - H^{n+\frac{1}{2}}_{scat,x}(i,j-1,k)}{\Delta y} - \frac{H^{n+\frac{1}{2}}_{scat,y}(i,j,k) - H^{n+\frac{1}{2}}_{scat,y}(i,j,k-1)}{\Delta z} \\
&+ (\varepsilon_0 - \varepsilon_x(i,j,k)) \frac{E^{n+1}_{inc,x}(i,j,k) + E^{n}_{inc,x}(i,j,k)}{\Delta t} \\
&- \sigma^e_x(i,j,k) \frac{E^{n+1}_{inc,x}(i,j,k) + E^{n}_{inc,x}(i,j,k)}{2},
\end{aligned} \tag{11.7}$$

Equation (11.7) can be arranged such that the new value of the scattered electric field component $E^{n+1}_{scat,x}(i,j,k)$ is calculated using other terms as

$$\begin{aligned} E^{n+1}_{scat,x}(i,j,k) = {} & C_{exe}(i,j,k) \times E^{n}_{scat,x}(i,j,k) \\ & + C_{exhz}(i,j,k) \times \left[H^{n+\frac{1}{2}}_{scat,z}(i,j,k) - H^{n+\frac{1}{2}}_{scat,z}(i,j-1,k)\right] \\ & + C_{exhy}(i,j,k) \times \left[H^{n+\frac{1}{2}}_{scat,y}(i,j,k) - H^{n+\frac{1}{2}}_{scat,y}(i,j,k-1)\right] \\ & + C_{exeic}(i,j,k) \times E^{n+1}_{inc,x}(i,j,k) + C_{exeip}(i,j,k) \times E^{n}_{inc,x}(i,j,k), \end{aligned} \tag{11.8}$$

where

$$\begin{aligned} C_{exe}(i,j,k) &= \frac{2\varepsilon_x(i,j,k) - \sigma^e_x(i,j,k)\Delta t}{2\varepsilon_x(i,j,k) + \sigma^e_x(i,j,k)\Delta t}, \\ C_{exhz}(i,j,k) &= \frac{2\Delta t}{\Delta y\left(2\varepsilon_x(i,j,k) + \sigma^e_x(i,j,k)\Delta t\right)}, \\ C_{exhy}(i,j,k) &= -\frac{2\Delta t}{\Delta z\left(2\varepsilon_x(i,j,k) + \sigma^e_x(i,j,k)\Delta t\right)}, \\ C_{exeic}(i,j,k) &= \frac{2(\varepsilon_0 - \varepsilon_x(i,j,k)) - \sigma^e_x(i,j,k)\Delta t}{2\varepsilon_x(i,j,k) + \sigma^e_x(i,j,k)\Delta t}, \\ C_{exeip}(i,j,k) &= -\frac{2(\varepsilon_0 - \varepsilon_x(i,j,k)) + \sigma^e_x(i,j,k)\Delta t}{2\varepsilon_x(i,j,k) + \sigma^e_x(i,j,k)\Delta t}. \end{aligned}$$

In this equation, the *ic* term in the subscript of the coefficient C_{exeic} indicates that this coefficient is multiplied with the *current* value of the *incident* field component, whereas the *ip* term in the subscript of the coefficient C_{exeip} indicates that this coefficient is multiplied with the *previous* value of the *incident* field component.

Similarly the updating equation can be written for $E^{n+1}_{scat,y}(i,j,k)$ as

$$\begin{aligned} E^{n+1}_{scat,y}(i,j,k) = {} & C_{eye}(i,j,k) \times E^{n}_{scat,y}(i,j,k) \\ & + C_{eyhx}(i,j,k) \times \left[H^{n+\frac{1}{2}}_{scat,x}(i,j,k) - H^{n+\frac{1}{2}}_{scat,x}(i,j,k-1)\right] \\ & + C_{eyhz}(i,j,k) \times \left[H^{n+\frac{1}{2}}_{scat,z}(i,j,k) - H^{n+\frac{1}{2}}_{scat,z}(i-1,j,k)\right] \\ & + C_{eyeic}(i,j,k) \times E^{n+1}_{inc,y}(i,j,k) + C_{eyeip}(i,j,k) \times E^{n}_{inc,y}(i,j,k), \end{aligned} \tag{11.9}$$

where

$$\begin{aligned} C_{eye}(i,j,k) &= \frac{2\varepsilon_y(i,j,k) - \sigma^e_y(i,j,k)\Delta t}{2\varepsilon_y(i,j,k) + \sigma^e_y(i,j,k)\Delta t}, \\ C_{eyhx}(i,j,k) &= \frac{2\Delta t}{\Delta z\left(2\varepsilon_y(i,j,k) + \sigma^e_y(i,j,k)\Delta t\right)}, \end{aligned}$$

$$C_{eyhz}(i,j,k) = -\frac{2\Delta t}{\Delta x\left(2\varepsilon_y(i,j,k) + \sigma_y^e(i,j,k)\Delta t\right)},$$

$$C_{eyeic}(i,j,k) = \frac{2\left(\varepsilon_0 - \varepsilon_y(i,j,k)\right) - \sigma_y^e(i,j,k)\Delta t}{2\varepsilon_y(i,j,k) + \sigma_y^e(i,j,k)\Delta t},$$

$$C_{eyeip}(i,j,k) = -\frac{2\left(\varepsilon_0 - \varepsilon_y(i,j,k)\right) + \sigma_y^e(i,j,k)\Delta t}{2\varepsilon_y(i,j,k) + \sigma_y^e(i,j,k)\Delta t}.$$

The updating equation can be written for $E^{n+1}_{scat,z}(i,j,k)$ as

$$\begin{aligned}
E^{n+1}_{scat,z}(i,j,k) = {} & C_{eze}(i,j,k) \times E^{n}_{scat,z}(i,j,k) \\
& + C_{ezhy}(i,j,k) \times \left[H^{n+\frac{1}{2}}_{scat,y}(i,j,k) - H^{n+\frac{1}{2}}_{scat,y}(i,j,k-1)\right] \\
& + C_{ezhx}(i,j,k) \times \left[H^{n+\frac{1}{2}}_{scat,x}(i,j,k) - H^{n+\frac{1}{2}}_{scat,x}(i-1,j,k)\right] \\
& + C_{ezeic}(i,j,k) \times E^{n+1}_{inc,z}(i,j,k) + C_{ezeip}(i,j,k) \times E^{n}_{inc,z}(i,j,k),
\end{aligned} \tag{11.10}$$

where

$$C_{eze}(i,j,k) = \frac{2\varepsilon_z(i,j,k) - \sigma_z^e(i,j,k)\Delta t}{2\varepsilon_z(i,j,k) + \sigma_z^e(i,j,k)\Delta t},$$

$$C_{ezhy}(i,j,k) = \frac{2\Delta t}{\Delta x(2\varepsilon_z(i,j,k) + \sigma_z^e(i,j,k)\Delta t)},$$

$$C_{ezhx}(i,j,k) = -\frac{2\Delta t}{\Delta y(2\varepsilon_z(i,j,k) + \sigma_z^e(i,j,k)\Delta t)},$$

$$C_{ezeic}(i,j,k) = \frac{2(\varepsilon_0 - \varepsilon_z(i,j,k)) - \sigma_z^e(i,j,k)\Delta t}{2\varepsilon_z(i,j,k) + \sigma_z^e(i,j,k)\Delta t},$$

$$C_{eyeip}(i,j,k) = -\frac{2(\varepsilon_0 - \varepsilon_z(i,j,k)) + \sigma_z^e(i,j,k)\Delta t}{2\varepsilon_z(i,j,k) + \sigma_z^e(i,j,k)\Delta t}.$$

Following a similar procedure, the updating equations for the magnetic field components can be obtained as follows. The updating equation for $H^{n+\frac{1}{2}}_{scat,x}(i,j,k)$ is

$$\begin{aligned}
H^{n+\frac{1}{2}}_{scat,x}(i,j,k) = {} & C_{hxh}(i,j,k) \times H^{n-\frac{1}{2}}_{scat,x}(i,j,k) \\
& + C_{hxez}(i,j,k) \times \left[E^{n}_{scat,z}(i,j+1,k) - E^{n}_{scat,z}(i,j,k)\right] \\
& + C_{hxey}(i,j,k) \times \left[E^{n}_{scat,y}(i,j,k+1) - E^{n}_{scat,y}(i,j,k)\right] \\
& + C_{hxhic}(i,j,k) \times H^{n+\frac{1}{2}}_{inc,x}(i,j,k) + C_{hxhip}(i,j,k) \times H^{n-\frac{1}{2}}_{inc,x}(i,j,k)
\end{aligned} \tag{11.11}$$

where

$$
\begin{aligned}
C_{hxh}(i,j,k) &= \frac{2\mu_x(i,j,k) - \sigma_x^m(i,j,k)\Delta t}{2\mu_x(i,j,k) + \sigma_x^m(i,j,k)\Delta t}, \\
C_{hxez}(i,j,k) &= -\frac{2\Delta t}{\Delta y\left(2\mu_x(i,j,k) + \sigma_x^m(i,j,k)\Delta t\right)}, \\
C_{hxey}(i,j,k) &= \frac{2\Delta t}{\Delta z\left(2\mu_x(i,j,k) + \sigma_x^m(i,j,k)\Delta t\right)}, \\
C_{hxhic}(i,j,k) &= \frac{2(\mu_0 - \mu_x(i,j,k)) - \sigma_x^m(i,j,k)\Delta t}{2\mu_x(i,j,k) + \sigma_x^m(i,j,k)\Delta t}, \\
C_{hxhip}(i,j,k) &= -\frac{2(\mu_0 - \mu_x(i,j,k)) + \sigma_x^m(i,j,k)\Delta t}{2\mu_x(i,j,k) + \sigma_x^m(i,j,k)\Delta t}.
\end{aligned}
$$

The updating equation for $H_{scat,y}^{n+\frac{1}{2}}(i,j,k)$ is

$$
\begin{aligned}
H_{scat,y}^{n+\frac{1}{2}}(i,j,k) = {} & C_{hyh}(i,j,k) \times H_{scat,y}^{n-\frac{1}{2}}(i,j,k) \\
& + C_{hyex}(i,j,k) \times \left[E_{scat,x}^{n}(i,j+1,k) - E_{scat,x}^{n}(i,j,k)\right] \\
& + C_{hyez}(i,j,k) \times \left[E_{scat,z}^{n}(i,j,k+1) - E_{scat,z}^{n}(i,j,k)\right] \\
& + C_{hyhic}(i,j,k) \times H_{inc,y}^{n+\frac{1}{2}}(i,j,k) + C_{hyhip}(i,j,k) \times H_{inc,y}^{n-\frac{1}{2}}(i,j,k),
\end{aligned}
\tag{11.12}
$$

where

$$
\begin{aligned}
C_{hyh}(i,j,k) &= \frac{2\mu_y(i,j,k) - \sigma_y^m(i,j,k)\Delta t}{2\mu_y(i,j,k) + \sigma_y^m(i,j,k)\Delta t}, \\
C_{hyex}(i,j,k) &= -\frac{2\Delta t}{\Delta z\left(2\mu_y(i,j,k) + \sigma_y^m(i,j,k)\Delta t\right)}, \\
C_{hyez}(i,j,k) &= \frac{2\Delta t}{\Delta x\left(2\mu_y(i,j,k) + \sigma_y^m(i,j,k)\Delta t\right)}, \\
C_{hyhic}(i,j,k) &= \frac{2\left(\mu_0 - \mu_y(i,j,k)\right) - \sigma_y^m(i,j,k)\Delta t}{2\mu_y(i,j,k) + \sigma_y^m(i,j,k)\Delta t}. \\
C_{hyhip}(i,j,k) &= -\frac{2\left(\mu_0 - \mu_y(i,j,k)\right) + \sigma_y^m(i,j,k)\Delta t}{2\mu_y(i,j,k) + \sigma_y^m(i,j,k)\Delta t},
\end{aligned}
$$

Finally, the updating equation for $H^{n+\frac{1}{2}}_{scat,z}(i,j,k)$ is

$$\begin{aligned} H^{n+\frac{1}{2}}_{scat,z}(i,j,k) &= C_{hzh}(i,j,k) \times H^{n-\frac{1}{2}}_{scat,z}(i,j,k) \\ &+ C_{hzey}(i,j,k) \times \left[E^{n}_{scat,y}(i,j+1,k) - E^{n}_{scat,y}(i,j,k)\right] \\ &+ C_{hzex}(i,j,k) \times \left[E^{n}_{scat,x}(i,j,k+1), -E^{n}_{scat,x}(i,j,k)\right] \\ &+ C_{hzhic}(i,j,k) \times H^{n+\frac{1}{2}}_{inc,z}(i,j,k) + C_{hzhip}(i,j,k) \times H^{n-\frac{1}{2}}_{inc,z}(i,j,k) \end{aligned} \tag{11.13}$$

where

$$C_{hzh}(i,j,k) = \frac{2\mu_z(i,j,k) - \sigma^m_z(i,j,k)\Delta t}{2\mu_z(i,j,k) + \sigma^m_z(i,j,k)\Delta t},$$

$$C_{hzey}(i,j,k) = -\frac{2\Delta t}{\Delta x\left(2\mu_z(i,j,k) + \sigma^m_z(i,j,k)\Delta t\right)},$$

$$C_{hzex}(i,j,k) = \frac{2\Delta t}{\Delta y\left(2\mu_z(i,j,k) + \sigma^m_z(i,j,k)\Delta t\right)},$$

$$C_{hzhic}(i,j,k) = \frac{2(\mu_0 - \mu_z(i,j,k)) - \sigma^m_z(i,j,k)\Delta t}{2\mu_z(i,j,k) + \sigma^m_z(i,j,k)\Delta t},$$

$$C_{hzhip}(i,j,k) = -\frac{2(\mu_0 - \mu_z(i,j,k)) + \sigma^m_z(i,j,k)\Delta t}{2\mu_z(i,j,k) + \sigma^m_z(i,j,k)\Delta t}.$$

Equations (11.8)–(11.13) are the updating equations with corresponding definitions for associated coefficients for the scattered field formulation. Comparing these equations with the set of general updating (1.26)–(1.31) one can realize that the forms of the equations and expressions of the coefficients are the same except that (11.8)–(11.13) include additional incident field terms.

11.3 Expressions for the incident plane waves

The aforementioned scattered field updating equations are derived to calculate scattered electric fields in a problem space in response to an incident field. The incident field in general can be any far-zone source that can be expressed by analytical expressions. In this context, incident plane waves are the most commonly used type of incident field. This section discusses the incident plane wave field expressions.

Figure 11.1 illustrates an incident plane wave traveling in the direction denoted by a unit vector $\hat{k}$. The incident electric field in general may have a θ component and a ϕ component when expressed in a spherical coordinate system denoting its polarization. The incident electric field can be expressed for a given point denoted by a position vector $\vec{r}$ as

$$\vec{E}_{inc} = \left(E_\theta\hat{\theta} + E_\phi\hat{\phi}\right) f\left(t - \frac{1}{c}\hat{k}\cdot\vec{r}\right), \tag{11.14}$$

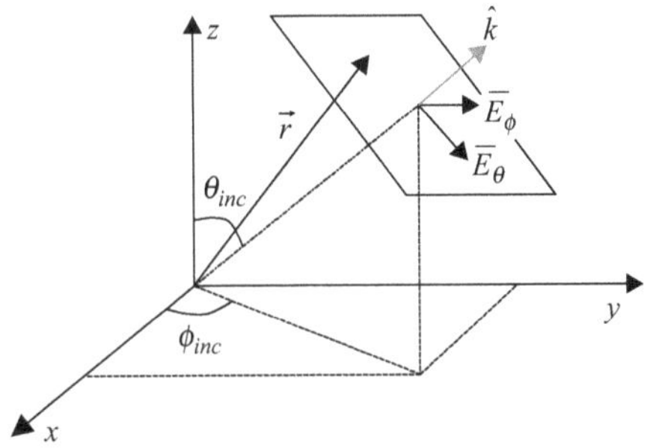

Figure 11.1 An incident plane wave.

where f is a function determining the waveform of the incident electric field. This equation can be rewritten by allowing a time delay t_0 and spatial shift l_0 as

$$\vec{E}_{inc} = \left(E_\theta\hat{\theta} + E_\phi\hat{\phi}\right) f\left((t - t_0) - \frac{1}{c}\left(\hat{k}\cdot\vec{r} - l_0\right)\right). \tag{11.15}$$

The vectors $\vec{r}$ and $\hat{k}$ can be written in Cartesian coordinates as

$$\vec{r} = \hat{x}x + \hat{y}y + \hat{z}z, \quad \hat{k} = \hat{x}\sin\theta_{inc}\cos\phi_{inc} + \hat{y}\sin\theta_{inc}\sin\phi_{inc} + \hat{z}\cos\theta_{inc}, \tag{11.16}$$

where θ_{inc} and ϕ_{inc} are the angles indicating the direction of propagation of the incident plane wave as shown in Figure 11.1. The components of the incident electric field E_θ and E_ϕ can also be expressed in Cartesian coordinates. Therefore, after using (11.16) in (11.15) and applying the spherical to Cartesian coordinates transformation, one can obtain the components of the incident electric field for any point (x, y, z) in space as

$$E_{inc,x} = (E_\theta\cos\theta_{inc}\cos\phi_{inc} - E_\phi\sin\phi_{inc})f_{wf}, \tag{11.17a}$$

$$E_{inc,y} = (E_\theta\cos\theta_{inc}\sin\phi_{inc} + E_\phi\cos\phi_{inc})f_{wf}, \tag{11.17b}$$

$$E_{inc,z} = (-E_\theta\sin\theta_{inc})f_{wf}, \tag{11.17c}$$

where f_{wf} is given by

$$f_{wf} = f\left((t - t_0) - \frac{1}{c}(x\sin\theta_{inc}\cos\phi_{inc} + y\sin\theta_{inc}\sin\phi_{inc} + z\cos\theta_{inc} - l_0)\right). \tag{11.18}$$

Once the expressions for the incident electric field components are obtained, the expressions for magnetic field components can be obtained using

$$\vec{H}_{inc} = \frac{1}{\eta_0}\hat{k}\times\vec{E}_{inc},$$

where n_0 is the intrinsic impedance of free space, and the $\times$ denotes the cross-product operator. Then the magnetic field components are

$$H_{inc,x} = \frac{-1}{\eta_0}(E_\phi\cos\theta_{inc}\cos\phi_{inc} + E_\theta\sin\phi_{inc})f_{wf}, \tag{11.19a}$$

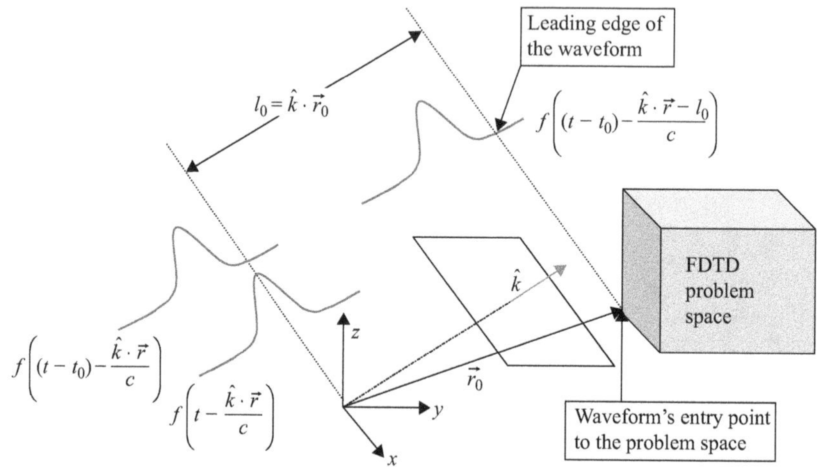

Figure 11.2 An incident plane wave delayed in time and shifted in space.

$$H_{inc,y} = \frac{-1}{\eta_0}(E_\phi \cos\theta_{inc} \sin\phi_{inc} - E_\theta \cos\phi_{inc})f_{wf}, \tag{11.19b}$$

$$H_{inc,z} = \frac{1}{\eta_0}(E_\phi \sin\theta_{inc})f_{wf}. \tag{11.19c}$$

Equations (11.17)–(11.19) express an incident field with a polarization determined by $(E_\theta\hat{\theta} + E_\phi\hat{\phi})$, propagating in a direction determined by the angles θ_{inc} and ϕ_{inc} and having a waveform described by f_{wf}. In general, the waveform can be a function that has the desired bandwidth characteristics as discussed in Chapter 5. However, there are two additional parameters in (11.18) we have not discussed yet: t_0 and l_0. These parameters are used to shift the given waveform in time and space such that when the FDTD iterations start the incident field in the problem space is zero, and as time proceeds the incident field propagates into the FDTD problem space.

Consider Figure 11.2, which illustrates a plane wave with a Gaussian waveform propagating in a direction $\hat{k}$ toward an FDTD problem space. This waveform is expressed by a function

$$f\left(t - \frac{1}{c}\left(\hat{k}\cdot\vec{r}\right)\right),$$

where the maximum of the waveform appears on a plane including the origin and normal to $\hat{k}$ at $t = 0$. Clearly the leading edge of this waveform is closer to the FDTD problem space. A time delay t_0 can be applied to this function such that

$$f\left((t - t_0) - \frac{1}{c}\left(\hat{k}\cdot\vec{r}\right)\right),$$

and the leading edge of the waveform coincides with the origin. The value of the waveform can be calculated as explained in Chapter 5. After applying the time delay, there is a distance between the leading edge of the waveform and the problem space that needs to be accounted for; otherwise, it may take considerable time after the simulation is started until the waveform enters to the problem space. The distance between the waveform and the problem space can be found as $l_0 = \hat{k} \cdot \vec{r}_0$, where $\vec{r}_0$ is the position vector indicating the closest point of the problem space to the waveform. A problem space has eight corners, and one of these corners will be closest to the waveform approaching from an arbitrary direction. If $\vec{r}_0$ indicates the corner points of the problem space, then the minimum distance for l_0 can be calculated by $l_0 = min\left[\hat{k} \cdot \vec{r}_0\right]$. After l_0 is determined, the waveform equation can be modified as

$$f\left((t - t_0) - \frac{1}{c}\left(\hat{k} \cdot \vec{r} - l_0\right)\right),$$

which leads to (11.18).

In some scattering applications, a scatterer is placed at the origin and the scattered field due to an incident plane wave is calculated. The direction of propagation of the plane wave must be defined appropriately in such applications. Consider Figure 11.3, where two plane waves are illustrated. Although these two plane waves are traveling in the same direction, one of them is illustrated as traveling toward the origin while the other one is traveling away from the origin. The equations given in this section are derived based on the wave traveling away from the origin; the angles determining the propagation, θ_{inc} and ϕ_{inc}, and amplitudes of the incident field components E_θ and E_ϕ are as shown in this figure. If it is desired to define the angles and field components based on the wave traveling toward the origin such as θ'_{inc}, ϕ'_{inc}, E'_θ, and E'_ϕ, then the following conversions need to be employed to use the aforementioned equations as is:

$$\theta_{inc} = \pi - \theta'_{inc}, \tag{11.20a}$$

$$\phi_{inc} = \pi + \phi'_{inc}, \tag{11.20b}$$

$$E_\theta = E'_\theta, \tag{11.20c}$$

$$E_\phi = -E'_\phi. \tag{11.20d}$$

One of the types of results that can be obtained from the scattered field due to an incident plane wave is the radar cross-section (RCS) of a scatterer. The near-field to far-field NF-FF transformation algorithm discussed in Chapter 9 is used for this purpose. In this case the scattered near fields calculated in a problem space are captured on an imaginary surface

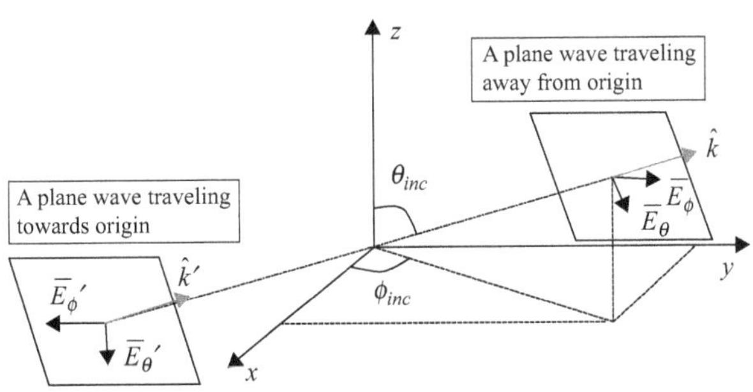

Figure 11.3 Two incident plane waves: one traveling toward the origin and the other traveling away from the origin.

enclosing the scatterer to determine the equivalent fictitious surface currents. These currents are transformed to the frequency domain while being captured. After the time-marching loop is completed, the far-field terms L_θ, L_ϕ, N_θ, and N_ϕ are calculated following the procedure described in Chapter 9.

At this point the components of the bistatic RCS can be calculated using

$$RCS_\theta = \frac{k^2}{8\pi\eta_0 P_{inc}} |L_\phi + \eta_0 N_\theta|^2 \tag{11.21a}$$

$$RCS_\phi = \frac{k^2}{8\pi\eta_0 P_{inc}} |L_\theta - \eta_0 N_\phi|^2, \tag{11.21b}$$

which is very similar to (9.39). The only difference is that P_{rad} is replaced by P_{inc}, which is the power carried by the incident plane wave. The P_{inc} can be calculated as

$$P_{inc} = \frac{1}{2\eta_0} |E_{inc}(\omega)|^2, \tag{11.22}$$

where $E_{inc}(\omega)$ is the discrete Fourier transform of the incident electric field waveform at the frequency for which the RCS calculation is sought.

11.4 MATLAB® implementation of the scattered field formulation

The scattered field formulation is presented in the previous sections. In this section the implementation of the scattered field formulation is discussed.

11.4.1 Definition of the incident plane wave

For a scattering problem it is necessary to define an incident field as the source. In Section 11.3 the equations necessary to construct a plane wave are presented, and the plane wave is used as the incident field in this discussion. In the FDTD program an incident plane wave is defined in the subroutine ***define_sources_and_lumped_elements*** as shown in Listing 11.1. In the given code first a new empty structure, **incident_plane_wave**, is defined. Then the parameters of the incident plane wave are defined as fields of this structure. Here **E_theta** denotes the maximum amplitude of the θ component, whereas **E_phi** denotes the maximum amplitude of the ϕ component of the electric field waveform. The fields **theta_incident** and **phi_incident** denote the angles indicating the direction of propagation, θ_{inc} and ϕ_{inc}, of the incident plane wave. It should be noted here that these parameters are based on the convention that the plane wave is propagating in a direction away from the origin rather than toward the origin. If the parameters are readily available for the convention wave traveling toward the origin, then the necessary transformation needs to be performed based on (11.20). Finally, the waveform of the incident plane wave need to be defined. The definition of the waveform for the incident plane wave follows the same procedure as the voltage source or the current source waveform. The types of waveforms are defined as structure **waveforms**, and the fields **waveform_type** and **waveform_index** are defined under **incident_plane_wave**, which point to a specific waveform defined under **waveforms**.

Listing 11.1 define_sources_and_lumped_elements

```
disp('defining sources and lumped element components');

voltage_sources = [];
current_sources = [];
diodes = [];
resistors = [];
inductors = [];
capacitors = [];
incident_plane_wave = [];

% define source waveform types and parameters
waveforms.gaussian(1).number_of_cells_per_wavelength = 0;
waveforms.gaussian(2).number_of_cells_per_wavelength = 15;

% Define incident plane wave, angles are in degrees
incident_plane_wave.E_theta = 1;
incident_plane_wave.E_phi = 0;
incident_plane_wave.theta_incident = 0;
incident_plane_wave.phi_incident = 0;
incident_plane_wave.waveform_type = 'gaussian';
incident_plane_wave.waveform_index = 1;
```

11.4.2 Initialization of the incident fields

The next step is the initialization of the incident fields and the updating coefficients. The initialization of the incident fields is performed in the subroutine ***initialize_sources_and_lumped_elements***. The section of the code listed in Listing 11.2 is added to this subroutine.

Listing 11.2 initialize_sources_and_lumped_elements

```
% initialize incident plane wave
if isfield(incident_plane_wave, 'E_theta')
    incident_plane_wave.enabled = true;
else
    incident_plane_wave.enabled = false;
end

if incident_plane_wave.enabled
    % create incident field arrays for current time step
    Hxic = zeros(nxp1, ny, nz);
    Hyic = zeros(nx, nyp1, nz);
    Hzic = zeros(nx, ny, nzp1);
    Exic = zeros(nx, nyp1, nzp1);
    Eyic = zeros(nxp1, ny, nzp1);
    Ezic = zeros(nxp1, nyp1, nz);
    % create incident field arrays for previous time step
    Hxip = zeros(nxp1, ny, nz);
    Hyip = zeros(nx, nyp1, nz);
    Hzip = zeros(nx, ny, nzp1);
```

```
Exip = zeros(nx,nyp1,nzp1);
Eyip = zeros(nxp1,ny,nzp1);
Ezip = zeros(nxp1,nyp1,nz);

% calculate the amplitude factors for field components
theta_incident = incident_plane_wave.theta_incident*pi/180;
phi_incident = incident_plane_wave.phi_incident*pi/180;
E_theta = incident_plane_wave.E_theta;
E_phi = incident_plane_wave.E_phi;
eta_0 = sqrt(mu_0/eps_0);
Exi0 = E_theta * cos(theta_incident) * cos(phi_incident) ...
    - E_phi * sin(phi_incident);
Eyi0 = E_theta * cos(theta_incident) * sin(phi_incident) ...
    + E_phi * cos(phi_incident);
Ezi0 = -E_theta * sin(theta_incident);
Hxi0 = (-1/eta_0)*(E_phi * cos(theta_incident) ...
    * cos(phi_incident) + E_theta * sin(phi_incident));
Hyi0 = (-1/eta_0)*(E_phi * cos(theta_incident) ...
    * sin(phi_incident) - E_theta * cos(phi_incident));
Hzi0 = (1/eta_0)*(E_phi * sin(theta_incident));

% Create position arrays indicating the coordinates of the nodes
x_pos = zeros(nxp1,nyp1,nzp1);
y_pos = zeros(nxp1,nyp1,nzp1);
z_pos = zeros(nxp1,nyp1,nzp1);
for ind = 1:nxp1
    x_pos(ind,:,:) = (ind - 1) * dx + fdtd_domain.min_x;
end
for ind = 1:nyp1
    y_pos(:,ind,:) = (ind - 1) * dy + fdtd_domain.min_y;
end
for ind = 1:nzp1
    z_pos(:,:,ind) = (ind - 1) * dz + fdtd_domain.min_z;
end

% calculate spatial shift, l_0, required for incident plane wave
r0 =[fdtd_domain.min_x fdtd_domain.min_y fdtd_domain.min_z;
fdtd_domain.min_x fdtd_domain.min_y fdtd_domain.max_z;
fdtd_domain.min_x fdtd_domain.max_y fdtd_domain.min_z;
fdtd_domain.min_x fdtd_domain.max_y fdtd_domain.max_z;
fdtd_domain.max_x fdtd_domain.min_y fdtd_domain.min_z;
fdtd_domain.max_x fdtd_domain.min_y fdtd_domain.max_z;
fdtd_domain.max_x fdtd_domain.max_y fdtd_domain.min_z;
fdtd_domain.max_x fdtd_domain.max_y fdtd_domain.max_z;];

k_vec_x = sin(theta_incident)*cos(phi_incident);
k_vec_y = sin(theta_incident)*sin(phi_incident);
k_vec_z = cos(theta_incident);

k_dot_r0 = k_vec_x * r0(:,1) ...
    + k_vec_y * r0(:,2) ...
    + k_vec_z * r0(:,3);
```

```
    l_0 = min(k_dot_r0)/c;

    % calculate k.r for every field component
    k_dot_r_ex = ((x_pos(1:nx,1:nyp1,1:nzp1)+dx/2) * k_vec_x ...
        + y_pos(1:nx,1:nyp1,1:nzp1) * k_vec_y ...
        + z_pos(1:nx,1:nyp1,1:nzp1) * k_vec_z)/c;

    k_dot_r_ey = (x_pos(1:nxp1,1:ny,1:nzp1) * k_vec_x ...
        + (y_pos(1:nxp1,1:ny,1:nzp1)+dy/2) * k_vec_y ...
        + z_pos(1:nxp1,1:ny,1:nzp1) * k_vec_z)/c;

    k_dot_r_ez = (x_pos(1:nxp1,1:nyp1,1:nz) * k_vec_x ...
        + y_pos(1:nxp1,1:nyp1,1:nz) * k_vec_y ...
        + (z_pos(1:nxp1,1:nyp1,1:nz)+dz/2) * k_vec_z)/c;

    k_dot_r_hx = (x_pos(1:nxp1,1:ny,1:nz) * k_vec_x ...
        + (y_pos(1:nxp1,1:ny,1:nz)+dy/2) * k_vec_y ...
        + (z_pos(1:nxp1,1:ny,1:nz)+dz/2) * k_vec_z)/c;

    k_dot_r_hy = ((x_pos(1:nx,1:nyp1,1:nz)+dx/2) * k_vec_x ...
        + y_pos(1:nx,1:nyp1,1:nz) * k_vec_y ...
        + (z_pos(1:nx,1:nyp1,1:nz)+dz/2) * k_vec_z)/c;

    k_dot_r_hz = ((x_pos(1:nx,1:ny,1:nzp1)+dx/2) * k_vec_x ...
        + (y_pos(1:nx,1:ny,1:nzp1)+dy/2) * k_vec_y ...
        + z_pos(1:nx,1:ny,1:nzp1) * k_vec_z)/c;

    % embed spatial shift in k.r
    k_dot_r_ex = k_dot_r_ex - l_0;
    k_dot_r_ey = k_dot_r_ey - l_0;
    k_dot_r_ez = k_dot_r_ez - l_0;
    k_dot_r_hx = k_dot_r_hx - l_0;
    k_dot_r_hy = k_dot_r_hy - l_0;
    k_dot_r_hz = k_dot_r_hz - l_0;

    % store the waveform
    wt_str = incident_plane_wave.waveform_type;
    wi_str = num2str(incident_plane_wave.waveform_index);
    eval_str = ['a_waveform = waveforms.' ...
        wt_str '(' wi_str ').waveform;'];
    eval(eval_str);
    incident_plane_wave.waveform = a_waveform;

    clear x_pos y_pos z_pos;
end
```

In this code the first step is to determine whether an incident plane wave is defined. If an incident plane wave is defined, then it indicates that the FDTD simulation is running a scattering problem and, therefore, the algorithm of the scattered field formulation will be employed. Here a logical parameter **incident_plane_wave.enabled** is assigned a value "***true***" or "***false***" indicating the mode of the FDTD simulation.

The calculation of the scattered field components using the scattered field updating equation is the same as the calculation of the *total* field components using the general updating equation, except for the additional incident field terms. Therefore, the parameters corresponding to the field arrays **Ex**, **Ey**, **Ez**, **Hx**, **Hy**, and **Hz** can be assumed as scattered fields if the mode of the simulation is scattering. Then the field arrays corresponding to incident fields must be defined. The new field arrays are thus defined as **Exi**, **Eyi**, **Ezi**, **Hxi**, **Hyi**, and **Hzi** and are initialized as zeros. One should notice that the sizes of these three-dimensional arrays are the same as their corresponding scattered field arrays.

In the next step the amplitudes of the electric and magnetic field waveforms are transformed from the spherical coordinates to Cartesian coordinates based on (11.17) and (11.19), yielding **Exi0**, **Eyi0**, **Ezi0**, **Hxi0**, **Hyi0**, and **Hzi0**.

The incident field components are computed at different positions in a three-dimensional space. The coordinates of these field components can be determined with respect to positions of the nodes of the FDTD grid. Three three-dimensional arrays are constructed, each of which stores the x, y, and z coordinates of the nodes.

As discussed in Section 11.3 the waveform of the incident field should be shifted in time and space to ensure that the leading edge of the waveform enters into the problem space after the simulation has started. These shifts are indicated as t_0 and l_0, respectively. The temporal shift t_0 is determined in the subroutine ***initialize_waveforms***, which has been called in ***initialize_sources_and_lumped_elements*** before. Thus, t_0 for a specific waveform is readily available. The spatial shift, however, needs to be calculated. Eight corners determine the boundaries of the problem space, and an incident plane wave will enter to the problem space from one of these corers. The coordinates of these corners are stored in an array denoted as **r0**. The value of l_0 is calculated by $l_0 = min[\hat{k} \cdot \vec{r}_0]$, where $\hat{k}$ is the propagation vector indicating the direction of propagation. The components of the vector $\hat{k}$ are calculated based on (11.16) and are stored in the parameters **k_vec_x**, **k_vec_y**, and **k_vec_z**. Then these parameters are used together with **r0** to calculate l_0 and to store it in the parameter **1_0**.

Another term that determines the three-dimensional distribution of the waveform is $\hat{k} \cdot \vec{r}$ as shown in (11.18). Here the vector $\vec{r}$ is the position vector. In the three-dimensional case this dot product would yield three-dimensional arrays. Since all field components are located at different positions, this dot product would yield a different three-dimensional array for each component. Therefore, the dot product is performed separately for each field and is stored in the arrays **k_dot_r_ex**, **k_dot_r_ey**, **k_dot_r_ez**, **k_dot_r_hx**, **k_dot_r_hy**, and **k_dot_r_hz**. Then the spatial shift l_0 appearing in (11.18) is embedded in these arrays.

Finally, the waveform of the incident plane wave is acquired from the structure **waveforms** and is stored in the structure **incident_plane_wave**. This waveform is used only for plotting purposes after the FDTD calculation is completed. The distribution of the waveform and incident plane wave has to be recalculated at every time step and stored in the three-dimensional incident field arrays for use in the scattered field updating equations.

11.4.3 Initialization of the updating coefficients

The scattered field formulation algorithm requires a new set of updating coefficients. As discussed in Section 11.2 the updating equations for the scattered field formulation are the same as the general updating equations except that the scattered field updating equations have additional terms. These terms consist of incident fields multiplied by incident field coefficients.

Listing 11.3 initialize_incident_field_updating_coefficients

```
% initialize incident field updating coefficients

if incident_plane_wave.enabled == false
    return;
end

% Coeffiecients updating Ex
Cexeic= (2*(1 - eps_r_x)*eps_0-dt*sigma_e_x) ...
    ./(2*eps_r_x*eps_0+dt*sigma_e_x);
Cexeip=-(2*(1-eps_r_x)*eps_0+dt*sigma_e_x) ...
    ./(2*eps_r_x*eps_0+dt*sigma_e_x);

% Coeffiecients updating Ey
Ceyeic= (2*(1 - eps_r_y)*eps_0-dt*sigma_e_y) ...
    ./(2*eps_r_y*eps_0+dt*sigma_e_y);
Ceyeip=-(2*(1-eps_r_y)*eps_0+dt*sigma_e_y) ...
    ./(2*eps_r_y*eps_0+dt*sigma_e_y);

% Coeffiecients updating Ez
Cezeic= (2*(1 - eps_r_z)*eps_0-dt*sigma_e_z) ...
    ./(2*eps_r_z*eps_0+dt*sigma_e_z);
Cezeip=-(2*(1-eps_r_z)*eps_0+dt*sigma_e_z) ...
    ./(2*eps_r_z*eps_0+dt*sigma_e_z);

% Coeffiecients updating Hx
Chxhic= (2*(1 - mu_r_x)*mu_0-dt*sigma_m_x)./(2*mu_r_x*mu_0+dt*sigma_m_x);
Chxhip=-(2*(1-mu_r_x)*mu_0+dt*sigma_m_x)./(2*mu_r_x*mu_0+dt*sigma_m_x);

% Coeffiecients updating Hy
Chyhic= (2*(1 - mu_r_y)*mu_0-dt*sigma_m_y)./(2*mu_r_y*mu_0+dt*sigma_m_y);
Chyhip=-(2*(1-mu_r_y)*mu_0+dt*sigma_m_y)./(2*mu_r_y*mu_0+dt*sigma_m_y);

% Coeffiecients updating Hz
Chzhic=(2*(1 - mu_r_z)*mu_0-dt*sigma_m_z)./(2*mu_r_z*mu_0+dt*sigma_m_z);
Chzhip=-(2*(1-mu_r_z)*mu_0+dt*sigma_m_z)./(2*mu_r_z*mu_0+dt*sigma_m_z);
```

The incident field coefficients are initialized in a subroutine ***initialize_incident_field_updating_coefficients***, the implementation of which is shown in Listing 11.3. This subroutine is called in ***initialize_updating_coefficients*** after all other updating coefficient subroutines are called. Here a set of new three-dimensional arrays is constructed representing the updating coefficients related to the incident field components based on (11.8)–(11.13).

11.4.4 Calculation of the scattered fields

As the iterations proceed in the FDTD time-marching algorithm, the incident plane wave enters the FDTD problem space and propagates such that at every time instant the values of the incident field components change. Therefore, the values of the incident field components need to be recalculated at every time step. A new subroutine named ***update_incident_fields*** is implemented for this purpose and is provided in Listing 11.4.

Listing 11.4 update_incident_fields

```
% update incident fields for the current time step
if incident_plane_wave.enabled == false
    return;
end

tm = current_time + dt/2;
te = current_time + dt;

% update incident fields for previous time step
Hxip = Hxic; Hyip = Hyic; Hzip = Hzic;
Exip = Exic; Eyip = Eyic; Ezip = Ezic;

wt_str = incident_plane_wave.waveform_type;
wi = incident_plane_wave.waveform_index;

% if waveform is Gaussian waveforms
if strcmp(incident_plane_wave.waveform_type,'gaussian')
    tau = waveforms.gaussian(wi).tau;
    t_0 = waveforms.gaussian(wi).t_0;
    Exic = Exi0 * exp(-((te - t_0 - k_dot_r_ex )/tau).^2);
    Eyic = Eyi0 * exp(-((te - t_0 - k_dot_r_ey )/tau).^2);
    Ezic = Ezi0 * exp(-((te - t_0 - k_dot_r_ez )/tau).^2);
    Hxic = Hxi0 * exp(-((tm - t_0 - k_dot_r_hx )/tau).^2);
    Hyic = Hyi0 * exp(-((tm - t_0 - k_dot_r_hy )/tau).^2);
    Hzic = Hzi0 * exp(-((tm - t_0 - k_dot_r_hz )/tau).^2);
end

% if waveform is derivative of Gaussian
if strcmp(incident_plane_wave.waveform_type,'derivative_gaussian')
    tau = waveforms.derivative_gaussian(wi).tau;
    t_0 = waveforms.derivative_gaussian(wi).t_0;
    Exic = Exi0 * (-sqrt(2*exp(1))/tau)*(te - t_0 - k_dot_r_ex) ...
           .*exp(-((te - t_0 - k_dot_r_ex)/tau).^2);
    Eyic = Eyi0 * (-sqrt(2*exp(1))/tau)*(te - t_0 - k_dot_r_ey) ...
           .*exp(-((te - t_0 - k_dot_r_ey)/tau).^2);
    Ezic = Ezi0 * (-sqrt(2*exp(1))/tau)*(te - t_0 - k_dot_r_ez) ...
           .*exp(-((te - t_0 - k_dot_r_ez)/tau).^2);
    Hxic = Hxi0 * (-sqrt(2*exp(1))/tau)*(tm - t_0 - k_dot_r_hx) ...
           .*exp(-((tm - t_0 - k_dot_r_hx)/tau).^2);
    Hyic = Hyi0 * (-sqrt(2*exp(1))/tau)*(tm - t_0 - k_dot_r_hy) ...
           .*exp(-((tm - t_0 - k_dot_r_hy)/tau).^2);
    Hzic = Hzi0 * (-sqrt(2*exp(1))/tau)*(tm - t_0 - k_dot_r_hz) ...
           .*exp(-((tm - t_0 - k_dot_r_hz)/tau).^2);
end

% if waveform is cosine modulated Gaussian
if strcmp(incident_plane_wave.waveform_type, ...
    'cosine_modulated_gaussian')
    f = waveforms.cosine_modulated_gaussian(wi).modulation_frequency;
    tau = waveforms.cosine_modulated_gaussian(wi).tau;
```

```
    t_0 = waveforms.cosine_modulated_gaussian(wi).t_0;
    Exic = Exi0 * cos(2*pi*f*(te - t_0 - k_dot_r_ex)) ...
           .*exp(-((te - t_0 - k_dot_r_ex)/tau).^2);
    Eyic = Eyi0 * cos(2*pi*f*(te - t_0 - k_dot_r_ey)) ...
           .*exp(-((te - t_0 - k_dot_r_ey)/tau).^2);
    Ezic = Ezi0 * cos(2*pi*f*(te - t_0 - k_dot_r_ez)) ...
           .*exp(-((te - t_0 - k_dot_r_ez)/tau).^2);
    Hxic = Hxi0 * cos(2*pi*f*(tm - t_0 - k_dot_r_hx)) ...
           .*exp(-((tm - t_0 - k_dot_r_hx)/tau).^2);
    Hyic = Hyi0 * cos(2*pi*f*(tm - t_0 - k_dot_r_hy)) ...
           .*exp(-((tm - t_0 - k_dot_r_hy)/tau).^2);
    Hzic = Hyi0 * cos(2*pi*f*(tm - t_0 - k_dot_r_hz)) ...
           .*exp(-((tm - t_0 - k_dot_r_hz)/tau).^2);
end
```

This subroutine is called in ***run_fdtd_time_marching_loop***, in the time-marching loop, before calling ***update_magnetic_fields*** and therefore before updating the magnetic field components. Since the waveform needs to be recalculated at every time step, this subroutine is composed of several sections, each of which pertains to a specific type of waveform. For instance, this implementation supports the Gaussian, derivative of Gaussian, and cosine-modulated Gaussian waveforms. Thus, at every iteration only the section pertaining to the type of the waveform of the incident plane wave is executed. At every time step all the six incident electric and magnetic field arrays are recalculated. One should notice that the electric and magnetic field components are calculated at time instants a half time step off from each other since the magnetic field components are evaluated at half-integer time steps while the electric field components are evaluated at integer time steps.

After ***update_incident_fields*** is called in ***run_fdtd_time_marching_loop***, the subroutine ***update_magnetic_fields*** is called in which the magnetic field components are updated for the current time step. The implementation of ***update_magnetic_fields*** is modified as shown in Listing 11.5 to accommodate the scattered field formulation. Here the general updating equation is kept as is; however, an additional section is added that will be executed if the mode of the simulation is scattering. In this additional section the incident field arrays are multiplied by the corresponding coefficients and are added to the magnetic field components, which in this case are assumed to be scattered fields. Similarly, the implementation of ***update_electric_fields*** is modified as shown in Listing 11.6 to accommodate the scattered field formulation for updating the scattered electric field components.

11.4.5 Postprocessing and simulation results

The scattered electric and magnetic field components at desired locations can be captured as the FDTD simulation is running by defining **sampled_electric_fields** and **sampled_magnetic_fields** in the subroutine ***define_output_parameters*** before the FDTD simulation is started. Another type of output from a scattering simulation is the RCS as discussed in Section 11.3. The calculation of RCS follows the same procedure as the NF–FF transformation. Only a slight modification of the code is required at the last stage of the NF–FF transformation.

Listing 11.5 update_magnetic_fields

```
% update magnetic fields

current_time = current_time + dt/2;

Hx = Chxh.*Hx+Chxey.*(Ey(1:nxp1,1:ny,2:nzp1)-Ey(1:nxp1,1:ny,1:nz)) ...
    + Chxez.*(Ez(1:nxp1,2:nyp1,1:nz)-Ez(1:nxp1,1:ny,1:nz));

Hy = Chyh.*Hy+Chyez.*(Ez(2:nxp1,1:nyp1,1:nz)-Ez(1:nx,1:nyp1,1:nz)) ...
    + Chyex.*(Ex(1:nx,1:nyp1,2:nzp1)-Ex(1:nx,1:nyp1,1:nz));

Hz = Chzh.*Hz+Chzex.*(Ex(1:nx,2:nyp1,1:nzp1)-Ex(1:nx,1:ny,1:nzp1)) ...
    + Chzey.*(Ey(2:nxp1,1:ny,1:nzp1)-Ey(1:nx,1:ny,1:nzp1));

if incident_plane_wave.enabled
    Hx = Hx + Chxhic .* Hxic + Chxhip .* Hxip;
    Hy = Hy + Chyhic .* Hyic + Chyhip .* Hyip;
    Hz = Hz + Chzhic .* Hzic + Chzhip .* Hzip;
end
```

Listing 11.6 update_electric_fields

```
% update electric fields except the tangential components
% on the boundaries

current_time = current_time + dt/2;

Ex(1:nx,2:ny,2:nz) = Cexe(1:nx,2:ny,2:nz).*Ex(1:nx,2:ny,2:nz) ...
                     + Cexhz(1:nx,2:ny,2:nz).*...
                     (Hz(1:nx,2:ny,2:nz)-Hz(1:nx,1:ny-1,2:nz)) ...
                     + Cexhy(1:nx,2:ny,2:nz).*...
                     (Hy(1:nx,2:ny,2:nz)-Hy(1:nx,2:ny,1:nz-1));

Ey(2:nx,1:ny,2:nz)=Ceye(2:nx,1:ny,2:nz).*Ey(2:nx,1:ny,2:nz) ...
                     + Ceyhx(2:nx,1:ny,2:nz).* ...
                     (Hx(2:nx,1:ny,2:nz)-Hx(2:nx,1:ny,1:nz-1)) ...
                     + Ceyhz(2:nx,1:ny,2:nz).* ...
                     (Hz(2:nx,1:ny,2:nz)-Hz(1:nx-1,1:ny,2:nz));

Ez(2:nx,2:ny,1:nz)=Ceze(2:nx,2:ny,1:nz).*Ez(2:nx,2:ny,1:nz) ...
                     + Cezhy(2:nx,2:ny,1:nz).* ...
                     (Hy(2:nx,2:ny,1:nz)-Hy(1:nx-1,2:ny,1:nz)) ...
                     + Cezhx(2:nx,2:ny,1:nz).*...
                     (Hx(2:nx,2:ny,1:nz)-Hx(2:nx,1:ny-1,1:nz));

if incident_plane_wave.enabled
    Ex = Ex + Cexeic .* Exic + Cexeip .* Exip;
    Ey = Ey + Ceyeic .* Eyic + Ceyeip .* Eyip;
    Ez = Ez + Cezeic .* Ezic + Cezeip .* Ezip;
end
```

However, before discussing this modification, it should be noted that if the calculation of the RCS is desired, the frequencies for which the RCS calculation is sought must be defined in the subroutine ***define_output_parameters*** as the parameter **farfield.frequencies(i)**. Furthermore, the parameter **farfield.number_of_cells_from_outer_boundary** must be defined as well.

The calculation of RCS from the captured NF–FF fictitious currents and display of results are performed in the subroutine ***calculate_and_display_farfields***. The initial lines of this subroutine are modified as shown in Listing 11.7. In the last part of the given code section it can be seen that if the mode of the simulation is scattering, then the plotted radiation patterns are RCS; otherwise, plots show the directivity of the field patterns. Another modification is that a new subroutine named ***calculate_incident_plane_wave_power*** is called after the subroutine ***calculate_radiated_power***. The RCS calculation is based on (11.21), which requires the value of the incident field power. Hence, the subroutine ***calculate_incident_plane_wave_power***, which is shown in Listing 11.8, is implemented to calculate the power of the incident plane wave at the desired frequency or frequencies based on (11.22).

The actual calculation of RCS is performed in the subroutine ***calculate_farfields_per_plane***. The last part of this subroutine is modified as shown in Listing 11.9. If the mode of the simulation is scattering then RCS is calculated using (11.21); otherwise, the directivity is calculated.

Finally, a section of code is added to subroutine ***display_transient_parameters***, as shown in Listing 11.10, to display the waveform of the incident plane wave that would be observed at the origin. It should be noted that the displayed waveform includes the temporal shift t_0 but not the spatial shift l_0.

Listing 11.7 calculate_and_display_farfields

```
if number_of_farfield_frequencies == 0
    return;
end

calculate_radiated_power;
calculate_incident_plane_wave_power;

j = sqrt(-1);
number_of_angles = 360;

% parameters used by polar plotting functions
step_size = 10;        % increment between the rings in the polar grid
Nrings = 4;            % number of rings in the polar grid
line_style1 = 'b-';    % line style for theta component
line_style2 = 'r--';   % line style for phi component
scale_type = 'dB';     % linear or dB

if incident_plane_wave.enabled == false
    plot_type = 'D';
else
    plot_type = 'RCS';
end
```

Listing 11.8 calculate_incident_plane_wave_power

```
% Calculate total radiated power
if incident_plane_wave.enabled == false
    return;
end
x = incident_plane_wave.waveform;
time_shift = 0;
[X] = time_to_frequency_domain(x,dt,farfield.frequencies,time_shift);
incident_plane_wave.frequency_domain_value = X;
incident_plane_wave.frequencies = frequency_array;
E_t = incident_plane_wave.E_theta;
E_p = incident_plane_wave.E_phi;
eta_0 = sqrt(mu_0/eps_0);
incident_plane_wave.incident_power = ...
    (0.5/eta_0) * (E_t^2 + E_p^2) * abs(X)^2;
```

Listing 11.9 calculate_farfields_per_plane

```
if incident_plane_wave.enabled == false
  % calculate directivity
  farfield_dataTheta(mi,:)=(k^2./(8*pi*eta_0*radiated_power(mi))) ...
      .* (abs(Lphi+eta_0*Ntheta).^2);
  farfield_dataPhi(mi,:)  =(k^2./(8*pi*eta_0*radiated_power(mi))) ...
      .* (abs(Ltheta-eta_0*Nphi).^2);
else
  % calculate radar cross-section
  farfield_dataTheta(mi,:)   = ...
      (k^2./(8*pi*eta_0*incident_plane_wave.incident_power(mi))) ...
      .* (abs(Lphi+eta_0*Ntheta).^2);
  farfield_dataPhi(mi,:)     = ...
      (k^2./(8*pi*eta_0*incident_plane_wave.incident_power(mi))) ...
      .* (abs(Ltheta-eta_0*Nphi).^2);
end
```

Listing 11.10 display_transient_parameters

```
% figure for incident plane wave
if incident_plane_wave.enabled == true
    figure;
    xlabel('time (ns)','fontsize',12);
    ylabel('(volt/meter)','fontsize',12);
    title('Incident electric field','fontsize',12);
    grid on; hold on;
    sampled_value_theta = incident_plane_wave.E_theta * ...
        incident_plane_wave.waveform;
    sampled_value_phi = incident_plane_wave.E_phi * ...
        incident_plane_wave.waveform;
    sampled_time = time(1:time_step)*1e9;
    plot(sampled_time, sampled_value_theta,'b-',...
        sampled_time, sampled_value_phi,'r:','linewidth',1.5);
    legend('E_{\theta,inc}','E_{\phi,inc}'); drawnow;
end
```

11.5 Simulation examples

In the previous sections we discussed scattered field algorithm and demonstrated its implementation using MATLAB® programs. In this section we provide examples of scattering problems.

11.5.1 Scattering from a dielectric sphere

In this section we present the calculation of the bistatic RCS of a dielectric sphere. Figure 11.4 shows an FDTD problem space including a dielectric sphere illuminated by an x polarized plane wave traveling in the positive z direction. The problem space is composed of cells with size $\Delta x = 0.75$ cm, $\Delta y = 0.75$ cm, and $\Delta z = 0.75$ cm. The dielectric sphere radius is 10 cm, relative permittivity is 3, and relative permeability is 2. The definition of the problem space is shown in Listing 11.11. The incident plane wave source is defined as shown in Listing 11.12. The waveform of the plane wave is Gaussian. The output of the FDTD simulation is defined as far field at 1 GHz, which indicates that the RCS will be calculated. Furthermore, the x component of the scattered electric field at the origin is defined as the output as shown in Listing 11.13.

The simulation is run for 2,000 time steps, and the sampled electric field at the origin and the RCS plots are obtained. Figure 11.5 shows the sampled scattered electric field captured at the origin and shows the waveform of the incident plane wave. Figures 11.6, 11.7, and 11.8 show the RCS of the dielectric sphere in the xy, xz, and yz planes, respectively.

To verify the accuracy of the calculated RCSs, the results are compared with the analytical solution presented in [44]. The θ component of the RCS in the xz plane calculated by the

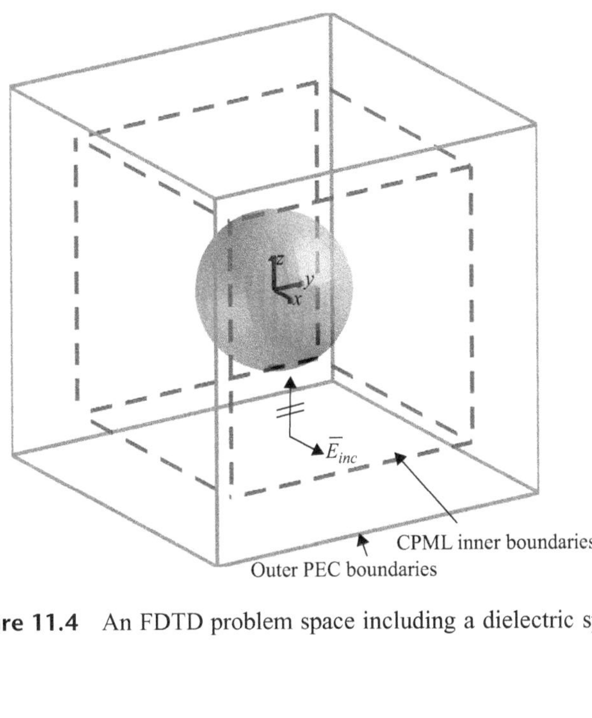

Figure 11.4 An FDTD problem space including a dielectric sphere.

Listing 11.11 define_geometry.m

```
disp('defining the problem geometry');

bricks  = [];
spheres = [];
thin_wires = [];

% define a sphere
spheres(1).radius = 100e-3;
spheres(1).center_x = 0;
spheres(1).center_y = 0;
spheres(1).center_z = 0;
spheres(1).material_type = 4;
```

Listing 11.12 define_sources_and_lumped_elements.m

```
disp('defining sources and lumped element components');

voltage_sources = [];
current_sources = [];
diodes = [];
resistors = [];
inductors = [];
capacitors = [];
incident_plane_wave = [];

% define source waveform types and parameters
waveforms.gaussian(1).number_of_cells_per_wavelength = 0;
waveforms.gaussian(2).number_of_cells_per_wavelength = 15;

% Define incident plane wave, angles are in degrees
incident_plane_wave.E_theta = 1;
incident_plane_wave.E_phi = 0;
incident_plane_wave.theta_incident = 0;
incident_plane_wave.phi_incident = 0;
incident_plane_wave.waveform_type = 'gaussian';
incident_plane_wave.waveform_index = 1;
```

FDTD simulation is compared with the analytical solution in the Cartesian plot in Figure 11.9, whereas the ϕ component of the RCS in the *yz* plane calculated by the FDTD simulation is compared with the analytical solution in the Cartesian plot in Figure 11.10. Both comparisons show very good agreement.

Listing 11.13 shows that an animation is also defined as an output. The scattered electric field is thus captured on the *xz* plane displayed during the simulation. Figure 11.11 shows the snapshots of the animation captured at time steps 120, 140, 160, and 180.

Listing 11.13 define_output_parameters.m

```
disp('defining output parameters');

sampled_electric_fields = [];
sampled_magnetic_fields = [];
sampled_voltages = [];
sampled_currents = [];
ports = [];
farfield.frequencies = [];

% figure refresh rate
plotting_step = 10;

% mode of operation
run_simulation = true;
show_material_mesh = true;
show_problem_space = true;

% far field calculation parameters
farfield.frequencies(1) = 1.0e9;
farfield.number_of_cells_from_outer_boundary = 13;

% frequency domain parameters
frequency_domain.start = 20e6;
frequency_domain.end   = 4e9;
frequency_domain.step  = 20e6;


% define sampled electric fields
% component: vector component íx í,í y í,í z í, or magnitude ímí
% display plot = true, in order to plot field during simulation
sampled_electric_fields(1).x = 0;
sampled_electric_fields(1).y = 0;
sampled_electric_fields(1).z = 0;
sampled_electric_fields(1).component = 'x';
sampled_electric_fields(1).display_plot = false;

% define animation
% field_type shall be 'e' or 'h'
% plane cut shall be 'xy', yz, or zx
% component shall be 'x', 'y', 'z', or 'm;
animation(1).field_type = 'e';
animation(1).component = 'm';
animation(1).plane_cut(1).type = 'zx';
animation(1).plane_cut(1).position  = 0;
animation(1).enable = true;
animation(1).display_grid = false;
animation(1).display_objects = true;
```

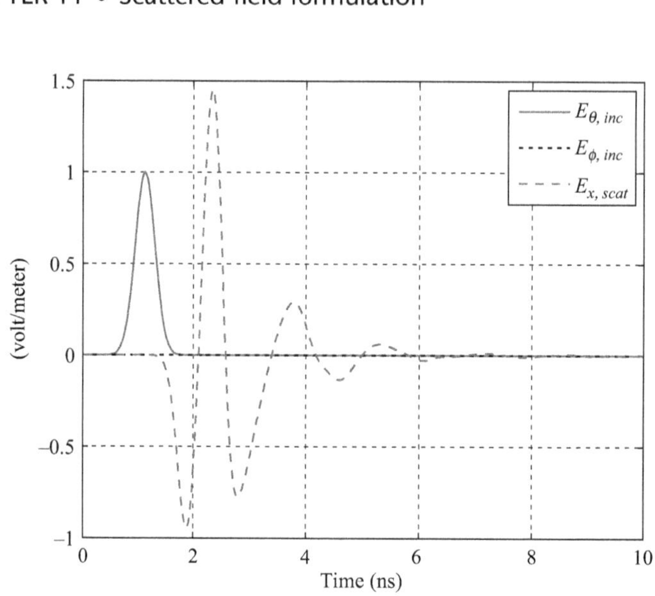

Figure 11.5 Scattered electric field captured at the origin and the incident field waveform.

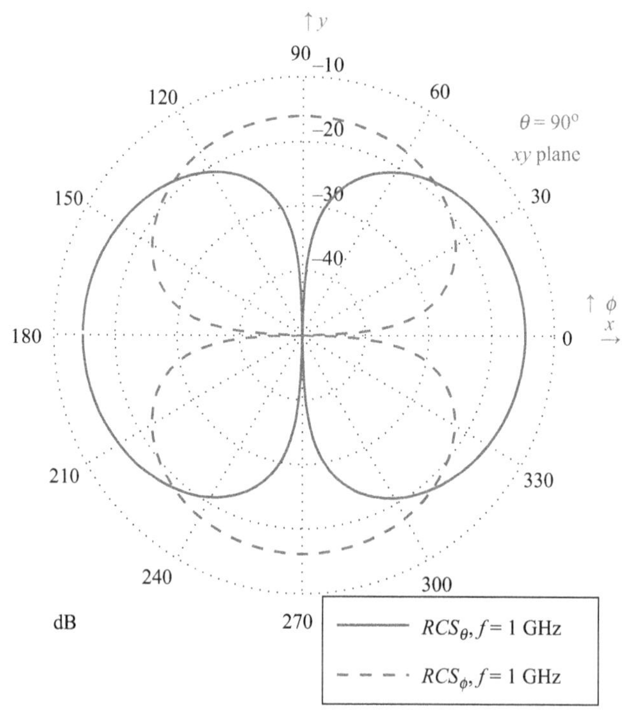

Figure 11.6 Bistatic RCS at 1 GHz in the *xy* plane.

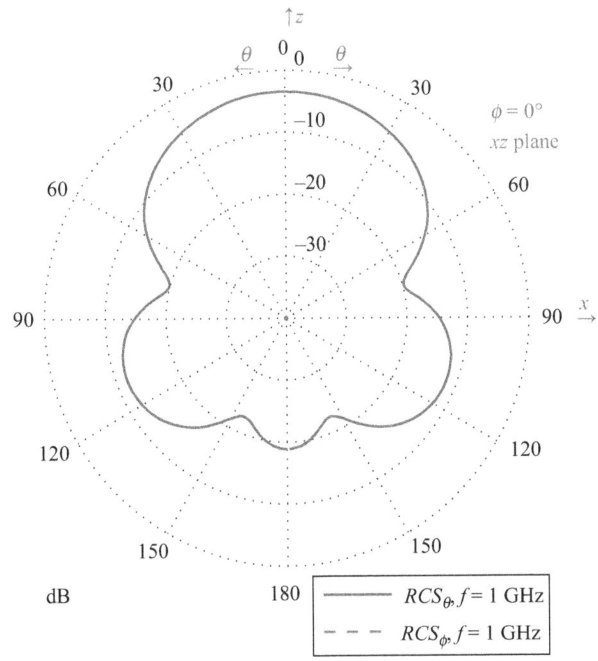

Figure 11.7 Bistatic RCS at 1 GHz in the *xz* plane.

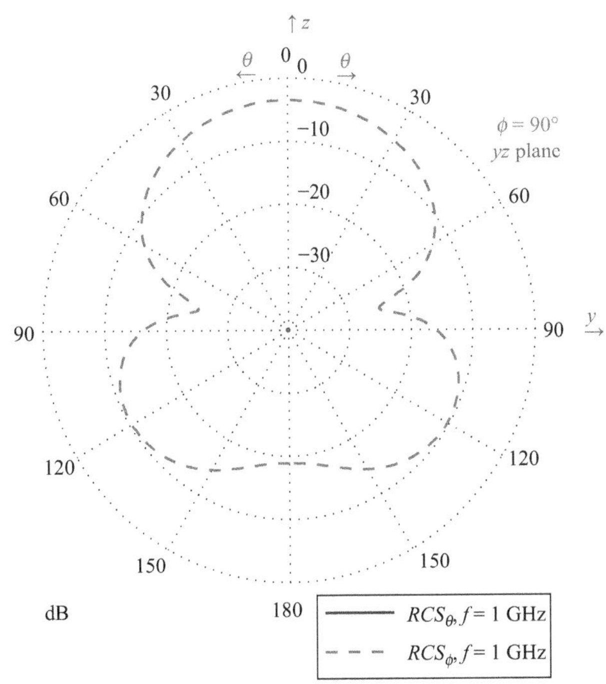

Figure 11.8 Bistatic RCS at 1 GHz in the *yz* plane.

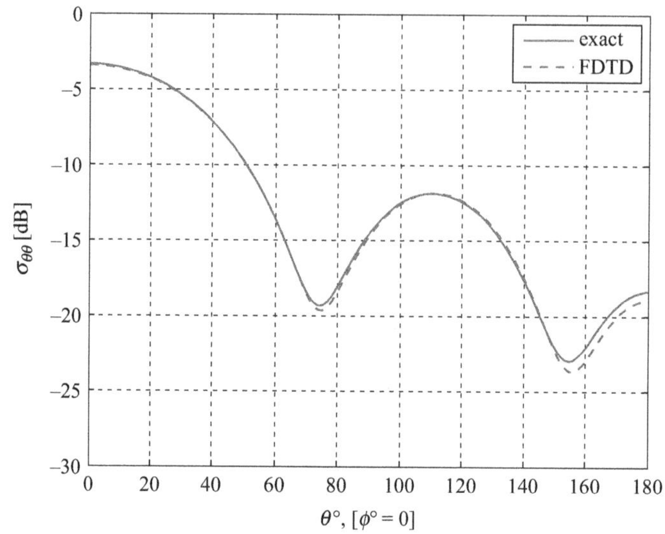

Figure 11.9 Calculated RCS_θ at 1 GHz in the xz plane compared with the analytical solution.

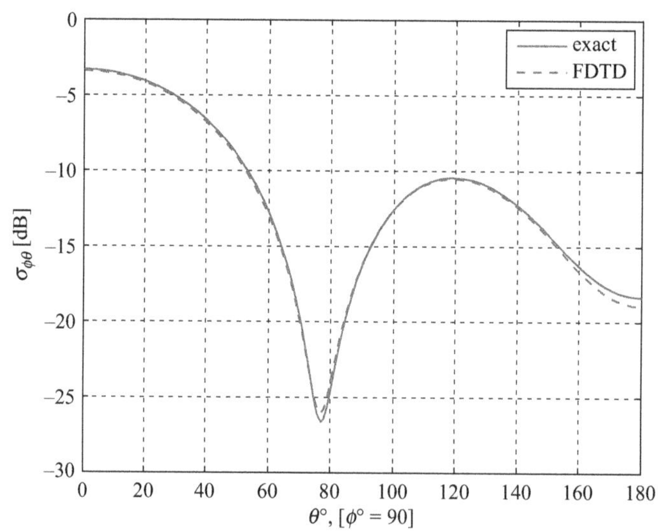

Figure 11.10 Calculated RCS_ϕ at 1 GHz in the yz plane compared with the analytical solution.

11.5.2 Scattering from a dielectric cube

In the previous section calculation of the bistatic RCS of a dielectric sphere is presented. In this section we present calculation of the RCS of a cube using the FDTD method and comparison with results obtained by a method of moments (MoM) solver [45]. In Figure 11.12 the FDTD problem space including a dielectric cube is illustrated. In the same figure the triangular meshing of the surface of the cube used by the MoM solver is shown as well.

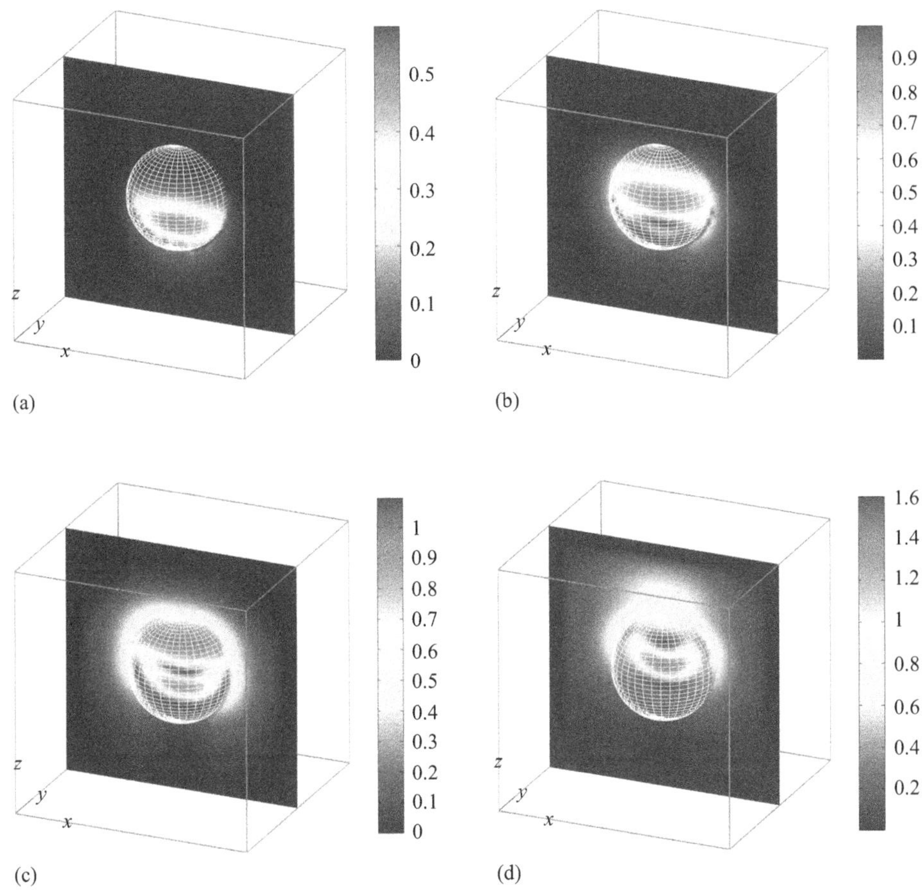

Figure 11.11 Scattered field due to a sphere captured on the *xz* plane cut: (a) at time step 120; (b) at time step 140; (c) at time step 160; and (d) at time step 180.

The FDTD problem space is composed of cubic cells 0.5 cm on a side. The definition of the problem geometry is shown in Listing 11.14. Each side of the cube is 16 cm. The relative permittivity of the cube is 5, and the relative permeability is 1. The incident plane wave is defined as shown in Listing 11.15. The incident plane wave illuminating the cube is θ polarized and propagates in a direction expressed by angles $\theta = 45^\circ$ and $\phi = 30^\circ$. The waveform of the plane wave is Gaussian. The far-field output is defined at 1 GHz as shown in Listing 11.16. Finally, a sampled electric field is defined at the origin to observe whether the fields in the problem space are sufficiently decayed.

After the FDTD time-marching loop is completed, the far-field patterns are obtained as shown in Figures 11.13, 11.14, and 11.15; for the RCS at 1 GHz in the *xy*, *xz*, and *yz* planes, respectively. The same problem is simulated at 1 GHz using the MoM solver. Figure 11.16 shows the compared RCS_θ in the *xz* plane, whereas Figure 11.17 shows the compared RCS_ϕ. The results show a good agreement.

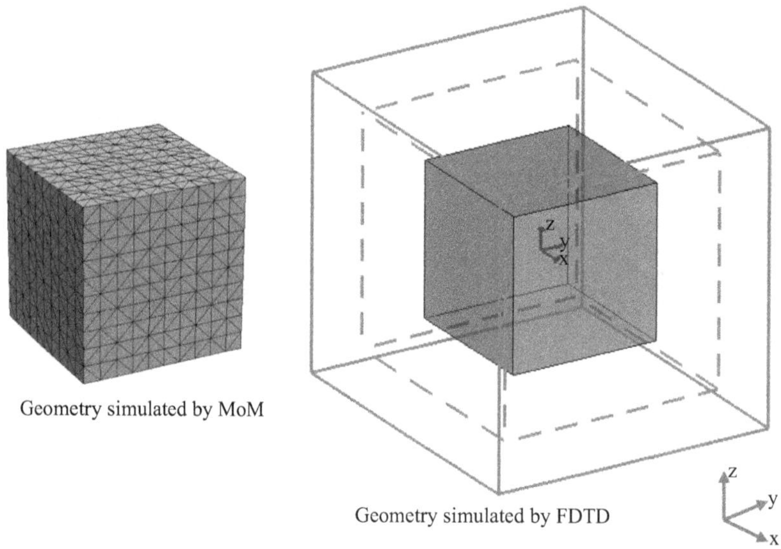

Figure 11.12 An FDTD problem space including a cube, and the surface mesh of the cube used by the MoM solver.

Listing 11.14 define_geometry.m

```
disp('defining the problem geometry');

bricks  = [];
spheres = [];
thin_wires = [];

% define dielectric
bricks(1).min_x = -80e-3;
bricks(1).min_y = -80e-3;
bricks(1).min_z = -80e-3;
bricks(1).max_x = 80e-3;
bricks(1).max_y = 80e-3;
bricks(1).max_z = 80e-3;
bricks(1).material_type = 4;
```

Listing 11.15 define_sources_and_lumped_elements.m

```
disp('defining sources and lumped element components');

voltage_sources = [];
current_sources = [];
diodes = [];
resistors = [];
inductors = [];
capacitors = [];
incident_plane_wave = [];

% define source waveform types and parameters
```

```
waveforms.gaussian(1).number_of_cells_per_wavelength = 0;
waveforms.gaussian(2).number_of_cells_per_wavelength = 15;

% Define incident plane wave, angles are in degrees
incident_plane_wave.E_theta = 1;
incident_plane_wave.E_phi = 0;
incident_plane_wave.theta_incident = 45;
incident_plane_wave.phi_incident = 30;
incident_plane_wave.waveform_type = 'gaussian';
incident_plane_wave.waveform_index = 1;
```

Listing 11.16 define_output_parameters.m

```
disp('defining output parameters');

sampled_electric_fields = [];
sampled_magnetic_fields = [];
sampled_voltages = [];
sampled_currents = [];
ports = [];
farfield.frequencies = [];

% figure refresh rate
plotting_step = 10;

% mode of operation
run_simulation = true;
show_material_mesh = true;
show_problem_space = true;

% far field calculation parameters
farfield.frequencies(1) = 1.0e9;
farfield.number_of_cells_from_outer_boundary = 13;

% frequency domain parameters
frequency_domain.start = 20e6;
frequency_domain.end   = 4e9;
frequency_domain.step  = 20e6;


% define sampled electric fields
% component: vector component íx í,í y í,í z í, or magnitude ímí
% display plot = true, in order to plot field during simulation
sampled_electric_fields(1).x = 0;
sampled_electric_fields(1).y = 0;
sampled_electric_fields(1).z = 0;
sampled_electric_fields(1).component = 'x';
sampled_electric_fields(1).display_plot = false;
```

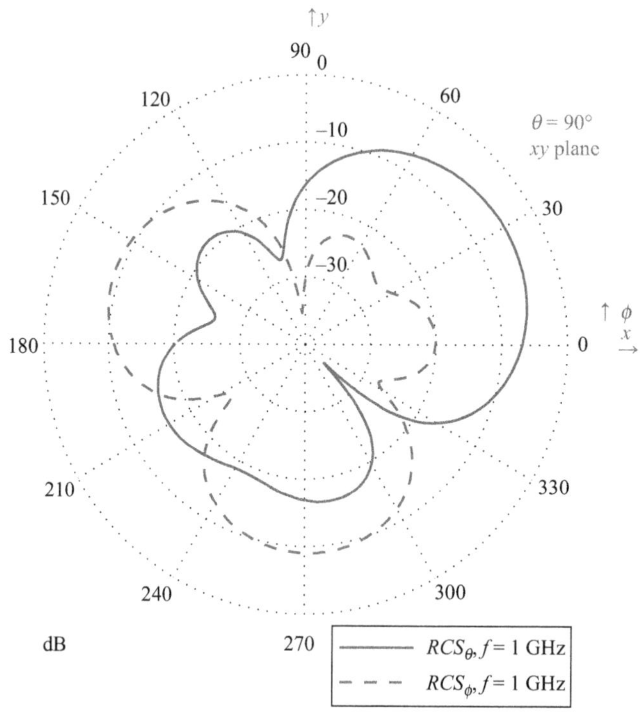

Figure 11.13 Bistatic RCS at 1 GHz in the *xy* plane.

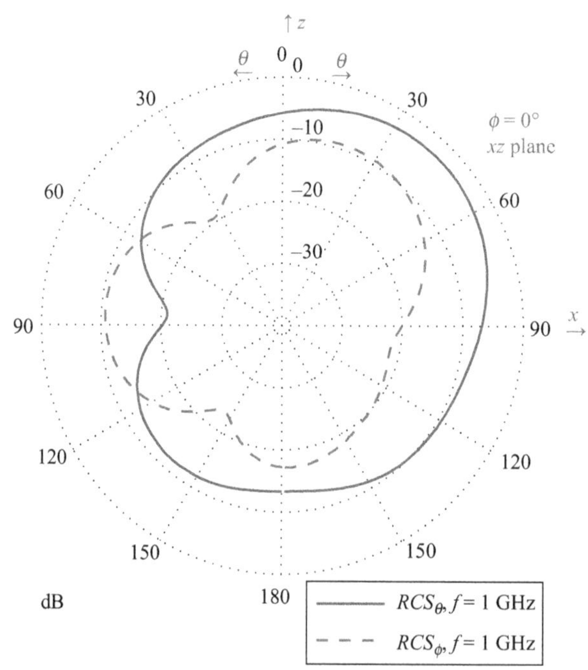

Figure 11.14 Bistatic RCS at 1 GHz in the *xz* plane.

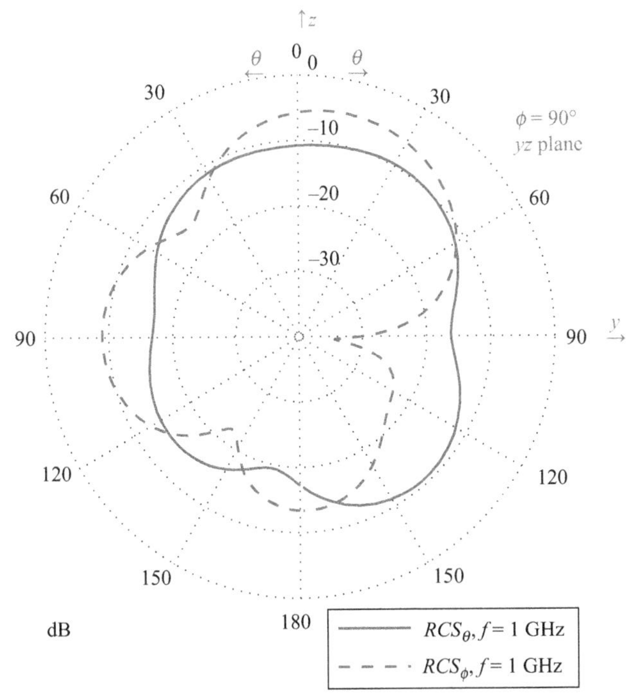

Figure 11.15 Bistatic RCS at 1 GHz in the *yz* plane.

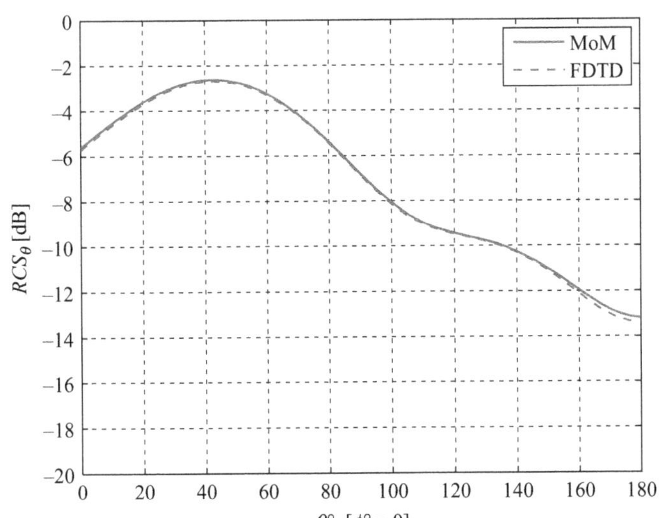

Figure 11.16 Calculated RCS_θ at 1 GHz in the *xz* plane compared with the MoM solution.

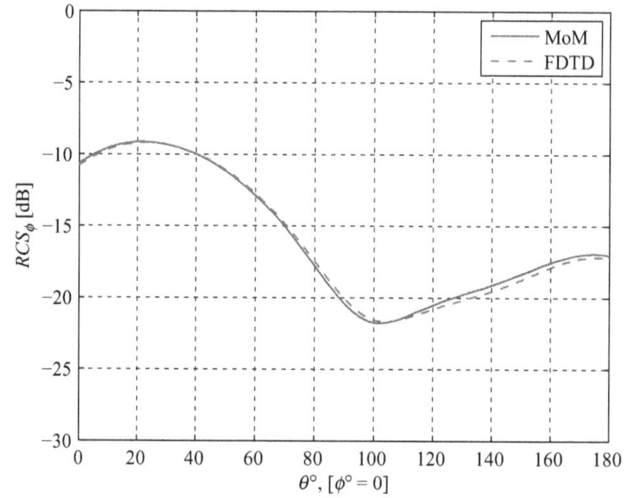

Figure 11.17 Calculated RCS_ϕ at 1 GHz in the xz plane compared with the MoM solution.

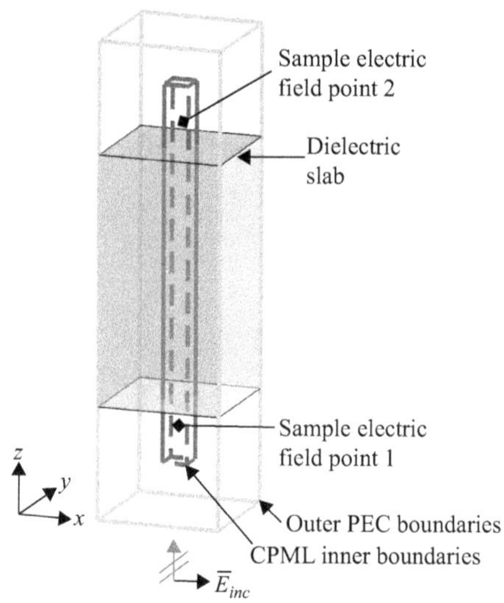

Figure 11.18 An FDTD problem space including a dielectric slab.

11.5.3 Reflection and transmission coefficients of a dielectric slab

Chapter 8 demonstrated that the objects with infinite length can be simulated by letting the objects penetrate into the convolutional perfectly matched layer (CPML) boundaries. In this example we show the simulation of a dielectric slab, which is essentially a one-dimensional problem. The availability of incident plane wave allows us to calculate the reflection and transmission coefficients of the slab.

The geometry of the problem in consideration is shown in Figure 11.18. A dielectric slab, having dielectric constant 4 and thickness 20 cm, penetrates into the CPML boundaries in *xn*, *xp*, *yn*, and *yp* directions. The problem space is composed of cells having 0.5 cm size on a side. The source of the simulation is an *x*-polarized incident field traveling in the positive *z* direction. Two sampled electric fields are defined to capture the *x* component of the electric field at two points, each 5 cm away from the slab – one below the slab and the other above the slab as illustrated in Figure 11.18. The definition of the problem space boundary conditions, incident field, the slab, and the sampled electric fields are shown in Listings 11.17–11.20, respectively. Notice that the CPML parameters have been modified for better performance.

In the previous sections, *scattered field formulation* has been demonstrated for calculating the *scattered field* when an *incident field* is used to excite the problem space. The sampled electric field therefore is the one being captured while the simulation is running.

Listing 11.17 define_problem_space_parameters.m

```
% ==< boundary conditions >========
% Here we define the boundary conditions parameters
% 'pec' : perfect electric conductor
% 'cpml' : conlvolutional PML
% if cpml_number_of_cells is less than zero
% CPML extends inside of the domain rather than outwards

boundary.type_xn = 'cpml';
boundary.air_buffer_number_of_cells_xn = 0;
boundary.cpml_number_of_cells_xn = -8;

boundary.type_xp = 'cpml';
boundary.air_buffer_number_of_cells_xp = 0;
boundary.cpml_number_of_cells_xp = -8;

boundary.type_yn = 'cpml';
boundary.air_buffer_number_of_cells_yn = 0;
boundary.cpml_number_of_cells_yn = -8;

boundary.type_yp = 'cpml';
boundary.air_buffer_number_of_cells_yp = 0;
boundary.cpml_number_of_cells_yp = -8;

boundary.type_zn = 'cpml';
boundary.air_buffer_number_of_cells_zn = 10;
boundary.cpml_number_of_cells_zn = 8;

boundary.type_zp = 'cpml';
boundary.air_buffer_number_of_cells_zp = 10;
boundary.cpml_number_of_cells_zp = 8;

boundary.cpml_order = 4;
boundary.cpml_sigma_factor = 1;
boundary.cpml_kappa_max = 1;
boundary.cpml_alpha_min = 0;
boundary.cpml_alpha_max = 0;
```

Listing 11.18 define_geometry.m

```
% define dielectric
bricks(1).min_x = -0.05;
bricks(1).min_y = -0.05;
bricks(1).min_z = 0;
bricks(1).max_x = 0.05;
bricks(1).max_y = 0.05;
bricks(1).max_z = 0.2;
bricks(1).material_type = 4;
```

Listing 11.19 define_sources_and_lumped_elements.m

```
% define source waveform types and parameters
waveforms.derivative_gaussian(1).number_of_cells_per_wavelength = 20;

% Define incident plane wave, angles are in degrees
incident_plane_wave.E_theta = 1;
incident_plane_wave.E_phi = 0;
incident_plane_wave.theta_incident = 0;
incident_plane_wave.phi_incident = 0;
incident_plane_wave.waveform_type = 'derivative_gaussian';
incident_plane_wave.waveform_index = 1;
```

Listing 11.20 define_output_parameters.m

```
% frequency domain parameters
frequency_domain.start = 2e6;
frequency_domain.end   = 2e9;
frequency_domain.step  = 1e6;

% define sampled electric fields
% component: vector component íx í,í y í,í z í, or magnitude ímí
% display plot = true, in order to plot field during simulation
sampled_electric_fields(1).x = 0;
sampled_electric_fields(1).y = 0;
sampled_electric_fields(1).z = -0.025;
sampled_electric_fields(1).component = 'x';
sampled_electric_fields(1).display_plot = false;

sampled_electric_fields(2).x = 0;
sampled_electric_fields(2).y = 0;
sampled_electric_fields(2).z = 0.225;
sampled_electric_fields(2).component = 'x';
sampled_electric_fields(2).display_plot = false;
```

The reflected field from the slab is the scattered field, and the ratio of the reflected field to incident field is required to calculate the reflection coefficient using

$$|\Gamma| = \frac{|\vec{E}_{scat}|}{|\vec{E}_{inc}|}, \tag{11.23}$$

Listing 11.21 capture_sampled_electric_fields.m

```
% Capturing the incident electric fields
if incident_plane_wave.enabled
    for ind=1:number_of_sampled_electric_fields
        is = sampled_electric_fields(ind).is;
        js = sampled_electric_fields(ind).js;
        ks = sampled_electric_fields(ind).ks;

        switch (sampled_electric_fields(ind).component)
            case 'x'
                sampled_value = 0.5 * sum(Exic(is-1:is,js,ks));
            case 'y'
                sampled_value = 0.5 * sum(Eyic(is,js-1:js,ks));
            case 'z'
                sampled_value = 0.5 * sum(Ezic(is,js,ks-1:ks));
            case 'm'
                svx = 0.5 * sum(Exic(is-1:is,js,ks));
                svy = 0.5 * sum(Eyic(is,js-1:js,ks));
                svz = 0.5 * sum(Ezic(is,js,ks-1:ks));
                sampled_value = sqrt(svx^2 + svy^2 + svz^2);
        end
        sampled_electric_fields(ind).incident_field_value(time_step) ...
                                                = sampled_value;
    end
end
```

where Γ is the reflection coefficient. Here the scattered field is the one sampled at a point below the slab, and the incident field shall be sampled at the same point as well. Therefore, the section of code shown in Listing 11.21 is added to the subroutine ***capture_sampled_electric_fields*** to sample the incident electric field as the time-marching loop proceeds.

The ratio of the transmitted field to incident field is required to calculate the transmission coefficient, where the transmitted field is the *total field* that is the sum of the incident and scattered fields. Then the transmission coefficient can be calculated using

$$|T| = \frac{|\vec{E}_{tot}|}{|\vec{E}_{inc}|} = \frac{|\vec{E}_{scat} + \vec{E}_{inc}|}{|\vec{E}_{inc}|}, \tag{11.24}$$

where T is the transmission coefficient. Here the scattered field is the one sampled at a point above the slab, and the incident field is sampled at the same point as well.

The simulation for this problem is executed, the incident and scattered fields are captured at two points above and below the slab and transformed to frequency domain, and the reflection and transmission coefficients are calculated using (11.23) and (11.24), respectively. The calculation results are plotted in Figure 11.19 together with the exact solution of the same problem. The exact solution is obtained using the program published in [46]. A very good agreement can be observed between the simulated results and exact solution up to 1 GHz.

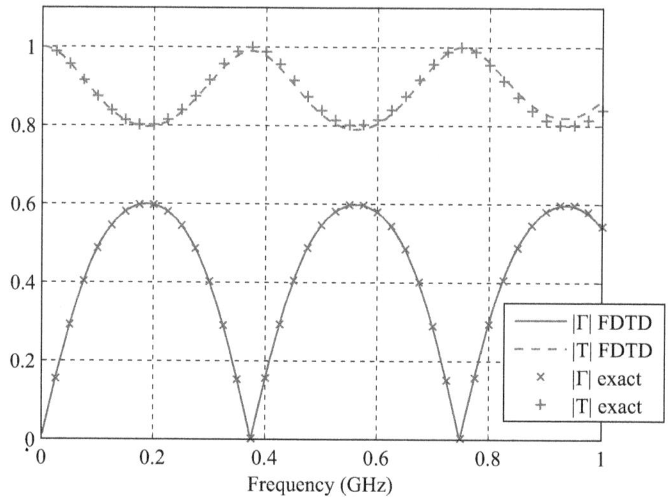

Figure 11.19 Reflection and transmission coefficients of the dielectric slab.

11.6 Exercises

11.1 Consider the problem demonstrated in Section 11.5.1, where the RCS of a dielectric sphere is calculated due to an incident field traveling in the positive z direction. In the given example the direction of incidence is defined as $\theta = 0°$ and $\phi = 0°$. Change the incident angle such that $\theta = 45°$ and $\phi = 0°$, run the simulation, and obtain the RCS at 1 GHz. Examine the RCS plots, and verify that the RCS in the xz plane is the same as the one shown in Figure 11.7, except that it is rotated by 45° about the y axis.

11.2 Consider the scattering problem constructed for Exercise 11.1, where the RCS of a dielectric sphere is calculated due to an incident field. The RCS calculations are based on the scattered fields; the fictitious electric and magnetic current densities used for NF–FF transformation are calculated using the scattered field electric and magnetic fields. The total field in the problem space is the sum of the incident field and the scattered field. Modify the program such that the RCS calculations are performed based on the total fields rather than on the scattered fields. It is necessary to modify the code of the subroutine ***calculate_JandM***. Run the simulation and obtain the RCS of the sphere. Verify that the RCS data are the same as the ones obtained from Exercise 11.1. Notice that the incident field does not contribute to far-field scattering at all.

11.3 Consider the dielectric slab problem demonstrated in Section 11.5.3. You can perform the reflection and transmission coefficient calculation for stacked slabs composed of multiple layers with different dielectric constants. For instance, create two slabs in the problem space, each 10 cm thick. Then set the dielectric constant of the bottom slab as 4 and the dielectric constant of the top slab as 2. Run the simulation, and then calculate the reflection and transmission coefficients. Compare your result with the exact solution of the same problem. For the exact solution you can refer to [46]. As an alternative, you can construct a one-dimensional FDTD code similar to the one presented in Chapter 1 based on the scattered field formulation to solve the same problem.

CHAPTER 12

Total field/scattered field formulation

12.1 Introduction

Another technique that introduces far-zone sources (plane wave) into the FDTD computational domain is the Total Field/Scattered Field (TF/SF) method presented by Umashankar and Taflove [47]. In the scattered field (SF) method, discussed before, the scattered fields are defined in the entire computational domain and the incident fields are imposed on the entire computational domain. Unlike the SF method, in the TF/SF method the computational domain is divided into two regions: a total fields region, in which all objects are placed, and a scattered fields region that surrounds the total fields region as shown in Figure 12.1. Then, the incident plane wave is imposed on the boundary between these two regions to compensate for the discontinuity that arises due to the difference in the field values on the two sides of the boundary. The concept will be illustrated for the one-dimensional case first, and then the discussion will be extended to the three-dimensional case in the following sections.

Figure 12.2 illustrates total and scattered field regions and the field components for one-dimensional computational space. The TF/SF boundaries are indicated with the lower and upper nodes, "*li*" and "*ui*," respectively. The field components inside the total field region are considered as the total field components, whereas the field components inside the scattered field region are considered as the scattered field components and they are indicated with a "scat" subscript. The scattered field region does not contain any objects, thus it is free space as material type and it is surrounded by absorbing boundaries such as CPML. Thus the scattered fields satisfy Maxwell's equations for free space, which is translated into a one-dimensional updating equations, for instance, as

$$E^{n+1}_{scat,z}(i) = E^{n}_{scat,z}(i) + C_{ezhy}(i)\left(H^{n+\frac{1}{2}}_{scat,y}(i) - H^{n+\frac{1}{2}}_{scat,y}(i-1)\right), \tag{12.1a}$$

$$H^{n+0.5}_{scat,y}(i) = H^{n-0.5}_{scat,y}(i) + C_{hyez}(i)\left(E^{n}_{scat,z}(i+1) - E^{n}_{scat,z}(i)\right), \tag{12.1b}$$

where

$$C_{hyez}(i) = \frac{\Delta t}{\mu_0 \Delta x}, \quad \text{and} \quad C_{ezhy}(i) = \frac{\Delta t}{\varepsilon_0 \Delta x}. \tag{12.2}$$

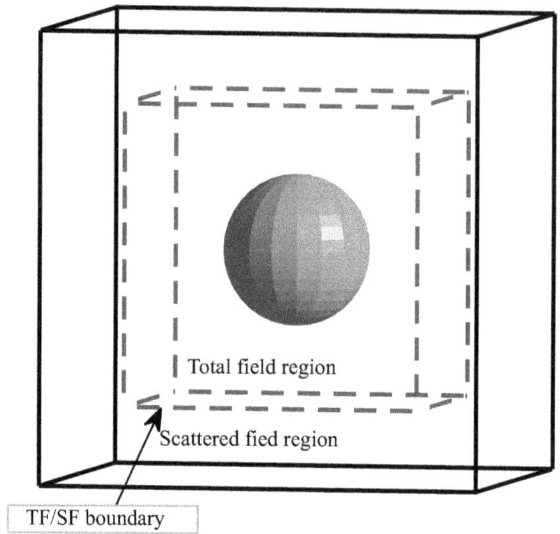

Figure 12.1 Total and scattered field regions for a three-dimensional FDTD computational domain.

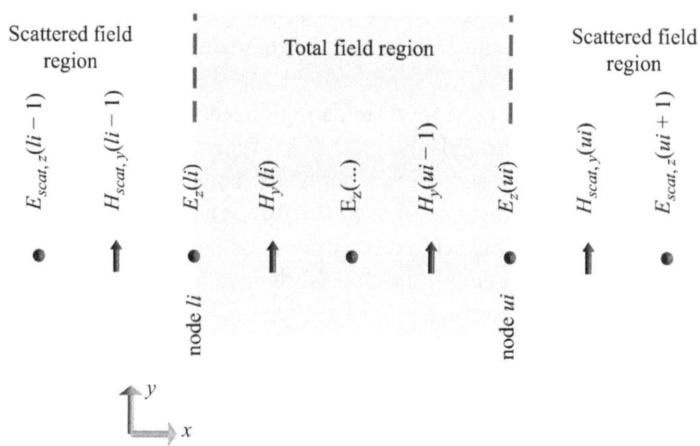

Figure 12.2 Total and scattered field components and regions for a one-dimensional FDTD computational domain.

One should notice that these updating equations are the same as the ones in (1.42) and (1.43), used for total fields, except that they are defined for free space. Since the total and scattered field components are distinct and complementary in space, and they follow the same form of the updating equations, there is no need to use separate arrays to store the total and scattered field components, as well as their respective coefficients. The updating coefficients in (1.42) and (1.43) can be simply set as the ones in (12.2) for respective scattered field components, and the field updates can be performed as usual based on (1.42) and (1.43), while implicitly considering that the field components that correspond to the scattered field region are the scattered field components and the other field components are the total field components.

After the field updates in the entire computational domain, one needs to compensate for the discontinuity between the scattered and total fields on the TF/SF boundary by using the incident field. For instance, when the field component $E_z(li)$ in Figure 12.2 is updated, the first step would be

$$E_z^{n+1}(li) = E_z^n(li) + \frac{\Delta t}{\varepsilon_0 \Delta x}\left(H_y^{n+\frac{1}{2}}(li) - H_{scat,y}^{n+\frac{1}{2}}(li-1)\right). \tag{12.3}$$

Here, the magnetic field component with index $(li - 1)$ is a scattered field component, since it is in the scattered field region, where a total field component, $H_y(li - 1)$, is needed to be used. Since the required $H_y(li - 1)$ is the sum of $H_{scat,y}(li - 1)$ and an incident field component, $H_{inc,y}(li - 1)$, which can be calculated analytically, the second step required to complete the $E_z^{n+1}(li)$ update would be

$$E_z^{n+1}(li) = E_z^{n+1}(li) - \frac{\Delta t}{\varepsilon_0 \Delta x} H_{inc,y}^{n+\frac{1}{2}}(li-1). \tag{12.4}$$

A similar compensation is needed for the magnetic component, $H_{scat,y}(li - 1)$, in the scattered field region. Regular field update performs

$$H_{scat,y}^{n+0.5}(li-1) = H_{scat,y}^{n-0.5}(li-1) + \frac{\Delta t}{\mu_0 \Delta x}\left(E_z^n(li) - E_{scat,z}^n(li-1)\right), \tag{12.5}$$

where $E_z^n(li)$ is used instead of the needed $E_{scat,z}^n(li)$ component. Similar to (12.4), an additional update as

$$H_{scat,y}^{n+0.5}(li-1) = H_{scat,y}^{n+0.5}(li-1) - \frac{\Delta t}{\mu_0 \Delta x} E_{inc,z}^n(li), \tag{12.6}$$

compensates for the needed field component.

The same kind of additional updates are required on the other (right) TF/SF boundary as well. In that case, the additional updates can be listed as

$$E_z^{n+1}(ui) = E_z^{n+1}(ui) + \frac{\Delta t}{\varepsilon_0 \Delta x} H_{inc,y}^{n+\frac{1}{2}}(ui), \tag{12.7}$$

$$H_{scat,y}^{n+0.5}(ui) = H_{scat,y}^{n+0.5}(ui) + \frac{\Delta t}{\mu_0 \Delta x} E_{inc,z}^n(ui). \tag{12.8}$$

As illustrated above, the values of the incident fields are needed only at the two sides of the TF/SF boundaries, unlike the scattered field formulation, which requires the values of the incident fields to be computed in the entire computational domain.

It is straightforward to extend the TF/SF concept to two- and three-dimensional domains. In the three-dimensional case, the incident field corrections are needed on the six faces of a TF/SF boundary surface that surrounds all scatterers in the computational space.

Figures 12.3 and 12.4 illustrate the field components that need compensation using the incident fields after performing the regular field updates. In these figures, the boundaries of the imaginary surface are indicated by the node indices *li*, *lj*, *lk*, *ui*, *uj*, and *uk*, where "*i*", "*j*", and "*k*" denote the indices in *x*, *y*, and *z* directions, respectively, while "*l*" and "*u*"

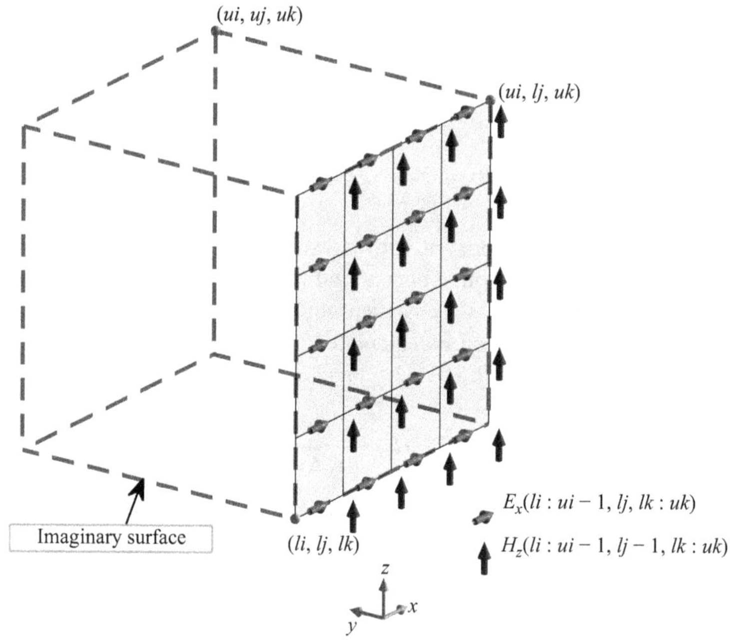

Figure 12.3 E_x and H_z field components that needs incident field compensation on the yn face of the imaginary surface.

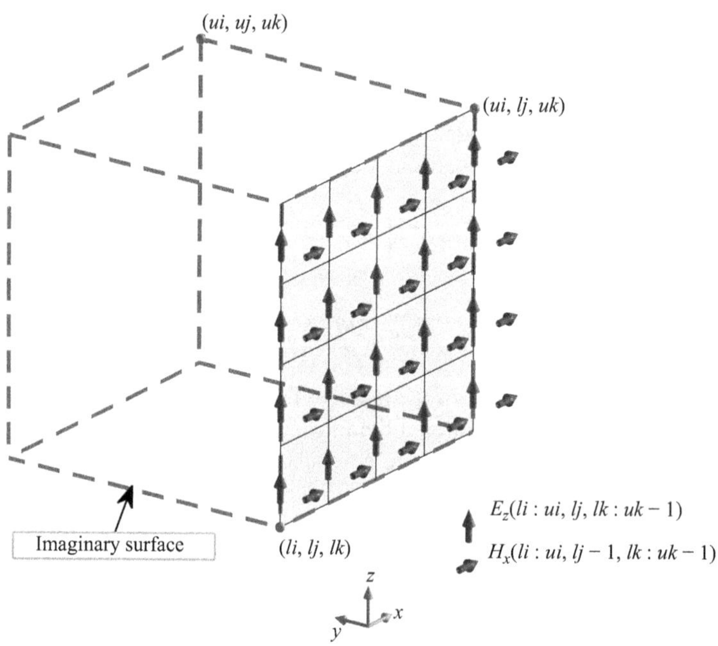

Figure 12.4 E_z and H_x field components that needs incident field compensation on the yn face of the imaginary surface.

denote lower and upper limits of the boundaries. One can visualize in Figure 12.3 that the additional field updates for the *yn* boundary would be as

$$\begin{aligned} E_x^{n+1}(li:ui-1,lj,lk:uk) = E_x^{n+1}(li:ui-1,lj,lk:uk) \\ -\frac{\Delta t}{\varepsilon_0 \Delta x} H_{inc,z}^{n+\frac{1}{2}}(li:ui-1,lj-1,lk:uk), \end{aligned} \tag{12.9}$$

and

$$\begin{aligned} H_{scat,z}^{n+0.5}(li:ui-1,lj-1,lk:uk) = H_{scat,z}^{n+0.5}(li:ui-1,lj-1,lk:uk) \\ -\frac{\Delta t}{\mu_0 \Delta x} E_{inc,x}^{n}(li:ui-1,lj,lk:uk), \end{aligned} \tag{12.10}$$

where ":" denotes the range of indices similar to its usage in FORTRAN and MATLAB®. Similarly, one can visualize in Figure 12.4 that the additional field updates would be

$$\begin{aligned} E_z^{n+1}(li:ui,lj,lk:uk-1) = E_z^{n+1}(li:ui,lj,lk:uk-1) \\ +\frac{\Delta t}{\varepsilon_0 \Delta x} H_{inc,x}^{n+\frac{1}{2}}(li:ui,lj-1,lk:uk-1), \end{aligned} \tag{12.11}$$

and

$$\begin{aligned} H_{scat,x}^{n+0.5}(li:ui,lj-1,lk:uk-1) = H_{scat,x}^{n+0.5}(li:ui,lj-1,lk:uk-1) \\ +\frac{\Delta t}{\mu_0 \Delta x} E_{inc,z}^{n}(li:ui,lj,lk:uk-1). \end{aligned} \tag{12.12}$$

Equations (12.9)–(12.12) can be modified for the *yp* boundary as

$$\begin{aligned} E_x^{n+1}(li:ui-1,uj,lk:uk) = E_x^{n+1}(li:ui-1,uj,lk:uk) \\ +\frac{\Delta t}{\varepsilon_0 \Delta x} H_{inc,z}^{n+\frac{1}{2}}(li:ui-1,uj,lk:uk), \end{aligned} \tag{12.13}$$

$$\begin{aligned} H_{scat,z}^{n+0.5}(li:ui-1,uj,lk:uk) = H_{scat,z}^{n+0.5}(li:ui-1,uj,lk:uk) \\ +\frac{\Delta t}{\mu_0 \Delta x} E_{inc,x}^{n}(li:ui-1,uj,lk:uk), \end{aligned} \tag{12.14}$$

$$\begin{aligned} E_z^{n+1}(li:ui,uj,lk:uk-1) = E_z^{n+1}(li:ui,uj,lk:uk-1) \\ -\frac{\Delta t}{\varepsilon_0 \Delta x} H_{inc,x}^{n+\frac{1}{2}}(li:ui,uj,lk:uk-1), \end{aligned} \tag{12.15}$$

$$\begin{aligned} H_{scat,x}^{n+0.5}(li:ui,uj,lk:uk-1) = H_{scat,x}^{n+0.5}(li:ui,uj,lk:uk-1) \\ -\frac{\Delta t}{\mu_0 \Delta x} E_{inc,z}^{n}(li:ui,uj,lk:uk-1). \end{aligned} \tag{12.16}$$

Equations for other boundaries can be obtained similarly.

12.2 MATLAB® implementation of the TF/SF formulation

In Section 11.4, the MATLAB implementation of a scattered field formulation was presented including the definition of an incident plane wave, initialization of the incident fields, initialization of updating coefficients and calculation of the scattered fields. Thus it will be easy to modify the code of the scattered field formulation and develop a code for the TF/SF formulation since there is a significant overlap among the elements of these methods. In particular, the way the incident field is defined and implemented is the same except that the incident fields are defined on the entire problem space in the former, while they are defined on only both sides of the TF/SF surface in the latter.

12.2.1 Definition and initialization of incident fields

A plane wave with a Gaussian waveform will be considered as the incident field in this implementation. The definition of an incident plane wave was illustrated in Listing 11.1. The same definition can be retained. However, the initialization of the incident fields will be significantly different for the implementation for the TF/SF formulation since the fields need to be known only around the six sides of the rectangular surface defining the TF/SF boundary.

The first task in the initialization of the incident plane wave is to determine the location of the TF/SF boundary. This surface should reside in the air gap between the objects and the absorbing boundary regions enclosing the problem space. Listing 12.1 shows the part of the initialization subroutine ***initialize_incident_plane_wave***, where this boundary is placed three cells away from CPML absorbing boundaries and the values of the node indices *li*, *lj*, *lk*, *ui*, *uj*, and *uk* are updated accordingly.

Listing 12.1 initialize_incident_plane_wave.m: Locate the TF/SF boundary

```
% leaving three cells between the outer boundary and the TF/SF boundary
li = 3;    lj = 3;    lk = 3;
ui = nx-3;  uj = ny-3;  uk = nz-3;
if strcmp(boundary.type_xn, 'cpml')
    li = boundary.cpml_number_of_cells_xn + 3;
end
if strcmp(boundary.type_yn, 'cpml')
    lj = boundary.cpml_number_of_cells_yn + 3;
end
if strcmp(boundary.type_zn, 'cpml')
    lk = boundary.cpml_number_of_cells_zn + 3;
end
if strcmp(boundary.type_xp, 'cpml')
    ui = nx-boundary.cpml_number_of_cells_xp - 3;
end
if strcmp(boundary.type_yp, 'cpml')
    uj = ny-boundary.cpml_number_of_cells_yp - 3;
end
if strcmp(boundary.type_zp, 'cpml')
    uk = nz-boundary.cpml_number_of_cells_zp - 3;
end
```

The next step is to allocate incident field arrays for the six sides of the TF/SF surface. For instance, as illustrated in Figures 12.3 and 12.4, arrays for $E_{inc,x}$, $E_{inc,z}$, $H_{inc,x}$, and $H_{inc,z}$ need to be defined for the *yn* boundary. Each of these arrays is three dimensional, while its size in the *y* dimension is equal to one, thus making it effectively a two-dimensional array. Similarly, four incident field arrays are needed for each of other sides, thus in total 24 arrays are defined as illustrated in Listing 12.2.

Then the amplitudes of the incident electric and magnetic field waveforms are transformed from spherical coordinates to Cartesian coordinates to yield **Exi0**, **Eyi0**, **Ezi0**, **Hxi0**, **Hyi0**, and **Hzi0** as shown in Listing 12.3. Moreover, the components of the propagation vector are calculated and stored in **k_vec_x**, **k_vec_y**, **k_vec_z** parameters. Also, the spatial shift required for the plane wave, which ensures that the wave is outside the computational domain when the time-marching loop starts and lets the wave enter to the computational domain on its leading front as the time-marching loop proceeds, is calculated and stored as the parameter **l_0**. It should be noted that these calculations are the same as the corresponding ones in Listing 11.2.

Listing 11.3 also includes the calculation of three-dimensional arrays **k_dot_r_ex**, **k_dot_r_ey**, **k_dot_r_ez**, **k_dot_r_hx**, **k_dot_r_hy**, and **k_dot_r_hz** as discussed in Section 11.4.2. These arrays need to be defined on the six faces of TF/SF boundary as two-dimensional arrays for the case of TF/SF formulation. For instance, Listing 12.4 shows the allocation and initialization of these arrays for the *yn* side of the TF/SF boundary.

Listing 12.2 initialize_incident_plane_wave.m: Allocate incident field arrays

```
Exi_yn = zeros(ui-li,1,uk-lk+1);
Exi_yp = zeros(ui-li,1,uk-lk+1);
Exi_zn = zeros(ui-li,uj-lj+1,1);
Exi_zp = zeros(ui-li,uj-lj+1,1);

Eyi_xn = zeros(1,uj-lj,uk-lk+1);
Eyi_xp = zeros(1,uj-lj,uk-lk+1);
Eyi_zn = zeros(ui-li+1,uj-lj,1);
Eyi_zp = zeros(ui-li+1,uj-lj,1);

Ezi_xn = zeros(1,uj-lj+1,uk-lk);
Ezi_xp = zeros(1,uj-lj+1,uk-lk);
Ezi_yn = zeros(ui-li+1,1,uk-lk);
Ezi_yp = zeros(ui-li+1,1,uk-lk);

Hzi_yn = zeros(ui-li,1,uk-lk+1);
Hzi_yp = zeros(ui-li,1,uk-lk+1);
Hyi_zn = zeros(ui-li,uj-lj+1,1);
Hyi_zp = zeros(ui-li,uj-lj+1,1);

Hzi_xn = zeros(1,uj-lj,uk-lk+1);
Hzi_xp = zeros(1,uj-lj,uk-lk+1);
Hxi_zn = zeros(ui-li+1,uj-lj,1);
Hxi_zp = zeros(ui-li+1,uj-lj,1);

Hyi_xn = zeros(1,uj-lj+1,uk-lk);
Hyi_xp = zeros(1,uj-lj+1,uk-lk);
Hxi_yn = zeros(ui-li+1,1,uk-lk);
Hxi_yp = zeros(ui-li+1,1,uk-lk);
```

Listing 12.3 initialize_incident_plane_wave.m: Calculate amplitudes and propagation vector

```
% calculate the amplitude factors for field components
theta_incident = incident_plane_wave.theta_incident*pi/180;
phi_incident = incident_plane_wave.phi_incident*pi/180;
E_theta = incident_plane_wave.E_theta;
E_phi = incident_plane_wave.E_phi;
eta_0 = sqrt(mu_0/eps_0);
Exi0 = E_theta * cos(theta_incident) * cos(phi_incident) ...
   - E_phi * sin(phi_incident);
Eyi0 = E_theta * cos(theta_incident) * sin(phi_incident) ...
   + E_phi * cos(phi_incident);
Ezi0 = -E_theta * sin(theta_incident);
Hxi0 = (-1/eta_0)*(E_phi * cos(theta_incident) ...
   * cos(phi_incident) + E_theta * sin(phi_incident));
Hyi0 = (-1/eta_0)*(E_phi * cos(theta_incident) ...
   * sin(phi_incident) - E_theta * cos(phi_incident));
Hzi0 = (1/eta_0)*(E_phi * sin(theta_incident));

% calculate spatial shift, l_0, required for incident plane wave
r0 =[fdtd_domain.min_x fdtd_domain.min_y fdtd_domain.min_z;
fdtd_domain.min_x fdtd_domain.min_y fdtd_domain.max_z;
fdtd_domain.min_x fdtd_domain.max_y fdtd_domain.min_z;
fdtd_domain.min_x fdtd_domain.max_y fdtd_domain.max_z;
fdtd_domain.max_x fdtd_domain.min_y fdtd_domain.min_z;
fdtd_domain.max_x fdtd_domain.min_y fdtd_domain.max_z;
fdtd_domain.max_x fdtd_domain.max_y fdtd_domain.min_z;
fdtd_domain.max_x fdtd_domain.max_y fdtd_domain.max_z;];

k_vec_x = sin(theta_incident)*cos(phi_incident);
k_vec_y = sin(theta_incident)*sin(phi_incident);
k_vec_z = cos(theta_incident);

k_dot_r0 = k_vec_x * r0(:,1) ...
   + k_vec_y * r0(:,2) ...
   + k_vec_z * r0(:,3);

l_0 = min(k_dot_r0)/c;
```

Listing 12.4 initialize_incident_plane_wave.m: Calculate $\hat{k} \cdot \bar{r}$ arrays

```
k_dot_r_ex_yn = zeros(ui-li,1,uk-lk+1);
k_dot_r_ez_yn = zeros(ui-li+1,1,uk-lk);
k_dot_r_hx_yn = zeros(ui-li+1,1,uk-lk);
k_dot_r_hz_yn = zeros(ui-li,1,uk-lk+1);

% calculate k.r for every field component on the TF/SF boundary
for mk=lk:uk
      for mi=li:ui-1
      mii = mi - li + 1;
      mjj = 1;
      mkk = mk - lk + 1;
```

```
 mj = lj;
        x = min_x + dx * (mi-0.5);
        y = min_y + dy * (mj-1);
        z = min_z + dz * (mk-1);

        k_dot_r_ex_yn(mii,mjj,mkk) = ...
            (x*k_vec_x + y*k_vec_y + z*k_vec_z)/c;

        y = min_y + dy * (mj-1.5);
        k_dot_r_hz_yn(mii,mjj,mkk) = ...
            (x*k_vec_x + y*k_vec_y + z*k_vec_z)/c;

    end
end

for mi=li:ui
    for mk=lk:uk-1
        mjj = 1;
        mii = mi - li + 1;
        mkk = mk - lk + 1;

        mj = lj;
        x = min_x + dx * (mi-1);
        y = min_y + dy * (mj-1);
        z = min_z + dz * (mk-0.5);
        k_dot_r_ez_yn(mii,mjj,mkk) = ...
            (x*k_vec_x + y*k_vec_y + z*k_vec_z)/c;

        y = min_y + dy * (mj-1.5);
        k_dot_r_hx_yn(mii,mjj,mkk) = ...
            (x*k_vec_x + y*k_vec_y + z*k_vec_z)/c;

    end
end

% embed spatial shift in k.r
k_dot_r_ex_yn = k_dot_r_ex_yn - l_0;
k_dot_r_ez_yn = k_dot_r_ez_yn - l_0;
k_dot_r_hx_yn = k_dot_r_hx_yn - l_0;
k_dot_r_hz_yn = k_dot_r_hz_yn - l_0;
```

12.2.2 Updating incident fields

The incident field arrays and other auxiliary arrays are allocated and initialized in the subroutine ***initialize_incident_plane_wave*** as discussed in the previous section. Now these arrays are ready for use in the time-marching loop of the FDTD method, which is implemented in the subroutine ***run_fdtd_time_marching_loop***. The incident electric and magnetic field components on the TF/SF boundary need to be recalculated at every iteration of the time-marching loop before they are used to update the total or scattered fields on both sides of the boundary. Listing 12.5 shows the subroutine ***update_incident_fields***, which is used to recalculate these boundary field components. First, the total electric and magnetic field

Listing 12.5 update_incident_fields.m

```
tm = current_time + dt/2;
te = current_time + dt;

% if waveform is Gaussian waveforms
tau = incident_plane_wave.tau;
t_0 = incident_plane_wave.t_0;

Exi_yn = Exi0 * exp(-((tm - t_0 - k_dot_r_ex_yn )/tau).^2);
Exi_yp = Exi0 * exp(-((tm - t_0 - k_dot_r_ex_yp )/tau).^2);
Exi_zn = Exi0 * exp(-((tm - t_0 - k_dot_r_ex_zn )/tau).^2);
Exi_zp = Exi0 * exp(-((tm - t_0 - k_dot_r_ex_zp )/tau).^2);

Eyi_zn = Eyi0 * exp(-((tm - t_0 - k_dot_r_ey_zn )/tau).^2);
Eyi_zp = Eyi0 * exp(-((tm - t_0 - k_dot_r_ey_zp )/tau).^2);
Eyi_xn = Eyi0 * exp(-((tm - t_0 - k_dot_r_ey_xn )/tau).^2);
Eyi_xp = Eyi0 * exp(-((tm - t_0 - k_dot_r_ey_xp )/tau).^2);

Ezi_xn = Ezi0 * exp(-((tm - t_0 - k_dot_r_ez_xn )/tau).^2);
Ezi_xp = Ezi0 * exp(-((tm - t_0 - k_dot_r_ez_xp )/tau).^2);
Ezi_yn = Ezi0 * exp(-((tm - t_0 - k_dot_r_ez_yn )/tau).^2);
Ezi_yp = Ezi0 * exp(-((tm - t_0 - k_dot_r_ez_yp )/tau).^2);

Hxi_yn = Hxi0 * exp(-((te - t_0 - k_dot_r_hx_yn )/tau).^2);
Hxi_yp = Hxi0 * exp(-((te - t_0 - k_dot_r_hx_yp )/tau).^2);
Hxi_zn = Hxi0 * exp(-((te - t_0 - k_dot_r_hx_zn )/tau).^2);
Hxi_zp = Hxi0 * exp(-((te - t_0 - k_dot_r_hx_zp )/tau).^2);

Hyi_zn = Hyi0 * exp(-((te - t_0 - k_dot_r_hy_zn )/tau).^2);
Hyi_zp = Hyi0 * exp(-((te - t_0 - k_dot_r_hy_zp )/tau).^2);
Hyi_xn = Hyi0 * exp(-((te - t_0 - k_dot_r_hy_xn )/tau).^2);
Hyi_xp = Hyi0 * exp(-((te - t_0 - k_dot_r_hy_xp )/tau).^2);

Hzi_xn = Hzi0 * exp(-((te - t_0 - k_dot_r_hz_xn )/tau).^2);
Hzi_xp = Hzi0 * exp(-((te - t_0 - k_dot_r_hz_xp )/tau).^2);
Hzi_yn = Hzi0 * exp(-((te - t_0 - k_dot_r_hz_yn )/tau).^2);
Hzi_yp = Hzi0 * exp(-((te - t_0 - k_dot_r_hz_yp )/tau).^2);
```

components are defined at time instants that are referred to as "*te*" and "*tm*," respectively, in Listing 12.5. One should notice that "*te*" used to calculate incident magnetic field components. Incident magnetic field components are needed at the time instants at which total/scattered electric field components are defined since these components are added to (or subtracted from) the total/scattered electric field components as (12.9), (12.11), (12.13), and (12.15). Similarly, "*tm*" used to calculate incident electric field components since these components are added to (or subtracted from) the total/scattered magnetic field components as (12.10), (12.12), (12.14), and (12.16).

12.2.3 Updating fields on both sides of the TF/SF boundary

In the FDTD time-marching loop, the magnetic field components in the entire problem space are updated in the subroutine ***update_magnetic_fields***, as shown in Listing 12.6,

Listing 12.6 update_magnetic_fields.m

```
% update magnetic fields

current_time = current_time + dt/2;

Hx = Chxh.*Hx+Chxey.*(Ey(1:nxp1,1:ny,2:nzp1)-Ey(1:nxp1,1:ny,1:nz)) ...
    + Chxez.*(Ez(1:nxp1,2:nyp1,1:nz)-Ez(1:nxp1,1:ny,1:nz));

Hy = Chyh.*Hy+Chyez.*(Ez(2:nxp1,1:nyp1,1:nz)-Ez(1:nx,1:nyp1,1:nz)) ...
    + Chyex.*(Ex(1:nx,1:nyp1,2:nzp1)-Ex(1:nx,1:nyp1,1:nz));

Hz = Chzh.*Hz+Chzex.*(Ex(1:nx,2:nyp1,1:nzp1)-Ex(1:nx,1:ny,1:nzp1)) ...
    + Chzey.*(Ey(2:nxp1,1:ny,1:nzp1)-Ey(1:nx,1:ny,1:nzp1));

if incident_plane_wave.enable
   xui = incident_plane_wave.ui;
   xuj = incident_plane_wave.uj;
   xuk = incident_plane_wave.uk;
   xli = incident_plane_wave.li;
   xlj = incident_plane_wave.lj;
   xlk = incident_plane_wave.lk;
    % Hx_yn
       Hx(xli:xui,xlj-1,xlk:xuk-1) = ...
       Hx(xli:xui,xlj-1,xlk:xuk-1) + dt/(mu_0*dy) * Ezi_yn(:,1,:);
    % Hx_yp
       Hx(xli:xui,xuj,xlk:xuk-1) = ...
       Hx(xli:xui,xuj,xlk:xuk-1) - dt/(mu_0*dy) * Ezi_yp(:,1,:);
    % Hx_zn
       Hx(xli:xui,xlj:xuj-1,xlk-1) = ...
       Hx(xli:xui,xlj:xuj-1,xlk-1) - dt/(mu_0*dz) * Eyi_zn(:,:,1);
    % Hx_zp
       Hx(xli:xui,xlj:xuj-1,xuk) = ...
       Hx(xli:xui,xlj:xuj-1,xuk) + dt/(mu_0*dz) * Eyi_zp(:,:,1);

    % Hy_xn
       Hy(xli-1,xlj:xuj,xlk:xuk-1) = ...
       Hy(xli-1,xlj:xuj,xlk:xuk-1) - dt/(mu_0*dx) * Ezi_xn(1,:,:);
    % Hy_xp
       Hy(xui,xlj:xuj,xlk:xuk-1) = ...
       Hy(xui,xlj:xuj,xlk:xuk-1) + dt/(mu_0*dx) * Ezi_xp(1,:,:);
    % Hy_zn
       Hy(xli:xui-1,xlj:xuj,xlk-1) = ...
       Hy(xli:xui-1,xlj:xuj,xlk-1) + dt/(mu_0*dz) * Exi_zn(:,:,1);
    % Hy_zp
       Hy(xli:xui-1,xlj:xuj,xuk) = ...
       Hy(xli:xui-1,xlj:xuj,xuk) - dt/(mu_0*dz) * Exi_zp(:,:,1);

    % Hz_yn
       Hz(xli:xui-1, xlj-1,xlk:xuk) = ...
       Hz(xli:xui-1, xlj-1,xlk:xuk) - dt/(mu_0*dy) * Exi_yn(:,1,:);
    % Hz_yp
       Hz(xli:xui-1, xuj,xlk:xuk) = ...
       Hz(xli:xui-1, xuj,xlk:xuk) + dt/(mu_0*dy) * Exi_yp(:,1,:);
    % Hz_xn
       Hz(xli-1,xlj:xuj-1,xlk:xuk) = ...
       Hz(xli-1,xlj:xuj-1,xlk:xuk) + dt/(mu_0*dx) * Eyi_xn(1,:,:);
```

```
    % Hz_xp
        Hz(xui,xlj:xuj-1,xlk:xuk) = ...
        Hz(xui,xlj:xuj-1,xlk:xuk) - dt/(mu_0*dx) * Eyi_xp(1,:,:);
end
```

using the usual updating equations. These field components will be implicitly regarded as total fields inside the TF/SF boundaries and scattered fields outside. Once the update is completed, the incident field components can be added or subtracted from the boundary components as illustrated in Listing 12.6. In the code the lines denoted as updating **Hz_yn, Hx_yn, Hz_yp,** and **Hx_yp** correspond to (12.10), (12.12), (12.14), and (12.16), respectively.

Similarly, the electric field components in the entire problem space are updated in the subroutine ***update_electric_fields***, as shown in Listing 12.7, using the usual updating equations. Once the update is completed, the incident field components can be added or

Listing 12.7 update_electric_fields.m

```
% update electric fields except the tangential components
% on the boundaries

current_time        =current_time + dt/2;

Ex(1:nx,2:ny,2:nz)  =Cexe(1:nx,2:ny,2:nz).*Ex(1:nx,2:ny,2:nz) ...
                     + Cexhz(1:nx,2:ny,2:nz).*...
                      (Hz(1:nx,2:ny,2:nz)-Hz(1:nx,1:ny-1,2:nz)) ...
                     + Cexhy(1:nx,2:ny,2:nz).*...
                      (Hy(1:nx,2:ny,2:nz)-Hy(1:nx,2:ny,1:nz-1));

Ey(2:nx,1:ny,2:nz)  =Ceye(2:nx,1:ny,2:nz).*Ey(2:nx,1:ny,2:nz) ...
                     + Ceyhx(2:nx,1:ny,2:nz).*  ...
                      (Hx(2:nx,1:ny,2:nz)-Hx(2:nx,1:ny,1:nz-1)) ...
                     + Ceyhz(2:nx,1:ny,2:nz).*  ...
                      (Hz(2:nx,1:ny,2:nz)-Hz(1:nx-1,1:ny,2:nz));

Ez(2:nx,2:ny,1:nz)  =Ceze(2:nx,2:ny,1:nz).*Ez(2:nx,2:ny,1:nz) ...
                     + Cezhy(2:nx,2:ny,1:nz).*  ...
                      (Hy(2:nx,2:ny,1:nz)-Hy(1:nx-1,2:ny,1:nz)) ...
                     + Cezhx(2:nx,2:ny,1:nz).*...
                      (Hx(2:nx,2:ny,1:nz)-Hx(2:nx,1:ny-1,1:nz));

if incident_plane_wave.enable
   xui = incident_plane_wave.ui;
   xuj = incident_plane_wave.uj;
   xuk = incident_plane_wave.uk;
   xli = incident_plane_wave.li;
   xlj = incident_plane_wave.lj;
   xlk = incident_plane_wave.lk;
    % Ex_yn
        Ex(xli:xui-1,xlj,xlk:xuk) = ...
        Ex(xli:xui-1,xlj,xlk:xuk) - dt/(eps_0*dy) * Hzi_yn(:,1,:);
```

```
    % Ex_yp
        Ex(xli:xui-1,xuj,xlk:xuk) = ...
        Ex(xli:xui-1,xuj,xlk:xuk) + dt/(eps_0*dy) * Hzi_yp(:,1,:);
    % Ex_zn
        Ex(xli:xui-1,xlj:xuj,xlk) = ...
        Ex(xli:xui-1,xlj:xuj,xlk) + dt/(eps_0*dz) * Hyi_zn(:,:,1);
    % Ex_zp
        Ex(xli:xui-1,xlj:xuj,xuk) = ...
        Ex(xli:xui-1,xlj:xuj,xuk) - dt/(eps_0*dz) * Hyi_zp(:,:,1);

    % Ey_xn
        Ey(xli,xlj:xuj-1,xlk:xuk) = ...
        Ey(xli,xlj:xuj-1,xlk:xuk) + dt/(eps_0*dx) * Hzi_xn(1,:,:);
    % Ey_xp
        Ey(xui,xlj:xuj-1,xlk:xuk) = ...
        Ey(xui,xlj:xuj-1,xlk:xuk) - dt/(eps_0*dx) * Hzi_xp(1,:,:);
    % Ey_zn
        Ey(xli:xui,xlj:xuj-1,xlk) = ...
        Ey(xli:xui,xlj:xuj-1,xlk) - dt/(eps_0*dz) * Hxi_zn(:,:,1);
    % Ey_zp
        Ey(xli:xui,xlj:xuj-1,xuk) = ...
        Ey(xli:xui,xlj:xuj-1,xuk) + dt/(eps_0*dz) * Hxi_zp(:,:,1);

    % Ez_xn
        Ez(xli,xlj:xuj,xlk:xuk-1) = ...
        Ez(xli,xlj:xuj,xlk:xuk-1) - dt/(eps_0*dx) * Hyi_xn(1,:,:);
    % Ez_xp
        Ez(xui,xlj:xuj,xlk:xuk-1) = ...
        Ez(xui,xlj:xuj,xlk:xuk-1) + dt/(eps_0*dx) * Hyi_xp(1,:,:);
    % Ez_yn
        Ez(xli:xui,xlj,xlk:xuk-1) = ...
        Ez(xli:xui,xlj,xlk:xuk-1) + dt/(eps_0*dy) * Hxi_yn(:,1,:);
    % Ez_yp
        Ez(xli:xui,xuj,xlk:xuk-1) = ...
        Ez(xli:xui,xuj,xlk:xuk-1) - dt/(eps_0*dy) * Hxi_yp(:,1,:);
end
```

subtracted from the boundary components as illustrated in Listing 12.7. In the code, the lines denoted as updating **Ex_yn**, **Ez_yn**, **Ex_yp**, and **Ez_yp** correspond to (12.9), (12.11), (12.13), and (12.15), respectively.

12.3 Simulation examples

The presented TF/SF formulation can be used to solve general scattering problems in which the objects float in free space, that is, objects do not touch the boundaries or do not penetrate into the absorbing boundaries of the problem space. Although the same types of problems can be solved by the scattered field formulation, discussed in Chapter 11, as well, solution using the TF/SF formulation is much faster while it also requires less memory. Two examples are presented in the following sections.

12.3.1 Fields in an empty problem space

The distribution of total and scattered fields in an empty problem space is illustrated in this example. A brick of size 20, 20, 100 mm and material type of air is surrounded by 10 cells of air gap and 8 cells of CPML in all directions. The cell size of the grid is 1 mm on a side. An incident field that propagates in $\theta = 30^\circ$ and $\phi = 90^\circ$ direction with a Gaussian waveform is defined as the source in the subroutine ***define_sources_and_lumped_elements*** as shown in Listing 12.8. The transient fields are captured and displayed using animation on a *yz* plane-cut. Figure 12.5 shows the magnitude of the electric fields at 130, 230, and 330 time steps. One can notice that a wave of Gaussian waveform travels in the $\theta = 30^\circ$ direction as the time-marching loop proceeds. The field displayed inside the TF/SF boundary is the total field and any field that would be displayed outside the TF/SF boundary is the scattered field. In this example, the scattered field does not exist due to the absence of scatterers in the problem space therefore the displayed field is actually the incident field, which is also the total field. One can also notice the discontinuity of the field on the TF/SF boundary; the scattered field is zero outside the TF/SF boundary.

Listing 12.8 define_sources_and_lumped_elements.m

```
% Define incident plane wave, angles are in degrees
incident_plane_wave.E_theta = 1;
incident_plane_wave.E_phi = 0;
incident_plane_wave.theta_incident = 30;
incident_plane_wave.phi_incident = 90;
incident_plane_wave.waveform_type = 'gaussian';
incident_plane_wave.number_of_cells_per_wavelength = 20;
incident_plane_wave.enable = true;
```

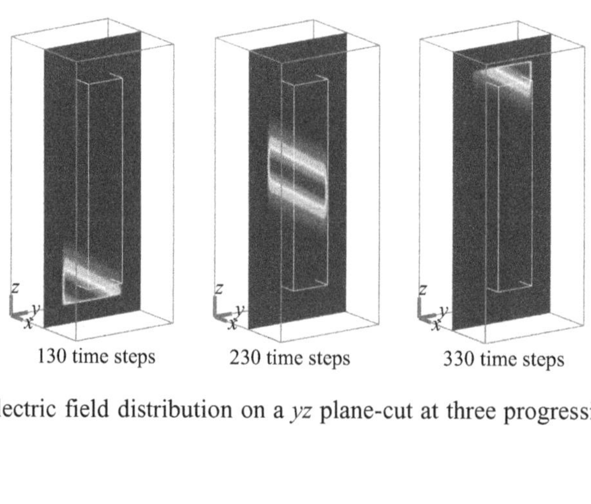

130 time steps   230 time steps   330 time steps

Figure 12.5 Electric field distribution on a *yz* plane-cut at three progressive time instants.

12.3.2 Scattering from a dielectric sphere

Scattering from a dielectric sphere is presented in Section 11.5.1, where bistatic RCS of a sphere is calculated using the scattered field formulation. The same example is repeated here by employing the TF/SF formulation. The RCS on *xy*, *xz*, and *yz* planes are obtained the same as the ones shown in Figures 11.6, 11.7, and 11.8, respectively, using the TF/SF formulation. This example is repeated using both formulations on a personal computer with Intel(R) Core(TM)2 Quad CPU Q9550 running at 2.83 GHz and calculation times are recorded as 10 min using the scattered field formulation while it is recorded as about 7 minutes using the TF/SF formulation.

The electric fields on a *xz* plane are captured and displayed in Figure 12.6 at progressive time steps of the simulation with TF/SF formulation. One can notice the discontinuity on the TF/SF boundary between the total and scattered fields in these images.

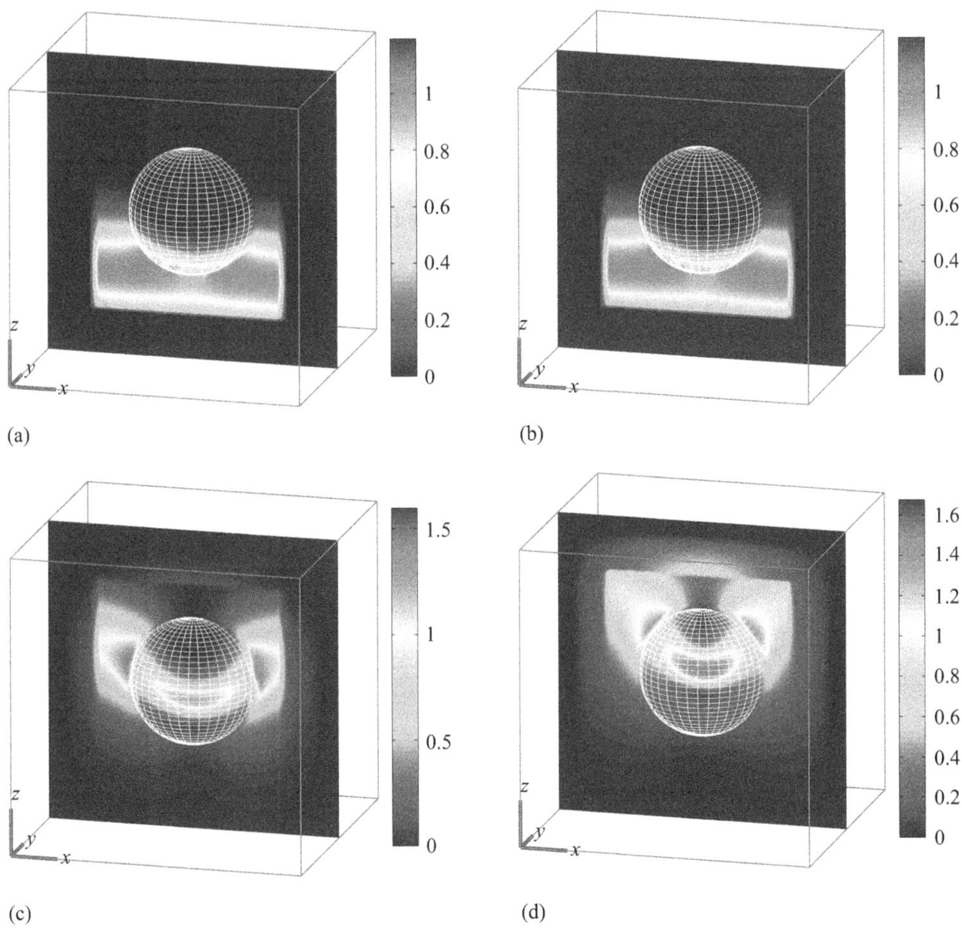

Figure 12.6 Electric field distribution on the *xz* plane: (a) at time step 120; (b) at time step 140; (c) at time step 160; and (d) at time step 180.

12.4 Exercises

12.1 Consider the problem in Section 12.3.2. The outputs definition file ***define_output_parameters*** includes the definition of far-field calculation parameters. The variable that indicates the position of the imaginary NF–FF surface **is number_of_cells_from_outer_boundary**. This variable can be set such that the NF–FF surface is inside the TF/SF boundary or outside the TF/SF boundary. If the NF–FF surface is inside the TF/SF boundary the total field is used for far-field calculations, otherwise, the scattered field is used for calculations. Run the simulation and obtain the RCS of the sphere at 1 GHz once using the total field and once using the scattered field. Verify that the RCS data are the same in both cases. You should obtain the same results as the ones in Figures 11.6–11.8. Make sure that the NF–FF surface does not overlap with the CPML boundaries. Notice that this exercise is similar to Exercise 11.1.

12.2 Consider the dipole antenna in Exercise 9.1. Replace the voltage source at the terminal of the antenna by 50°Ohms resistor. Run the simulation using a theta polarized plane wave that propagates in the theta = 90° direction. Capture sampled voltage at the terminal of the dipole and store it as V_{90°. Then, rerun the simulation using a theta polarized plane wave that propagates in the theta = 45° direction, capture sampled voltage at the terminal of the dipole and store it as V_{45°. Transform the sampled voltages to frequency domain and find their ratio $V_R = V_{45^\circ}/V_{90^\circ}$.

At very low frequencies, the dipole antenna acts like an ideal dipole. Verify that the voltage ratio V_R is $\sin(\theta)$, the radiation pattern expression of an ideal dipole, where $\theta = 45^\circ$.

At 4.5 GHz, the dipole antenna acts like a half-wave dipole. Verify that the voltage ratio V_R is $\cos\left(\frac{\pi}{2}\cos\theta\right)/\sin(\theta)$, the radiation pattern expression of a half-wave dipole, where $\theta = 45^\circ$.

CHAPTER 13

Dispersive material modeling

One of the attractive features of the finite-difference time-domain (FDTD) method is that it allows for modeling behavior of complex media such as dispersion and nonlinearity. Accurate algorithms can be developed to model the electromagnetic properties of these media and to integrate within the discrete time domain solution. In this chapter we will present the simulation of dispersive materials using the FDTD method.

The electromagnetic material properties – permittivity, permeability, and conductivities – were treated as constants for a medium while developing the updating equations throughout the previous chapters. In many applications these parameters are relatively constant over the frequency band of interest and simulations based on constant parameter assumption yield sufficiently accurate results in many applications. On the other hand, in some applications, the values of material parameters vary significantly as functions of frequency in the frequency band of interest. For instance, electromagnetic properties of biological tissues, earth, and artificial metamaterials are highly dependent on frequency and their dispersive nature needs to be modeled accurately in FDTD simulations.

Dispersion modeling has been one of the key topics in FDTD research over the years and many algorithms have been introduced by researchers. Most of these algorithms are classified mainly under the auxiliary differential equation (ADE) [48–52], recursive convolution (RC) [53–57], and Z-transform [58,59] techniques. An in-depth review of these techniques for modeling dispersive and other complex media in the FDTD method is presented in [60].

In general, frequency dependence of permittivity, permeability, or conductivity can be expressed as a sum of rational functions of angular frequency, ω. The most common types of these rational functions are referred to as Debye, Lorentz, or Drude model. Debye models are commonly used to approximate the frequency behavior of biological tissues and soil permittivities. Lorentz models describe the frequency behavior of some metamaterials such as Double-Negative media close to resonances, while Drude models are useful in describing the behavior of metals at optical frequencies, and they can sometimes also be augmented by Lorentz terms [60]. It should be noted that some applications employ some other dispersion models described in the literature besides these three models. For instance, Cole-Cole is another model, yet a more general one than Debye, for biological applications, while Condon is a model used to describe chiral media.

In the following sections, we will present integration of Debye, Lorentz, and Drude models into FDTD by using the ADE technique. Then, we will illustrate an implementation of the resulting algorithms in the MATLAB® FDTD code.

13.1 Modeling dispersive media using ADE technique

13.1.1 Modeling Debye medium using ADE technique

In this section we illustrate an algorithm to model Debye medium based on the ADE technique as presented by [51]. Here we consider dispersive permittivity, while development of an algorithm for dispersive permeability or conductivity follows the same procedure.

Ampere's law is expressed for a dispersive medium with P poles as

$$\nabla \times \tilde{H} = j\omega\varepsilon_0\varepsilon_\infty\tilde{E} + \sigma\tilde{E} + \sum_{k=1}^{P}\tilde{J}_k \tag{13.1}$$

in frequency domain, while in the time domain

$$\nabla \times \overline{H} = \varepsilon_0\varepsilon_\infty\frac{\partial}{\partial t}\overline{E} + \sigma\overline{E} + \sum_{k=1}^{P}\overline{J}_k, \tag{13.2}$$

where ε_∞ is the relative permittivity of the medium at infinite frequencies, and $\overline{J}_k$ is the kth polarization current. In the frequency domain, the polarization current $\tilde{J}_k$ can be written for Debye model as

$$\tilde{J}_k = j\omega\varepsilon_0\frac{A_k(\varepsilon_s - \varepsilon_\infty)}{1 + j\omega\tau_k}\tilde{E} = j\omega\frac{\zeta_k}{1 + j\omega\tau_k}\tilde{E}. \tag{13.3}$$

where ε_s is the static relative permittivity, A_k is the amplitude of the kth term, and τ_k is the relaxation time of the kth term. Here we define $\zeta_k = A_k\varepsilon_0(\varepsilon_s - \varepsilon_\infty)$ for brevity. While developing the FDTD updating equations, as the first step, we need to express (13.2) in discrete time at time step $n + 0.5$ as

$$\nabla \times \overline{H}^{n+0.5} = \varepsilon_0\varepsilon_\infty\frac{\partial}{\partial t}\overline{E}^{n+0.5} + \sigma\overline{E}^{n+0.5} + \sum_{k=1}^{P}\overline{J}_k^{n+0.5}, \tag{13.4}$$

which requires the value of $\overline{J}_k^{n+0.5}$. This value can be retrieved as follows. One can arrange (13.3) as

$$j\omega\tau_k\tilde{J}_k + \tilde{J}_k = j\omega\zeta_k\tilde{E}, \tag{13.5}$$

which, then, can be transformed to time domain as a differential equation

$$\tau_k\frac{\partial}{\partial t}\overline{J}_k + \overline{J}_k = \zeta_k\frac{\partial}{\partial t}\overline{E}. \tag{13.6}$$

This equation is called an ADE, which can be used to retrieve a value for $\overline{J}_k^{n+0.5}$. It should be noted that various approaches can be employed to retrieve a value for the polarization current term in (13.6). The following approach is employed in [51].

Equation (13.6) can be represented in discrete time at time step $n + 0.5$ as

$$\tau_k \frac{\overline{J}_k^{n+1} - \overline{J}_k^n}{\Delta t} + \frac{\overline{J}_k^{n+1} + \overline{J}_k^n}{2} = \zeta_k \frac{\overline{E}^{n+1} - \overline{E}^n}{\Delta t}, \tag{13.7}$$

which leads to

$$\overline{J}_k^{n+1} = \frac{(2\tau_k - \Delta t)}{(2\tau_k + \Delta t)} \overline{J}_k^n + \frac{2\zeta_k}{(2\tau_k + \Delta t)} \left(\overline{E}^{n+1} - \overline{E}^n \right). \tag{13.8}$$

This equation yields $\overline{J}_k^{n+1}$ if the values of $\overline{J}_k^n$, $\overline{E}^{n+1}$, and $\overline{E}^n$ are known, however, the value at $n + 0.5$, $\overline{J}_k^{n+0.5}$, is needed in (13.4). One can write

$$\overline{J}_k^{n+0.5} = \frac{\overline{J}_k^{n+1} + \overline{J}_k^n}{2}, \tag{13.9}$$

which can be expressed, using $\overline{J}_k^{n+1}$ from (13.8), as

$$\overline{J}_k^{n+0.5} = \frac{2\tau_k}{(2\tau_k + \Delta t)} \overline{J}_k^n + \frac{\zeta_k}{(2\tau_k + \Delta t)} \left(E^{n+1} - E^n \right) \tag{13.10}$$

Now, (13.10) can be integrated into (13.4) as

$$\begin{aligned} \nabla \times \overline{H}^{n+0.5} &= \varepsilon_0 \varepsilon_\infty \frac{\overline{E}^{n+1} - \overline{E}^n}{\Delta t} + \sigma \frac{\overline{E}^{n+1} + \overline{E}^n}{2} \\ &+ \sum_{k=1}^{P} \left(\frac{2\tau_k}{(2\tau_k + \Delta t)} \overline{J}_k^n + \frac{\zeta_k}{(2\tau_k + \Delta t)} \left(E^{n+1} - E^n \right) \right). \end{aligned} \tag{13.11}$$

which can be arranged, by moving the $\overline{E}^{n+1}$ term to left-hand side, to obtain an updating equation for the electric field as

$$\begin{aligned} \overline{E}^{n+1} &= \frac{2\Delta t}{(2\varepsilon_0 \varepsilon_\infty + \sigma \Delta t + \xi)} \nabla \times \overline{H}^{n+0.5} + \frac{2\varepsilon_0 \varepsilon_\infty - \sigma \Delta t + \xi}{2\varepsilon_0 \varepsilon_\infty + \sigma \Delta t + \xi} \overline{E}^n \\ &- \frac{2\Delta t}{(2\varepsilon_0 \varepsilon_\infty + \sigma \Delta t + \xi)} \sum_{k=1}^{P} \frac{2\tau_k}{(2\tau_k + \Delta t)} \overline{J}_k^n, \end{aligned} \tag{13.12}$$

where

$$\xi = \sum_{k=1}^{P} \frac{2\Delta t \zeta_k}{(2\tau_k + \Delta t)}. \tag{13.13}$$

An algorithm can be constructed using (13.8) and (13.12) as illustrated in Figure 13.1. At every time step, magnetic field components are updated as usual. Next, electric field components are updated using the past values of electric and magnetic field components, as well as the polarization current components following (13.12). Then, the polarization

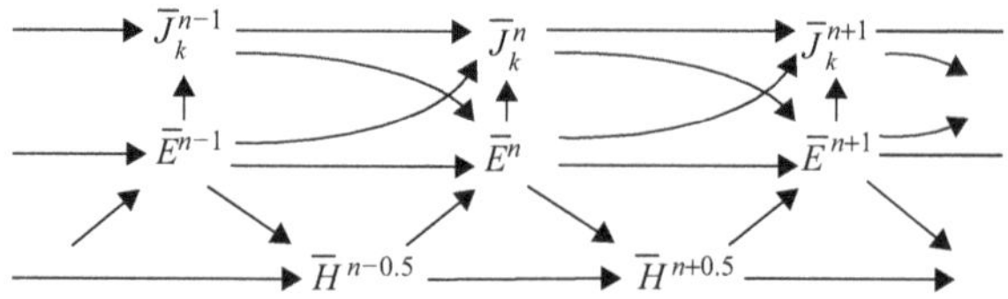

Figure 13.1 Update sequence of fields in the presented Debye modeling algorithm.

current components are calculated using the current and past values of electric field components and the past values of the polarization current components following (13.8). One should notice that this formulation requires an additional array to store $\overline{E}^n$ in order to implement (13.8).

13.1.2 Modeling Lorentz medium using ADE technique

In this section we illustrate an algorithm to model Lorentz medium based on the ADE technique. It should be noted that the procedure to develop an ADE and find a solution to the developed ADE is not unique. Here we discuss a procedure presented by [52] as an alternative to the one introduced by [51] that is presented in the previous section.

Ampere's law can be expressed for a dispersive medium with P poles, alternatively, as

$$\nabla \times \tilde{H} = j\omega\varepsilon_0\varepsilon_\infty\tilde{E} + \sigma\tilde{E} + j\omega\sum_{k=1}^{P}\tilde{Q}_k \tag{13.14}$$

in frequency domain, while in time domain

$$\nabla \times \overline{H} = \varepsilon_0\varepsilon_\infty\frac{\partial}{\partial t}\overline{E} + \sigma\overline{E} + \sum_{k=1}^{P}\frac{\partial}{\partial t}\overline{Q}_k. \tag{13.15}$$

Comparing (13.14) with (13.1), one can notice that they are the same when $\tilde{J}_k = j\omega\tilde{Q}_k$. The difference is that $\tilde{J}_k$ terms are used in [51] to develop ADEs for Debye and Lorentz dispersion models while $\tilde{Q}_k$ terms are used in [52] instead.

In the frequency domain, the term $\tilde{Q}_k$ can be written for Lorentz model as

$$\tilde{Q}_k = \frac{A_k\varepsilon_0(\varepsilon_s - \varepsilon_\infty)\omega_k^2}{\omega_k^2 + 2j\omega\delta_k - \omega^2}\tilde{E} = \frac{\psi_k}{\omega_k^2 + 2j\omega\delta_k - \omega^2}\tilde{E}, \tag{13.16}$$

where ω_k is the pole location and δ_k is the damping factor of the kth term. Here we define $\psi_k = A_k\varepsilon_0(\varepsilon_s - \varepsilon_\infty)\omega_k^2$ for brevity. An ADE can be constructed from (13.16) as

$$\left(\frac{\partial^2}{\partial t^2} + 2\delta_k\frac{\partial}{\partial t} + \omega_k^2\right)\overline{Q}_k = \psi_k\overline{E},$$

which can be expressed in discrete time using central difference approximation at time step n as

$$\frac{\overline{Q}_k^{n+1} - 2\overline{Q}_k^n + \overline{Q}_k^{n-1}}{\Delta t^2} + \delta_k \frac{\overline{Q}_k^{n+1} - \overline{Q}_k^{n-1}}{\Delta t} + \omega_k^2 \overline{Q}_k^n = \psi_k \overline{E}^n. \tag{13.17}$$

Equation (13.17) can be arranged to calculate the future value of $\overline{Q}_k$, such that

$$\overline{Q}_k^{n+1} = \frac{2 - (\Delta t)^2 \omega_k^2}{(\delta_k \Delta t + 1)} \overline{Q}_k^n + \frac{(\delta_k \Delta t - 1)}{(\delta_k \Delta t + 1)} \overline{Q}_k^{n-1} + \frac{(\Delta t)^2 \psi_k}{(\delta_k \Delta t + 1)} \overline{E}^n. \tag{13.18}$$

Meanwhile, (13.15) can be represented in discrete time at time step $n + 0.5$ as

$$\nabla \times \overline{H}^{n+0.5} = \varepsilon_0 \varepsilon_\infty \frac{\overline{E}^{n+1} - \overline{E}^n}{\Delta t} + \sigma \frac{\overline{E}^{n+1} + \overline{E}^n}{2} + \frac{1}{\Delta t} \sum_{k=1}^{P} \left(Q_k^{n+1} - Q_k^n\right), \tag{13.19}$$

which can be arranged, by moving the $\overline{E}^{n+1}$ term to left-hand side, to obtain an updating equation for the electric field as

$$\overline{E}^{n+1} = \frac{2\Delta t}{2\varepsilon_0 \varepsilon_\infty + \Delta t \sigma} \nabla \times \overline{H}^{n+0.5} + \frac{2\varepsilon_0 \varepsilon_\infty - \Delta t \sigma}{2\varepsilon_0 \varepsilon_\infty + \Delta t \sigma} \overline{E}^n - \frac{2}{2\varepsilon_0 \varepsilon_\infty + \Delta t \sigma} \sum_{k=1}^{P} \left(Q_k^{n+1} - Q_k^n\right) \tag{13.20}$$

An algorithm can be constructed using (13.18) and (13.20) as illustrated in Figure 13.2. At every time step, magnetic field components are updated as usual. Next, the new values of $\overline{Q}_k$ are calculated using their past values and the past values of the electric field components following (13.18). Then, electric field components are updated using the past values of electric and magnetic field components as well as the new and past values of $\overline{Q}_k$ following (13.20). One should notice that this formulation requires additional arrays to store $\overline{Q}_k^n$ and $\overline{Q}_k^{n-1}$ in order to implement (13.18).

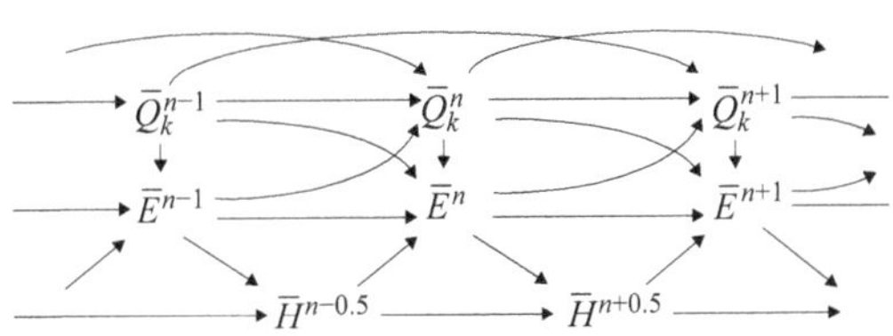

Figure 13.2 Update sequence of fields in the presented Lorentz and Drude modeling algorithms.

13.1.3 Modeling Drude medium using ADE technique

The ADE technique introduced by [52], discussed above, was employed to model Drude medium in [61]. In this section we illustrate the formulation presented in [61].

Here we consider (13.15) as the starting point to develop an algorithm for Drude medium. In the frequency domain, the term $\tilde{Q}_k$ can be written for Drude model as

$$\tilde{Q}_k = \frac{\varepsilon_0 \omega_k^2}{\omega^2 - j\omega\gamma_k}\tilde{E}, \tag{13.21}$$

where ω_k is the plasma frequency, and γ_k is the inverse of the pole relaxation time. An ADE can be constructed from (13.21) as

$$\left(-\frac{\partial^2}{\partial t^2} - \gamma_k \frac{\partial}{\partial t}\right)\overline{Q}_k = \varepsilon_0 \omega_k^2 \overline{E}. \tag{13.22}$$

Then, (13.22) can be expressed in discrete time using central difference approximation at time step n, and rearranged to calculate the future value of $\overline{Q}_k$ as

$$\overline{Q}_k^{n+1} = \frac{4}{\Delta t\gamma_k + 2}\overline{Q}_k^n + \frac{\Delta t\gamma_k - 2}{\Delta t\gamma_k + 2}\overline{Q}_k^{n-1} - \frac{2\Delta t^2 \varepsilon_0 \omega_k^2}{\Delta t\gamma_k + 2}\overline{E}^n. \tag{13.23}$$

Equation (13.23) is in the same form as (13.18), while its coefficients are different. Therefore, we can use (13.23) and (13.20) to construct an ADE algorithm to model Drude medium, which is essentially the same as the one constructed for Lorentz medium as shown in Figure 13.2: at every time step, magnetic field components are updated as usual. Next, the new values of $\overline{Q}_k$ are calculated following (13.23). Then, electric field components are updated following (13.20).

13.2 MATLAB® implementation of ADE algorithm for Lorentz medium

The presented Lorentz medium modeling ADE algorithm is implemented into the MATLAB FDTD code that also includes the TF/SF formulation. While a Lorentz model of a single term ($P = 1$ in (13.14)) is considered in the presented implementation, it is possible to extend the implementation to accommodate multiple terms as well, where $P > 1$. The details of the implementation are presented in the following subsections.

13.2.1 Definition of Lorentz material parameters

A dispersive material is described using additional parameters along with the standard permittivity, permeability, and conductivity parameters. In the case of the Lorentz model infinite relative permittivity ε_∞, static relative permittivity ε_s, amplitude A_k, pole location ω_k, and damping factor δ_k of the kth term are the parameters that need to be defined for an FDTD simulation.

In the MATLAB FDTD code, the material parameters are defined in the subroutine ***define_problem_space_parameters***, where a structure array, referred to as **material_types**, is constructed to store the electromagnetic parameters of materials that are used in the simulation. In the presented implementation, new fields are added to the **material_types** structure to support Lorentz media modeling. Listing 13.1 shows an element of the **material_types** array including the Lorentz parameters. Here, **lorentz_eps_s** refers to the static relative permittivity, **lorentz_A** refers to the amplitude, **lorentz_omega** refers to the pole location, and **lorentz_delta** refers to the damping factor. The parameter **lorentz_dispersive** is used to identify the material type if it is Lorentz dispersive or not. As for the infinite permittivity, one can realize in (13.1) and (13.2) that it is actually the same as the relative permittivity of a nondispersive medium. Therefore, the parameter **eps_r** is used to store the relative infinite permittivity if the material is of Lorentz type.

Listing 13.1 define_problem_sec_parameters.m

```
% lorentz material
material_types(4).eps_r      =2; % eps_inf for dispersive material
material_types(4).mu_r       =1;
material_types(4).sigma_e    =0;
material_types(4).sigma_m    =0;
material_types(4).color      =[0 0 1];
material_types(4).lorentz_dispersive = true;
material_types(4).lorentz_eps_s = 5;
material_types(4).lorentz_A = 1;
material_types(4).lorentz_omega = 2*pi*2e9;
material_types(4).lorentz_delta = pi*2e9;
```

13.2.2 Material grid construction for Lorentz objects

The FDTD material grid is constructed in the subroutine ***initialize_fdtd_material_grid***, where three-dimensional arrays are constructed for permittivity, permeability, and conductivity distributions in the problem space. Later, these arrays are used to calculate updating coefficients. In order to account for Lorentz modeling, a new subroutine, referred to as ***create_dispersive_objects***, is implemented and called at the end of the subroutine ***initialize_fdtd_material_grid***.

Listing 13.2 shows the subroutine ***create_dispersive_objects***. At the first part of the code, three-dimensional arrays are created to store the x, y, and z coordinates of the electric field components. These arrays are then used to calculate the distances of the field components from the center of a sphere, and identify the indices of the field components that are located inside the sphere. Once the indices of these field components are identified, corresponding elements in three-dimensional arrays **dispersive_Ex**, **dispersive_Ey**, and **dispersive_Ez** are assigned with the material index of the sphere if the sphere is dispersive. Then similarly, the field components that are located in a brick are identified and the corresponding elements in **dispersive_Ex**, **dispersive_Ey**, and **dispersive_Ez** are assigned with the material index of the brick if the brick is dispersive. Once these calculations are completed, these three arrays will include distribution of the material type indices of the dispersive material types.

Next, a new structure array, referred to as **lorentz**, is constructed. Each element in this array corresponds to a dispersive material type in the **material_types** structure array. Here, for a dispersive material type, the **dispersive_Ex**, **dispersive_Ey**, and **dispersive_Ez** arrays are

Listing 13.2 create_dispersive_objects.m

```
% Three dimensional arrays to store field coordinates
% Used to identify field components that are inside a sphere
Ex_x_coordinates = zeros(nx, nyp1, nzp1);
Ex_y_coordinates = zeros(nx, nyp1, nzp1);
Ex_z_coordinates = zeros(nx, nyp1, nzp1);

Ey_x_coordinates = zeros(nxp1, ny, nzp1);
Ey_y_coordinates = zeros(nxp1, ny, nzp1);
Ey_z_coordinates = zeros(nxp1, ny, nzp1);

Ez_x_coordinates = zeros(nxp1, nyp1, nz);
Ez_y_coordinates = zeros(nxp1, nyp1, nz);
Ez_z_coordinates = zeros(nxp1, nyp1, nz);

% Arrays to store material indices distribution of dispersive objects
dispersive_Ex = zeros(nx, nyp1, nzp1);
dispersive_Ey = zeros(nxp1, ny, nzp1);
dispersive_Ez = zeros(nxp1, nyp1, nz);

% calculate field coordinates
for ind = 1:nx
    Ex_x_coordinates(ind,:,:) = (ind - 0.5) * dx + fdtd_domain.min_x;
end
for ind = 1:nyp1
    Ex_y_coordinates(:,ind,:) = (ind - 1) * dy + fdtd_domain.min_y;
end
for ind = 1:nzp1
    Ex_z_coordinates(:,:,ind) = (ind - 1) * dz + fdtd_domain.min_z;
end

for ind = 1:nxp1
    Ey_x_coordinates(ind,:,:) = (ind - 1) * dx + fdtd_domain.min_x;
end
for ind = 1:ny
    Ey_y_coordinates(:,ind,:) = (ind - 0.5) * dy + fdtd_domain.min_y;
end
for ind = 1:nzp1
    Ey_z_coordinates(:,:,ind) = (ind - 1) * dz + fdtd_domain.min_z;
end

for ind = 1:nxp1
    Ez_x_coordinates(ind,:,:) = (ind - 1) * dx + fdtd_domain.min_x;
end
for ind = 1:nyp1
    Ez_y_coordinates(:,ind,:) = (ind - 1) * dy + fdtd_domain.min_y;
end
for ind = 1:nz
    Ez_z_coordinates(:,:,ind) = (ind - 0.5) * dz + fdtd_domain.min_z;
end

disp('creating dispersive spheres');
for ind=1:number_of_spheres
    mat_ind = spheres(ind).material_type;
    material = material_types(mat_ind);
    if material.lorentz_dispersive
```

```
        % distance of the Ex components from the center of the sphere
        distance = sqrt((spheres(ind).center_x - Ex_x_coordinates).^2 ...
             + (spheres(ind).center_y - Ex_y_coordinates).^2 ...
             + (spheres(ind).center_z - Ex_z_coordinates).^2);
        I = find(distance<=spheres(ind).radius);
        dispersive_Ex(I) = mat_ind;

        % distance of the Ey components from the center of the sphere
        distance = sqrt((spheres(ind).center_x - Ey_x_coordinates).^2 ...
             + (spheres(ind).center_y - Ey_y_coordinates).^2 ...
             + (spheres(ind).center_z - Ey_z_coordinates).^2);
        I = find(distance<=spheres(ind).radius);
        dispersive_Ey(I) = mat_ind;

        % distance of the Ez components from the center of the sphere
        distance = sqrt((spheres(ind).center_x - Ez_x_coordinates).^2 ...
             + (spheres(ind).center_y - Ez_y_coordinates).^2 ...
             + (spheres(ind).center_z - Ez_z_coordinates).^2);
        I = find(distance<=spheres(ind).radius);
        dispersive_Ez(I) = mat_ind;
    end
end

disp('creating dispersive bricks');
for ind = 1:number_of_bricks
    mat_ind = bricks(ind).material_type;
    material = material_types(mat_ind);
    if material.lorentz_dispersive
        % convert brick end coordinates to node indices
        ni = get_node_indices(bricks(ind), fdtd_domain);
        is = ni.is; js = ni.js; ks = ni.ks;
        ie = ni.ie; je = ni.je; ke = ni.ke;

        dispersive_Ex (is:ie-1, js:je, ks:ke) = mat_ind;
        dispersive_Ey (is:ie, js:je-1, ks:ke) = mat_ind;
        dispersive_Ez (is:ie, js:je, ks:ke-1) = mat_ind;
    end
end

% create a new structure array to store indices of field components
% that need to be updated using Lorentz formulation
number_of_lorentz = 0;
for ind = 1:number_of_material_types
    material = material_types(ind);
    if material.lorentz_dispersive
        number_of_lorentz = number_of_lorentz + 1;

        lorentz(number_of_lorentz).material = ind;

        I = find(dispersive_Ex == ind);
        lorentz(number_of_lorentz).Ex_indices = I;
        eps_r_x(I) = material.eps_r;
        sigma_e_x(I) = material.sigma_e;
```

```
        I = find(dispersive_Ey == ind);
        lorentz(number_of_lorentz).Ey_indices = I;
        eps_r_y(I) = material.eps_r;
        sigma_e_y(I) = material.sigma_e;

        I = find(dispersive_Ez == ind);
        lorentz(number_of_lorentz).Ez_indices = I;
        eps_r_z(I) = material.eps_r;
        sigma_e_z(I) = material.sigma_e;
    end
end

clear Ex_x_coordinates Ex_y_coordinates Ex_z_coordinates
clear Ey_x_coordinates Ey_y_coordinates Ey_z_coordinates
clear Ez_x_coordinates Ez_y_coordinates Ez_z_coordinates
clear dispersive_Ex dispersive_Ey dispersive_Ez
```

queried for the index of the material type in consideration, and the indices of the elements associated with this material type are identified. Then these indices are stored as linear arrays **Ex_indices**, **Ey_indices**, and **Ez_indices** in the **lorentz** structure. The indices of the field components that need to be updated using Lorentz formulation are now known through the arrays **Ex_indices**, **Ey_indices**, and **Ez_indices**.

13.2.3 Initialization of updating coefficients

The updating coefficients are initialized in the subroutine ***initialize_updating_coefficients*** in the MATLAB FDTD code. A new subroutine, ***initialize_dispersive_coefficients***, is implemented and called after the calculation of general updating coefficients in ***initialize_updating_coefficients***.

It should be noted that (13.18) and (13.20) are vector equations and they need to decomposed as scalar equations in terms of x, y, and z components. For instance, (13.18) can be expressed for the x, y, and z components as

$$Q_{k,x}^{n+1}(i,j,k) = C_{qq}Q_{k,x}^{n}(i,j,k) + C_{qqm}Q_{k,x}^{n-1}(i,j,k) + C_{qe}E_x^n(i,j,k), \tag{13.24a}$$

$$Q_{k,y}^{n+1}(i,j,k) = C_{qq}Q_{k,y}^{n}(i,j,k) + C_{qqm}Q_{k,y}^{n-1}(i,j,k) + C_{qe}E_y^n(i,j,k), \tag{13.24b}$$

$$Q_{k,z}^{n+1}(i,j,k) = C_{qq}Q_{k,z}^{n}(i,j,k) + C_{qqm}Q_{k,z}^{n-1}(i,j,k) + C_{qe}E_z^n(i,j,k). \tag{13.24c}$$

Here, C_{qq}, C_{qqm}, and C_{qe} are coefficients associated with the respective field components and they are defined as

$$C_{qq} = \frac{2-(\Delta t)^2\omega_k^2}{(\delta_k\Delta t+1)}, \quad C_{qqm} = \frac{(\delta_k\Delta t-1)}{(\delta_k\Delta t+1)}, \quad C_{qe} = \frac{(\Delta t)^2\psi_k}{(\delta_k\Delta t+1)}.$$

Similarly, (13.20) can be expressed for E_x^{n+1} as

$$
\begin{aligned}
E_x^{n+1}(i,j,k) = {} & C_{exe}(i,j,k)E_x^n(i,j,k) \\
& + C_{exhz}(i,j,k)\left(H_z^{n+0.5}(i,j,k) - H_z^{n+0.5}(i,j-1,k)\right) \\
& + C_{exhy}(i,j,k)\left(H_y^{n+0.5}(i,j,k) - H_y^{n+0.5}(i,j,k-1)\right) \\
& - \frac{\Delta y}{\Delta t} C_{exhz}(i,j,k)\sum_{k=1}^{P}\left(Q_{k,x}^{n+1} - Q_{k,x}^{n}\right).
\end{aligned}
\tag{13.25}
$$

One can notice that C_{exe}, C_{exhz}, and C_{exhy} are the same as the general updating coefficients (i.e., Chapter 1 (1.26)). Also, the coefficient that is multiplied by the summation of the differences of $Q_{k,x}^{n+1}$ and $Q_{k,x}^{n}$ terms is simply C_{exhz} scaled by $\frac{\Delta y}{\Delta t}$. Equation (13.25) is essentially the same as (1.26) in Chapter 1, except that it has an additional last term due to the dispersion.

Similarly, (13.20) can be expressed for E_y^{n+1} as

$$
\begin{aligned}
E_y^{n+1}(i,j,k) = {} & C_{eye}(i,j,k)E_y^n(i,j,k) \\
& + C_{eyhx}(i,j,k)\left(H_x^{n+0.5}(i,j,k) - H_x^{n+0.5}(i,j,k-1)\right) \\
& + C_{eyhz}(i,j,k)\left(H_z^{n+0.5}(i,j,k) - H_z^{n+0.5}(i-1,j,k)\right) \\
& - \frac{\Delta z}{\Delta t} C_{eyhx}(i,j,k)\sum_{k=1}^{P}\left(Q_{k,y}^{n+1} - Q_{k,y}^{n}\right),
\end{aligned}
\tag{13.26}
$$

and for E_z^{n+1} as

$$
\begin{aligned}
E_z^{n+1}(i,j,k) = {} & C_{eze}(i,j,k)E_z^n(i,j,k) \\
& + C_{ezhy}(i,j,k)\left(H_y^{n+0.5}(i,j,k) - H_y^{n+0.5}(i-1,j,k)\right) \\
& + C_{ezhx}(i,j,k)\left(H_x^{n+0.5}(i,j,k) - H_x^{n+0.5}(i,j-1,k)\right) \\
& - \frac{\Delta x}{\Delta t} C_{ezhy}(i,j,k)\sum_{k=1}^{P}\left(Q_{k,z}^{n+1} - Q_{k,z}^{n}\right).
\end{aligned}
\tag{13.27}
$$

Listing 13.3 shows the implementation of ***initialize_dispersive_coefficients***. First, for each Lorentz medium, arrays are constructed to store the additional terms $\overline{Q}_k^{n+1}$ and $\overline{Q}_k^{n}$ required by the Lorentz ADE algorithm and initialized to zero. Then, the coefficients C_{qq}, C_{qqm}, and C_{qe} in (13.24) are calculated and stored. Since the coefficients in (13.25) are the

Listing 13.3 initialize_dispersive_coefficients.m

```
% initialize Lorentz updating coefficients

for ind = 1:number_of_lorentz

    material = material_types(lorentz(ind).material);

    n_elements = size(lorentz(ind).Ex_indices,2);
    lorentz(number_of_lorentz).Qxp = zeros(1, n_elements);
    lorentz(number_of_lorentz).Qx  = zeros(1, n_elements);

    n_elements = size(lorentz(ind).Ey_indices,2);
    lorentz(number_of_lorentz).Qyp = zeros(1, n_elements);
    lorentz(number_of_lorentz).Qy  = zeros(1, n_elements);

    n_elements = size(lorentz(ind).Ez_indices,2);
    lorentz(number_of_lorentz).Qzp = zeros(1, n_elements);
    lorentz(number_of_lorentz).Qz  = zeros(1, n_elements);

    psi = eps_0 * material.lorentz_A ...
       * (material.lorentz_eps_s - material.eps_r) ...
       * material.lorentz_omega^2;

    den = material.lorentz_delta * dt + 1;
    lorentz(number_of_lorentz).Cqq =  (2 - dt^2 ...
       * material.lorentz_omega^2)/den;
    lorentz(number_of_lorentz).Cqqm =  ...
       (material.lorentz_delta * dt - 1)/den;
    lorentz(number_of_lorentz).Cqe =  ...
       dt^2 * psi / den;
end
```

same as the already calculated general updating coefficients, they do not need to be stored separately.

13.2.4 Field updates in time-marching loop

Electric fields are updated in the subroutine ***update_electric_fields*** at every time step of the FDTD time-marching loop. This subroutine is modified based on the sequence of field updates illustrated in Figure 13.2.

Listing 13.4 shows the modified section of the code in ***update_electric_fields***. In the first part of the code the $\overline{Q}_k^{n+1}$ terms are calculated using $\overline{Q}_k^n$, $\overline{Q}_k^{n-1}$, and $\overline{E}^n$ following (13.18). Then regular field updates are performed, where the components of $\overline{E}^{n+1}$ are calculated for the current time step. Then, the difference of Lorentz terms $\overline{Q}_k^{n+1}$ and $\overline{Q}_k^n$ is multiplied by the associated coefficients and added to the electric field components following (13.20). One can notice in the code that the arrays **Ex_indices**, **Ey_indices**, and **Ez_indices** are used to access to the electric field components that are associated with the Lorentz materials in the problem space.

Listing 13.4 update_electric_fields.m

```
% update the additional Lorents terms
for ind = 1:number_of_lorentz

    Cqq  = lorentz(number_of_lorentz).Cqq;
    Cqqm = lorentz(number_of_lorentz).Cqqm;
    Cqe  = lorentz(number_of_lorentz).Cqe;

    Qx = lorentz(ind).Qxp;
    lorentz(ind).Qxp = ...
          Cqq  * lorentz(ind).Qxp ...
        + Cqqm * lorentz(ind).Qx  ...
        + Cqe  * Ex(lorentz(ind).Ex_indices);
    lorentz(ind).Qx = Qx;

    Qy = lorentz(ind).Qyp;
    lorentz(ind).Qyp = ...
          Cqq  * lorentz(ind).Qyp ...
        + Cqqm * lorentz(ind).Qy  ...
        + Cqe  * Ey(lorentz(ind).Ey_indices);
    lorentz(ind).Qy = Qy;

    Qz = lorentz(ind).Qzp;
    lorentz(ind).Qzp = ...
          Cqq  * lorentz(ind).Qzp ...
        + Cqqm * lorentz(ind).Qz  ...
        + Cqe  * Ez(lorentz(ind).Ez_indices);
    lorentz(ind).Qz = Qz;
end

% Regular field updates
Ex(1:nx,2:ny,2:nz) = Cexe(1:nx,2:ny,2:nz).*Ex(1:nx,2:ny,2:nz) ...
                   + Cexhz(1:nx,2:ny,2:nz).*...
                   (Hz(1:nx,2:ny,2:nz)-Hz(1:nx,1:ny-1,2:nz)) ...
                   + Cexhy(1:nx,2:ny,2:nz).*...
                   (Hy(1:nx,2:ny,2:nz)-Hy(1:nx,2:ny,1:nz-1));

Ey(2:nx,1:ny,2:nz) = Ceye(2:nx,1:ny,2:nz).*Ey(2:nx,1:ny,2:nz) ...
                   + Ceyhx(2:nx,1:ny,2:nz).* ...
                   (Hx(2:nx,1:ny,2:nz)-Hx(2:nx,1:ny,1:nz-1)) ...
                   + Ceyhz(2:nx,1:ny,2:nz).* ...
                   (Hz(2:nx,1:ny,2:nz)-Hz(1:nx-1,1:ny,2:nz));

Ez(2:nx,2:ny,1:nz) = Ceze(2:nx,2:ny,1:nz).*Ez(2:nx,2:ny,1:nz) ...
                   + Cezhy(2:nx,2:ny,1:nz).* ...
                   (Hy(2:nx,2:ny,1:nz)-Hy(1:nx-1,2:ny,1:nz)) ...
                   + Cezhx(2:nx,2:ny,1:nz).*...
                   (Hx(2:nx,2:ny,1:nz)-Hx(2:nx,1:ny-1,1:nz));

% Add additional Lorentz terms to electric field components
for ind = 1:number_of_lorentz
    indices = lorentz(ind).Ex_indices;
    Ex(indices) = Ex(indices) ...
        - (dy/dt) * Cexhz(indices) .* (lorentz(ind).Qxp - lorentz(ind).Qx);
```

```
    indices = lorentz(ind).Ey_indices;
    Ey(indices) = Ey(indices) ...
        - (dz/dt) * Ceyhx(indices) .* (lorentz(ind).Qyp - lorentz(ind).Qy);

    indices = lorentz(ind).Ez_indices;
    Ez(indices) = Ez(indices) ...
        - (dx/dt) * Cezhy(indices) .* (lorentz(ind).Qzp - lorentz(ind).Qz);
end
```

13.3 Simulation examples

13.3.1 Scattering from a dispersive sphere

In this first example, we consider a single-term Lorentz dispersive sphere with a radius of 10 cm. The Lorentz parameters of the sphere are $\varepsilon_\infty = 2$, $\varepsilon_s = 5$, $A_1 = 1$, $\omega_1 = 4\pi \times 10^9$, and $\delta_1 = 2\pi \times 10^9$. The definition of these parameters in the FDTD code is shown in Listing 13.1. Examining (13.14) and (13.16), one can describe an equivalent complex relative permittivity for a single term Lorentz medium as

$$\varepsilon_r(\omega) = \varepsilon_\infty + \frac{\sigma}{j\omega\varepsilon_0} + \frac{A_1(\varepsilon_s - \varepsilon_\infty)\omega_1^2}{\omega_1^2 + 2j\omega\delta_1 - \omega^2}. \tag{13.28}$$

The relative permittivity is calculated using (13.28) for the above parameter values as a function of frequency and plotted in Figure 13.3.

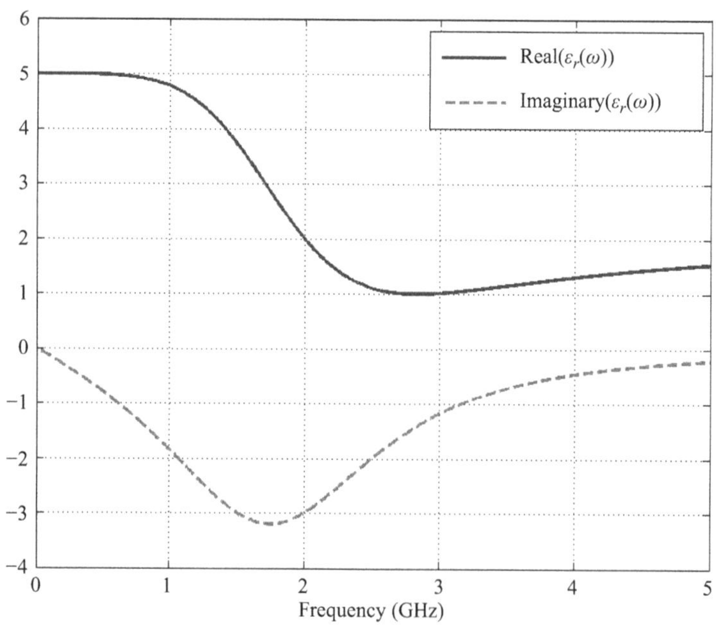

Figure 13.3 Complex relative permittivity versus frequency.

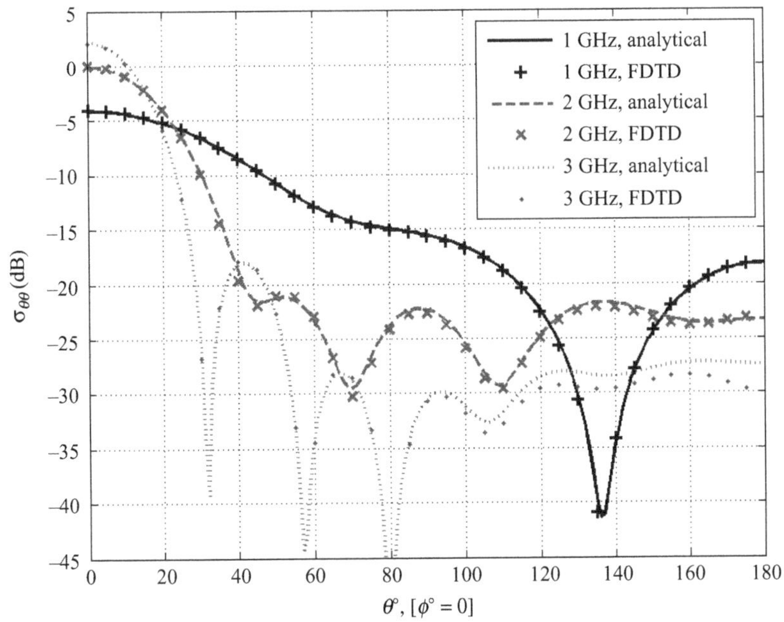

Figure 13.4 Co-polarized RCS of a dispersive sphere at three frequencies.

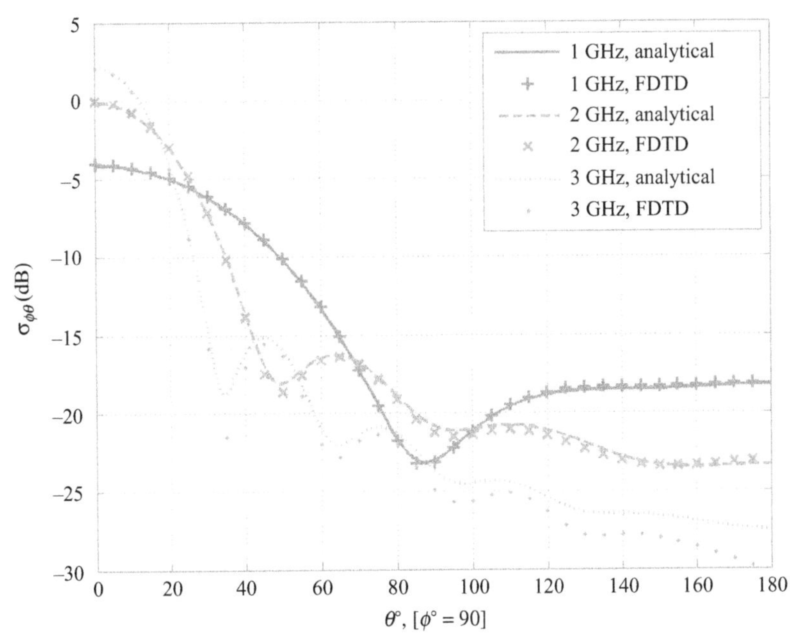

Figure 13.5 Cross-polarized RCS of a dispersive sphere at three frequencies.

The sphere is illuminated by an x-polarized plane wave with a Gaussian waveform that propagates in the positive z direction. The problem space is modeled using a cell size of 5 mm on a side and the simulation is performed for 3,000 time steps. The bistatic radar cross-section of the sphere is calculated at frequencies 1, 2, and 3 GHz. At these frequencies, the relative permittivity values are $\varepsilon_r(1\text{ GHz}) = 4.769 - 1.846j$, $\varepsilon_r(2\text{ GHz}) = 2 - 3j$, and $\varepsilon_r(3\text{ GHz}) = 1.016 - 1.18j$, as can be verified from Figure 13.3.

The results for $\sigma_{\theta\theta}$ in the xz-plane are shown in Figure 13.4, whereas the results for $\sigma_{\phi\theta}$ in the yz-plane are shown in Figure 13.5 along with the analytical solutions obtained using the program presented in [62]. A very good agreement can be observed between the FDTD simulation results and the analytical solutions except for the slight differences at the back-scatter angles at highest frequency. The simulation accuracy can be improved by using smaller cells or by employing an averaging scheme for media properties on the boundary between the sphere surface and surrounding media. For instance, such a method to determine effective permittivity at the interface of dispersive dielectrics is presented in [63].

13.4 Exercises

13.1 Consider the problem in Section 13.3.1. The dispersive sphere material has the relative permittivity of $\varepsilon_r = 4.769 - 1.846j$ at 1 GHz. A lossy and nondispersive medium can be imagined to have this particular relative permittivity value 1 GHz. For instance, the complex relative permittivity of such a medium can be expressed as $\varepsilon_{rc} = \varepsilon_r' - j\frac{\sigma}{\omega\varepsilon_0}$. Hence, $\varepsilon_r' = 4.769$ and $\sigma = 1.846\ \omega\varepsilon_0 = 0.1027$ at 1 GHz. Model the sphere as a lossy and nondispersive by using these relative permittivity and conductivity values and run the simulation to calculate the co- and cross-polarized RCS of the sphere. Verify that the results match with the ones in 13.4 and 13.5 at 1 GHz. Note that the results will match only at 1 GHz.

13.2 Consider the dielectric cube example in Section 11.5.2, and assume that it is made of a dispersive material described by the one in Figure 13.3. Run the simulation and obtain the RCS of the cube at 1, 2, and 3 GHz.

One can employ the procedure discussed in Exercise 13.1 to verify that the calculated results are correct: run the simulation for each of these frequencies and each time model the cube as a lossy and nondispersive object that has the corresponding complex permittivity at the frequency of interest. Result of the dispersive media simulation and the equivalent lossy and nondispersive media simulation should be the same at a frequency of interest.

Follow this approach and verify the results of the dispersive cube simulation. Notice that the complex relative permittivity values at 1, 2, and 3 GHz are listed in Section 13.3.1.

CHAPTER 14

Analysis of periodic structures

Many electromagnetic applications require the use of periodic structures such as frequency selective surfaces (FSS), electromagnetic band gap structures, corrugated surfaces, phased antenna arrays, periodic absorbers or negative index materials. These structures extend to several wavelengths in size therefore their analyses are time-consuming and memory-extensive using the conventional FDTD method. To overcome the limitation of the conventional FDTD method, first, a periodic structure is considered as infinitely periodic. For instance, Figure 14.1 illustrates a unit cell of a two-dimensional periodic structure on a substrate that extends to infinity in the x and y directions. Then the assumption of infinite periodicity is utilized to develop FDTD algorithms that analyze only one unit cell of the periodicity instead of the entire structure and obtain results for the entire infinite size structure or structures composed of large number of unit cells with approximations. These algorithms mainly deal with the boundaries of a unit cell, therefore they are referred to as periodic boundary conditions (PBC).

14.1 Periodic boundary conditions

The PBC algorithms are divided into two main categories: field transformation methods and direct field methods [64]. The field transformation methods introduce auxiliary fields to eliminate the need for time-advanced data. The transformed field equations are then discretized and solved using FDTD techniques. The split-field method [65] and multispatial grid method [66] are two approaches in this category. The direct field category methods work directly with Maxwell's equations, and hence there is no need for any field transformation. The sine-cosine method [67] is an example of this category. In this method the structure is excited simultaneously with sine and cosine waveforms, therefore it is a single frequency method that does not maintain the wide-band capability of FDTD.

A simple and efficient PBC algorithm that belongs to the direct field category, referred to as constant horizontal wavenumber approach, is introduced in [68–70] and yet has a wideband capability. In this chapter, we will demonstrate the constant horizontal wavenumber approach and its MATLAB® implementation to calculate reflection and transmission coefficients of an infinite periodic structure.

It should be noted that in most applications that use periodic structures the reflection and/or transmission properties are of interest due to incident plane waves with a number of incident angles and a number of frequencies. One would seek to obtain the reflection and/or

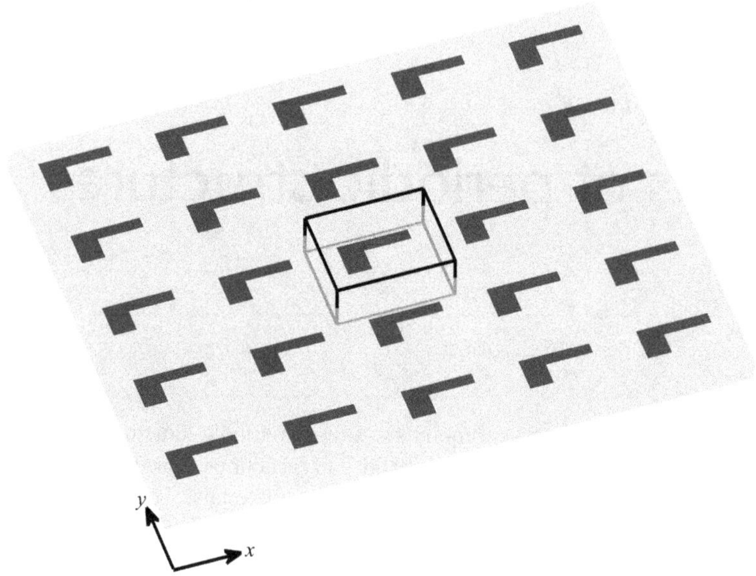

Figure 14.1 A periodic structure and a unit cell of two-dimension periodicity.

transmission coefficient distribution as a two-dimensional figure on which the axes are the frequency and angle of incidence. For instance, Figure 14.2 shows the magnitude of reflection coefficient of a dielectric slab with a thickness of 1 cm and relative permittivity of 4 due to obliquely incident TE plane waves. In this example the figure displays data up to 20 GHz for incident angles range from 0 to 90°. Here the figure is constructed using $N_f \times N_a$ data points where $N_f = 200$ is the number of frequencies and $N_a = 90$ is the number of incident angles for which the reflection coefficients are calculated.

The sine-cosine method can be used to perform calculations at a single frequency and a single angle, therefore, to obtain an angle-frequency plane figure such as the one in Figure 14.2 it is needed to run the PBC FDTD program $N_f \times N_a$ times, which makes it inefficient. For instance, Figure 14.2 shows a data point calculated by the sine-cosine method for 12 GHz and 60° using a single run.

On the other hand, for instance, the split-field method preserves the wideband capability and one can obtain wideband results for a wide frequency band in a single program run for one incident angle. Therefore, to obtain an angle-frequency plane figure one needs to run the split-field PBC FDTD program N_a times. For instance, Figure 14.2 shows the line on which data points, calculated by a single run of split-field method when the angle of incidence is 30°.

The split-field method and multispatial grid method have the wideband capability; however, there are two main limitations with these methods. First, the transformed equations have additional terms that require special handling such as splitting the field or using a multigrid algorithm to implement the FDTD, which increases the complexity of the problem. Second, as the angle of incidence increases from normal incidence (0°) to grazing incidence (90°), the stability factor needs to be reduced, so the FDTD time step decreases significantly [64]. As a result, smaller time steps are needed for oblique incidence to generate stable results, which increases the computational time for such cases [71].

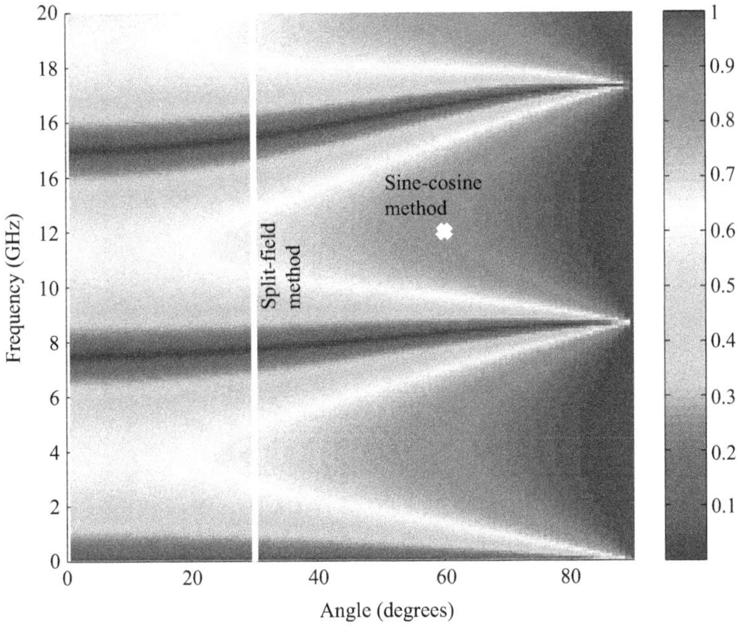

Figure 14.2 Magnitude of reflection coefficient in the angle-frequency plane.

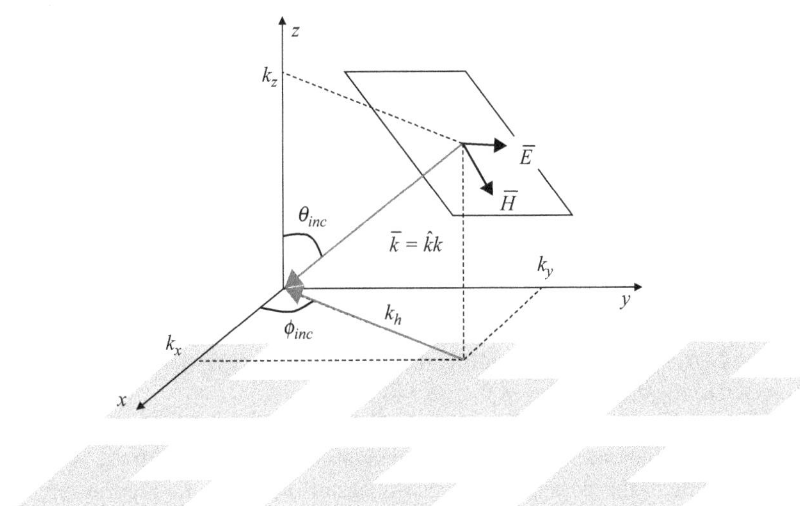

Figure 14.3 Propagation vector of an incident plane wave and its horizontal component.

While one way to display reflection and transmission coefficient distributions is to plot them on angle-frequency planes, an alternative way, which essentially shows the same data, is to plot them on horizontal wavenumber-frequency planes. Figure 14.3 illustrates the propagation vector, $\hat{k}$, of an incident plane wave. One can imagine an infinitely periodic structure in the x and y directions with z direction as the normal to the interface between the

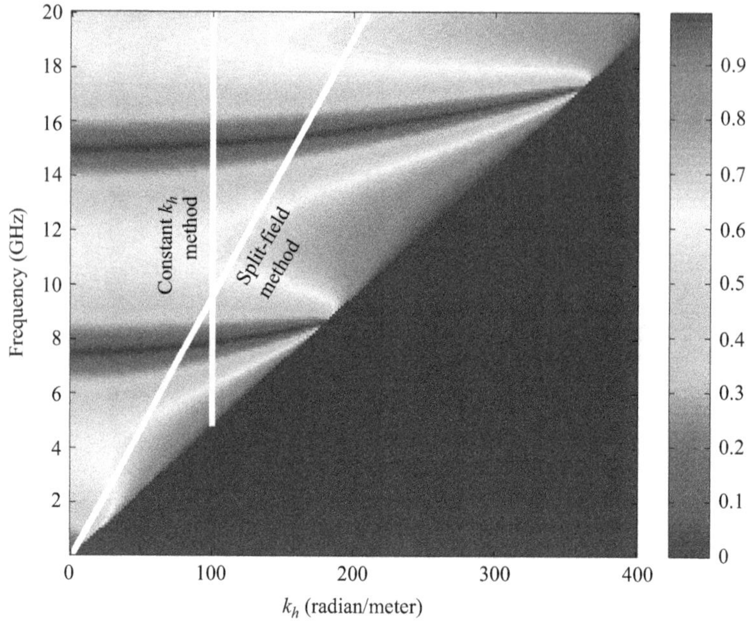

Figure 14.4 Magnitude of reflection coefficient in the horizontal wavenumber-frequency plane.

periodic structure and free space. Also shown in the figure is the horizontal component of the propagation vector, denoted with length k_h, such that $k_h = \sqrt{k_x^2 + k_y^2}$. The relation between the wavenumber, k, and its horizontal component, k_h, is fixed through the angle θ_{inc} such that

$$k_h = k \sin\theta_{inc} = \omega\sqrt{\mu_0\varepsilon_0}\sin\theta_{inc} = \frac{\omega}{c}\sin\theta_{inc}. \tag{14.1}$$

where ω is the angular frequency of the wave propagating in free space. Therefore, one can use this fixed relationship between θ_{inc} and k_h to display reflection and transmission coefficient distributions on horizontal wavenumber-frequency planes instead of the angle-frequency planes. For instance, the data in Figure 14.2 is displayed on a horizontal wavenumber-frequency plane in Figure 14.4. The set of data points calculated using the split-field method in Figure 14.2 is also projected to the data display in Figure 14.4.

It is possible to imagine many incident plane waves all with the same horizontal wavenumber. Figure 14.5 illustrates a number of propagation vectors all with the same horizontal wavenumber. Although the horizontal wavenumbers of these vectors are the same, but their wavenumbers, angles of incidence, as well as frequencies are different. In the constant horizontal wavenumber PBC algorithm an FDTD simulation is performed by setting a constant horizontal wavenumber instead of a specific angle of incidence. Therefore, it yields results supporting a number of angles each with a different frequency in a single PBC FDTD run. For instance, Figure 14.4 shows the data points on a vertical line that can be obtained with a single run of this method for the constant horizontal wavenumber of 100 radian/meter. To obtain the data distribution on the entire horizontal wavenumber-frequency plane, the simulation needs to be repeated for a number of constant horizontal wavenumbers, N_{chw}.

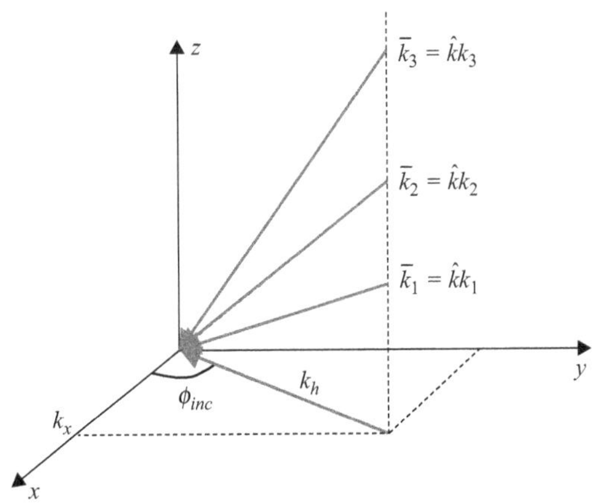

Figure 14.5 Propagation vectors with the same horizontal wavenumber.

The idea of using a constant wavenumber in FDTD was originated from guided wave analysis and eigenvalue problems in [72], and it was extended to the plane wave scattering problems in [73–75]. The approach offers many advantages, such as implementation simplicity, stability condition, and numerical errors similar to the conventional FDTD, computational efficiency near the grazing incident angles, and the wide-band capability [71]. The formulation of the constant horizontal wavenumber method and its MATLAB implementation are presented in the following sections.

14.2 Constant horizontal wavenumber method

Consider the example shown in Figure 14.6, where a unit cell of a periodic structure is shown. The periodic structure is a dielectric slab and a rectangular PEC patch is placed on top of the slab in this case. A time harmonic plane wave with angular frequency ω is incident on the periodic structure with oblique incidence. The direction of incoming wave can be denoted by θ_{inc} and ϕ_{inc}. The unit cell is terminated by CPML boundaries at the top and bottom sides. The other sides will be treated as periodic boundaries. Also shown in the figure below, the upper CPML region is a plane, referred to as the source plane, where incident field is injected into the computational space.

Figure 14.7 shows the field components on an xy plane-cut of the grid of this problem space. The size of the problem space is $P_x = N_x\Delta x$ in the x direction and $P_y = N_y\Delta y$ in the y direction. At steady state, a field component on the right boundary of the grid is a time-delayed equivalent of a field on the left boundary. For instance, one can write

$$E_y(x = P_x, y, z, t) = E_y\left(x = 0, y, z, t - \frac{P_x}{c}\sin\theta_{inc}\right), \tag{14.2}$$

and

$$H_z\left(x = P_x - \frac{\Delta x}{2}, y, z, t\right) = H_z\left(x = -\frac{\Delta x}{2}, y, z, t - \frac{P_x}{c}\sin\theta_{inc}\right). \tag{14.3}$$

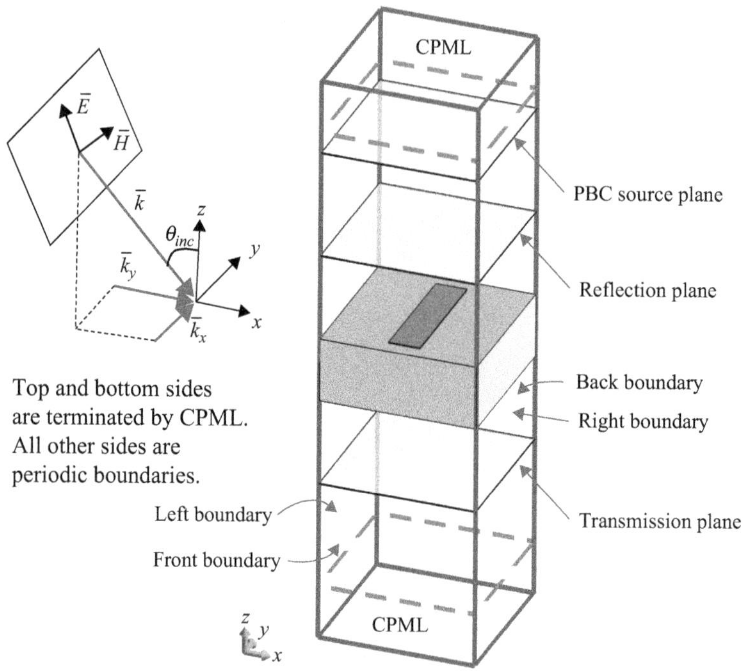

Figure 14.6 An incident plane wave and a unit cell.

Similarly, a field component on the back boundary of the grid is a time-delayed equivalent of a field on the front boundary such that

$$E_x(x, y = P_y, z, t) = E_x\left(x, y = 0, z, t - \frac{P_y}{c}\sin\theta_{inc}\right), \tag{14.4}$$

and

$$H_z\left(x, y = P_y - \frac{\Delta y}{2}, z, t\right) = H_z\left(x, y = -\frac{\Delta y}{2}, z, t - \frac{P_y}{c}\sin\theta_{inc}\right). \tag{14.5}$$

When transformed from time-domain to frequency-domain (14.2) can be written as

$$E_y(x = P_x, y, z, t) = E_y(x = 0, y, z, t)e^{-j\frac{\omega}{c}\sin\theta_{inc}P_x}, \tag{14.6}$$

which can be expressed using (14.1) as

$$E_y(x = P_x, y, z) = E_y(x = 0, y, z)e^{-jk\sin\theta_{inc}P_x} = E_y(x = 0, y, z)e^{-jk_xP_x}. \tag{14.7}$$

This is how the fields on a periodic structure can be expressed according to the Floquet theory, and it implies that any field component in the problem space of a periodic structure is a phase shifted equivalent of another field component that is distant by P_x in the x direction by a phase shift of k_xP_x or $-k_xP_x$. Similarly, any field component is a phase shifted equivalent of another field component that is distant by P_y in the y direction by a phase shift

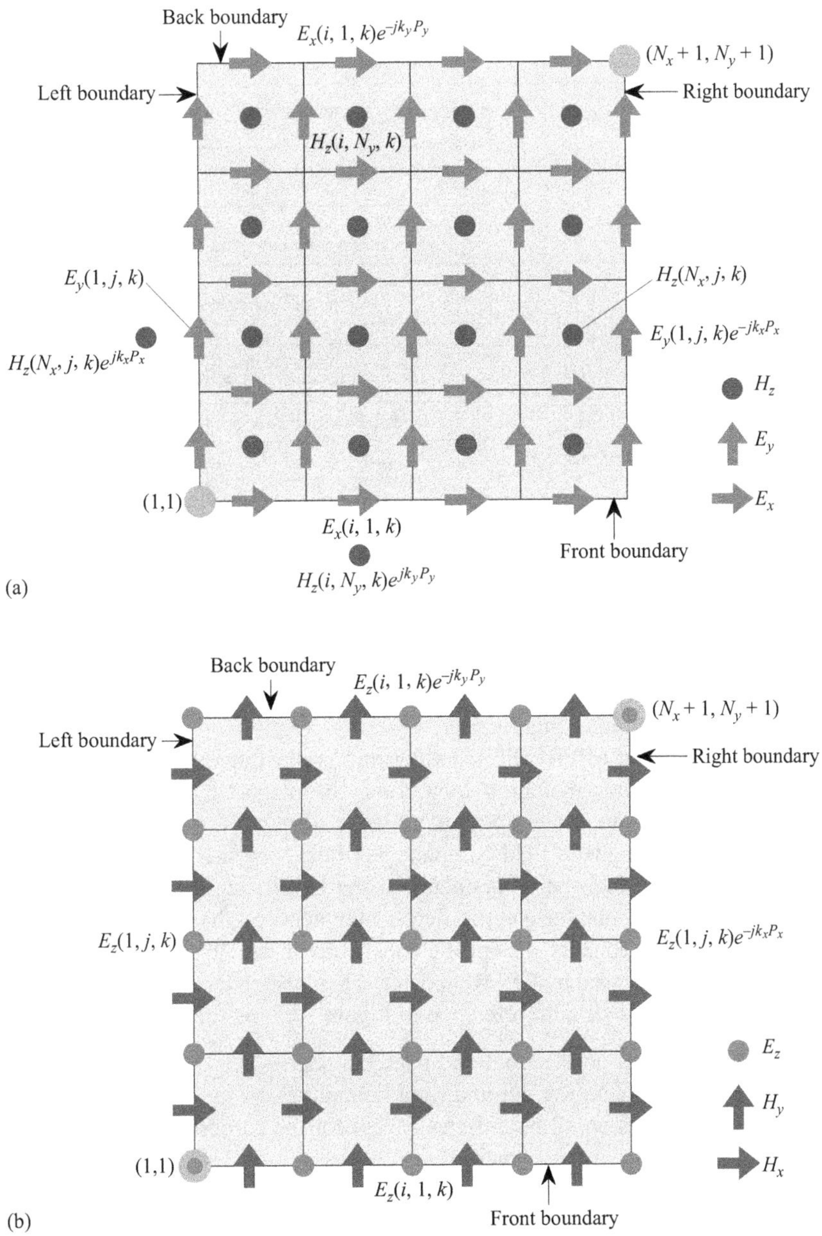

Figure 14.7 Field components on an xy plane-cut of a unit cell in a three-dimensional domain: (a) E_x, E_y, and H_z components and (b) H_x, H_y, and E_z components.

of k_yP_y or $-k_yP_y$. We can make use of this relationship between the field components and develop a PBC algorithm: while updating a field component on the boundary of a unit cell during an iteration of the FDTD time-marching loop, if we need to use the value of a neighboring field component for which the value is not readily available we can use a phase

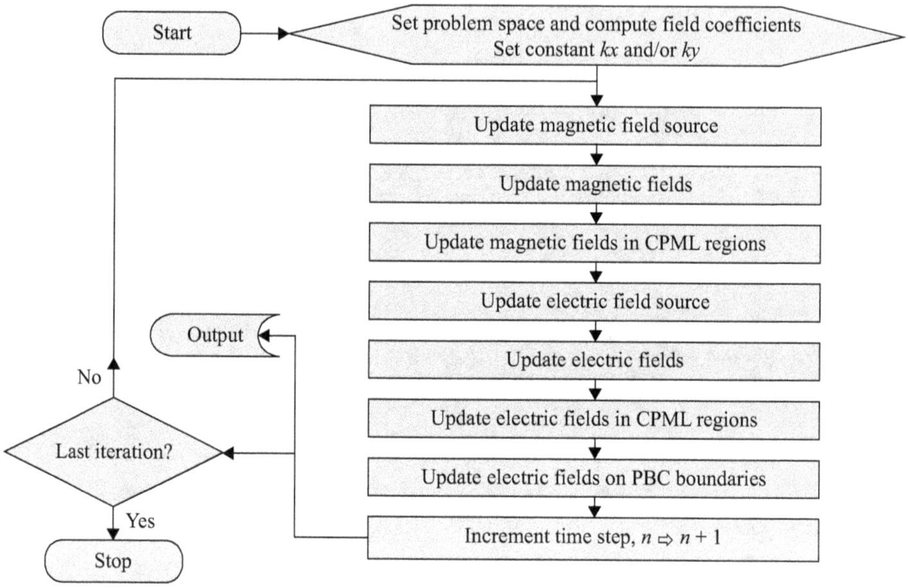

Figure 14.8 Flow chart of PBC FDTD algorithm.

shifted equivalent of another field component for which the value is known. This procedure will be elaborated in details in this section.

The general flow chart of PBC FDTD algorithm is illustrated in Figure 14.8. During an iteration of the FDTD time-marching loop, first, the magnetic field components on the source plane are updated to excite an incident wave, details of which are discussed in the next section. Then, all magnetic field components in the computational domain are updated for the new time step. Also, the components in the CPML regions are updated using the CPML equations. Afterwards, the electric field components on the source plane are updated, and all electric field components, except the ones lying on the PBC boundaries, are updated for the new time step followed by CPML updates. Then, the electric field components on the PBC boundaries are updated, with reference to Figure 14.7, as follows:

1. Update E_x components on the front and back boundaries.
2. Update E_y components on the left and right boundaries.
3. Update E_z components on all boundaries except for the corners.
4. Update E_z components at the corners.

Figure 14.7 shows the field components of a unit cell, the shaded regions, as well as the neighboring magnetic field components outside the unit cell. In order to calculate an electric field component on the front boundary, that is indexed as $E_x^{n+1}(i,1,k)$, one needs the value of the magnetic field component below it, that could be indexed as $H_z^{n+\frac{1}{2}}(i,0,k)$. While $H_z^{n+\frac{1}{2}}(i,0,k)$ is not in the computational space of the unit cell in consideration and its value is not known, a phase shifted equivalent of it, $H_z^{n+\frac{1}{2}}(i, N_y, k)$, can be used instead following the Floquet theory as

$$H_z^{n+\frac{1}{2}}(i,0,k) = H_z^{n+\frac{1}{2}}(i, N_y, k)e^{jk_yP_y}. \tag{14.8}$$

Then the electric field updating equation can be written for $E_x^{n+1}(i, 1, k)$ as

$$E_x^{n+1}(i, 1, k) = C_{exe}(i, 1, k) \times E_x^n(i, 1, k) + C_{exhz}(i, 1, k) \times \left[H_z^{n+\frac{1}{2}}(i, 1, k) - H_z^{n+\frac{1}{2}}(i, N_y, k)e^{jk_yP_y}\right] + C_{exhy}(i, 1, k) \times \left[H_y^{n+\frac{1}{2}}(i, 1, k) - H_y^{n+\frac{1}{2}}(i, 1, k-1)\right]. \quad (14.9)$$

Similarly, to update an electric field component on the back boundary, $E_x^{n+1}(i, N_y+1, k)$, the value of $H_z^{n+\frac{1}{2}}(i, N_y+1, k)$ is needed, which is not available. One can notice that $E_x^{n+1}(i, N_y+1, k)$ is a phase shifted equivalent of $E_x^{n+1}(i, 1, k)$, which has already been calculated using (14.9). Therefore, as a shortcut, $E_x^{n+1}(i, N_y+1, k)$ can be updated as

$$E_x^{n+1}(i, N_y+1, k) = E_x^{n+1}(i, 1, k)e^{-jk_yP_y}. \quad (14.10)$$

The E_y components on the left boundary can be updated using the periodicity in the x direction as

$$E_y^{n+1}(1, j, k) = C_{eye}(1, j, k) \times E_y^n(1, j, k) + C_{eyhx}(1, j, k) \times \left[H_x^{n+\frac{1}{2}}(1, j, k) - H_x^{n+\frac{1}{2}}(1, j, k-1)\right] + C_{eyhz}(1, j, k) \times \left[H_z^{n+\frac{1}{2}}(1, j, k) - H_z^{n+\frac{1}{2}}(N_x, j, k) \times e^{jk_xP_x}\right]. \quad (14.11)$$

Then, $E_y^{n+1}(1, j, k)$ can be used to calculate the electric field component on the right boundary as

$$E_y^{n+1}(N_x+1, j, k) = E_y^{n+1}(1, j, k)e^{-jk_yP_y}. \quad (14.12)$$

The same procedure can be repeated for the E_z components as well. The field components on the left boundary, except the corners, can be updated as

$$E_z^{n+1}(1, j, k) = C_{eze}(1, j, k) \times E_z^n(1, j, k) + C_{ezhy}(1, j, k) \times \left[H_y^{n+\frac{1}{2}}(1, j, k) - H_y^{n+\frac{1}{2}}(N_x, j, k) \times e^{jk_xP_x}\right] + C_{ezhx}(1, j, k) \times \left[H_x^{n+\frac{1}{2}}(1, j, k) - H_x^{n+\frac{1}{2}}(1, j-1, k)\right], \quad (14.13)$$

while the components on the right boundary, except the corners, can be updated as

$$E_z^{n+1}(N_x+1, j, k) = E_z^{n+1}(1, j, k) \times e^{-jk_xP_x}. \quad (14.14)$$

For the components on the front boundary, except the corners, one can write

$$E_z^{n+1}(i, 1, k) = C_{eze}(i, 1, k) \times E_z^n(i, 1, k) + C_{ezhy}(i, 1, k) \times \left[H_y^{n+\frac{1}{2}}(i, 1, k) - H_y^{n+\frac{1}{2}}(i-1, 1, k)\right] + C_{ezhx}(i, 1, k) \times \left[H_x^{n+\frac{1}{2}}(i, 1, k) - H_x^{n+\frac{1}{2}}(i, N_y, k) \times e^{jk_yP_y}\right] \quad (14.15)$$

while for the components on the back boundary, except the corners,

$$E_z^{n+1}(i, N_y + 1, k) = E_z^{n+1}(i, 1, k) \times e^{-jk_yP_y}. \tag{14.16}$$

The component of E_z at the front left corner, $E_z^{n+1}(1, 1, k)$, needs the values of $H_x^{n+\frac{1}{2}}(1, 0, k)$ and $H_y^{n+\frac{1}{2}}(0, 1, k)$ that are unknown. Therefore, instead, following the Floquet theory, one can write

$$\begin{aligned} E_z^{n+1}(1, 1, k) &= C_{eze}(1, 1, k) \times E_z^n(1, 1, k) \\ &+ C_{ezhy}(1, 1, k) \times \left[H_y^{n+\frac{1}{2}}(1, 1, k) - H_y^{n+\frac{1}{2}}(N_x, 1, k) \times e^{jk_xP_x}\right] \\ &+ C_{ezhx}(1, 1, k) \times \left[H_x^{n+\frac{1}{2}}(1, 1, k) - H_x^{n+\frac{1}{2}}(1, N_y, k) \times e^{jk_yP_y}\right]. \end{aligned} \tag{14.17}$$

Once the component of E_z at the front left corner is known, the components at the other corners can be calculated as

$$E_z^{n+1}(N_x + 1, 1, k) = E_z^{n+1}(1, 1, k) \times e^{-jk_xP_x}, \tag{14.18}$$

$$E_z^{n+1}(1, N_y + 1, k) = E_z^{n+1}(1, 1, k) \times e^{-jk_yP_y}, \tag{14.19}$$

$$E_z^{n+1}(N_x + 1, N_y + 1, k) = E_z^{n+1}(1, 1, k) \times e^{-jk_yP_y} \times e^{-jk_xP_x}. \tag{14.20}$$

The above PBC update equations are derived for obliquely incident plane waves. Thus, in the case of normal incidence, the horizontal wavenumbers are $k_x = k_y = 0$ radian/meter, therefore all exponential terms in the field component update equations become equal to one.

14.3 Source excitation

In Section 14.1 we discussed that many incident waves with different frequencies can have the same horizontal wavenumber, while each of these waves will have a different angle of incidence. Therefore, one can imagine that if a wave with a time waveform which has a wide frequency band is incident on a unit cell, all these frequency components of the wave can have the same horizontal wavenumber, i.e., k_x and k_y, while each component will have a different angle of incidence that satisfies (14.1). If we excite a computational domain with such a waveform and apply the presented PBC conditions, we can obtain the results for the frequencies in the spectrum of the incident wave at their associated angles of incidence in one FDTD simulation. The results of such a simulation will lie on a vertical line on the horizontal wavenumber-frequency plane as shown in Figure 14.4. In Figure 14.4, a vertical data line starts from a minimum frequency: At grazing angle, where $\theta = 90°$, (14.1) becomes

$$k_h = k = \frac{\omega}{c}, \tag{14.21}$$

which implies that ω is at its minimum value, $\omega_{\min} = ck_h$, therefore the minimum frequency that can be calculated is

$$f_{\min} = \frac{ck_h}{2\pi}. \tag{14.22}$$

Consequently, a wideband waveform that is constructed to excite the FDTD computational space should have a minimum frequency of $f_{\min}$ in its spectrum. This also avoids the problem of horizontal resonance in which the fields do not decay to zero over time. Therefore, for instance, a cosine-modulated Gaussian waveform with a proper bandwidth, BW, can be considered as a source waveform, which has a modulation frequency of

$$f_c = \frac{k_h c}{2\pi} + \frac{BW}{2}. \tag{14.23}$$

Some considerations that need to be taken into account while determining the bandwidth and the modulation frequency are discussed in Section 14.5.2.

The incident waveform can be injected into a computational domain at the location of a source plane, shown in Figure 14.6. An easy way for this implementation is to calculate the tangential field components of the incident wave on the source plane and add them to the field components on the source plane at every time step. As a result the source plane will radiate the incident field in the lower direction toward the periodic structure. The source plane will radiate a mirror image of the incident field in the upper direction as well, however, since the upper boundary is terminated by CPML, the mirror image field will be absorbed by CPML and it will not interfere with the fields below the source plane.

This incident waveform can be constructed to simulate TE, TM, or TEM mode. It is straightforward to excite the TEM mode: the tangential electric field components on the source plane, i.e., $E_{inc,x}$ and $E_{inc,y}$, are uniform over the source plane, therefore all components of incident $E_{inc,x}$ will be the same as well as all components of incident $E_{inc,y}$ will have the same value, which will be added to the respective field components on the source plane.

For the TE mode with the oblique incidence, and the incident field satisfies the periodicity condition such that

$$E_{inc,x}(i, N_y + 1, ks) = E_{inc,x}(i, 1, ks) \times e^{-jk_y P_y}, \tag{14.24a}$$

$$E_{inc,y}(N_x + 1, j, ks) = E_{inc,x}(1, j, ks) \times e^{-jk_x P_x}, \tag{14.24b}$$

where ks is the k index of the source plane. To excite the incident field, we can choose a point, for instance, node (1,1) at the lower left corner shown in Figure 14.7, as a reference and apply the phase shift to all $E_{inc,x}$ and $E_{inc,y}$ components on the source plane with reference to this point. For instance, if the value of x component of the incident field is E_{0x}^n at time step n then,

$$E_{inc,x}^n(i, j, ks) = E_{0x}^n \times e^{-jk_x(i-0.5)\Delta x} \times e^{-jk_y(j-1)\Delta y}. \tag{14.25}$$

Similarly, if the value of y component of the incident field is E_{0y}^n at time step n then,

$$E_{inc,y}^n(i, j, ks) = E_{0y}^n \times e^{-jk_x(i-1)\Delta x} \times e^{-jk_y(j-0.5)\Delta y}. \tag{14.26}$$

For the TM mode, we can add incident magnetic field component to the respective tangential magnetic field components on the source plane. In this case, the source plane can be chosen to be on the magnetic field grid. To excite the incident field, again choose the point at node (1,1) as a reference and apply the phase shift to all $H_{inc,x}$ and $H_{inc,y}$ components on the source plane with reference to this point. For instance, if the value of x component of the incident field is $H_{0x}^{n+\frac{1}{2}}$ at time step $n+\frac{1}{2}$ then,

$$H_{inc,x}^{n+\frac{1}{2}}(i,j,ks) = H_{0x}^{n+\frac{1}{2}} \times e^{-jk_x(i-1)\Delta x} \times e^{-jk_y(j-0.5)\Delta y}. \tag{14.27}$$

Similarly, if the value of y component of the incident field is $H_{0y}^{n+\frac{1}{2}}$ at time step $n+\frac{1}{2}$ then,

$$H_{inc,y}^{n+\frac{1}{2}}(i,j,ks) = H_{0y}^{n+\frac{1}{2}} \times e^{-jk_x(i-0.5)\Delta x} \times e^{-jk_y(j-1)\Delta y}. \tag{14.28}$$

It should be noted that the presented constant horizontal wavenumber method integrates complex exponential phase terms into a time domain simulation, therefore, unlike a conventional FDTD algorithm, the fields values are all complex rather than real. Hence, while the simulation results are not meaningful in time domain, the results are meaningful on a wide frequency-band in frequency domain.

14.4 Reflection and transmission coefficients

The reflection and/or transmission properties of periodic structures due to obliquely incident plane waves are sought in many applications. Figure 14.6 shows two planes, referred to as reflection and transmission planes, on which the fields can be captured and they can be used to calculate reflection and transmission coefficients. In this section we will discuss the equations to be used for the calculation of these coefficients.

It is possible to calculate co-polarized reflection, cross-polarized reflection, co-polarized transmission, as well as the cross-polarized transmission coefficients. Equations for these coefficients depend on the mode of excitation, i.e., TE, TM, or TEM modes.

Reflection coefficient of a periodic structure is usually presented in the frequency domain. The value of captured field components at a single point is sufficient to calculate the reflection coefficient, while the field components can be captured anywhere on the reflection plane illustrated in Figure 14.6. In the MATLAB implementation of the PBC FDTD, that will be discussed in the subsequent sections, an average of field components on the reflection plane is captured and used for reflection coefficient calculations. It should be noted that the field components on the reflection plane as well are subject to the phase shift discussed above. Before averaging, this phase shifts need to be adjusted so that all field components are at the same phase. Similar to the source excitation, we can choose the point at node (1,1) as a reference and update the phase shifts with reference to this point. The field components used for averaging after phase adjustments are represented as

$$E_{rp,x}^{n}(i,j,kr) = E_{x}^{n}(i,j,kr) \times e^{jk_x(i-0.5)\Delta x} \times e^{jk_y(j-1)\Delta y}, \tag{14.29a}$$

$$E_{rp,y}^{n}(i,j,kr) = E_{y}^{n}(i,j,kr) \times e^{jk_x(i-1)\Delta x} \times e^{jk_y(j-0.5)\Delta y}, \tag{14.29b}$$

$$E_{rp,z}^{n}(i,j,kr) = 0.5\left(E_{z}^{n}(i,j,kr-1) + E_{z}^{n}(i,j,kr)\right) \times e^{jk_x(i-1)\Delta x} \times e^{jk_y(j-1)\Delta y}, \tag{14.29c}$$

$$H_{rp,x}^{n+\frac{1}{2}}(i,j,kr) = 0.5\left(H_{x}^{n+\frac{1}{2}}(i,j,kr) + H_{x}^{n+\frac{1}{2}}(i,j,kr-1)\right) \times e^{jk_x(i-1)\Delta x} \times e^{jk_y(j-0.5)\Delta y}, \tag{14.29d}$$

$$H_{rp,y}^{n+\frac{1}{2}}(i,j,kr) = 0.5\left(H_y^{n+\frac{1}{2}}(i,j,kr) + H_y^{n+\frac{1}{2}}(i,j,kr-1)\right) \times e^{jk_x(i-0.5)\Delta x} \times e^{jk_y(j-1)\Delta y}, \quad (14.29e)$$

$$H_{rp,z}^{n+\frac{1}{2}}(i,j,kr) = H_z^{n+\frac{1}{2}}(i,j,kr) \times e^{jk_x(i-0.5)\Delta x} \times e^{jk_y(j-0.5)\Delta y}, \quad (14.29f)$$

where the subscript "*rp*" denotes the captured fields on the reflection plane. It should be noted that the reflection plane is considered to be an xy plane-cut of a unit cell that includes E_x, E_y, and H_z components as illustrated in Figure 14.7(a) for the above equations, where the plane-cut is indicated with the index kr. Therefore, E_x, E_y, and H_z components with kr index are used in (14.29a), (14.29b), and (14.29f), respectively, to calculate the respective phase adjusted fields. The components of E_z, H_x, and H_y do not lie on the same plane cut as the reflection plane. To obtain equivalent values for these field components on the reflection plane, the components above (with index kr) and below (with index $kr - 1$) the reflection plane are averaged and used in (14.29c), (14.29d), and (14.29e), respectively.

Similarly, the fields on the transmission plane can be captured and phase adjusted as well. After phase reversals, the field components are averaged and stored versus time in terms of the time step. After the FDTD time-marching loop is completed, these transient fields are transformed to frequency domain by discrete Fourier transform (DFT) and then the reflection and transmission coefficient calculations are performed.

One should notice that the fields that are captured on the reflection plane are total fields, such that

$$E_{rp} = E_{inc} + E_{ref}, \quad (14.30)$$

while the incident and reflected fields are needed for reflection coefficient calculations. Moreover, in particular, the co- and cross-polarized components of incident and reflected fields are needed for co- and cross-polarized reflection coefficient calculations.

14.4.1 TE mode reflection and transmission coefficients

Figure 14.9 illustrates the incident field in the TE mode. For the TE mode, the co-polarized total fields can be determined using the x and y components of the captured fields as

$$E_{rp,co} = E_{rp,y}\frac{k_x}{k_h} - E_{rp,x}\frac{k_y}{k_h}, \quad (14.31a)$$

$$H_{rp,co} = H_{rp,x}\frac{k_x}{k_h} + H_{rp,y}\frac{k_y}{k_h}, \quad (14.31b)$$

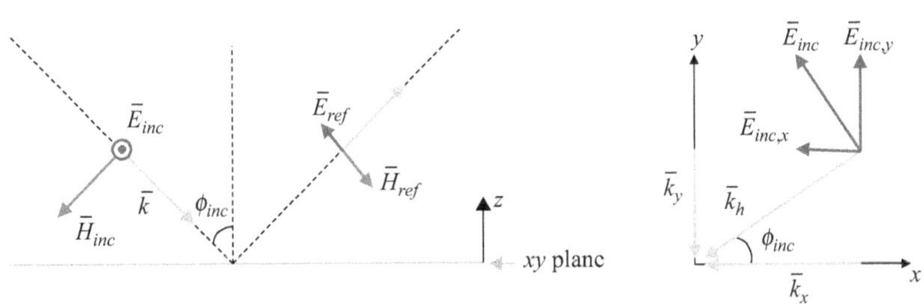

Figure 14.9 Illustration of incident fields in the TE case.

while the cross-polarized total electric field is the same as the cross-polarized reflected electric field, therefore

$$E_{ref,cr} = E_{rp,cr} = E_{rp,x}\frac{k_x}{k_h} + E_{rp,y}\frac{k_y}{k_h}. \tag{14.32}$$

The next step is to construct incident electric field such that

$$E_{inc} = \frac{1}{2}\left(E_{rp,co} + \eta_0 H_{rp,co}\frac{k}{k_z}\right), \tag{14.33}$$

where $k_z = \sqrt{k^2 - k_h^2}$. Once the incident electric field is computed, it can be subtracted from the co-polarized total electric to find the co-polarized reflected electric field as

$$E_{ref,co} = E_{rp,co} - E_{inc}, \tag{14.34}$$

which can be used to calculate the co-polarized reflection coefficient as

$$\Gamma_{co} = \frac{E_{ref,co}}{E_{inc}}. \tag{14.35}$$

Then, the cross-polarized reflected electric field can be used to determine the cross-polarized reflection coefficient as

$$\Gamma_{cr} = \frac{E_{ref,cr}}{E_{inc}}. \tag{14.36}$$

The fields captured on the transmission plane are the co- and cross-polarized components of the transmitted electric field which are similar to (14.31) and (14.32) such that

$$E_{tra,co} = E_{tra,y}\frac{k_x}{k_h} - E_{tra,x}\frac{k_y}{k_h}, \tag{14.37a}$$

$$E_{tra,cr} = E_{tra,x}\frac{k_x}{k_h} + E_{tra,y}\frac{k_y}{k_h}. \tag{14.37b}$$

Then, $E_{tra,co}$ and $E_{tra,cr}$ can be used together with E_{inc} to calculate the co- and cross-polarized transmission coefficients, respectively. However, one should notice that E_{inc} is calculated on the reflection plane while $E_{tra,co}$ and $E_{tra,cr}$ are obtained on the transmission plane. To calculate the transmission coefficients with the correct phases, E_{inc} should be evaluated on the transmission plane. If the distance between the refection and transmission planes is d_{rt}, the incident field on the transmission plane can be obtained by shifting the phase of previously defined E_{inc} by $k_z d_{rt}$, and hence co- and cross-polarized transmission coefficients can be calculated as

$$T_{co} = \frac{E_{tra,co}}{E_{inc}e^{jk_z d_{rt}}}, \tag{14.38}$$

$$T_{cr} = \frac{E_{tra,cr}}{E_{inc}e^{jk_z d_{rt}}}. \tag{14.39}$$

14.4.2 TM mode reflection and transmission coefficients

Figure 14.10 shows the incident field components for the TM mode. For the TM mode, the co-polarized total fields can be determined using the x and y components of the captured fields as

$$E_{rp,co} = -E_{rp,x}\frac{k_x}{k_h} - E_{rp,y}\frac{k_y}{k_h}, \tag{14.40a}$$

$$H_{rp,co} = H_{rp,y}\frac{k_x}{k_h} - H_{rp,x}\frac{k_y}{k_h}, \tag{14.40b}$$

while the cross-polarized total magnetic field is the same as the cross-polarized reflected magnetic field, therefore

$$H_{ref,cr} = H_{rp,cr} = H_{rp,x}\frac{k_x}{k_h} + H_{rp,y}\frac{k_y}{k_h}. \tag{14.41}$$

The next step is to construct the incident magnetic field such that

$$H_{inc} = \frac{1}{2}\left(H_{rp,co} + E_{rp,co}\frac{k}{\eta_0 k_z}\right), \tag{14.42}$$

Once the incident magnetic field is known, it can be subtracted from the co-polarized total magnetic field to find the co-polarized reflected magnetic field as

$$H_{ref,co} = H_{rp,co} - H_{inc}, \tag{14.43}$$

which can be used to calculate the co-polarized reflection coefficient as

$$\Gamma_{co} = \frac{H_{ref,co}}{H_{inc}}. \tag{14.44}$$

Then, the cross-polarized reflected magnetic field can be used to determine the cross-polarized reflection coefficient as

$$\Gamma_{cr} = \frac{H_{ref,cr}}{H_{inc}}. \tag{14.45}$$

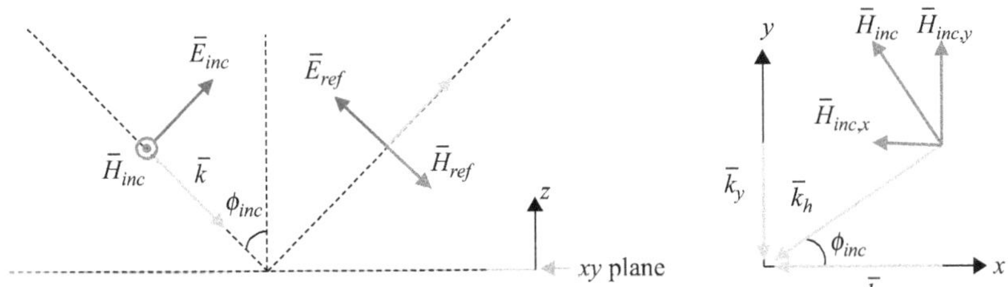

Figure 14.10 Illustration of incident fields in the TM case.

The fields captured on the transmission plane are transmitted fields, and the co- and cross-polarized components of the transmitted magnetic field can be found as

$$H_{tra,co} = H_{tra,y}\frac{k_x}{k_h} - H_{tra,x}\frac{k_y}{k_h}, \tag{14.46a}$$

$$H_{tra,cr} = H_{tra,x}\frac{k_x}{k_h} + H_{tra,y}\frac{k_y}{k_h}. \tag{14.46b}$$

Then, the co- and cross-polarized transmission coefficients can be written as

$$T_{co} = \frac{H_{tra,co}}{H_{inc}e^{jk_z d_{rt}}}, \tag{14.47}$$

$$T_{cr} = \frac{H_{tra,cr}}{H_{inc}e^{jk_z d_{rt}}}. \tag{14.48}$$

14.4.3 TEM mode reflection and transmission coefficients

Figure 14.11 illustrates the incident field in the TEM mode. For the TEM mode, the co-polarized total fields can be determined using the x and y components of the captured fields as

$$E_{rp,co} = E_{rp,x}\cos\phi + E_{rp,y}\sin\phi, \tag{14.49a}$$

$$H_{rp,co} = H_{rp,x}\sin\phi - H_{rp,y}\cos\phi, \tag{14.49b}$$

while the cross-polarized total electric field is the same as the cross-polarized reflected electric field, therefore

$$E_{ref,cr} = E_{rp,x}\sin\phi - E_{rp,y}\cos\phi. \tag{14.50}$$

Here, ϕ is the angle between the incident electric field vector and the x axis, which can be expressed as

$$\phi = \tan^{-1}\left(\frac{E_{inc,y}}{E_{inc,x}}\right). \tag{14.51}$$

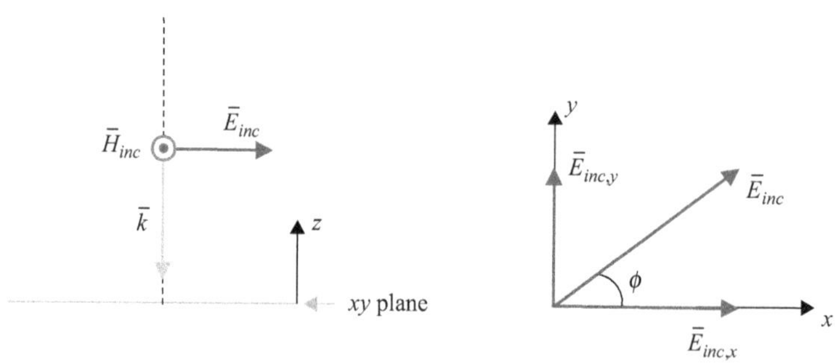

Figure 14.11 Illustration of incident fields in the TEM case.

The next step is the extraction of incident field as

$$E_{inc} = \frac{1}{2}\left(E_{rp,co} + \eta_0 H_{rp,co}\right). \tag{14.52}$$

Once the incident electric field is known, it can be subtracted from the co-polarized total electric field to find the co-polarized reflected electric field as

$$E_{ref,co} = E_{rp,co} - E_{inc}, \tag{14.53}$$

which can be used to calculate the co-polarized reflection coefficient as

$$\Gamma_{co} = \frac{E_{ref,co}}{E_{inc}}. \tag{14.54}$$

Then, the cross-polarized reflected electric field can be used to find the cross-polarized reflection coefficient as

$$\Gamma_{cr} = \frac{E_{ref,cr}}{E_{inc}}. \tag{14.55}$$

The fields captured on the transmission plane are transmitted fields, and the co- and cross-polarized components of the transmitted electric field can be found as

$$E_{tra,co} = E_{tra,x}\cos\phi + E_{tra,y}\sin\phi, \tag{14.56a}$$

$$E_{tra,cr} = E_{tra,x}\sin\phi - E_{tra,y}\cos\phi. \tag{14.56b}$$

Then, the co- and cross-polarized transmission coefficients can be written as

$$T_{co} = \frac{E_{tra,co}}{E_{inc}e^{jk_z d_{rt}}}, \tag{14.57}$$

$$T_{cr} = \frac{E_{tra,cr}}{E_{inc}e^{jk_z d_{rt}}}. \tag{14.58}$$

14.5 MATLAB® implementation of PBC FDTD algorithm

The presented constant horizontal wavenumber PBC algorithm is implemented into the base MATLAB FDTD code. The details of implementation are presented in the following subsections.

14.5.1 Definition of a PBC simulation

The parameters of a PBC simulation are specified in the subroutine ***define_problem_space_parameters***, as shown in Listing 14.1. Here, a parameter named as **periodic_boundary** is defined to store the several fields to specify various simulation parameters. First, a field mode is used to specify whether the mode of excitation will be TE, TM, or TEM. Depending on the mode, the relevant fields of **periodic_boundary** are further used to

Listing 14.1 define_problem_space_parameters.m

```
% ==<periodic boundary simulation parameters>========
disp('Defining periodic boundary simulation');

periodic_boundary.mode = 'TE'; % TE, TM, or TEM
periodic_boundary.E_phi = 1;
periodic_boundary.H_phi = 0;
periodic_boundary.E_x = 0;
periodic_boundary.E_y = 0;
periodic_boundary.kx = 100;
periodic_boundary.ky = 0;
periodic_boundary.source_z = 0.025;
periodic_boundary.reflection_z = 0.02;
periodic_boundary.transmission_z = -0.005;

% ==<boundary conditions>========
% Here we define the boundary conditions parameters
% 'pec' : perfect electric conductor
% 'cpml' : convolutional PML
% 'pbc' : Periodic Boundary Condition
% if cpml_number_of_cells is less than zero
% CPML extends inside of the domain rather than outwards

boundary.type_xn = 'pbc';
boundary.air_buffer_number_of_cells_xn = 0;
boundary.cpml_number_of_cells_xn = 8;

boundary.type_xp = 'pbc';
boundary.air_buffer_number_of_cells_xp = 0;
boundary.cpml_number_of_cells_xp = 8;

boundary.type_yn = 'pbc';
boundary.air_buffer_number_of_cells_yn = 0;
boundary.cpml_number_of_cells_yn = 10;

boundary.type_yp = 'pbc';
boundary.air_buffer_number_of_cells_yp = 0;
boundary.cpml_number_of_cells_yp = 8;

boundary.type_zn = 'cpml';
boundary.air_buffer_number_of_cells_zn = 20;
boundary.cpml_number_of_cells_zn = 8;

boundary.type_zp = 'cpml';
boundary.air_buffer_number_of_cells_zp = 60;
boundary.cpml_number_of_cells_zp = 8;
```

set simulation settings. For instance, if the mode is TEM, then the fields **E_x** and **E_y** are set with the values of the x and y components of the incident plane wave. If the mode is TE, then the field **E_phi** is the magnitude of the incident electric field, while parameters **k_x** and **k_y** are x and y components, k_x and k_y, of the constant horizontal wavenumber respectively. Similarly, if the mode is TM, then the field **H_phi** is the magnitude of the incident magnetic field, while parameters **k_x** and **k_y** are x and y components k_x and k_y. The parameters

source_z, **reflection_z**, and **transmission_z** are the z coordinates of the source, reflection, and transmission planes, respectively.

The presented PBC formulation assumes that four sides of the problems space – namely xn, xp, yn, and yp sides – are periodic boundaries while the other two sides, zn and zp, are CPML. These boundaries as well are specified in the ***define_problem_space_parameters***, as shown in Listing 14.1.

14.5.2 Initialization of PBC

A subroutine called as ***initialize_periodic_boundary_conditions*** is implemented to perform initialization for the PBC algorithm before the FDTD time-matching loop starts. Listing 14.2 shows the details of this subroutine.

First, the k indices of the source, reflection, and transmission planes are calculated and stored in respective parameters. Also, one-dimensional arrays are created for all six field components that will be used to store the transient fields captured on the reflection and transmission planes.

As discussed in the previous section, the field components need to be phase adjusted before they are averaged on the reflection and transmission planes. Since the phase correction is performed at every time step, it may be more practical to calculate and store the phase

Listing 14.2 initialize_periodic_boundary_conditions.m

```
periodic_boundary.source_ks = ...
    1+round((periodic_boundary.source_z - ...
    fdtd_domain.min_z)/fdtd_domain.dz);

if isfield(periodic_boundary,'reflection_z')
    periodic_boundary.calculate_reflection = true;
    periodic_boundary.reflection_ks = ...
    1+round((periodic_boundary.reflection_z -
    fdtd_domain.min_z)/fdtd_domain.dz);
    periodic_boundary.reflection_ex = zeros(1,number_of_time_steps);
    periodic_boundary.reflection_ey = zeros(1,number_of_time_steps);
    periodic_boundary.reflection_ez = zeros(1,number_of_time_steps);
    periodic_boundary.reflection_hx = zeros(1,number_of_time_steps);
    periodic_boundary.reflection_hy = zeros(1,number_of_time_steps);
    periodic_boundary.reflection_hz = zeros(1,number_of_time_steps);
else
        periodic_boundary.calculate_reflection = false;
end

if isfield(periodic_boundary,'transmission_z')
    periodic_boundary.calculate_transmission = true;
    periodic_boundary.transmission_ks = ...
    1+round((periodic_boundary.transmission_z -
    fdtd_domain.min_z)/fdtd_domain.dz);
    periodic_boundary.transmission_ex = zeros(1,number_of_time_steps);
    periodic_boundary.transmission_ey = zeros(1,number_of_time_steps);
    periodic_boundary.transmission_ez = zeros(1,number_of_time_steps);
    periodic_boundary.transmission_hx = zeros(1,number_of_time_steps);
    periodic_boundary.transmission_hy = zeros(1,number_of_time_steps);
    periodic_boundary.transmission_hz = zeros(1,number_of_time_steps);
    periodic_boundary.reflection_transmission_distance = ...
        dz*(periodic_boundary.reflection_ks - ...
        periodic_boundary.transmission_ks);
else
    periodic_boundary.calculate_transmission = false;
end
```

```
if strcmp(periodic_boundary.mode,'TEM')
    periodic_boundary.kx = 0;
    periodic_boundary.ky = 0;
end

kx = periodic_boundary.kx;
ky = periodic_boundary.ky;

pex = zeros(nx,ny);
pey = zeros(nx,ny);
pez = zeros(nx,ny);
phz = zeros(nx,ny);

for mi=1:nx
    for mj=1:ny
        pex(mi,mj) = exp(j*kx*(mi-0.5)*dx)*exp(j*ky*(mj-1)*dy);
        pey(mi,mj) = exp(j*kx*(mi-1)*dx)*exp(j*ky*(mj-0.5)*dy);
        pez(mi,mj) = exp(j*kx*(mi-1)*dx)*exp(j*ky*(mj-1)*dy);
        phz(mi,mj) = exp(j*kx*(mi-0.5)*dx)*exp(j*ky*(mj-0.5)*dy);
    end
end

periodic_boundary.phase_correction_ex = pex;
periodic_boundary.phase_correction_ey = pey;
periodic_boundary.phase_correction_ez = pez;

periodic_boundary.phase_correction_hx = pey;
periodic_boundary.phase_correction_hy = pex;
periodic_boundary.phase_correction_hz = phz;

periodic_boundary.Px = nx*dx;
periodic_boundary.Py = ny*dy;

initialize_plane_wave_source_for_pbc;
```

terms of the field components on a plane in two-dimensional arrays before the time-marching loop, and use these arrays at every time step. Two-dimensional phase arrays are defined and set following (14.29) as shown in Listing 14.2. One should notice that the phase array of H_x is the same as the phase array of E_y, since components of these fields are at the same coordinates in the xy plane, thus they experience the same amount of phase delay. Similarly, the phase array of H_y is the same as the phase array of E_x.

Finally, in Listing 14.2, another subroutine, ***initialize_plane_wave_source_for_pbc***, shown in Listing 14.3, is called to initialize the plane wave source. First, E_x and E_y, or H_x and H_y, components of the incident field are determined based on the mode of excitation. Then the excitation waveform, which is cosine-modulated Gaussian, is constructed. As discussed before in Section 5.1.4, one needs to determine mainly the center frequency and the bandwidth to construct a cosine-modulated Gaussian waveform. The minimum frequency in the spectrum of the waveform can be calculated using (22). The maximum frequency can be calculated using a predefined number of cells per wavelength at the highest frequency, as discussed in Section 5.1.2. In Listing 14.3, 40 cells per wavelength is used to determine the maximum frequency.

In Sections 5.1.2–5.1.4, the time constant τ was determined such that at the minimum and maximum frequencies the value of the waveform would be 10% of the maximum

Listing 14.3 initialize_plane_wave_source_for_pbc.m

```
% Here the incident plane wave is assumed to propagate in the -z direction
kx = periodic_boundary.kx;
ky = periodic_boundary.ky;
periodic_boundary.kxy = sqrt(kx^2+ky^2);
phi_incident = atan2(ky,kx);

if strcmp(periodic_boundary.mode,'TE')
    phi_incident = atan2(ky,kx);
    E_phi = periodic_boundary.E_phi;
    periodic_boundary.Exi0 = -E_phi * sin(phi_incident);
    periodic_boundary.Eyi0 = E_phi * cos(phi_incident);
    Exi = zeros(nx, nyp1);
    Eyi = zeros(nxp1, ny);
end
if strcmp(periodic_boundary.mode,'TM')
    phi_incident = atan2(ky,kx);
    H_phi = periodic_boundary.H_phi;
    periodic_boundary.Hxi0 = -H_phi * sin(phi_incident);
    periodic_boundary.Hyi0 =  H_phi * cos(phi_incident);
    Hxi = zeros(nxp1, ny);
    Hyi = zeros(nx, nyp1);
end
if strcmp(periodic_boundary.mode,'TEM')
    phi_incident = 0;
    periodic_boundary.Exi0 = periodic_boundary.E_x;
    periodic_boundary.Eyi0 = periodic_boundary.E_y;
    Exi = zeros(nx, nyp1);
    Eyi = zeros(nxp1, ny);
end

periodic_boundary.phi_incident = phi_incident;
f_min = c*periodic_boundary.kxy/(2*pi);
f_max = c/(40*max([dx,dy,dz]));%40 cells per wavelength
bandwidth = f_max-f_min;
frequency = (f_max+f_min)/2;
tau = (2*4.29/pi)./bandwidth;  %for edge of Gaussian 40 dB below max
t_0 = 4.5 * tau;
periodic_boundary.frequency_start = f_min;
periodic_boundary.frequency_end = f_max;
periodic_boundary.modulation_frequency = frequency;
periodic_boundary.bandwidth = bandwidth;
periodic_boundary.tau = tau;
periodic_boundary.t_0 = t_0;
periodic_boundary.waveform = ...
    cos(2*pi*frequency*(time - t_0)).*exp(-((time - t_0)/tau).^2);
```

value, in other words 20 dB less than the maximum value, in the frequency spectrum of the waveform. The problem of horizontal resonance, in which the fields do not decay to zero over time, needs to be avoided in a PBC simulation. The horizontal resonance would occur due to the frequencies lower than $f_{\min}$ in the spectrum. To filter out these lower frequencies better a more conservative figure than 20 dB is preferred and τ is calculated using 40 dB as a reference.

14.5.3 PBC updates in time-marching loop

Once the initializations are completed the marching loop can be modified for the PBC implementation.

14.5.3.1 Updating magnetic field PBC source on the source plane

As illustrated in Figure 14.8, first, magnetic fields are updated on the source plane to excite the source waveform if the excitation mode is TM. Listing 14.4 shows the subroutine ***update_magnetic_field_PBC_source***. Here, $H_{inc,x}$ and $H_{inc,y}$ are calculated and phase shifted following (14.27) and (14.28). Then, these field components are added to the corresponding field components on the source plane.

It should be noted that, the same shift is applied to field components at every time step, therefore, the phase shift terms could be stored as two-dimensional arrays and they could be used at every time step without recalculation, as done for the phase correction terms discussed in Section 14.5.2. In Listing 14.4 the phase shifts are recalculated to explicitly demonstrate the connection to (14.27) and (14.28).

14.5.3.2 Updating electric field PBC source on the source plane

As the next step in the time-marching loop, magnetic fields are updated in the entire computational domain followed by CPML updates. Afterwards, the electric field components on the source plane are updated in a subroutine ***update_electric_field_PBC_source*** as shown in

Listing 14.4 update_magnetic_field_PBC_source.m

```
% update magnetic field source for TM mode PBC
if strcmp(periodic_boundary.mode,'TM')

    ks = periodic_boundary.source_ks;

    Hxi(:,:) = periodic_boundary.Hxi0 ...
        * periodic_boundary.waveform(time_step);
    Hyi(:,:) = periodic_boundary.Hyi0 ...
        * periodic_boundary.waveform(time_step);

    for m = 1:nx
        Hyi(m,:) = Hyi(m,:) * exp(-j*kx*(m-0.5)*dx);
        Hxi(m,:) = Hxi(m,:) * exp(-j*kx*(m-1)*dx);
    end
    Hxi(nxp1,:) = Hxi(nxp1,:)*exp(-j*kx*nx*dx);

    for m = 1:ny
        Hyi(:,m) = Hyi(:,m) * exp(-j*ky*(m-1)*dy);
        Hxi(:,m) = Hxi(:,m) * exp(-j*ky*(m-0.5)*dy);
    end
    Hyi(:,nyp1) = Hyi(:,nyp1) * exp(-j*ky*ny*dy);

    Hx(:,:,ks) = Hx(:,:,ks) +  Hxi;
    Hy(:,:,ks) = Hy(:,:,ks) +  Hyi;

end
```

Listing 14.5 update_electric_field_PBC_source.m

```
% update electric field source for TE and TEM mode PBC

if strcmp(periodic_boundary.mode,'TE')
   ks = periodic_boundary.source_ks;

   Exi(:,:) = periodic_boundary.Exi0 ...
       * periodic_boundary.waveform(time_step);
   Eyi(:,:) = periodic_boundary.Eyi0 ...
       * periodic_boundary.waveform(time_step);

   for m = 1:nx
       Eyi(m,:) = Eyi(m,:) * exp(-j*kx*(m-1)*dx);
       Exi(m,:) = Exi(m,:) * exp(-j*kx*(m-0.5)*dx);
   end
   Eyi(nxp1,:) = Eyi(nxp1,:) * exp(-j*kx*nx*dx);

   for m = 1:ny
       Eyi(:,m) = Eyi(:,m) * exp(-j*ky*(m-0.5)*dy);
       Exi(:,m) = Exi(:,m) * exp(-j*ky*(m-1)*dy);
   end
   Exi(:,nyp1) = Exi(:,nyp1) * exp(-j*ky*ny*dy);

   Ex(:,:,ks) = Ex(:,:,ks) +  Exi;
   Ey(:,:,ks) = Ey(:,:,ks) +  Eyi;

end
if strcmp(periodic_boundary.mode,'TEM')
   ks = periodic_boundary.source_ks;

   Exi(:,:) = periodic_boundary.Exi0 ...
       * periodic_boundary.waveform(time_step);
   Eyi(:,:) = periodic_boundary.Eyi0 ...
       * periodic_boundary.waveform(time_step);

   Ex(:,:,ks) = Ex(:,:,ks) +  Exi;
   Ey(:,:,ks) = Ey(:,:,ks) +  Eyi;

end
```

Listing 14.5. Here, $E_{inc,x}$ and $E_{inc,y}$ are calculated and phase shifted following (14.25) and (14.26). Then, these field components are added to the corresponding field components on the source plane.

14.5.3.3 Updating electric field on PBC boundaries

The electric fields are then updated in the entire computational domain, except for the components lying on the PBC boundaries, followed by CPML updates. Afterwards, the electric field components are updated on the PBC boundaries in a subroutine ***update_electric_field_PBC_ABC*** as shown in Listing 14.6. Here, the components of E_x, E_y, and E_z are updated based on (14.9)–(14.20).

Listing 14.6 update_electric_field_PBC_ABC.m

```
if strcmp(periodic_boundary.mode,'TEM')
    Ex(1:nx,1,2:nz) = Cexe(1:nx,1,2:nz).*Ex(1:nx,1,2:nz) ...
        + Cexhz(1:nx,1,2:nz).*...
            (Hz(1:nx,1,2:nz)-Hz(1:nx,ny,2:nz)) ...
        + Cexhy(1:nx,1,2:nz).*...
            (Hy(1:nx,1,2:nz)-Hy(1:nx,1,1:nz-1));

    Ex(1:nx,nyp1,2:nz) = Ex(1:nx,1,2:nz);

    Ey(1,1:ny,2:nz) = Ceye(1,1:ny,2:nz).*Ey(1,1:ny,2:nz) ...
        + Ceyhx(1,1:ny,2:nz).* ...
            (Hx(1,1:ny,2:nz)-Hx(1,1:ny,1:nz-1)) ...
        + Ceyhz(1,1:ny,2:nz).* ...
            (Hz(1,1:ny,2:nz)-Hz(nx,1:ny,2:nz));

    Ey(nxp1,1:ny,2:nz) = Ey(1,1:ny,2:nz);

    Ez(1,2:ny,1:nz) = Ceze(1,2:ny,1:nz).*Ez(1,2:ny,1:nz) ...
        + Cezhy(1,2:ny,1:nz).* ...
            (Hy(1,2:ny,1:nz)-Hy(nx,2:ny,1:nz)) ...
        + Cezhx(1,2:ny,1:nz).*...
            (Hx(1,2:ny,1:nz)-Hx(1,1:ny-1,1:nz));

    Ez(nxp1,2:ny,1:nz) = Ez(1,2:ny,1:nz); ...

    Ez(2:nx,1,1:nz) = Ceze(2:nx,1,1:nz).*Ez(2:nx,1,1:nz) ...
        + Cezhy(2:nx,1,1:nz).* ...
            (Hy(2:nx,1,1:nz)-Hy(1:nx-1,1,1:nz)) ...
        + Cezhx(2:nx,1,1:nz).*...
            (Hx(2:nx,1,1:nz)-Hx(2:nx,ny,1:nz));

    Ez(2:nx,nyp1,1:nz) = Ez(2:nx,1,1:nz);

    Ez(1,1,1:nz) = Ceze(1,1,1:nz).*Ez(1,1,1:nz) ...
        + Cezhy(1,1,1:nz).* ...
            (Hy(1,1,1:nz)-Hy(nx,1,1:nz)) ...
        + Cezhx(1,1,1:nz).*...
            (Hx(1,1,1:nz)-Hx(1,ny,1:nz));

    Ez(nxp1,1,1:nz) = Ez(1,1,1:nz);

    Ez(1,nyp1,1:nz) = Ez(1,1,1:nz);

    Ez(nxp1,nyp1,1:nz) = Ez(1,1,1:nz);
end

if strcmp(periodic_boundary.mode,'TE') ...
        || strcmp(periodic_boundary.mode,'TM')

    kx = periodic_boundary.kx;
    ky = periodic_boundary.ky;
    Px = periodic_boundary.Px;
    Py = periodic_boundary.Py;

    Ex(1:nx,1,2:nz) = Cexe(1:nx,1,2:nz).*Ex(1:nx,1,2:nz) ...
        + Cexhz(1:nx,1,2:nz).*...
          (Hz(1:nx,1,2:nz)-Hz(1:nx,ny,2:nz)*exp(j*ky*Py)) ...
        + Cexhy(1:nx,1,2:nz).*...
          (Hy(1:nx,1,2:nz)-Hy(1:nx,1,1:nz-1));

    Ex(1:nx,nyp1,2:nz) = Ex(1:nx,1,2:nz)*exp(-j*ky*Py);
```

```
    Ey(1,1:ny,2:nz) = Ceye(1,1:ny,2:nz).*Ey(1,1:ny,2:nz) ...
        + Ceyhx(1,1:ny,2:nz).* ...
            (Hx(1,1:ny,2:nz)-Hx(1,1:ny,1:nz-1)) ...
        + Ceyhz(1,1:ny,2:nz).* ...
            (Hz(1,1:ny,2:nz)-Hz(nx,1:ny,2:nz)*exp(j*kx*Px));

    Ey(nxp1,1:ny,2:nz) = Ey(1,1:ny,2:nz)*exp(-j*kx*Px);

    Ez(1,2:ny,1:nz) = Ceze(1,2:ny,1:nz).*Ez(1,2:ny,1:nz) ...
        + Cezhy(1,2:ny,1:nz).* ...
            (Hy(1,2:ny,1:nz)-Hy(nx,2:ny,1:nz)*exp(j*kx*Px)) ...
        + Cezhx(1,2:ny,1:nz).*...
            (Hx(1,2:ny,1:nz)-Hx(1,1:ny-1,1:nz));

    Ez(nxp1,2:ny,1:nz) = Ez(1,2:ny,1:nz)*exp(-j*kx*Px); ...

    Ez(2:nx,1,1:nz) = Ceze(2:nx,1,1:nz).*Ez(2:nx,1,1:nz) ...
        + Cezhy(2:nx,1,1:nz).* ...
            (Hy(2:nx,1,1:nz)-Hy(1:nx-1,1,1:nz)) ...
        + Cezhx(2:nx,1,1:nz).*...
            (Hx(2:nx,1,1:nz)-Hx(2:nx,ny,1:nz)*exp(j*ky*Py));

    Ez(2:nx,nyp1,1:nz) = Ez(2:nx,1,1:nz)*exp(-j*ky*Py);

    Ez(1,1,1:nz) = Ceze(1,1,1:nz).*Ez(1,1,1:nz) ...
        + Cezhy(1,1,1:nz).* ...
            (Hy(1,1,1:nz)-Hy(nx,1,1:nz)*exp(j*kx*Px)) ...
        + Cezhx(1,1,1:nz).*...
            (Hx(1,1,1:nz)-Hx(1,ny,1:nz)*exp(j*ky*Py));

    Ez(nxp1,1,1:nz) = Ez(1,1,1:nz)*exp(-j*kx*Px);

    Ez(1,nyp1,1:nz) = Ez(1,1,1:nz)*exp(-j*ky*Py);

    Ez(nxp1,nyp1,1:nz) = Ez(1,1,1:nz)*exp(-j*kx*Px)*exp(-j*ky*Py);
end
```

14.5.3.4 Capturing fields on the reflection and transmission planes

Listing 14.7 shows the implementation of a subroutine, ***capture_fields_for_PBC***, used to capture fields on the reflection and transmission planes after the field updates at every time step. First, all field components on the reflection plane are captured and multiplied by two-dimensional phase correction arrays based on (14.29) as discussed in Section 14.4. Then the results are averaged and stored in transient reflection arrays. The same calculations are repeated to capture transmitted fields on the transmission plane.

14.5.3.5 Calculation of reflection and transmission coefficients

The transient fields on reflection and transmission planes are available in one-dimensional arrays after the FDTD time-marching loop is completed. These transients are used to calculate the reflection and transmission coefficients during the post-processing phase of the FDTD simulation in a subroutine ***calculate_reflection_and_transmission_for_PBC***, shown in Listing 14.8. The transient fields are transformed to frequency domain using DFT. Then, co- and cross-polarized reflection and transmission coefficients are calculated based on the

Listing 14.7 capture_fields_for_PBC.m

```
nxy = nx*ny;

if periodic_boundary.calculate_reflection

    ks = periodic_boundary.reflection_ks;

    ex = Ex(1:nx,1:ny,ks).*periodic_boundary.phase_correction_ex;
    ey = Ey(1:nx,1:ny,ks).*periodic_boundary.phase_correction_ey;
    ez = 0.5*(Ez(1:nx,1:ny,ks)+Ez(1:nx,1:ny,ks-1)) ...
         .*periodic_boundary.phase_correction_ez;

    hx = 0.5*(Hx(1:nx,1:ny,ks)+Hx(1:nx,1:ny,ks-1)) ...
       .*periodic_boundary.phase_correction_hx;
    hy = 0.5*(Hy(1:nx,1:ny,ks)+Hy(1:nx,1:ny,ks-1)) ...
       .*periodic_boundary.phase_correction_hy;
    hz = Hz(1:nx,1:ny,ks).*periodic_boundary.phase_correction_hz;

    periodic_boundary.reflection_ex(time_step) = sum(sum(ex))/nxy;
    periodic_boundary.reflection_ey(time_step) = sum(sum(ey))/nxy;
    periodic_boundary.reflection_ez(time_step) = sum(sum(ez))/nxy;

    periodic_boundary.reflection_hx(time_step) = sum(sum(hx))/nxy;
    periodic_boundary.reflection_hy(time_step) = sum(sum(hy))/nxy;
    periodic_boundary.reflection_hz(time_step) = sum(sum(hz))/nxy;
end

if periodic_boundary.calculate_transmission
    ks = periodic_boundary.transmission_ks;

    ex = Ex(1:nx,1:ny,ks).*periodic_boundary.phase_correction_ex;
    ey = Ey(1:nx,1:ny,ks).*periodic_boundary.phase_correction_ey;
    ez = 0.5*(Ez(1:nx,1:ny,ks)+Ez(1:nx,1:ny,ks-1)) ...
         .*periodic_boundary.phase_correction_ez;

    hx = 0.5*(Hx(1:nx,1:ny,ks)+Hx(1:nx,1:ny,ks-1)) ...
       .*periodic_boundary.phase_correction_hx;
    hy = 0.5*(Hy(1:nx,1:ny,ks)+Hy(1:nx,1:ny,ks-1)) ...
       .*periodic_boundary.phase_correction_hy;
    hz = Hz(1:nx,1:ny,ks).*periodic_boundary.phase_correction_hz;

    periodic_boundary.transmission_ex(time_step) = sum(sum(ex))/nxy;
    periodic_boundary.transmission_ey(time_step) = sum(sum(ey))/nxy;
    periodic_boundary.transmission_ez(time_step) = sum(sum(ez))/nxy;

    periodic_boundary.transmission_hx(time_step) = sum(sum(hx))/nxy;
    periodic_boundary.transmission_hy(time_step) = sum(sum(hy))/nxy;
    periodic_boundary.transmission_hz(time_step) = sum(sum(hz))/nxy;
end
```

Listing 14.8 calculate_reflection_and_transmission_for_PBC.m

```
frequencies = linspace(periodic_boundary.frequency_start, ...
    periodic_boundary.frequency_end, 200);

time_shift = 0;
reflection_ex_dft = time_to_frequency_domain( ...
    periodic_boundary.reflection_ex, dt, frequencies, time_shift);
reflection_ey_dft = time_to_frequency_domain( ...
    periodic_boundary.reflection_ey, dt, frequencies, time_shift);
reflection_ez_dft = time_to_frequency_domain( ...
    periodic_boundary.reflection_ez, dt, frequencies, time_shift);
time_shift = -dt/2;
reflection_hx_dft = time_to_frequency_domain( ...
    periodic_boundary.reflection_hx, dt, frequencies, time_shift);
reflection_hy_dft = time_to_frequency_domain( ...
    periodic_boundary.reflection_hy, dt, frequencies, time_shift);
reflection_hz_dft = time_to_frequency_domain( ...
    periodic_boundary.reflection_hz, dt, frequencies, time_shift);

if periodic_boundary.calculate_transmission
    time_shift = 0;
    transmission_ex_dft = time_to_frequency_domain( ...
        periodic_boundary.transmission_ex, dt, frequencies, time_shift);
    transmission_ey_dft = time_to_frequency_domain( ...
        periodic_boundary.transmission_ey, dt, frequencies, time_shift);
    transmission_ez_dft = time_to_frequency_domain( ...
        periodic_boundary.transmission_ez, dt, frequencies, time_shift);
    time_shift = -dt/2;
    transmission_hx_dft = time_to_frequency_domain( ...
        periodic_boundary.transmission_hx, dt, frequencies, time_shift);
    transmission_hy_dft = time_to_frequency_domain( ...
        periodic_boundary.transmission_hy, dt, frequencies, time_shift);
    transmission_hz_dft = time_to_frequency_domain( ...
        periodic_boundary.transmission_hz, dt, frequencies, time_shift);
end

% wavenumber components
k = frequencies*2*pi/c;
kx = periodic_boundary.kx;
ky = periodic_boundary.ky;
kh = sqrt(kx^2+ky^2);
kz = sqrt(k.^2-kh^2);

phi_incident = atan2(ky,kx);

eta_0 = sqrt(mu_0/eps_0);

if strcmp(periodic_boundary.mode,'TE')
    Erp_co = (reflection_ey_dft*kx/kh-reflection_ex_dft*ky/kh);
    Hrp_co = (reflection_hx_dft*kx/kh+reflection_hy_dft*ky/kh);
    Eref_cr = (reflection_ex_dft*kx/kh+reflection_ey_dft*ky/kh);

    Einc = (Erp_co+eta_0*Hrp_co.*k./kz)/2;
    Eref_co = Erp_co-Einc;

    Gamma_co = Eref_co./Einc;
    aGamma_co = abs(Gamma_co);              % Magnitude
    pGamma_co = angle(Gamma_co)/pi*180;     % Phase in degree

    Gamma_cr  = Eref_cr./Einc;
    aGamma_cr = abs(Gamma_cr);              % Magnitude
    pGamma_cr = angle(Gamma_cr)/pi*180;     % Phase in degree
```

```
    if periodic_boundary.calculate_transmission
        Etra_co = (transmission_ey_dft*kx/kh-transmission_ex_dft*ky/kh);
        Etra_cr = (transmission_ex_dft*kx/kh+transmission_ey_dft*ky/kh);

        Einc = Einc .* ...
            exp(j*kz*periodic_boundary.reflection_transmission_distance);

        T_co = Etra_co./Einc;
        aT_co = abs(T_co);                 % Magnitude
        pT_co = angle(T_co)/pi*180;        % Phase in degree

        T_cr = Etra_cr./Einc;
        aT_cr = abs(T_cr);                 % Magnitude
        pT_cr = angle(T_cr)/pi*180;        % Phase in degree
    end
end

if strcmp(periodic_boundary.mode,'TM')
    Erp_co  = (-reflection_ex_dft*kx/kh-reflection_ey_dft*ky/kh);
    Hrp_co  = (reflection_hy_dft*kx/kh-reflection_hx_dft*ky/kh);
    Href_cr = (reflection_hx_dft*kx/kh+reflection_hy_dft*ky/kh);

    Hinc = (Hrp_co+Erp_co.*k./(eta_0*kz))/2;
    Href_co = Hrp_co-Hinc;

    Gamma_co = Href_co./Hinc;
    aGamma_co = abs(Gamma_co);                % Magnitude
    pGamma_co = angle(Gamma_co)/pi*180;       % Phase in degree

    Gamma_cr = Href_cr./Hinc;
    aGamma_cr = abs(Gamma_cr);                % Magnitude
    pGamma_cr = angle(Gamma_cr)/pi*180;       % Phase in degree

    if periodic_boundary.calculate_transmission
        Htra_co = (transmission_hy_dft*kx/kh-transmission_hx_dft*ky/kh);
        Htra_cr = (transmission_hx_dft*kx/kh+transmission_hy_dft*ky/kh);

        Hinc = Hinc .* ...
            exp(j*kz*periodic_boundary.reflection_transmission_distance);

        T_co = Htra_co./Hinc;
        aT_co = abs(T_co);                 % Magnitude
        pT_co = angle(T_co)/pi*180;        % Phase in degree

        T_cr = Htra_cr./Hinc;
        aT_cr = abs(T_cr);                 % Magnitude
        pT_cr = angle(T_cr)/pi*180;        % Phase in degree
    end
end

if strcmp(periodic_boundary.mode,'TEM')

    Exi0 = periodic_boundary.Exi0;
    Eyi0 = periodic_boundary.Eyi0;
    Ei0 = sqrt(Exi0^2+Eyi0^2);
    phi = atan2(Eyi0,Exi0);

    Erp_co  = reflection_ex_dft*cos(phi) + reflection_ey_dft*sin(phi);
    Hrp_co  = reflection_hx_dft*sin(phi) - reflection_hy_dft*cos(phi);
    Eref_cr = reflection_ex_dft*sin(phi) - reflection_ey_dft*cos(phi);

    Einc = (Erp_co+eta_0*Hrp_co)/2;
    Eref_co = Erp_co-Einc;
```

```
    Gamma_co = Eref_co./Einc;
    aGamma_co = abs(Gamma_co);                % Magnitude
    pGamma_co = angle(Gamma_co)/pi*180;       % Phase in degree

    Gamma_cr  = Eref_cr./Einc;
    aGamma_cr = abs(Gamma_cr);                % Magnitude
    pGamma_cr = angle(Gamma_cr)/pi*180;       % Phase in degree

    if periodic_boundary.calculate_transmission
        Etra_co = transmission_ex_dft*cos(phi) + transmission_ey_dft*sin(phi);
        Etra_cr = transmission_ex_dft*sin(phi) - transmission_ey_dft*cos(phi);

        Einc = Einc .* ...
            exp(j*kz*periodic_boundary.reflection_transmission_distance);

        T_co = Etra_co./Einc;
        aT_co = abs(T_co);                    % Magnitude
        pT_co = angle(T_co)/pi*180;           % Phase in degree

        T_cr = Etra_cr./Einc;
        aT_cr = abs(T_cr);                    % Magnitude
        pT_cr = angle(T_cr)/pi*180;           % Phase in degree
    end
end

fGHz = frequencies*1e-9;

figure;
plot(fGHz,aGamma_co,'b-',fGHz,aGamma_cr,'r--','linewidth',1.5);
    legend('|\Gamma_{co}|','|\Gamma_{cr}|');
if periodic_boundary.calculate_transmission
    hold on;
    plot(fGHz,aT_co,'g-.',fGHz,aT_cr,'m:','linewidth',1.5);
    legend('|\Gamma_{co}|','|\Gamma_{cr}|','|T_{co}|', '|T_{cr}|');
end
grid on;
xlabel('frequency [GHz]','fontsize',12);
ylabel('magnitude','fontsize',12);

figure;
plot(fGHz,pGamma_co,'b-',fGHz,pGamma_cr,'r--','linewidth',1.5);
legend('|\Gamma_{co}|','|\Gamma_{cr}|');
if periodic_boundary.calculate_transmission
    hold on;
    plot(fGHz,pT_co,'g-.',fGHz,pT_cr,'m:','linewidth',1.5);
    legend('|\Gamma_{co}|','|\Gamma_{cr}|','|T_{co}|', '|T_{cr}|');
end
grid on;
xlabel('frequency [GHz]','fontsize',12);
ylabel('phase [degrees]','fontsize',12);
set(gca,'fontsize',12);

figure;
timens = time*1e9;
plot(timens, abs(periodic_boundary.reflection_ex),'b-', ...
     timens, abs(periodic_boundary.reflection_ey),'r--', ...
     timens, abs(periodic_boundary.reflection_ez),'k-.', ...
    'linewidth',1.5);
xlabel('time [ns]','fontsize',12);
ylabel('magnitude','fontsize',12);
grid on;
legend('reflection E_{x}','reflection E_{y}','reflection E_{z}');
set(gca,'fontsize',12);
```

```
if periodic_boundary.calculate_transmission
    figure;
    plot(timens, abs(periodic_boundary.transmission_ex),'b-', ...
        timens, abs(periodic_boundary.transmission_ey),'r--', ...
        timens, abs(periodic_boundary.transmission_ez),'k-.', ...
       'linewidth',1.5);
   xlabel('time [ns]','fontsize',12);
   ylabel('magnitude','fontsize',12);
   grid on;
   legend('transmission E_{x}','transmission E_{y}','transmission E_{z}');
   set(gca,'fontsize',12);
end
```

equations discussed in Sections 14.4.1–14.4.3 for TE, TM, and TEM modes, respectively. Once the reflection and transmission coefficients are calculated, their magnitudes and phases are displayed in separate figures.

14.6 Simulation examples

14.6.1 Reflection and transmission coefficients of a dielectric slab

A simple example that can be verified against analytical solutions is simulation of a dielectric slab. Here, we consider a dielectric slab case with thickness of 1 cm and relative permittivity of 4, for which reflection coefficient distribution is shown in Figures 14.2 and 14.4. The slab is illuminated by a TE waveform with horizontal wavenumber of $k_x = 100$ radian/meter and $k_y = 0$ radian/meter, as shown in Listing 14.1. Figure 14.12 shows magnitudes of the

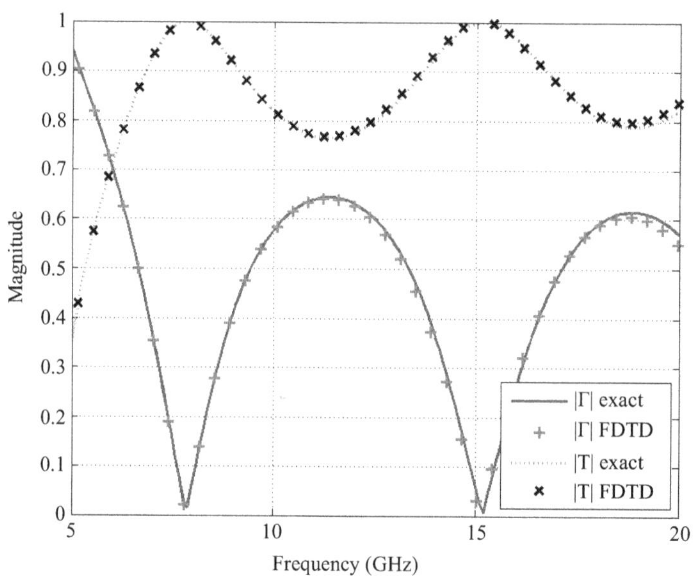

Figure 14.12 Reflection and transmission coefficients of a dielectric slab.

reflection and transmission coefficients calculated by FDTD and compared with the analytical solution. It should be noted that the minimum frequency that can be calculated using the presented PBC FDTD code is 4.8 GHz based on (14.22), hence the results are shown starting from 5 GHz in Figure 14.12.

14.6.2 Reflection and transmission coefficients of a dipole FSS

Figure 14.6 illustrates a unit cell in which a PEC rectangular patch placed on a dielectric slab. A periodic structure built on this kind of a unit cell is referred to as a Dipole Frequency Selective Surface (DFSS). In this example we consider a DFSS with a PEC patch of size 3 mm by 12 mm, on a 6 mm thick slab with 2.2 dielectric constant. The unit cell is 15 mm wide in x and y directions. The source plane is located 18 mm above the slab and it excites TE mode waves with horizontal wavenumbers $k_x = 20$ radian/meter and $k_y = 7.8$ radian/meter. The fields are captured 16 mm above and 3 mm below the slab as reflected and transmitted fields, respectively. FDTD simulation is run for 5000 time steps using a cell size of 0.5 mm on a side. Figure 14.13 shows the co- and cross-polarized reflection coefficients whereas Figure 14.14 shows the transmission coefficients calculated over the frequency range 2–15 GHz compared with solutions obtained from Ansoft Designer (AD) [76].

Next, the DFSS is simulated using TM mode waves with horizontal wavenumbers $k_x = 20$ radian/meter and $k_y = 7.8$ radian/meter as the excitation. To simulate the TM mode one can simply change the value of **periodic_boundary.mode** as 'TM' and set **periodic_boundary.H_phi** as 1 in Listing 14.1. Figure 14.15 shows the reflection coefficients whereas Figure 14.16 shows the transmission coefficients calculated by FDTD and compared with solutions obtained from Ansoft Designer.

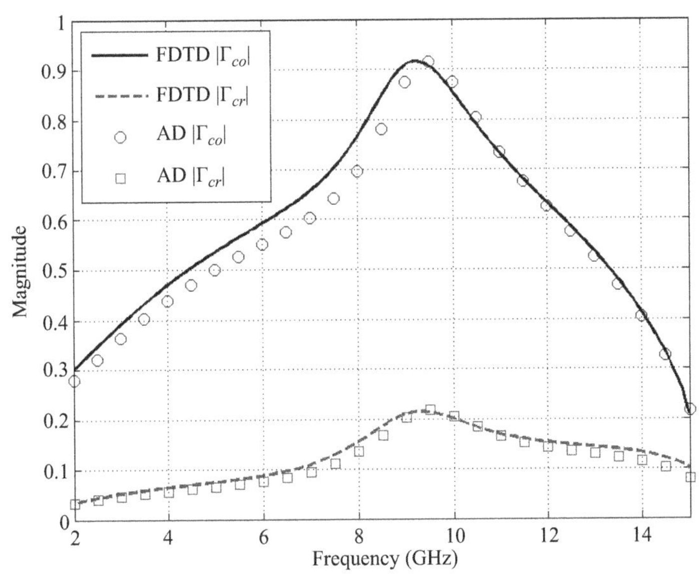

Figure 14.13 TE mode reflection coefficients of a DFSS: FDTD vs Ansoft Designer.

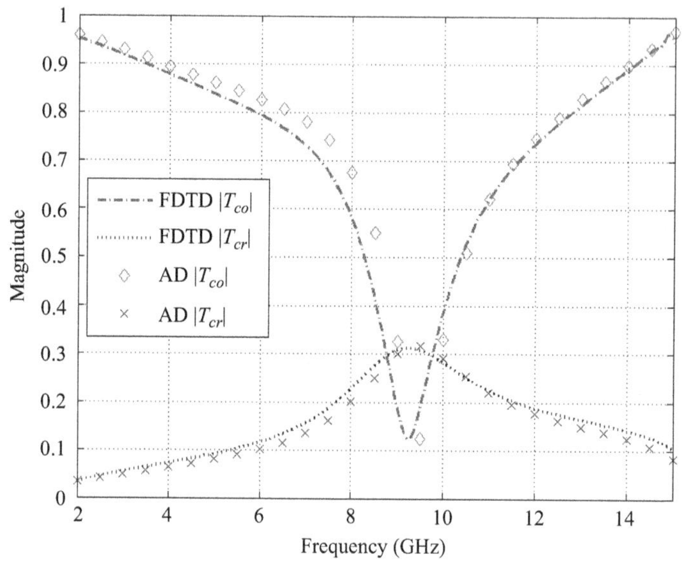

Figure 14.14 TE mode transmission coefficients of a DFSS: FDTD vs Ansoft Designer.

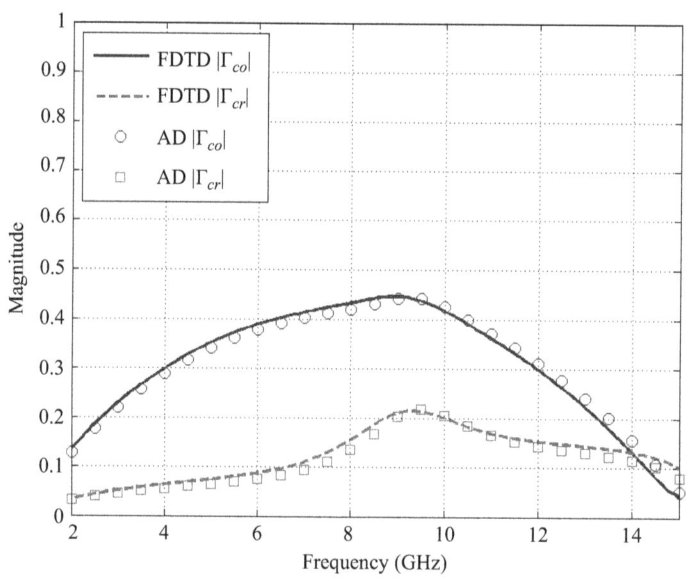

Figure 14.15 TM mode reflection coefficients of a DFSS: FDTD vs Ansoft Designer.

14.6.3 Reflection and transmission coefficients of a Jarusalem-cross FSS

In this example we consider a Jarusalem-Cross Frequency Selective Surface (JC-FSS) dimensions of which are illustrated in Figure 14.17. The PEC cross patch is in free-space

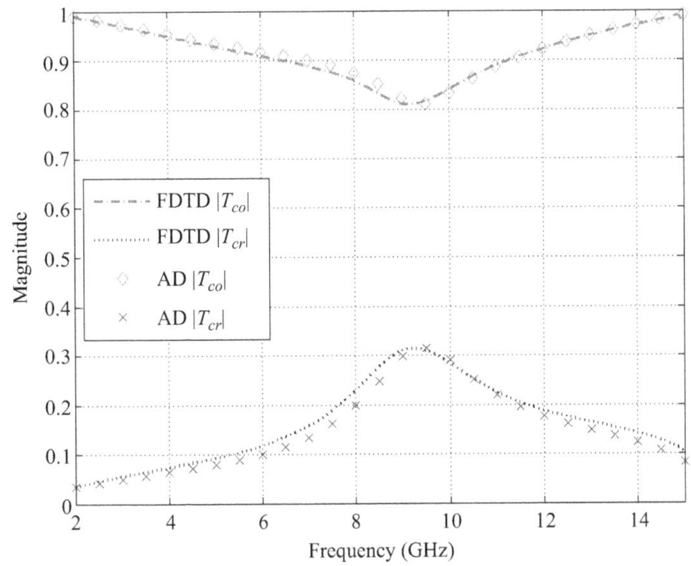

Figure 14.16 TM mode transmission coefficients of a DFSS: FDTD vs Ansoft Designer.

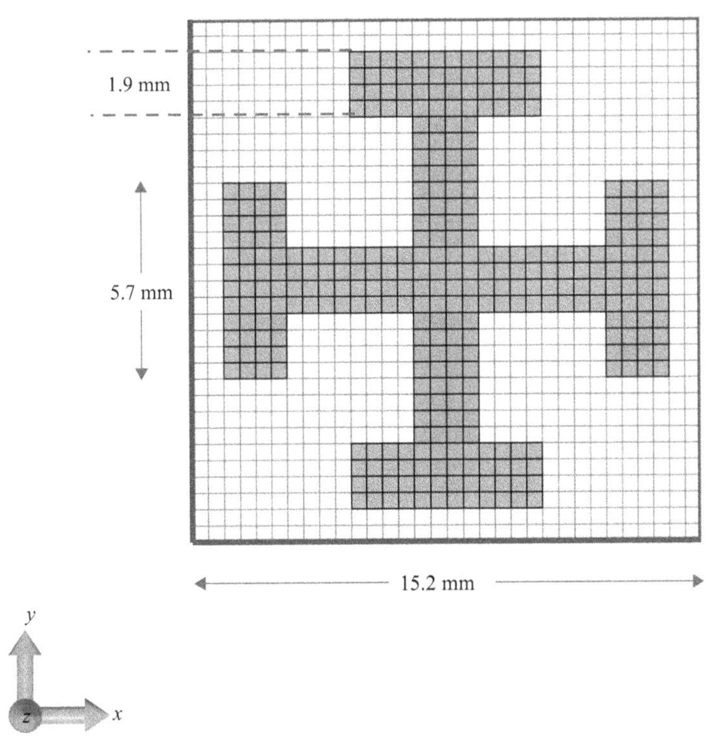

Figure 14.17 A Jarusalem-cross frequency selective circuit.

Listing 14.9 define_problem_space_parameters.m

```
% ==<periodic boundary simulation parameters>========
disp('Defining periodic boundary simulation');

periodic_boundary.mode = 'TEM';
periodic_boundary.E_x = 0;
periodic_boundary.E_y = 1;
periodic_boundary.source_z = 8e-3;
periodic_boundary.reflection_z = 6e-3;
periodic_boundary.transmission_z = -3e-3;
```

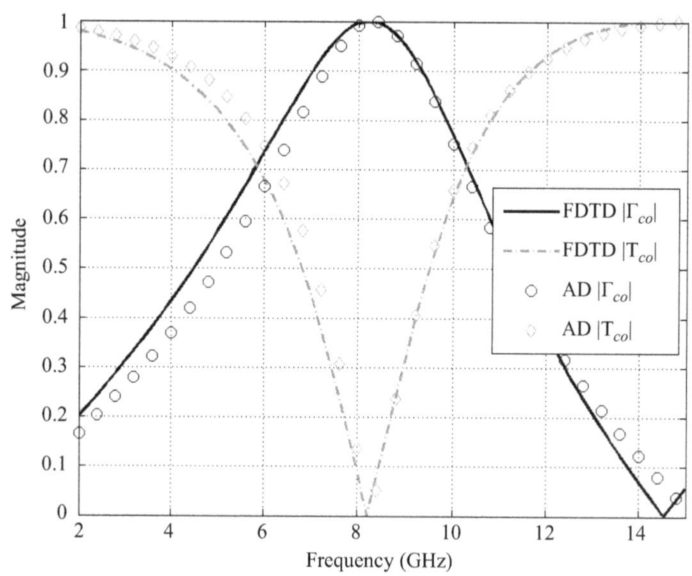

Figure 14.18 Reflection and transmission coefficients of a Jarusalem-cross FSS: FDTD vs Ansoft Designer.

without a dielectric slab support. The unit cell is 15.2 mm wide in x and y directions. The source plane is located 8 mm above the slab and it excites TEM mode waves with y polarization, as shown in Listing 14.9. The fields are captured 6 mm above and 3 mm below the patch as reflected and transmitted fields, respectively. FDTD simulation is run for 3,000 time steps using a cell size of 0.475 mm in x and y directions while 0.5 mm in z direction. Figure 14.18 shows the co- and cross-polarized reflection and transmission coefficients calculated over the frequency range 2–15 GHz compared with solutions obtained from Ansoft Designer.

CHAPTER 15

Nonuniform grid

15.1 Introduction

Choice of the FDTD grid cell size is a trade-off between reduced solution time and improved accuracy. For problems in which there are no fine geometrical details or any fine details that affect the solution accuracy, this trade-off can be well balanced by choosing a cell size which is not too small to keep reasonable solution time and accuracy. If fine details are expected to affect the accuracy, some advanced techniques can be used. For instance, subgridding techniques can be employed within the uniform FDTD grid, or a nonuniform grid can be used instead of the uniform grid.

One of the methods to model fine geometrical details in an FDTD problem, while keeping the solution time reasonable, is nonuniform grids: Instead of using the same cell size all over the computational domain, different cell sizes can be used at different regions. For instance consider the problem shown in Figure 15.1(a), where a coarse uniform grid is used. In this geometry there are two fine strips which are narrower than a cells size. Moreover, some parts of the geometry do not conform to the underlying grid. An approximate geometry that would conform to this grid will be used in the conventional FDTD, hence the loss of modeling accuracy will reflect into the results of the calculations.

To model the geometry more accurately, one can use a fine uniform grid, as shown in Figure 15.1(b), which will increase the problem size – in number of cells – significantly, and increase the calculation time much more. For instance, if the cell size is halved for a three-dimensional domain, the calculations will take about 16 times longer.

Figure 15.1(c) illustrates a nonuniform grid where coarse cells are used in the background of the problem space and fine cells are used only at the fine geometrical regions. Transitions between fine cell regions and coarse cell regions are made smooth by gradually changing the cell sizes between these regions, thus numerical reflections that might occur due to transitions can be minimized. Moreover, by appropriately shifting the mesh lines, the grid is made to conform to the objects completely. Compared with the uniform fine grid, the total number of cells in this grid is much less, thus the total simulation time will be less while maintaining a level of accuracy comparable to the one obtained with the fine uniform grid.

15.2 Transition between fine and coarse grid subregions

It is essential to set a smooth transition between the fine grid regions and coarse grid regions to minimize numerical reflections that would be generated due to the transition.

In this chapter, a transition scheme will be discussed in which the cell sizes in the transition region are gradually increased starting with the fine cell size on one end and ending with the coarse cell size at the other end. While this scheme will be illustrated for a one-dimensional grid, it applies to all directions of two and three-dimensional grids as well.

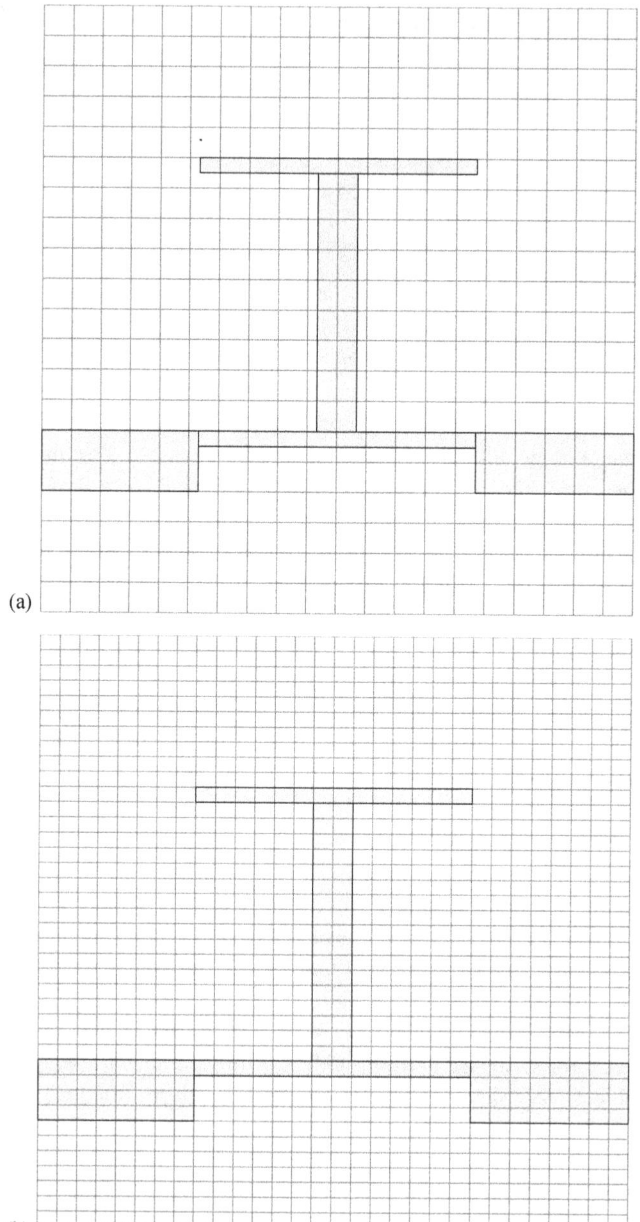

Figure 15.1 (a) An FDTD computational domain with coarse uniform grid; (b) an FDTD computational domain with fine uniform grid; and (c) an FDTD computational domain with nonuniform grid.

(c) 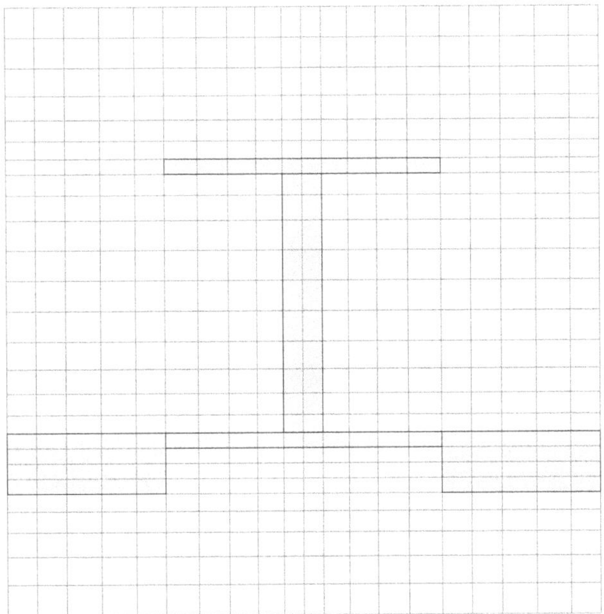

Figure 15.1 *(Continued)*

In what is called the uniform case all cells in the problem space have the same size in each coordinate direction Δx, Δy, or Δz. While each cell can have a different size in the nonuniform case, it is better to define subregions in the problem space, where needed, with uniform cells and establish nonuniform transition between the uniform subregions.

Figure 15.2 shows a nonuniform transition subregion (ΔT) between two uniform subregions. While performing the transition, the cell sizes, i.e., the distances between the subsequent electric field components, can be gradually changed. One can start with a cell size, shown as Δs length in Figure 15.2, just before the transition subregion. The length of first cell in the transition subregion can be set as

$$\Delta 1 = R\Delta s, \tag{15.1}$$

where R is rate of change between subsequent half cells. Thus the lengths of the cells in the transition subregion can be expressed as

$$\Delta M = R^M \Delta s, \tag{15.2}$$

where M is an index denoting a cell in the transition subregion. At the end of the transition the length of the first cell of the next uniform subregion is

$$\Delta e = R^{N_{nu}+1} \Delta s, \tag{15.3}$$

where N_{nu} is the number of cells in the transition subregion. If the transition is desired to happen on a given transition length ΔT, then it is needed to determine the rate R and the number of cells N_{nu}.

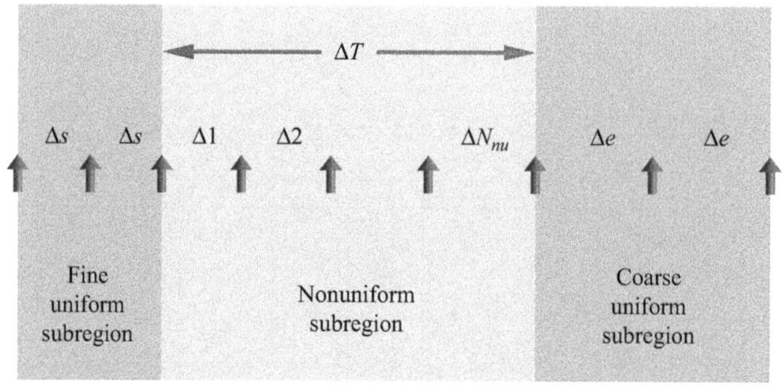

Figure 15.2 Uniform and nonuniform subregions.

The total length of the transition subregion, in addition to one cell at both ends from the uniform regions is

$$\Delta T + \Delta s + \Delta e = \sum_{k=0}^{N_{nu}+1} \Delta s R^k = \frac{\Delta s - \Delta s R^{N_{nu}+2}}{1-R} = \frac{\Delta s - \Delta e R}{1-R}. \tag{15.4}$$

Equation (15.4) can be used to find R as

$$R = \frac{\Delta T + \Delta e}{\Delta T + \Delta s}. \tag{15.5}$$

Then, the number of cells, N_{nu}, can be determined using (15.3) as

$$N_{nu} = \frac{\log(\Delta e/\Delta s)}{\log(R)} - 1. \tag{15.6}$$

One should notice that N_{nu} may not be found as an integer number by (15.6), however, it has to be an integer number. In this case, N_{nu} can be rounded to the closest appropriate integer. It should be noted that, if N_{nu} is rounded to a larger integer number, the transition cell sizes will be smaller than they are supposed to be and consequently a transition cell may become smaller than the fine uniform subregion cell size. Similarly, if N_{nu} is rounded to a smaller integer number, the cell sizes will be larger and a transition cell may get larger than the coarse uniform subregion cell size. In either case, there may be an abrupt transition between the uniform and nonuniform subregions, which may enhance the numerical error due to the transition. A smoother transition may be achieved if the transition region is selected to be appropriately long.

Since in a Yee cell the electric and magnetic field components are located at distinct positions, we can define a grid of electric field components as wells as a grid of magnetic field components. Figure 15.3 illustrates these two grids for a uniform one-dimensional space of N_x cells in the x direction. In the figure "nc_xe" denotes the coordinates of electric field grid points, whereas "nc_xh" denotes the coordinates of magnetic field points. The nonuniform electric field grid can be set up by determining the sizes of the cells (i.e., distances between the neighboring electric field components) as discussed above. Once the positions of the electric field components are determined, magnetic field components can be placed at the centers of the cells of the electric field grid and the magnetic field grid can be constructed as shown in Figure 15.4. A representation of a fine grid with nonuniform surrounding regions is illustrated in Figure 15.5.

This kind of field positioning on a dual grid, where magnetic fields are located at the midpoints of electric field grid cells, has been discussed and analyzed in [77] in terms of its accuracy. Since magnetic fields are at the midpoints, as it is also the case in a uniform grid, the corresponding updating equations that update magnetic fields are second order accurate. The electric fields are off-centered between the magnetic field components, which leads to

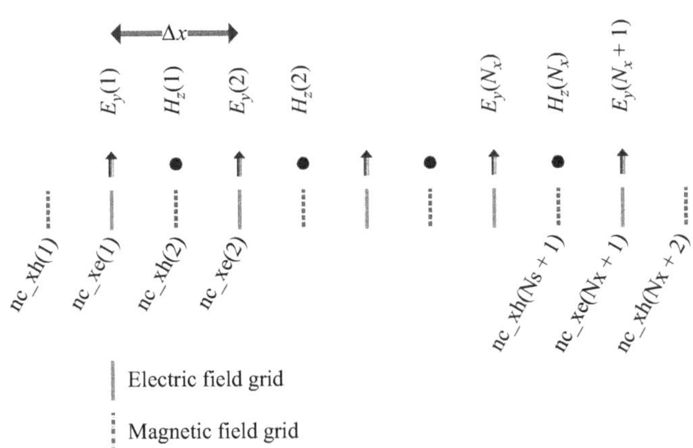

Figure 15.3 Uniform electric and magnetic field grids in a one-dimensional space.

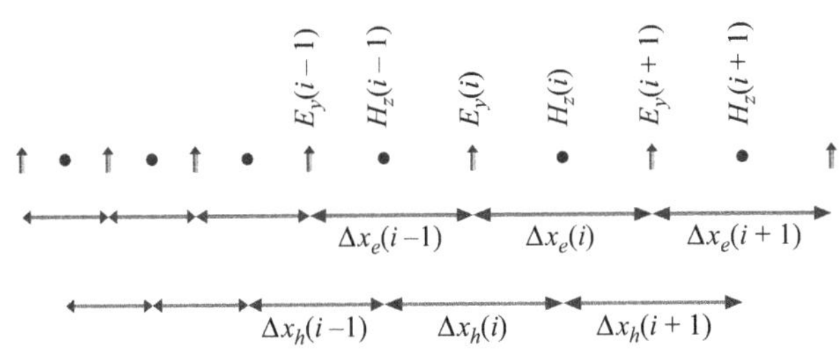

Figure 15.4 Electric and magnetic field positions in a nonuniform grid.

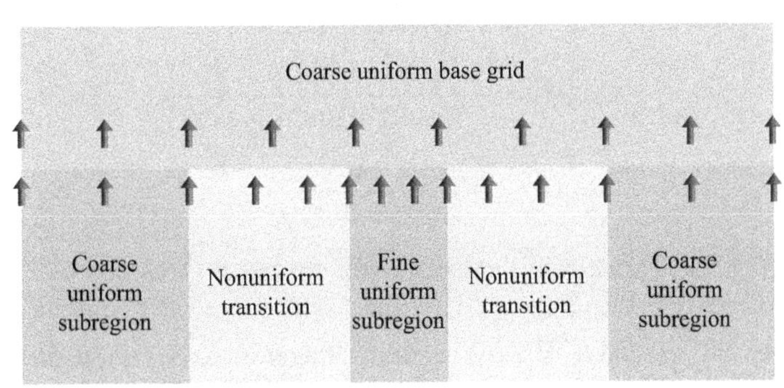

Figure 15.5 A fine grid subregion embedded in a coarse base grid.

first order accurate updating equations. However, it has been shown in [77] that, despite being first order accurate locally, Yee's scheme is second-order convergent regardless of mesh nonuniformity.

15.3 FDTD updating equations for the nonuniform grids

From Figure 15.4, one should notice that electric field components are located about the middle of the magnetic field grid cells, while the magnetic field components are located at the middle of the electric field grid cells. The cells in these two grids are different in size in the nonuniform regions, thus the cell sizes are denoted as Δx_e and Δx_h for electric and magnetic grids, respectively. Similarly, for a three-dimensional case we would have Δy_e, Δy_h, Δz_e, and Δz_h for the respective directions.

It is straightforward to construct general FDTD updating equations using the field and cell indexing scheme shown in Figure 15.4 for the one-dimensional case. Here, the E_y and H_z updating equations become

$$E_y^{n+1}(i) = C_{eye}(i) \times E_y^n(i) + C_{eyhz}(i) \times \left(H_z^{n+\frac{1}{2}}(i) - H_z^{n+\frac{1}{2}}(i-1)\right) + C_{eyj}(i) \times J_{iy}^{n+\frac{1}{2}}(i), \quad (15.7)$$

where

$$C_{eye}(i) = \frac{2\varepsilon_y(i) - \Delta t \sigma_y^e(i)}{2\varepsilon_y(i) + \Delta t \sigma_y^e(i)}$$

$$C_{eyhz}(i) = -\frac{2\Delta t}{\left(2\varepsilon_y(i) + \Delta t \sigma_y^e(i)\right)\Delta x_h(i)}$$

$$C_{eyj}(i) = -\frac{2\Delta t}{2\varepsilon_y(i) + \Delta t \sigma_y^e(i)}$$

and

$$H_z^{n+\frac{1}{2}}(i) = C_{hzh}(i) \times H_z^{n-\frac{1}{2}}(i) + C_{ezhy}(i) \times \left(E_y^n(i+1) - E_y^n(i)\right) + C_{hzm}(i) \times M_{iz}^n(i), \quad (15.8)$$

where

$$C_{hzh}(i) = \frac{2\mu_z(i) - \Delta t \sigma_z^m(i)}{2\mu_z(i) + \Delta t \sigma_z^m(i)}$$

$$C_{hzey}(i) = -\frac{2\Delta t}{(2\mu_z(i) + \Delta t \sigma_z^m(i))\Delta x_e(i)}$$

$$C_{hzm}(i) = -\frac{2\Delta t}{2\mu_z(i) + \Delta t \sigma_z^m(i)}.$$

Comparing (15.7) and (15.8) with those of the uniform case in (1.40) and (1.41) one can notice that the only difference is the usage of $\Delta x_e(i)$ and $\Delta x_h(i)$ instead of a constant Δx; The spatial distance between $H_z^{n+\frac{1}{2}}(i)$ and $H_z^{n+\frac{1}{2}}(i-1)$ is $\Delta x_h(i)$, therefore $\Delta x_h(i)$ is used in (15.7), whereas the spatial distance between $E_y^n(i+1)$ and $\Delta x_e(i)$ is $\Delta x_e(i)$, therefore $\Delta x_e(i)$ is used in (15.8).

Updating equations for nonuniform two and three-dimensional cases can be obtained easily by modifying the equations of the uniform case. For instance, the updating equation for E_x in the three-dimensional case (see Chapter 1 (1.26)) can be written as

$$\begin{aligned} E_x^{n+1}(i,j,k) &= C_{exe}(i,j,k) \times E_x^n(i,j,k) \\ &+ C_{exhz}(i,j,k) \times \left(H_z^{n+\frac{1}{2}}(i,j,k) - H_z^{n+\frac{1}{2}}(i,j-1,k)\right) \\ &+ C_{exhy}(i,j,k) \times \left(H_y^{n+\frac{1}{2}}(i,j,k) - H_y^{n+\frac{1}{2}}(i,j,k-1)\right) \\ &+ C_{exj}(i,j,k) \times J_{ix}^{n+\frac{1}{2}}(i,j,k), \end{aligned} \tag{15.9}$$

where

$$\begin{aligned} C_{exe}(i,j,k) &= \frac{2\varepsilon_x(i,j,k) - \Delta t\sigma_x^e(i,j,k)}{2\varepsilon_x(i,j,k) + \Delta t\sigma_x^e(i,j,k)} \\ C_{exhz}(i,j,k) &= \frac{2\Delta t}{\left(2\varepsilon_x(i,j,k) + \Delta t\sigma_x^e(i,j,k)\right)\Delta y_h(j)} \\ C_{exhy}(i,j,k) &= -\frac{2\Delta t}{\left(2\varepsilon_x(i) + \Delta t\sigma_x^e(i)\right)\Delta z_h(k)} \\ C_{exj}(i,j,k) &= -\frac{2\Delta t}{2\varepsilon_x(i,j,k) + \Delta t\sigma_x^e(i,j,k)}. \end{aligned}$$

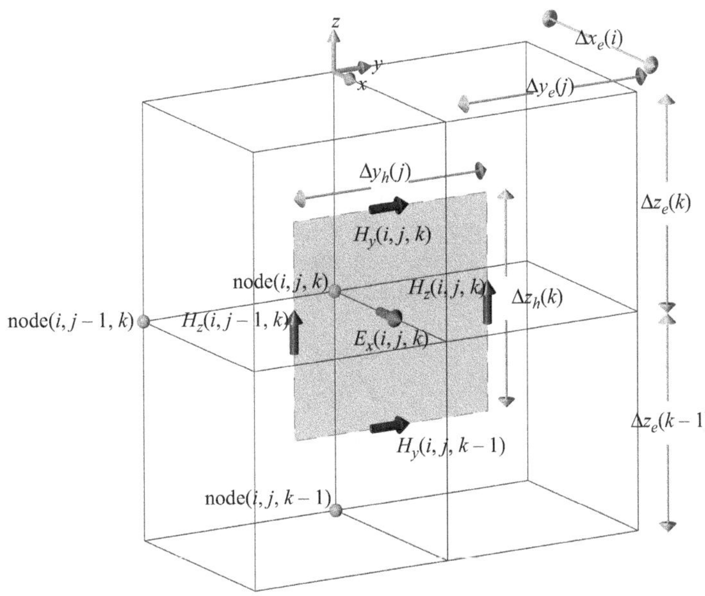

Figure 15.6 Electric field grid around $E_x(i,j,k)$, and magnetic grid cell sizes used to update $E_x(i,j,k)$.

Here the cell sizes $\Delta y_h(j)$ and $\Delta z_h(k)$ are used instead of Δy and Δz of the equation for the uniform grid. Figure 15.6 illustrates how the magnetic field components that surround the electric field component simulate Ampere's law and the path along which magnetic field curls conforms to the magnetic field grid.

Similarly, the updating equation for H_x in the three-dimensional case (Chapter 1, (1.29)) can be written as

$$\begin{aligned} H_x^{n+\frac{1}{2}}(i,j,k) = {} & C_{hxh}(i,j,k) \times H_x^{n-\frac{1}{2}}(i,j,k) \\ & + C_{hxey}(i,j,k) \times \left(E_y^n(i,j,k+1) - E_y^n(i,j,k)\right) \\ & + C_{hxez}(i,j,k) \times \left(E_z^n(i,j+1,k) - E_z^n(i,j,k)\right) \\ & + C_{hxm}(i,j,k) \times M_{ix}^n(i,j,k), \end{aligned} \tag{15.10}$$

where

$$C_{hxh}(i,j,k) = \frac{2\mu_x(i,j,k) - \Delta t \sigma_x^m(i,j,k)}{2\mu_x(i,j,k) + \Delta t \sigma_x^m(i,j,k)}$$

$$C_{hxey}(i,j,k) = \frac{2\Delta t}{\left(2\mu_x(i,j,k) + \Delta t \sigma_x^m(i,j,k)\right)\Delta z_e(k)}$$

$$C_{hxez}(i,j,k) = -\frac{2\Delta t}{\left(2\mu_x(i) + \Delta t \sigma_x^m(i)\right)\Delta y_e(j)}$$

$$C_{hxm}(i,j,k) = -\frac{2\Delta t}{2\mu_x(i,j,k) + \Delta t \sigma_x^e(i,j,k)}.$$

15.4 Active and passive lumped elements

In the previous section, it is shown that updating equations for nonuniform grid can be obtained by slightly modifying the updating equations of the uniform grid by using the right cell sizes. Readily available updating equations that model lumped elements on a uniform grid also can be modified appropriately and equations for the nonuniform case can be obtained by replacing the constant cell sizes by the indexed cell sizes of electric or magnetic field grids. For instance, for a voltage source in the z-direction the equation that updates E_z becomes (see Chapter 4 (4.10))

$$\begin{aligned} E_z^{n+1}(i,j,k) = {} & C_{eze}(i,j,k) \times E_z^n(i,j,k) \\ & + C_{ezhy}(i,j,k) \times \left(H_y^{n+\frac{1}{2}}(i,j,k) - H_y^{n+\frac{1}{2}}(i-1,j,k)\right) \\ & + C_{ezhx}(i,j,k) \times \left(H_x^{n+\frac{1}{2}}(i,j,k) - H_x^{n+\frac{1}{2}}(i,j-1,k)\right) \\ & + C_{ezs}(i,j,k) \times V_s^{n+\frac{1}{2}}(i,j,k), \end{aligned} \tag{15.11}$$

where

$$C_{eze}(i,j,k) = \frac{2\varepsilon_z(i,j,k) - \Delta t\sigma_z^e(i,j,k) - \dfrac{\Delta z_e(k)\Delta t}{R_s\Delta x_h(i)\Delta y_h(j)}}{2\varepsilon_z(i,j,k) + \Delta t\sigma_z^e(i,j,k) + \dfrac{\Delta z_e(k)\Delta t}{R_s\Delta x_h(i)\Delta y_h(j)}}$$

$$C_{ezhy}(i,j,k) = \frac{2\Delta t}{\left(2\varepsilon_z(i,j,k) + \Delta t\sigma_z^e(i,j,k) + \dfrac{\Delta z_e(k)\Delta t}{R_s\Delta x_h(i)\Delta y_h(j)}\right)\Delta x_h(i)}$$

$$C_{ezhx}(i,j,k) = -\frac{2\Delta t}{\left(2\varepsilon_z(i) + \Delta t\sigma_z^e(i) + \dfrac{\Delta z_e(k)\Delta t}{R_s\Delta x_h(i)\Delta y_h(j)}\right)\Delta y_h(j)}$$

$$C_{ezs}(i,j,k) = -\frac{2\Delta t}{\left(2\varepsilon_z(i,j,k) + \Delta t\sigma_z^e(i,j,k) + \dfrac{\Delta z_e(k)\Delta t}{R_s\Delta x_h(i)\Delta y_h(j)}\right)R_s\Delta x_h(i)\Delta y_h(j)}.$$

In (15.11), the cell size Δz_e in the electric field grid is used since the voltage source is along the z direction and conforming to the electric field grid. Other cell sizes Δx_h and Δy_h arise from the finite difference approximations of partial derivatives with respect to x and y, respectively, as discussed before.

Setting the source term in (15.11) to zero results in the updating equation for a resistor with resistance R as (see Chapter 4 (4.16))

$$\begin{aligned} E_z^{n+1}(i,j,k) = {} & C_{eze}(i,j,k) \times E_z^n(i,j,k) \\ & + C_{ezhy}(i,j,k) \times \left(H_y^{n+\frac{1}{2}}(i,j,k) - H_y^{n+\frac{1}{2}}(i-1,j,k)\right) \\ & + C_{ezhx}(i,j,k) \times \left(H_x^{n+\frac{1}{2}}(i,j,k) - H_x^{n+\frac{1}{2}}(i,j-1,k)\right), \end{aligned} \tag{15.12}$$

where

$$C_{eze}(i,j,k) = \frac{2\varepsilon_z(i,j,k) - \Delta t\sigma_z^e(i,j,k) - \dfrac{\Delta z_e(k)\Delta t}{R\Delta x_h(i)\Delta y_h(j)}}{2\varepsilon_z(i,j,k) + \Delta t\sigma_z^e(i,j,k) + \dfrac{\Delta z_e(k)\Delta t}{R\Delta x_h(i)\Delta y_h(j)}}$$

$$C_{ezhy}(i,j,k) = \frac{2\Delta t}{\left(2\varepsilon_z(i,j,k) + \Delta t\sigma_z^e(i,j,k) + \dfrac{\Delta z_e(k)\Delta t}{R\Delta x_h(i)\Delta y_h(j)}\right)\Delta x_h(i)}$$

$$C_{ezhx}(i,j,k) = -\frac{2\Delta t}{\left(2\varepsilon_z(i) + \Delta t\sigma_z^e(i) + \dfrac{\Delta z_e(k)\Delta t}{R\Delta x_h(i)\Delta y_h(j)}\right)\Delta y_h(j)}.$$

The equation for a capacitor with capacitance C can be obtained as (see Chapter 4 (4.20))

$$
\begin{aligned}
E_z^{n+1}(i,j,k) &= C_{eze}(i,j,k) \times E_z^n(i,j,k) \\
&+ C_{ezhy}(i,j,k) \times \left(H_y^{n+\frac{1}{2}}(i,j,k) - H_y^{n+\frac{1}{2}}(i-1,j,k)\right) \\
&+ C_{ezhx}(i,j,k) \times \left(H_x^{n+\frac{1}{2}}(i,j,k) - H_x^{n+\frac{1}{2}}(i,j-1,k)\right),
\end{aligned} \tag{15.13}
$$

where

$$
C_{eze}(i,j,k) = \frac{2\varepsilon_z(i,j,k) - \Delta t \sigma_z^e(i,j,k) - \dfrac{2C\Delta z_e(k)}{\Delta x_h(i)\Delta y_h(j)}}{2\varepsilon_z(i,j,k) + \Delta t \sigma_z^e(i,j,k) + \dfrac{2C\Delta z_e(k)}{\Delta x_h(i)\Delta y_h(j)}}
$$

$$
C_{ezhy}(i,j,k) = \frac{2\Delta t}{\left(2\varepsilon_z(i,j,k) + \Delta t \sigma_z^e(i,j,k) + \dfrac{2C\Delta z_e(k)}{\Delta x_h(i)\Delta y_h(j)}\right)\Delta x_h(i)}
$$

$$
C_{ezhx}(i,j,k) = -\frac{2\Delta t}{\left(2\varepsilon_z(i) + \Delta t \sigma_z^e(i) + \dfrac{2C\Delta z_e(k)}{\Delta x_h(i)\Delta y_h(j)}\right)\Delta y_h(j)}.
$$

The equation for an inductor with an inductance L becomes (see Chapter 4 (4.24))

$$
\begin{aligned}
E_z^{n+1}(i,j,k) &= C_{eze}(i,j,k) \times E_z^n(i,j,k) \\
&+ C_{ezhy}(i,j,k) \times \left(H_y^{n+\frac{1}{2}}(i,j,k) - H_y^{n+\frac{1}{2}}(i-1,j,k)\right) \\
&+ C_{ezhx}(i,j,k) \times \left(H_x^{n+\frac{1}{2}}(i,j,k) - H_x^{n+\frac{1}{2}}(i,j-1,k)\right) \\
&+ C_{ezj}(i,j,k) \times J_{iz}^{n+\frac{1}{2}}(i,j,k),
\end{aligned} \tag{15.14}
$$

where

$$
C_{eze}(i,j,k) = \frac{2\varepsilon_z(i,j,k) - \Delta t \sigma_z^e(i,j,k)}{2\varepsilon_z(i,j,k) + \Delta t \sigma_z^e(i,j,k)}
$$

$$
C_{ezhy}(i,j,k) = \frac{2\Delta t}{(2\varepsilon_z(i,j,k) + \Delta t \sigma_z^e(i,j,k))\Delta x_h(i)}
$$

$$
C_{ezhx}(i,j,k) = -\frac{2\Delta t}{(2\varepsilon_z(i) + \Delta t \sigma_z^e(i))\Delta y_h(j)}
$$

$$
C_{ezj}(i,j,k) = -\frac{2\Delta t}{(2\varepsilon_z(i,j,k) + \Delta t \sigma_z^e(i,j,k))}.
$$

It should be noted that during the FDTD time-marching iteration, at every time step the new value of $J_{iz}^{n+\frac{1}{2}}(i,j,k)$ should be calculated using

$$J_{iz}^{n+\frac{1}{2}}(i,j,k) = J_{iz}^{n-\frac{1}{2}}(i,j,k) + \frac{\Delta t \Delta z_e(k)}{L\Delta x_h(i)\Delta y_h(j)} E_z^n(i,j,k). \tag{15.15}$$

15.5 Defining objects snapped to the electric field grid

In Chapter 3, Section 3.4, the definition of material parameters is discussed for the case where all objects in the FDTD space are assumed to snap to the uniform electric field grid. The subcell averaging schemes discussed in Section 3.3 can be employed for the case of nonuniform grid as well. Figure 15.7 shows four Yee cells, each of which has a different size and a different permittivity. Since the material component $\varepsilon_z(i,j,k)$ is tangential to the four surrounding media types the equivalent permittivity for $\varepsilon_z(i,j,k)$ can be given as

$$\varepsilon_z(i,j,k) = \frac{\Delta x_1 \Delta y_1 \varepsilon(i-1,j-1,k) + \Delta x_1 \Delta y_2 \varepsilon(i-1,j,k) + \Delta x_2 \Delta y_1 \varepsilon(i,j-1,k) + \Delta x_2 \Delta y_2 \varepsilon(i,j,k)}{(\Delta x_1 + \Delta x_2) \times (\Delta y_1 + \Delta y_2)}. \tag{15.16}$$

Similarly the electric conductivity material components are defined at the same positions as the permittivity components, the same averaging scheme can be employed to get the equivalent electric conductivity $\sigma_z^e(i,j,k)$ as

$$\sigma_z^e(i,j,k) = \frac{\Delta x_1 \Delta y_1 \sigma^e(i-1,j-1,k) + \Delta x_1 \Delta y_2 \sigma^e(i-1,j,k) + \Delta x_2 \Delta y_1 \sigma^e(i,j-1,k) + \Delta x_2 \Delta y_2 \sigma^e(i,j,k)}{(\Delta x_1 + \Delta x_2) \times (\Delta y_1 + \Delta y_2)}. \tag{15.17}$$

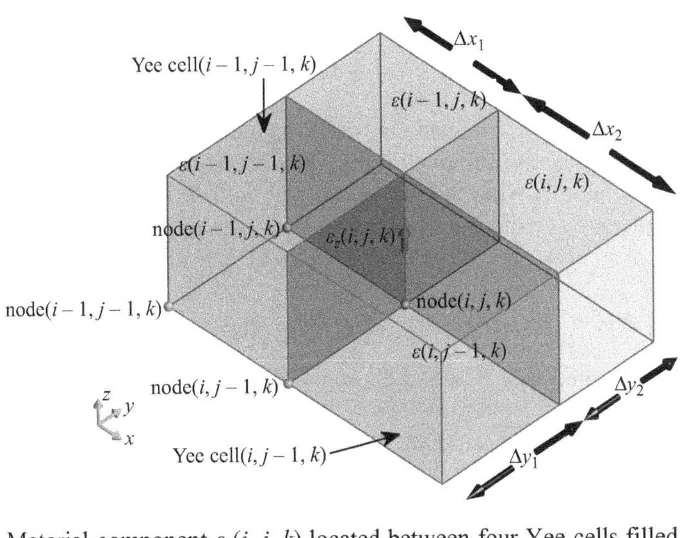

Figure 15.7 Material component $\varepsilon_z(i,j,k)$ located between four Yee cells filled with four different material types.

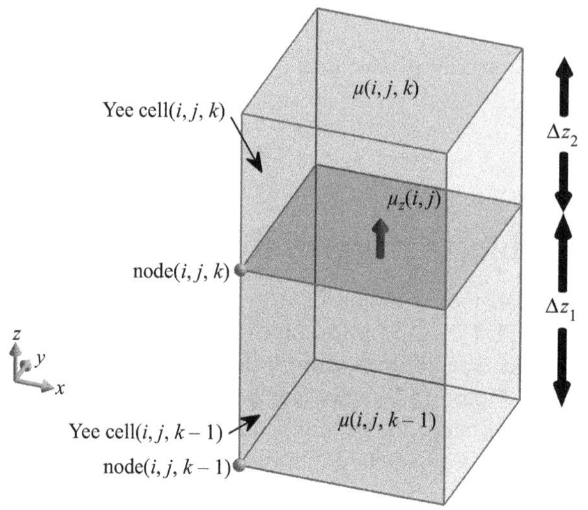

Figure 15.8 Material component $\mu_z(i, j, k)$ located between two Yee cells filled with two different material types.

Since all objects are assumed to snap to the electric field grid, then the material components associated with magnetic field components, permeability and magnetic conductivity, are located between two cells and are oriented normal to the cell boundaries as illustrated in Figure 15.8. In this case, the permeability $\mu_z(i, j, k)$ can be given as

$$\mu_z(i, j, k) = \frac{\mu(i, j, k-1)\mu(i, j, k) \times (\Delta z_1 \Delta z_2)}{\mu(i, j, k-1)\Delta z_2 + \mu(i, j, k) \times \Delta z_1}, \tag{15.18}$$

similarly, the magnetic conductivity $\sigma_z^m(i, j, k)$ becomes

$$\sigma_z^m(i, j, k) = \frac{\sigma^m(i, j, k-1)\sigma^m(i, j, k) \times (\Delta z_1 \Delta z_2)}{\sigma^m(i, j, k-1)\Delta z_2 + \sigma^m(i, j, k) \times \Delta z_1}. \tag{15.19}$$

So far we have discussed the development of the updating equations and construction of the material grids for the case of nonuniform cell sizes. It is straightforward to develop the required equations for other types of calculations within FDTD, such as near-field to far-field transformation, plane-wave incidence etc., since the same principles apply.

15.6 MATLAB® implementation of nonuniform grids

The previously presented base MATLAB program, which uses constant values of Δx, Δy, and Δz over the entire problem space, can be modified to support nonuniform grids. In the following sections, an implementation of nonuniform grids will be presented, which assumes that a problem space is initially constructed based on a uniform coarse grid that uses constant

Δx, Δy, and Δz values as the base cell sizes. Then fine uniform grid subregions are inserted into the main coarse grid and nonuniform transitions are established between the coarse and fine subregions as illustrated in Figure 15.5.

15.6.1 Definition of subregions

In the presented implementation, first it is assumed that the problem space is constructed based on a uniform grid and the base uniform grid cell size is defined as usual by assigning values to Δx, Δy, and Δz in the ***define_problem_space_parameters*** subroutine, as illustrated in Listing 15.1. Then subregions that will be inserted into the base grid are defined in the same file. Listing 15.1 shows definition of five subregions: three along the x direction while the other two along the y direction using a structure named as **subregions**. The field **cell_size**

Listing 15.1 define_problem_space_parameters.m

```
% Dimensions of a unit cell in x, y, and z directions (meters)
dx = 0.002;
dy = 0.002;
dz = 0.00095;

% direction can be 'x', 'y', or 'z'
subregions(1).direction = 'x';
subregions(1).cell_size = 1e-3;
subregions(1).region_start = 27e-3;
subregions(1).region_end   =  33e-3;
subregions(1).transition_length = 5e-3;

subregions(2).direction = 'y';
subregions(2).cell_size = 1e-3;
subregions(2).region_start = 7e-3;
subregions(2).region_end   =  13e-3;
subregions(2).transition_length = 5e-3;

subregions(3).direction = 'x';
subregions(3).cell_size = 1e-3;
subregions(3).region_start = -2e-3;
subregions(3).region_end   =  4e-3;
subregions(3).transition_length = 5e-3;

subregions(4).direction = 'x';
subregions(4).cell_size = 1e-3;
subregions(4).region_start = 56e-3;
subregions(4).region_end   =  62e-3;
subregions(4).transition_length = 5e-3;

subregions(5).direction = 'y';
subregions(5).cell_size = 1e-3;
subregions(5).region_start = 28e-3;
subregions(5).region_end   =  32e-3;
subregions(5).transition_length = 5e-3;
```

defines the size of a cell of the uniform fine subregion along the respective direction. These cells will fill the region determined by fixed coordinates **region_start** and **region_end**. The **transition_length** is the length on which the cell sizes are gradually increased or decreased to establish a smooth transition between the base cell size and the subregion cell size.

15.6.2 Initialization of subregions

As discussed in Chapter 3, one of the first tasks is to determine the size of a problem space, based on the objects in the domain and the boundary definitions, in the subroutine ***initialize_fdtd_material_grid*** by calling the subroutine ***calculate_domain_size***. Then another subroutine, named as ***initialize_subregions***, is called to initialize the subregions.

Listing 15.2 shows implementation of ***initialize_subregions***. Here, the base uniform grid is constructed using the problem space coordinates and dimensions, and the base cell sizes: three arrays are constructed to store the coordinates of the nodes of the electric field grid along x, y, and z directions (**node_coordinates_xe**, **node_coordinates_ye**, and

Listing 15.2 initialize_subregions.m

```
number_of_subregions = 0;
if exist('subregions','var')
    number_of_subregions = size(subregions,2);
end

% number of cells in x, y, and z directions
nx = round(fdtd_domain.size_x/dx);
ny = round(fdtd_domain.size_y/dy);
nz = round(fdtd_domain.size_z/dz);
node_coordinates_xe = (0:nx)*dx + fdtd_domain.min_x;
node_coordinates_ye = (0:ny)*dy + fdtd_domain.min_y;
node_coordinates_ze = (0:nz)*dz + fdtd_domain.min_z;

node_coordinates_xh = (0:nx+1)*dx + fdtd_domain.min_x - dx/2;
node_coordinates_yh = (0:ny+1)*dy + fdtd_domain.min_y - dy/2;
node_coordinates_zh = (0:nz+1)*dz + fdtd_domain.min_z - dz/2;

for ind = 1:number_of_subregions
    switch (subregions(ind).direction)
    case 'x'
    [node_coordinates_xe, node_coordinates_xh] ...
     = insert_subregion_into_domain ...
    (node_coordinates_xe, node_coordinates_xh, subregions(ind), dx);
    case 'y'
    [node_coordinates_ye, node_coordinates_yh] ...
     = insert_subregion_into_domain ...
    (node_coordinates_ye, node_coordinates_yh, subregions(ind), dy);
    case 'z'
    [node_coordinates_ze, node_coordinates_zh] ...
    = insert_subregion_into_domain ...
    (node_coordinates_ze, node_coordinates_zh, subregions(ind), dz);
    end
end
```

```
if isfield(boundary,'is_nonuniform_cpml')
    if (boundary.is_nonuniform_cpml)
        set_nonuniform_cpml_grid;
    end
end

% reset number of cells
nx = size(node_coordinates_xe, 2)-1;
ny = size(node_coordinates_ye, 2)-1;
nz = size(node_coordinates_ze, 2)-1;

% create cell size arrays
cell_sizes_xe = node_coordinates_xe(2:nx+1) - node_coordinates_xe(1:nx);
cell_sizes_ye = node_coordinates_ye(2:ny+1) - node_coordinates_ye(1:ny);
cell_sizes_ze = node_coordinates_ze(2:nz+1) - node_coordinates_ze(1:nz);

% create cell size arrays conforming the magnetic field grid
cell_sizes_xh = node_coordinates_xh(2:nx+2) - node_coordinates_xh(1:nx+1);
cell_sizes_yh = node_coordinates_yh(2:ny+2) - node_coordinates_yh(1:ny+1);
cell_sizes_zh = node_coordinates_zh(2:nz+2) - node_coordinates_zh(1:nz+1);

fdtd_domain.node_coordinates_xe = node_coordinates_xe;
fdtd_domain.node_coordinates_ye = node_coordinates_ye;
fdtd_domain.node_coordinates_ze = node_coordinates_ze;

fdtd_domain.node_coordinates_xh = node_coordinates_xh;
fdtd_domain.node_coordinates_yh = node_coordinates_yh;
fdtd_domain.node_coordinates_zh = node_coordinates_zh;

% some frequently used auxiliary parameters
nxp1 = nx+1;   nyp1 = ny+1;   nzp1 = nz+1;
nxm1 = nx-1;   nxm2 = nx-2;   nym1 = ny-1;
nym2 = ny-2;   nzm1 = nz-1;   nzm2 = nz-2;
```

node_coordinates_ze), whereas three other arrays are constructed to store the coordinates of the nodes of the magnetic field grid (**node_coordinates_xh**, **node_coordinates_yh**, and **node_coordinates_zh**). Then a function, ***insert_subregion_into_domain***, is called to insert nonuniform subregions into the base grid and node coordinate arrays are determined accordingly. After node coordinates are obtained, the distances between the consecutive grid nodes are evaluated and stored as cell size arrays. For instance, **cell_sizes_xe** corresponds to the cells sizes Δx_e in Figure 15.4, whereas, **cell_sizes_xh** corresponds to the cells sizes Δx_h.

Listing 15.3 shows the implementation of ***insert_subregion_into_domain***. In this implementation, first, the cell size of the uniform part of the subregion is adjusted: if the distance between **region_start** and **region_end** is not an integer multiple of the **subregion.cell_size**, then the **subregion.cell_size** is adjusted such that the cells fit between the **region_start** and **region_end**. Then, the length of the transition region that is before the uniform subregion is adjusted: The node of the base electric field grid that is closest to the transition start point is set as the new transition start point. Then, the transition cell sizes are calculated according to the procedure described in Section 15.2.

Listing 15.3 insert_subregion_into_domain.m

```
function [node_coordinates_e, node_coordinates_h] ...
    = insert_subregion_into_domain ...
    (node_coordinates_e, node_coordinates_h, subregion, base_cell_size)

% adjust subregion cell size
subregion_length = ...
    subregion.region_end - subregion.region_start;
n_subregion_cells = ...
    round(subregion_length/subregion.cell_size);
subregion.cell_size = subregion_length/n_subregion_cells;

bcs  = base_cell_size;
srcs = subregion.cell_size;

n_nodes_e = size(node_coordinates_e,2);

% transition before the subregion (indicated with bsr)
% adjust transition length
tr_start_position = (subregion.region_start - ...
    subregion.transition_length);
[mval, I] = min(abs(node_coordinates_e - tr_start_position));
I = find(node_coordinates_e == node_coordinates_e(I(1)));
tr_start_node = I;
tr_length = subregion.region_start ...
    - node_coordinates_e(tr_start_node);

% find number of cells for transition
R = (tr_length + bcs)/(tr_length + srcs);
number_of_transition_cells = ...
    floor(log10(bcs/srcs)/log10(R))-1;

% find transition cell sizes
transition_cell_sizes = ...
    srcs*R.^(1:number_of_transition_cells);

% adjust cell sizes so that total length is equal to
% transition length
transition_cell_sizes = transition_cell_sizes ...
    * (tr_length / sum(transition_cell_sizes));

bsr_transition_cell_sizes = fliplr(transition_cell_sizes);
bsr_tr_start_node = tr_start_node;

% transition after the subregion (indicated with asr)
% adjust transition length
tr_end_position = (subregion.region_end + ...
    subregion.transition_length);
[mval,I] = min(abs(node_coordinates_e - tr_end_position));
I = find(node_coordinates_e == node_coordinates_e(I(1)));
tr_end_node = I(1);
tr_length = node_coordinates_e(tr_end_node) ...
    - subregion.region_end;
```

```
% find number of cells for transition
R = (tr_length + bcs)/(tr_length+srcs);
number_of_transition_cells = ...
    floor(log10(bcs/srcs)/log10(R))-1;

% find transition cell sizes
transition_cell_sizes = ...
    srcs*R.^(1:number_of_transition_cells);

% adjust cell sizes so that total length is equal to
% transition length
transition_cell_sizes = transition_cell_sizes ...
    * (tr_length / sum(transition_cell_sizes));

asr_transition_cell_sizes = transition_cell_sizes;
asr_tr_end_node = tr_end_node;

% create cell size array for subregion
subregion_cell_sizes_e = ...
    srcs * ones(1, n_subregion_cells);

% adjust node coordinates
sr_all_cs_e = [bsr_transition_cell_sizes ...
    subregion_cell_sizes_e asr_transition_cell_sizes];

nodes_before_subregion_e = node_coordinates_e(1:bsr_tr_start_node);
nodes_after_subregion_e = node_coordinates_e(asr_tr_end_node:end);
a_node = node_coordinates_e(bsr_tr_start_node);

n_subregion_cells = size(sr_all_cs_e,2);
subregion_node_coordinates_e = zeros(1, n_subregion_cells-1);

for si = 1:n_subregion_cells-1
    a_node = a_node + sr_all_cs_e(si);
    subregion_node_coordinates_e(si) = a_node;
end

node_coordinates_e = [nodes_before_subregion_e ...
    subregion_node_coordinates_e nodes_after_subregion_e];

n_nodes = size(node_coordinates_e, 2);

% magnetic field components are placed at the centers of the cells
node_coordinates_h = ...
    0.5*(node_coordinates_e(1:n_nodes-1)+node_coordinates_e(2:n_nodes));

node_coordinates_h = [node_coordinates_e(1)-bcs/2 node_coordinates_h ...
    node_coordinates_e(n_nodes)+bcs/2];
```

Similarly, the transition cell sizes are calculated for the transition region after the uniform region part as well. Once the cell sizes in the entire nonuniform subregion are obtained, corresponding nodes are inserted in the positions, where nonuniform subregion overlaps with the base grid, and the electric field grid array is updated. After the coordinates of the nodes in the electric field grid are determined, the coordinates of the nodes of the magnetic field grid are calculated as the center points of the electric field grid cells. It should be noted that two more cells are padded to the magnetic field grid, each on one side of the grid. Therefore, the first and the last electric field components are located in the middle of these two cells, and the lengths of these two cells can be used to update the corresponding electric field components.

15.6.3 Initialization of updating coefficients

For nonuniform cell sizes the FDTD program code needs to be modified accordingly wherever cell sizes are used. In this section, we will discuss the modification of the basic updating equations to illustrate the application of nonuniform grids. The same logic can be employed to modify other parts of the code as well wherever needed.

The modified implementation of the subroutine ***initialize_updating_coefficients*** is shown in Listing 15.4. For instance, the updating coefficient **Ceyhz** requires a division by Δx.

Listing 15.4 initialize_updating_coefficients.m

```
% General electric field updating coefficients
% Coeffiecients updating Ex
[DX, DY, DZ] = ndgrid(cell_sizes_xe, cell_sizes_yh, cell_sizes_zh);
Cexe  =  (2*eps_r_x*eps_0 - dt*sigma_e_x)./(2*eps_r_x*eps_0 + dt*sigma_e_x);
Cexhz =  (2*dt/DY)./(2*eps_r_x*eps_0 + dt*sigma_e_x);
Cexhy = -(2*dt/DZ)./(2*eps_r_x*eps_0 + dt*sigma_e_x);

% Coeffiecients updating Ey
[DX, DY, DZ] = ndgrid(cell_sizes_xh, cell_sizes_ye, cell_sizes_zh);
Ceye  =  (2*eps_r_y*eps_0 - dt*sigma_e_y)./(2*eps_r_y*eps_0 + dt*sigma_e_y);
Ceyhx =  (2*dt/DZ)./(2*eps_r_y*eps_0 + dt*sigma_e_y);
Ceyhz = -(2*dt/DX)./(2*eps_r_y*eps_0 + dt*sigma_e_y);

% Coeffiecients updating Ez
[DX, DY, DZ] = ndgrid(cell_sizes_xh, cell_sizes_yh, cell_sizes_ze);
Ceze  =  (2*eps_r_z*eps_0 - dt*sigma_e_z)./(2*eps_r_z*eps_0 + dt*sigma_e_z);
Cezhy =  (2*dt/DX)./(2*eps_r_z*eps_0 + dt*sigma_e_z);
Cezhx = -(2*dt/DY)./(2*eps_r_z*eps_0 + dt*sigma_e_z);

% General magnetic field updating coefficients
% Coeffiecients updating Hx
[DX, DY, DZ] = ndgrid(cell_sizes_xh, cell_sizes_ye, cell_sizes_ze);
Chxh  =  (2*mu_r_x*mu_0 - dt*sigma_m_x)./(2*mu_r_x*mu_0 + dt*sigma_m_x);
Chxez = -(2*dt/DY)./(2*mu_r_x*mu_0 + dt*sigma_m_x);
Chxey =  (2*dt/DZ)./(2*mu_r_x*mu_0 + dt*sigma_m_x);

% Coeffiecients updating Hy
[DX, DY, DZ] = ndgrid(cell_sizes_xe, cell_sizes_yh, cell_sizes_ze);
Chyh  =  (2*mu_r_y*mu_0 - dt*sigma_m_y)./(2*mu_r_y*mu_0 + dt*sigma_m_y);
Chyex = -(2*dt/DZ)./(2*mu_r_y*mu_0 + dt*sigma_m_y);
Chyez =  (2*dt/DX)./(2*mu_r_y*mu_0 + dt*sigma_m_y);

% Coeffiecients updating Hz
[DX, DY, DZ] = ndgrid(cell_sizes_xe, cell_sizes_ye, cell_sizes_zh);
Chzh  =  (2*mu_r_z*mu_0 - dt*sigma_m_z)./(2*mu_r_z*mu_0 + dt*sigma_m_z);
Chzey = -(2*dt/DX)./(2*mu_r_z*mu_0 + dt*sigma_m_z);
Chzex =  (2*dt/DY)./(2*mu_r_z*mu_0 + dt*sigma_m_z);
```

Ceyhz is used to update **Ey**, which requires the distance between the **Hz** components before and after **Ey** as illustrated in Figure 15.4. These **Hz** components are on the magnetic field grid, thus the distance between them is denoted as Δx_h, which corresponds to the **cell_sizes_xh** array in the code. Then, **Ceyhz** can be constructed as a three-dimensional array in open form as

```
for i=1:nx+1
for j=1:ny
        for k=1:nz+1
                Ceyhz(i,j,k) = -(2*dt/cell_sizes_xh(i)) ...
                        ./(2*eps_r_y(i,j,k)*eps_0+dt*sigma_e_y(i,j,k));
                end
        end
end
```

Here, if **cell_sizes_xh** can be reshaped as a three-dimensional array of size $(nx + 1, ny, nz + 1)$, which varies only in the x direction, then the for loops in the above code can be eliminated. For instance, a three-dimensional array, named as **DX**, that has the size $(nx + 1, ny, nz + 1)$ and values of **cell_sizes_xh** in the x direction is used below to replace the above code:

```
Ceyhz(i,j,k) = -(2*dt/DX)./(2*eps_r_y*eps_0 + dt*sigma_e_y);
```

Similarly, a three-dimensional array, named as **DZ**, that has the size $(nx + 1, ny, nz + 1)$ and values of **cell_sizes_zh** in the z direction can be used to update **Ceyhx**:

```
Ceyhx = (2*dt/DZ)./(2*eps_r_y*eps_0 + dt*sigma_e_y);
```

The three-dimensional arrays **DX** and **DZ**, used to update **Ceyhz** and **Ceyhx** above, can be constructed simultaneously by using MATLAB function ***ndgrid*** as

```
[DX, DY, DZ] = ndgrid(cell_sizes_xh, cell_sizes_ye, cell_sizes_zh);
```

Here, ***ndgrid*** generates three-dimensional arrays as **DX**, **DY**, and **DZ**, using the size and data of the one-dimensional arrays **cell_sizes_xh**, **cell_sizes_ye**, **cell_sizes_zh**. For instance, the length of **cell_sizes_xh** is $N_x + 1$, the length of **cell_sizes_ye** is N_y, and the length of **cell_sizes_zh** is $N_z + 1$. Then each of **DX**, **DY**, and **DZ** has the size of $(N_x + 1) \times N_y \times (N_z + 1)$. **DX** contains the data of **cell_sizes_xh** listed along its first dimension that is copied $N_y \times (N_z + 1)$ times on other dimensions. **DY** contains the data of **cell_sizes_ye** listed along its second dimension that is copied $(N_x + 1) \times (N_z + 1)$ times on other dimensions. Similarly, **DZ** contains the data of **cell_sizes_zh** listed along its third dimension that is copied $(N_x + 1) \times N_y$ times on other dimensions. One can refer to MATLAB help for further details of the ***ndgrid*** function.

Listing 15.5 initialize_fdtd_parameters_and_arrays.m

```
% Duration of a time step in seconds
min_dx = min(cell_sizes_xe);
min_dy = min(cell_siazes_ye);
min_dz = min(cell_sizes_ze);

dt = 1/(c*sqrt((1/min_dx^2)+(1/min_dy^2)+(1/min_dz^2)));
dt = courant_factor*dt;
```

Similarly, the updating coefficient of a magnetic field component requires the distances between the neighboring electric field components, which lie on the electric field grid, as illustrated in Figure 15.4. Thus, the cell sizes of the electric field grid are used in the updating equations of the magnetic field components as shown in Listing 15.4.

15.6.4 Initialization of time step duration

The Courant-Friedrichs-Lewy (CFL) condition requires that the time increment Δt has a value given by (2.6) as discussed in Section 2.1. If nonuniform grid is used, then the dimensions of the smallest cell need to be used in (2.6) to determine Δt. The subroutine ***initialize_fdtd_parameters_and_arrays*** can be modified as shown in Listing 15.5 such that the minimum values of Δx,Δy, and Δz are obtained from the respective cell size arrays and are used to determine **dt**.

15.7 Simulation examples

Two simulation examples will be presented here. These simulations were executed on personal computer with Intel(R) Core(TM)2 Quad CPU Q9550 running at 2.83 GHz.

15.7.1 Microstrip patch antenna

In this demonstration, the microstrip rectangular patch antenna discussed in Exercise 9.3 is simulated by using a fine grid, a coarse grid, and a nonuniform grid to evaluate the accuracy and computational advantage of the nonuniform grid. Table 15.1 shows the details of these three grids. It should be noted that the base grid of the nonuniform grid simulation is the same as the grid of the coarse grid simulation, which has Δx and Δy as 2 mm. In the non-uniform grid simulation, a fine grid subregion with a cell size of Δx and Δy equal to 1 mm is used around the port and along the four edges of the microstrip patch. Using finer cells improve the accuracy of the input impedance as well as the return loss calculations. Figure 15.9 illustrates the cell view of the patch antenna based on the nonuniform grid, where the fine regions and the nonuniform transitions to the coarse underlying grid can be observed. The definitions of the nonuniform subregions for this particular example are shown in Listing 15.1. In the fine grid simulation, the entire problem space has a cell size of Δx and Δy equal to 1 mm.

Table 15.1 Comparison of fine grid, coarse grid, and nonuniform grid simulations of a microstrip rectangular patch antenna.

Grid type	(Δx, Δy, Δz) (mm)	($N_x \times N_y \times N_z$)	Total number of cells	Number of time steps	Simulation time (minutes)
Fine	(1,1,0.95)	(96 × 76 × 38)	277,248	8,000	33.0
Coarse	(2,2,0.95)	(66 × 56 × 38)	140,448	5,755	10.6
Nonuniform	(2,2,0.95)	(77 × 62 × 38)	181,412	8,000	19.6

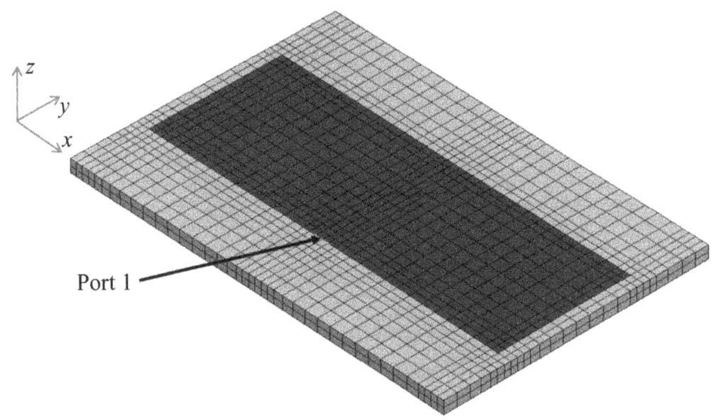

Figure 15.9 Cell view of the microstrip rectangular patch antenna.

The fine grid and the nonuniform grid simulations are performed for 8,000 time steps, while the coarse grid simulation is performed for 5,755 time steps using a Gaussian pulse with $\tau = 3.336 \times 10^{-11}$ as a source. It should be noted that in the fine grid and the nonuniform grid simulations the minimum cell sizes are the same, thus the duration of a time step, Δt, that is dictated by the CFL stability condition, is the same. Since the minimum cell size in the coarse grid simulation is larger, the time step is larger. The different numbers of time steps chosen in these simulations are necessary in order to simulate the transient response of these three cases for the same duration, which is 13.62 ns.

Each of these three cases is simulated and computation time is recorded in Table 15.1. Compared with the coarse grid simulation, the computational overhead of nonuniform grid is significantly less than that of the fine grid.

Figure 15.10 shows the comparison of the power reflection coefficient of fine grid, coarse grid, and nonuniform grid simulations of this patch antenna. Here we can use the fine grid simulation result as a reference for accuracy comparison with the coarse and the nonuniform grids. The results imply that accuracy of simulations can be improved by modeling critical areas of a geometry using fine grid subregions, thus employing nonuniform grids, compared with coarse grid simulations with the overhead of increased memory requirement and computation time, which are still significantly less than the fine grid requirements.

15.7.2 Three-pole microstrip low-pass filter

The second example is a three-pole low-pass filter presented in [78]. Dimensions of this filter are shown in Figure 15.11. The boundaries of the problem space are terminated by PEC on

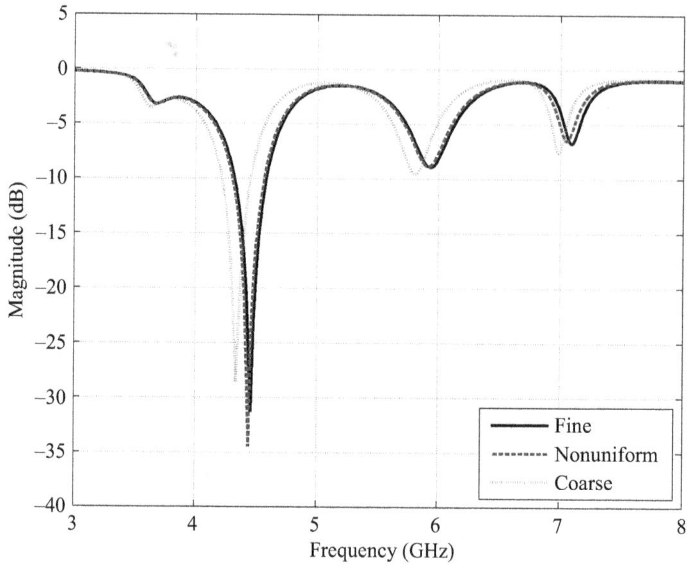

Figure 15.10 Comparison of power reflection coefficient of fine grid, coarse grid, and nonuniform grid simulations of a microstrip rectangular patch antenna.

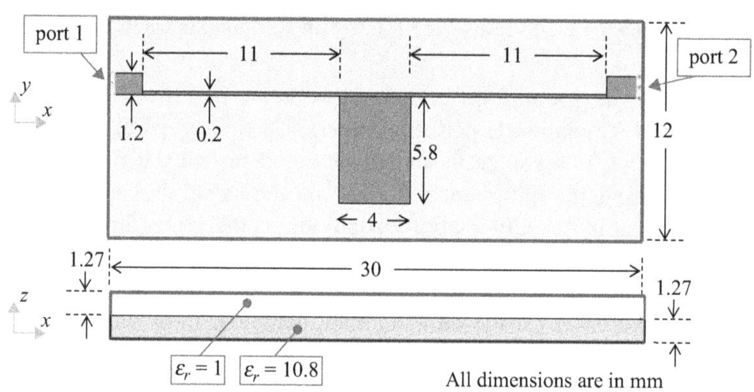

Figure 15.11 Geometry and dimensions of a three-pole low-pass filter.

all sides. The most critical dimension in this circuit is the 0.2 mm wide microstrip lines that connect the center stub to the wide feeding lines on each side. Similar to the previous example, the circuit is simulated by using a fine grid, a coarse grid, and a nonuniform grid to evaluate the performance of the nonuniform grid.

Table 15.2 shows the details of these three grids. Here, the base grid of the nonuniform grid simulation is the same as the grid of the coarse grid simulation. In the nonuniform grid simulation, as shown in Listing 15.6, a fine grid subregion with a cell size of Δy equal to 0.1 mm is used around the narrow strip to model that region more accurately.

Table 15.2 Comparison of fine grid, coarse grid, and nonuniform grid simulations of a three-pole low-pass microstrip filter.

Grid type	(Δx, Δy, Δz) (mm)	($N_x \times N_y \times N_z$)	Total number of cells	Number of time steps	Simulation time (minutes)
Fine	(0.1,0.1,0.127)	($300 \times 120 \times 20$)	720, 000	200,000	1,800
Coarse	(0.2,0.2,0.127)	($150 \times 60 \times 20$)	180,000	130,764	275
Nonuniform	(0.2,0.2,0.127)	($150 \times 76 \times 20$)	228,000	168,966	460

Listing 15.6 define_problem_space_parameters.m

```
% Dimensions of a unit cell in x, y, and z directions (meters)
dx = 0.2e-3;
dy = 0.2e-3;
dz = 0.127e-3;

% direction can be 'x', 'y', or 'z'
subregions(1).direction = 'y';
subregions(1).cell_size = 0.1e-3;
subregions(1).region_start = 7.4e-3;
subregions(1).region_end   = 9.4e-3;
subregions(1).transition_length = 2e-3;
```

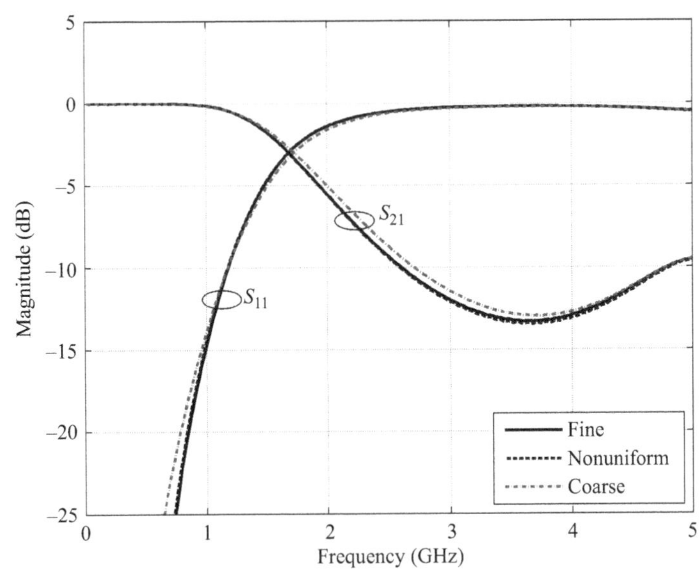

Figure 15.12 Comparison of scattering parameters of fine grid, coarse grid, and nonuniform grid simulations of a three-pole low-pass microstrip filter.

Each of these three cases is simulated for 37 ns using a Gaussian pulse with $\tau = 4.2362 \times 10^{-12}$ as a source. The number of time steps and total computation times are recorded in Table 15.2. Time results show that although the nonuniform simulation takes longer time than the coarse simulation, it takes much shorter time than the fine grid simulation. Figure 15.12 shows the scattering parameters of these three cases compared with each other, which further justifies the benefit of appropriate use of fine grid subregions through nonuniform grids where needed despite the slightly increased computational overhead.

CHAPTER 16

Graphics processing unit acceleration of finite-difference time-domain method

Recent developments in the design of graphics processing units (GPUs) have been occurring at a much greater pace than with central processor units (CPUs) and very powerful processing units have been designed solely for the processing of computer graphics. For instance, the Tesla® K40 GPU includes 2,880 cores and 12 GB memory, and it can reach to 4.29 Tflops peak single precision floating point performance. Due to this potential in faster computations, the GPUs have received the attention of the scientific computing community. Initially these cards were designed for computer graphics and floating precision arithmetic has been sufficient for such applications. Due to the demand of higher precision arithmetic from the scientific community, the vendors have started to develop graphics cards that support double precision arithmetic as well by introducing a new generation of graphical computation cards.

The computational electromagnetics community as well has started to utilize the computational power of graphics cards, and in particular, several implementations of finite-difference time-domain (FDTD) [79–89] method have been reported in the literature. For instance, Brook is used as the programming language in [79–85], High Level Shader Language (HLSL) is reported in [86], while the FDTD implementations in [87–89] are based on OpenGL. Relatively recently, the introduction of the Compute Unified Device Architecture (CUDA) [90] development environment from NVIDIA made GPU computing much easier. CUDA is a general purpose parallel computing architecture. To program the CUDA architecture, developers can use C, which can then be run at great performance on a CUDA-enabled processor [91]. CUDA has been reported as the programming environment for implementation of FDTD in several publications, while [92] and [93] illustrate methods to improve the efficiency of FDTD using CUDA, which can be used as guidelines while programming FDTD using CUDA.

It should be noted that OpenCL [94] has been introduced recently as a programming platform to develop codes on parallel devices. It enables programming on parallel architectures including GPU. It has been used to develop FDTD implementations as well [95]. Although OpenCL provides an alternative platform to CUDA, there is a large community that uses CUDA and the technical support for CUDA is well established, therefore, CUDA will keep its position as the prominent GPU programming platform for the near future.

16.1 GPU programming using CUDA

CUDA, as other languages or programming platforms, has its advantages and disadvantages. As the major advantages: CUDA is easier to learn compared with other alternatives and NVIDIA provides extensive support to developers and users. One major disadvantage is that CUDA can run only on CUDA-enabled NVIDIA cards. While OpenCL and DirectCompute are becoming the new alternatives to CUDA as programming languages on modern GPU-based architectures, CUDA may keep its popularity for scientific computing due to vast learning resources available for developers. In this chapter we will illustrate an implementation of a two-dimensional FDTD program using CUDA.

In order to start programming with CUDA, one need to install CUDA drivers, CUDA Toolkit that would include C/C++ compiler, Visual Profiler, and GPU-accelerated BLAS, FFT, Sparse Matrix, and RNG libraries. Another crucial component is GPU Computing Software Development Kit (SDK), which includes several tools and code samples. All these components are available at NVIDIA's web portal for Windows, Linux, and Mac OS X operating systems.

In order to program using CUDA, one needs to know the CUDA terminology and architecture. NVIDIA CUDA Programming Guide [96], available at NVIDIA's web portal, is a good resource to learn CUDA. In this section, brief descriptions of some concepts in CUDA are summarized from the Programming Guide in order to prepare the reader for the discussions that follow.

16.1.1 Host and device

CUDA provides an extended version of C language for programming. Thus it is relatively easy for a C programmer to develop codes for CUDA. In CUDA's programming model the part of the computer architecture which includes CPU is referred to as "host", while the GPU based card is referred to as "device". A CUDA program would consist of parts that would run on the host and the device: the main C program would run on the host, while the code including parallel computations would execute on the device. Basically, the "device" is a physically separate device on which the CUDA threads execute. These threads can be launched by the program running on the host.

Both the host and the device maintain their own DRAM, referred to as host memory and device memory, respectively. Generally, for data parallel computations, data is transferred from the host memory to the device memory, computations are performed on the device, and results are transferred back to host memory. For instance, Listing 16.1 shows a small program which adds two vectors. First, vectors A, B, and C are allocated on host (CPU) memory, and A and B are initialized with some numbers. Then corresponding vectors are allocated on device (GPU) memory. Vectors A and B are copied to device memory. Then a function is executed on the device which adds these vectors and stores the results in a vector on the device memory. The resulting vector is then copied back to vector C on the host memory to provide the expected results.

Listing 16.1 Addition of two vectors in CUOA.

```
#include <stdlib.h>
#include <stdio.h>

// Kernel definition
__global__ void AddVectors(int* A, int* B, int* C)
{
        int i = threadIdx.x;
        C[i] = A[i] + B[i];
}

int main()
{
        int N = 256;
        int mem_size = N*sizeof(int);

        // Allocate memory for 1D arrays on host memory
        int* A = (int*) malloc(mem_size);
        int* B = (int*) malloc(mem_size);
        int* C = (int*) malloc(mem_size);

        // initialize arrays
        for (int i=0; i<N; i++) A[i] = 2*i;
        for (int i=0; i<N; i++) B[i] = 3*i;

        // allocate device memory
    int* d_A;
    int* d_B;
    int* d_C;
        cudaMalloc( (void**) &d_A, mem_size);
        cudaMalloc( (void**) &d_B, mem_size);
        cudaMalloc( (void**) &d_C, mem_size);

    // copy host memory to device
        cudaMemcpy( d_A, A, mem_size, cudaMemcpyHostToDevice);
        cudaMemcpy( d_B, B, mem_size, cudaMemcpyHostToDevice);

        // Add vectors on device
        AddVectors<<<1, N>>>(d_A, d_B, d_C);

    // copy result from device to host
    cudaMemcpy( C, d_C, mem_size, cudaMemcpyDeviceToHost);

        // dsiplay results
        for (int i=0;i<N;i++) printf("%d ",C[i]);

    // cleanup memory
    free(A);    free(B);        free(C);
        cudaFree(d_A);
        cudaFree(d_B);
        cudaFree(d_C);
}
```

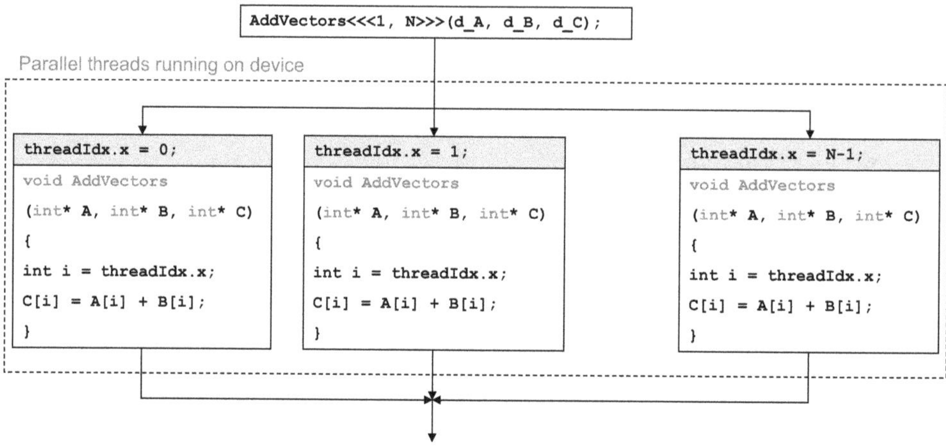

Figure 16.1 Parallel threads executing the same code on different set of data identified by the threadIdx.x value.

16.1.1.1 Kernels

A C function defined to run on the device is called a kernel. A kernel is defined using the __global__ declaration specifier as shown in Listing 16.1. A kernel, when called, as illustrated in Figure 16.1, executes N times in parallel by N different CUDA threads, as opposed to only once like regular C functions. Each CUDA thread executes the same kernel code, but for a different section of data accessed by an index identifying the thread. The number of CUDA threads for each call is specified using <<<...>>> syntax. In the example code, <<<1, N>>> indicates that N threads are executed in parallel. In each thread a pair-wise addition of elements in arrays A and B are performed. Here each thread has a unique thread ID, which is an integer number between 0 and $N-1$ in the given example. The thread ID is accessible within the kernel through the built-in threadIdx variable. A thread uses its specific thread ID, and uses it to access the array elements with the same index as the thread ID, and perform calculations on them. If the number of threads is the same as the number of array elements, then the entire array is processed through one-to-one mapping between array elements and the threads.

16.1.2 Thread hierarchy

The thread ID threadIdx can be used to map threads to data elements in arrays that these threads process. This mapping enables data parallel computations, i.e., running the same code for different sets of data. The variable threadIdx is a 3-component vector, so that threads can be identified using a one-dimensional, two-dimensional, or three-dimensional thread index, forming a one-dimensional, two-dimensional, or three-dimensional thread block. This provides a natural way to invoke computation across the elements in a domain such as a vector (one-dimensional array), matrix (two-dimensional array), or field (three-dimensional array). As an example, the following code in Listing 16.2 adds two three-dimensional arrays A and B of size $M \times N \times P$ and stores the result into array C. Here there is a one-to-one mapping between array elements and the threads in a three-dimensional thread block.

Listing 16.2 Addition of three dimensional arrays.

```
__global__ void AddArrays3D (int A[M][N][P], int B[M][N][P], int
C[M][N][P])
{
        int i = threadIdx.x;
        int j = threadIdx.y;
        int k = threadIdx.z;
        C[i][j][k] = A[i][j][k] + B[i][j][k];
}
int main()
{
...
        int M = 8;
        int N = 8;
        int P = 4;
        dim3 block_size(M, N, P);
        AddArrays3D<<<1, block_size>>>(A, B, C);
...
}
```

In CUDA, a number of threads form a thread block. For instance, in this example code, a thread block of M × N × P threads is defined by the statement "`dim3 block_size (M, N, P);`". However, there is a limit for the number of threads that can form a block: in NVIDIA graphics cards with compute capability 1.x, the maximum number of threads in a block is 512, while it is 1024 in cards with compute capability 2.x. In order to map larger number of threads, one can define a grid which consists of a number of equally-shaped thread blocks, so that the total number of threads in the grid is equal to the number of threads per block times the number of blocks. The grid of thread blocks can be one-, two-, or three-dimensional. For instance, Figure 16.2 illustrates a two-dimensional grid of 2×2 blocks, where each block is a two-dimensional array of 2×2 threads. The given grid thus includes 16 threads.

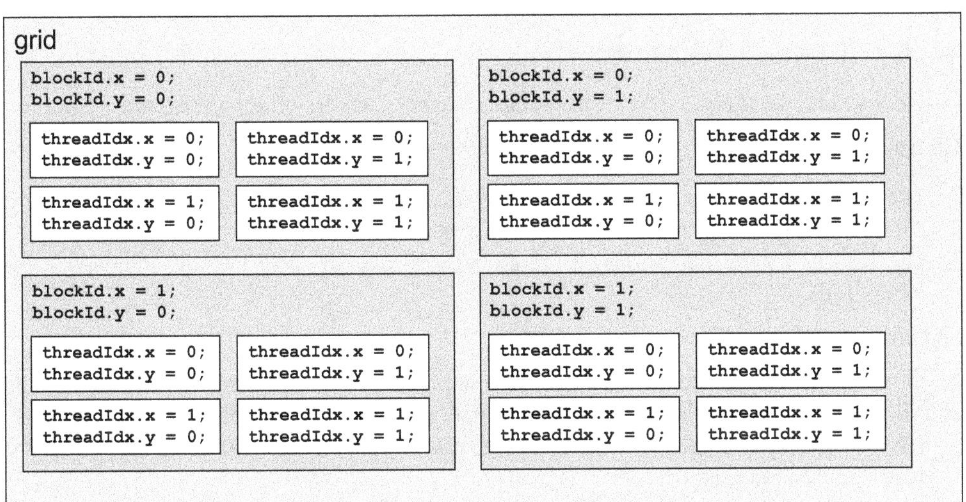

Figure 16.2 Grid of thread blocks.

Listing 16.3 Addition of two dimensional arrays.

```
__global__ void AddArrays2D(int A[M][N], int B[M][N], int C[M][N])
{
int i = blockIdx.x * blockDim.x + threadIdx.x;
int j = blockIdx.y * blockDim.y + threadIdx.y;
C[i][j] = A[i][j] + B[i][j];
}
int main()
{
...
        dim3 grid_size(3, 4);
        dim3 block_size(16,32);
        AddArrays2D<<< grid_size, block_size>>>(A, B, C);
...
}
```

The dimension of the grid is specified by the first parameter of the <<<...>>> syntax. Listing 16.3 shows a code that performs pair-wise addition of two-dimensional arrays. In this example code below, a 3×4 grid is defined, where each thread block in the grid includes 16×32 threads. Thus the threads can be easily mapped to two-dimensional 48×128 size arrays. Each block within the grid can be identified by a one-dimensional or two-dimensional index accessible within the kernel through the built-in blockIdx variable. The dimension of the thread block is accessible within the kernel through the built-in blockDim variable. It is straightforward to map threads to data elements in arrays using the blockDim, blockIdx, and threadIdx internal variables as shown in the kernel of Listing 16.3.

It should be noted that thread blocks execute independent from each other; therefore, one should consider this independence while developing an algorithm in CUDA. The number of thread blocks in a grid is generally determined by the size of the data being processed.

For instance, if an array of size 256×128 is to be processed by thread blocks of 16×16, one can use a grid of size 16×8.

16.1.3 Memory hierarchy

CUDA threads may access data from multiple memory spaces on the GPU card.

Global memory:

1. The main memory space on the device is called global memory.
2. All threads can access the global memory to read and write data.
3. The read and write operations to the global memory are long latency, thus should be minimized.

Shared memory:

1. This is a memory space that is available to a block throughout the lifetime as the block.
2. All threads in the block can access the shared memory of the block.
3. Shared memory is a cached memory space; it is computationally much more efficient to access the shared memory.
4. It is better to copy data from global memory to shared memory and process the data while it resides on the shared memory if the data are reused.

Two other memory spaces are called "constant memory" and "texture memory", which are read-only memory spaces on the device. For instance, if threads will access some data only in read-only mode, the data can be copied to the "constant memory" before the execution of the threads, and can be accessed by the threads during execution. Moreover, each thread has a private local memory, which is a set of local 32-bit registers. It is important to use these memory spaces effectively according to the capabilities they provide in order to achieve an algorithm with high computational efficiency in CUDA.

16.1.4 Performance optimization in CUDA

One should be familiar with the CUDA architecture to some extent in order to develop a program with optimum performance. CUDA Best Practices Guide, available at NVIDIA's web portal, is a good reference for programmers; it provides recommendations for optimization and a list of best practices for programming with CUDA. While not all of these recommendations are applicable to the case of FDTD programming; the following list of recommendations is taken into consideration to develop an FDTD implementation discussed in the next section:

(R1) structure the algorithm in a way that exposes as much data parallelism as possible. Once the parallelism of the algorithm has been exposed, it needs to be mapped to the hardware as efficiently as possible.
(R2) ensure global memory accesses are coalesced whenever possible.
(R3) minimize the use of global memory. Prefer shared memory access where possible.
(R4) use shared memory to avoid redundant transfers from global memory.
(R5) hide latency arising from register dependencies, maintain at least 25 percent occupancy on devices with CUDA compute capability 1.1 and lower, and 18.75 percent occupancy on later devices.
(R6) use a multiple of 32 threads for the number of threads per block as this provides optimal computing efficiency and facilitates coalescing.

16.1.5 Achieving parallelism

At every iteration of the FDTD loop, new values of three magnetic field components are calculated at every cell simultaneously using the past values of electric field components. After magnetic field updates are completed, values of three electric field components are updated at every cell simultaneously in a separate function using the past values of magnetic field components. Since the calculations for each cell can be performed independent from the other cells, a CUDA algorithm can be developed by assigning each cell calculation to a separate thread, and the highest level of parallelism can be achieved to satisfy the recommendation R1. Therefore, the data parallelism required to benefit from CUDA is inherently satisfied by FDTD.

16.2 CUDA implementation of two-dimensional FDTD

In this section we will present an implementation of a two-dimensional FDTD program in CUDA. The program consists of two parts: the first is a MATLAB® code that is based on the two-dimensional FDTD code that has been illustrated in Chapters 7 and 8 while discussing absorbing boundary conditions, and the second is a C code using CUDA that is developed to run FDTD time-marching loop calculations on a graphics card. These codes are available on the CD accompanying this book in folders "fdtd_2d_with_cuda" and "fdtd_2d_cuda_code", respectively.

The two-dimensional FDTD code that has been presented in Chapter 8 is modified such that problem definitions, initializations, and results postprocessing is done in MATLAB as usual, while the program launches an executable that runs the time-marching loop on a graphics card. The executable, "fdtd2d.exe", is implemented in C and it is compiled and built using CUDA's nvcc compiler. The C code can perform FDTD time-marching loop in one of two modes: either on CPU or on GPU. A parameter, named as **computation_platform**, is added to ***define_problem_space_parameters_2d*** to set the computation platform as shown in Listing 16.4. If this parameter is set as 1, the program will run as usual in MATLAB. If the parameter is set as 2 or 3, then the "fdtd2d.exe" will be launched. If the parameter is 2, the executable will perform FDTD computations on CPU, otherwise on GPU. Listing 16.5 illustrates the modified code in the subroutine ***run_fdtd_time_marching_loop_2d***.

Listing 16.6 shows the subroutine ***launch_executable***. Since the parameters and arrays that FDTD calculations require are initialized in MATLAB, they need to be transferred to the executable for use. This transfer is performed through a binary file defined as "`exe_data_file_name`" in Listing 16.6: All required arrays and parameters are adjusted and written to this data file, the data file is read by "fdtd2d.exe", and FDTD calculations are performed either on CPU or on GPU. The intermediate data file is accessed in binary format, instead of ASCII format, to transfer data as is without loss of precision between the MATLAB and the C programs, and binary files require less space for storage. Once the

Listing 16.4 define_problem_space_parameters_2d

```
% the platform on which the calculations will be performed
% 1: matlab on cpu, 2: C executable on cpu, 3: C executable on gpu
computation_platform = 3;
```

Listing 16.5 run_fdtd_time_marching_loop_2d

```
if computation_platform ~=1
    launch_executable;
else
    for time_step = 1:number_of_time_steps
        update_magnetic_fields_2d;
        update_impressed_M;
        update_magnetic_fields_for_CPML_2d;
        capture_sampled_magnetic_fields_2d;
        update_electric_fields_2d;
        update_impressed_J;
        update_electric_fields_for_CPML_2d;
        capture_sampled_electric_fields_2d;
        display_sampled_parameters_2d;
    end
end
```

Listing 16.6 launch_executable

```
% save the data to a file to process on gpu
exe_data_file_name = 'exe_data_file.dat';
save_project_data_to_file;

cmd = ['fdtd2d_on_gpu.exe ' exe_data_file_name];
status = system(cmd);
if status
   disp('Calculations are not successful!');
   return;
end

time_step = number_of_time_steps;

result_data_file_name = ['result_' exe_data_file_name];
fid = fopen(result_data_file_name, 'r');

for ind = 1:number_of_sampled_electric_fields
   sampled_electric_fields(ind).sampled_value = ...
      fread(fid, number_of_time_steps, 'float32').';
end

for ind = 1:number_of_sampled_magnetic_fields
   sampled_magnetic_fields(ind).sampled_value = ...
      fread(fid, number_of_time_steps, 'float32').';
end

fclose(fid);
```

calculations are completed by "fdtd2d.exe" results are written to an output data file, which then is read by MATLAB for postprocessing and displaying the results. In Listing 16.6, MATLAB ***system*** command is used to launch the executable.

16.2.1 Coalesced global memory access

In CUDA, memory instructions are the instructions that read from or write to shared, constant, or global memory. When accessing global memory, there are 400 to 600 clock cycles of memory latency. Much of this global memory latency can be hidden by the thread scheduler if there are sufficient independent arithmetic instructions that can be issued while waiting for the global memory access to complete [96]. Unfortunately when FDTD calculations are performed on a computer the operations are dominated by memory accesses rather than arithmetic instructions. Hence, the memory access inefficiency is the main bottle neck that reduces efficiency of FDTD on GPU. Global memory bandwidth is used most efficiently when the simultaneous memory accesses by threads in a half-warp (during the execution of a single read or write instruction) can be coalesced into a single memory transaction of 32, 64, or 128 bytes [96].

If the size of two-dimensional arrays, thus the size of the FDTD domain in number of cells, in the x and y directions, is a multiple of 16, then the coalesced memory access is ensured. In general an FDTD domain size would be an arbitrary number. In order to achieve

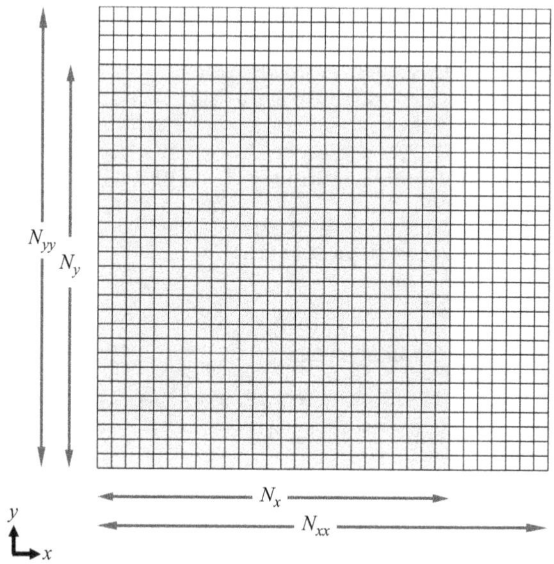

Figure 16.3 An extended two-dimensional computational domain.

coalesced memory access, the FDTD domain can be extended by padded cells such that the number of cells in x and y directions is an integer multiple of 16 as illustrated in Figure 16.3. Although, padding these cells increases the amount of memory needed to store arrays, it improves the efficiency of the kernel functions tremendously. Thus the recommendation R2 is satisfied.

It should be noted that, these padded cells are beyond the boundaries of the original domain and they should be electrically isolated from the original domain. As long as the original domain boundaries are kept as PEC, such as in the case of CPML boundaries, the fields in the original domain will be isolated from the padded cells, thus padded cells can be assumed to be filled with any type of material. The easiest way is to set these padded cells as free space.

If the size of a two-dimensional FDTD domain is $Nx \times Ny$ cells, where Nx and Ny are the numbers of cells in x and y directions, respectively, then the modified size of the FDTD domain becomes $Nxx \times Nyy$, where Nxx and Nyy are the new numbers of cells. In the presented example code, the extended size of the problem space is determined in the subroutine ***save_project_data_to_file*** as

```
% extend the domain and adjust number of cells for gpu
nxx = (floor(nx/16)+1)*16;
nyy = (floor(ny/16)+1)*16;
```

Then the two-dimensional arrays of fields and coefficients can be extended to $Nxx \times Nyy$. However, in fact, the two-dimensional arrays are extended to $Nxx \times (Nyy + 2)$ in the subroutine ***save_project_data_to_file*** before they are copied to the binary data file. The reason for the additional padded cells in the y direction will be explained in the subsequent sections.

16.2.2 Thread to cell mapping

The "fdtd2d_on_gpu.exe" program starts with reading the binary data file; first arrays are allocated on the CPU memory for two-dimensional arrays, then the coefficient and field data are read into these arrays. These arrays initially reside on the host (CPU) memory and they need to be copied to device (GPU) global memory. The `cudaMalloc()` function of CUDA is used to allocate memory on the device global memory for these arrays and `cudaMemcpy()` function is used to copy the data to the global memory. Once these arrays are ready on the global memory, they are ready for processing on the graphics card by kernel functions.

Listing 16.7 shows a function that performs an iteration of the time-marching loop on GPU. Similarly, Listing 16.8 shows a function that performs an iteration on CPU. It can be seen that main components of an iteration, such as electric and magnetic field updates, source updates, boundary condition updates, etc., are performed by calls to associated functions. We will discuss the electric and magnetic field updating functions in detail for the TMz case to illustrate an implementation using CUDA.

Listing 16.7 FDTD iteration on GPU

```
bool fdtdIterationOnGpu()
{
        int n_bx = (nxx/TILE_SIZE) + (nxx%TILE_SIZE == 0 ? 0 : 1);
        int n_by = (nyy/TILE_SIZE) + (nyy%TILE_SIZE == 0 ? 0 : 1);
        dim3 threads = dim3(TILE_SIZE, TILE_SIZE, 1);
        dim3 grid = dim3(n_bx, n_by, 1);

        dim3 threads_sef = dim3(number_of_sampled_electric_fields, 1, 1);
        dim3 threads_smf = dim3(number_of_sampled_magnetic_fields, 1, 1);
        dim3 grid_sef = dim3(1, 1, 1);
        dim3 grid_smf = dim3(1, 1, 1);

        if (is_TEz)
             update_magnetic_fields_on_gpu_TEz<<<grid,
threads>>>(dvChzh, dvChzex, dvChzey, dvHz, dvEx, dvEy, nxx);

        if (is_TMz)
             update_magnetic_fields_on_gpu_TMz<<<grid,
threads>>>(dvChxh, dvChxez, dvChyh, dvChyez, dvHx,  dvHy, dvEz, nxx);

        update_impressed_magnetic_currents_on_gpu(time_step);

        update_magnetic_fields_for_CPML_on_gpu();

        capture_sampled_magnetic_fields_on_gpu<<<grid_smf, threads_smf>>>
               (dvHx, dvHy, dvHz, dvsampled_magnetic_fields_component,
dvsampled_magnetic_fields_is,
               dvsampled_magnetic_fields_js,
dvsampled_magnetic_fields_sampled_value, time_step,
number_of_time_steps, nxx);

        if (is_TEz)
        update_electric_fields_on_gpu_TEz<<<grid, threads>>>
(dvCexe, dvCexhz, dvCeye, dvCeyhz,  dvEx, dvEy, dvHz, nxx);

        if (is_TMz)
        update_electric_fields_on_gpu_TMz<<<grid, threads>>>
(dvCeze, dvCezhy, dvCezhx, dvHx,  dvHy, dvEz, nxx);
```

```
        update_impressed_electric_currents_on_gpu(time_step);

        update_electric_fields_for_CPML_on_gpu();

        capture_sampled_electric_fields_on_gpu<<<grid_sef, threads_sef>>>
                (dvEz, dvEy, dvEz, dvsampled_electric_fields_component,
dvsampled_electric_fields_is,
                dvsampled_electric_fields_js,
dvsampled_electric_fields_sampled_value, time_step,
number_of_time_steps, nxx);

        printf("timestep: %d \n", time_step);
        time_step++;

        return true;
}
```

Listing 16.8 FDTD iteration on CPU

```
bool fdtdIterationOnCpu()
{
        if (is_TEz) update_magnetic_fields_on_cpu_TEz();

        if (is_TMz) update_magnetic_fields_on_cpu_TMz();

        update_impressed_magnetic_currents_on_cpu(time_step);

        update_magnetic_fields_for_CPML_on_cpu();

        capture_sampled_magnetic_fields_on_cpu(time_step);

        if (is_TEz)     update_electric_fields_on_cpu_TEz();

        if (is_TMz)     update_electric_fields_on_cpu_TMz();

        update_impressed_electric_currents_on_cpu(time_step);

        update_electric_fields_for_CPML_on_cpu();

        capture_sampled_electric_fields_on_cpu(time_step);

        printf("timestep: %d \n", time_step);
        time_step++;

        return true;
}
```

The kernel function that updates the magnetic field components is called

```
update_magnetic_fields_on_gpu_TMz<<<grid, threads>>>
(dvChxh, dvChxez, dvChyh, dvChyez, dvHx, dvHy, dvEz, nxx);
```

Here the parameter names preceded with "dv" are pointers to respective arrays on the global memory of the graphics card. The CUDA notation `<<<grid, threads>>>`

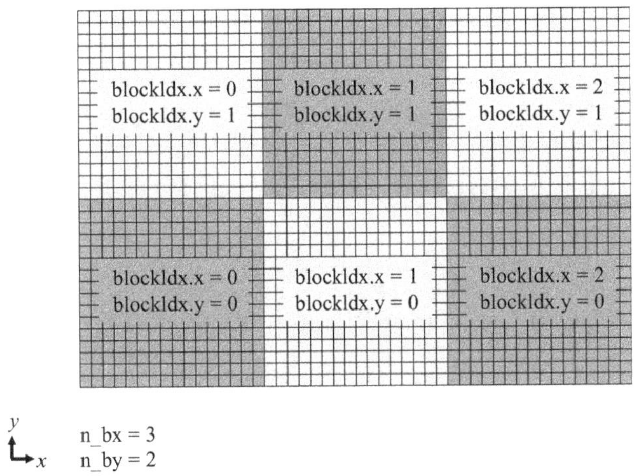

Figure 16.4 A grid of thread blocks that spans a 48×32 cells two-dimensional problem space.

describes the grid of threads that will be launched for this kernel. The parameter `threads` is initialized as

```
dim3 threads = dim3(TILE_SIZE, TILE_SIZE, 1);
```

which means that a thread block consists of `TILE_SIZExTILE_SIZE` square shaped array of threads. In this implementation the `TILE_SIZE` is set as 16, thus the recommendation R6 in Section 16.1 is satisfied. The parameter `grid` is initialized as

```
int n_bx = (nxx/TILE_SIZE) + (nxx%TILE_SIZE == 0 ? 0 : 1);
int n_by = (nyy/TILE_SIZE) + (nyy%TILE_SIZE == 0 ? 0 : 1);
dim3 grid = dim3(n_bx, n_by, 1);
```

which means that the grid is composed of $n\_bx \times n\_by$ thread blocks. This thread grid scheme helps to conveniently map threads to cells. Figure 16.4 illustrates such a grid that spans a problem space.

Listing 16.9 shows the kernel function that updates the magnetic field components for the TMz case, whereas Listing 16.10 shows the kernel that updates the electric field components. Respective functions that perform magnetic and electric field updates on CPU are shown in Listings 16.11 and 16.12 for comparison. In this implementation for GPU each cell is processed by an associated thread. The CUDA internal variables `threadIdx` and `blockIdx` are used to identify the threads and data elements which they will update. The variable `blockIdx` is used to identify the thread blocks as illsutrated in Figure 16.4. Similarly, the variable `threadIdx` is used to identify threads within a thread block as shown in Figure 16.5. Thus, one can identify any thread, and also cell, in a two-dimensional array with indices i and j such that

```
int i = blockIdx.x * blockDim.x + threadIdx.x;
int j = blockIdx.y * blockDim.y + threadIdx.y;
```

At this point it should be noted that although the arrays that are being processed are two-dimensional, they are stored in device (GPU) global memory as one-dimensional arrays and elements of these arrays are accessed in kernel functions in a linear fashion. Thus a conversion from i and j indices to the linear index, denoted as "ci" in the Listing 16.4, is required.

Listing 16.9 update_magnetic_fields_on_gpu_TMz

```
__global__ void
update_magnetic_fields_on_gpu_TMz(float* Chxh, float* Chxez, float*
Chyh, float* Chyez, float* Hx,  float* Hy, float* Ez, int nxx)
{
        __shared__ float sEz[TILE_SIZE][2*TILE_SIZE+1];

        int tx = threadIdx.x;
        int ty = threadIdx.y;
        int i = blockIdx.x * blockDim.x + tx;
        int j = blockIdx.y * blockDim.y + ty;

        int ci = (j+1)*nxx+i;

        sEz[ty][tx] = Ez[ci];
        sEz[ty][tx+TILE_SIZE] = Ez[ci+TILE_SIZE];

        __syncthreads();

        Hx[ci] = Chxh[ci] * Hx[ci] + Chxez[ci] * (Ez[ci+nxx]-sEz[ty][tx]);
        Hy[ci] = Chyh[ci] * Hy[ci] + Chyez[ci] * (sEz[ty][tx+1] - sEz[ty][tx])
}
```

Listing 16.10 update_electric_fields_on_gpu_TEz

```
__global__ void
update_electric_fields_on_gpu_TEz(float* Cexe, float* Cexhz, float*
Ceye, float* Ceyhz,  float* Ex, float* Ey, float* Hz, int nxx)
{
        __shared__ float sHz[TILE_SIZE][2*TILE_SIZE+1];

        int tx = threadIdx.x;
        int ty = threadIdx.y;
        int i = blockIdx.x * blockDim.x + tx;
        int j = blockIdx.y * blockDim.y + ty;

        int ci = (j+1)*nxx+i;

        sHz[ty][tx] = Hz[ci-TILE_SIZE];
        sHz[ty][tx+TILE_SIZE] = Hz[ci];

        __syncthreads();

        Ex[ci] = Cexe[ci] * Ex[ci] + Cexhz[ci] * (Hz[ci]-Hz[ci-nxx]);
        Ey[ci] = Ceye[ci] * Ey[ci] + Ceyhz[ci] * (sHz[ty][tx+TILE_SIZE]-
sHz[ty][tx+TILE_SIZE-1]);
  }
```

Listing 16.11 update_magnetic_fields_on_gpu_TMz

```
void update_magnetic_fields_on_cpu_TMz()
{
        for (int i=0;i<number_of_cells-nxx;i++)
         Hx[i] = Chxh[i] * Hx[i] + Chxez[i] * (Ez[i+nxx]-Ez[i]);

        for (int i=0;i<number_of_cells-nxx;i++)
         Hy[i] = Chyh[i] * Hy[i] + Chyez[i] * (Ez[i+1] - Ez[i]);
}
```

Listing 16.12 update_electric_fields_on_gpu_TMz

```
void update_electric_fields_on_cpu_TMz()
{
         for (int i=nxx;i<number_of_cells;i++)
                 Ez[i] = Ceze[i] * Ez[i] + Cezhy[i] * (Hy[i]-Hy[i-1]) +
Cezhx[i] * (Hx[i]-Hx[i-nxx]);
}
```

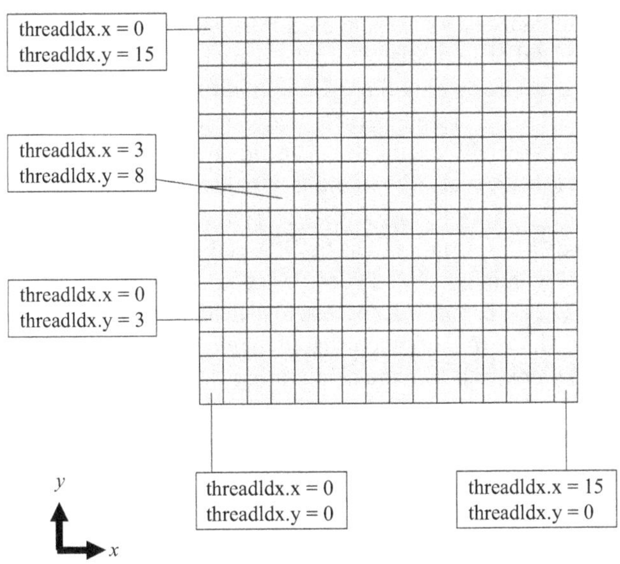

Figure 16.5 Threads in a thread block.

In MATLAB the index i runs faster than the index j, i.e., the data stored at (i, j) is adjacent to $(i+1, j)$ on the physical memory, thus the conversion to linear index can be done simply as

```
int ci = j*nxx+i;
```

However, in Listings 16.9 and 16.10 one can notice that the linear index is calculated as

```
int ci = (j+1)*nxx+i;
```

The reason is as follows: Consider the equation used to update E_z as shown in (1.36). In order to update $E_z(i,j)$ one needs $H_x(i,j-1)$, hence one cannot update $E_z(i,1)$ since $H_x(i,0)$ does not exist. Here MATLAB array indexing scheme is used which starts from 1, unlike C, which starts from 0. The update of $E_z(i,j)$, therefore, shall be performed starting from $j=2$, which can be controlled in the kernel code using an `if` statement. An `if` statement would reduce the efficiency of a kernel function considerably, thus to avoid such a situation, in the given FDTD algorithm one layer of cells is padded to the two-dimensional arrays in the $-y$ direction. This means that all field components that are being processed are shifted by one cell in the $+y$ direction, so final results would not be affected. Hence the calculation of `ci` is adjusted. Similar situation happens when updating H_z. In order to update $H_z(i,nyy)$ one needs $E_x(i,nyy+1)$, which does not exist on the physical memory if the arrays are of size $nxx \times nyy$. A straightforward solution is to pad another layer of cells to the two-dimensional arrays in the $+y$ direction. These two additional layers of cells are appended to the arrays in the subroutine ***save_project_data_to_file***.

16.2.3 Use of shared memory

Because it is on the GPU chip, the access to shared memory is much faster than the local and global memory. Parameters that reside in the shared memory space of a thread block have the lifetime of the block, and are accessible from all the threads within the block [96]. Therefore if a data block on global memory is going to be used frequently in a kernel, it is better to load the data to shared memory and reuse the data from the shared memory. Here it should be reminded that, shared memory is available only through the lifetime of a thread, thus the relevant field data have to be reloaded to the shared memory every time a kernel function is called.

Shared memory is especially useful when threads need to access to unaligned data. For instance, examining (1.38) reveals that in order to calculate $H_y(i,j)$, a thread mapped to the cell (i,j) needs E_z at (i,j) as well as E_z at $(i+1,j)$. In the kernel code the index of a thread that is processing the cell (i,j) is indexed as “ci”. A cell with index $(i+1,j)$ can be accessed by `ci+1`, while a cell with index $(i,j+1)$ can be accessed by `ci+nxx`. Access to $(i,j+1)$ is coalesced, however $(i+1,j)$ is not. If an access to a field component at a neighboring cell in the x direction is needed, i.e. $(i+1,j)$, then shared memory can be used to load the data block mapped by the thread block, and then the neighboring field value is accessed from the shared memory. At this point one need to use the CUDA function `__syncthreads()` to ensure that all threads in the block are synchronized; thus all necessary data are loaded to the shared memory before they are used by the neighboring threads.

As discussed above, uncoalesced memory accesses can be prevented by using shared memory. However, a problem arises when accessing the neighboring cells' data through shared memory. While loading the shared memory, each thread copies one element from the global memory to the shared memory. If the thread on the boundary of the thread block needs to access the data in the neighboring cell, this data will not be available since it has not been loaded to the shared memory. One way to overcome this problem is to load another set of data, which includes the neighboring cell's data, to the shared memory. In the presented implementation shown in Listing 16.9, two square blocks of data are copied from global memory to the shared memory as

```
sEz[ty][tx] = Ez[ci];
sEz[ty][tx+TILE_SIZE] = Ez[ci+TILE_SIZE];
```

Then E_z at $(i+1,j)$ is safely accessed from the shared memory to update $H_y(i,j)$ as

```
Hy[ci] = Chyh[ci] * Hy[ci] + Chyez[ci] * (sEz[ty][tx+1] -
sEz[ty][tx]);
```

A similar treatment is shown in Listing 16.10, where $E_y(i,j)$ is updated by $H_z(i-1,j)$ through an effective use of shared memory: First two blocks of data is from global memory to the shared memory as

```
sHz[ty][tx] = Hz[ci-TILE_SIZE];
sHz[ty][tx+TILE_SIZE] = Hz[ci];
```

Then $E_y(i,j)$ is updated as

```
Ey[ci] = Ceye[ci] * Ey[ci] + Ceyhz[ci] * (sHz[ty][tx+TILE_
SIZE]-sHz[ty][tx+TILE_SIZE-1]);
```

16.2.4 Optimization of number of threads

As pointed out in recommendations R5 and R6 in Section 16.1, occupancy of the microprocessors and number of threads in a block are two other important parameters that affect the performance of a CUDA program. Number of threads and occupancy are tightly connected. It is possible to set the number of threads as a desired value while it may not be possible to control the occupancy; it is a function of number of threads, number of registers used in the kernel, amount of shared memory used by the kernel, compute capability of the device, etc. A good practice is to optimize the number of threads while keeping the occupancy at a reasonable value. As a rule of thumb, it is better to keep kernel functions small such that they do not use many registers.

CUDA Visual Profiler is a graphical user interface based profiling tool provided by NVIDIA that can be used to measure performance and find potential opportunities for optimization in order to achieve maximum performance of kernels in a CUDA program. It can show information such as total GPU and CPU times of a thread, memory read-write throughputs, global memory load and store efficiencies, occupancy, etc. Therefore, the kernel functions can be profiled using the Visual Profiler and, if not satisfactory, their efficiencies can be improved.

In the presented code the block size is chosen as $16 \times 16 = 256$. One can choose a different configuration and test with Visual Profiler to evaluate the efficiency. It should also be noted that the presented grid scheme, i.e., two-dimensional array of square thread blocks, is not the only way to map threads to cells. It is possible to use other grid configurations, such as using one-dimensional thread blocks in a one-dimensional grid, develop algorithms based on these configurations, and evaluate if the developed algorithms are superior.

16.3 Performance of two-dimensional FDTD on CUDA

The main purpose of developing a code to run FDTD program on a graphics card is to run it faster by utilizing the parallel processing capability of GPU. The performance of the developed CUDA code is examined as a function of problem size. The analysis is performed on an NVIDIA® Tesla™ C1060 Computing Processor installed on a 64-bit Windows XP computer. This card has 240 streaming processor cores operating at 1.3 GHz. Size of a two-dimensional FDTD problem domain has been swept and the throughputs of the program are calculated as number of million cells processed per second (*NMCPS*) as a measure of the performance of the CUDA program. Number of million cells is calculated as [97]

$$NMCPS = \frac{n_{steps} \times Nxx \times Nyy}{t_s} \times 10^{-6}, \quad (16.1)$$

where n_{steps} is the number of time steps the program has been run and t_s is the total time of program run in seconds. Here, the problem that is discussed in Section 7.5.2, illustrated in Figure 7.15, is simulated for performance evaluation. The problem space is terminated by PEC boundaries. The results, shown in Figure 16.6, reveal that as the problem size is increased, the throughput of the CUDA calculations increase to about 900 million cells per second. Figure 16.6 also shows the throughput of the FDTD program when all calculations are performed on the CPU. The average throughput on CPU is about 25 million cells per second. For the presented implementations, GPU calculations are more than 36 times faster than the CPU calculations.

CUDA and OpenGL are two software platforms both of which operate on the GPU hardware, while their intended use are different; CUDA is to improve the performance of data parallel computations, while OpenGL is to manipulate data to produce 2D and 3D computer graphics. While running an FDTD simulation, it is possible to capture electromagnetic fields and display them as an on-the-fly animation. If the FDTD calculations are performed on a graphics card, which is used also to perform the OpenGL operations to display the field, one can copy the field data from graphics card memory (device global memory) to computer's main memory that is processed by CPU (host memory), process the data, and copy back to GPU memory to display via OpenGL. Thanks to CUDA/OpenGL

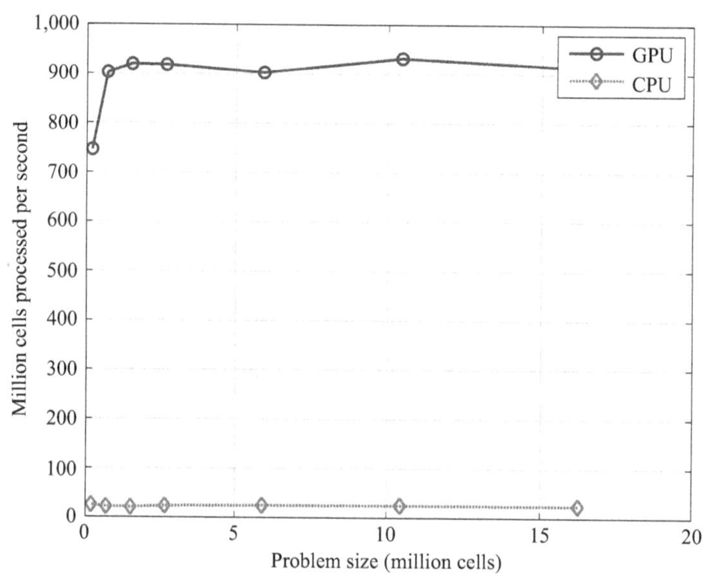

Figure 16.6 Throughput of CUDA FDTD calculations on a Tesla C1060 card.

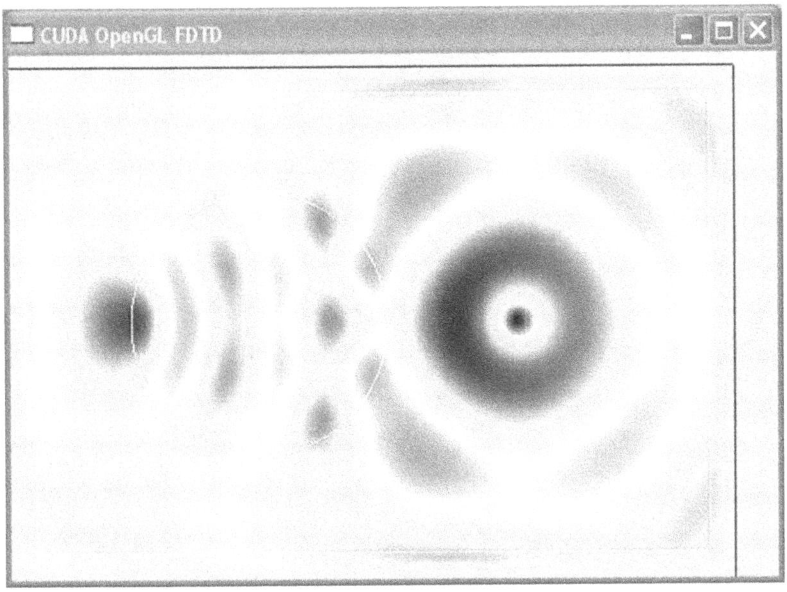

Figure 16.7 A snapshot of electric field distribution scattered from a dielectric cylinder due to a line source with sinusoidal excitation.

interoperability provided by CUDA, it is possible to avoid the back and forth data transfer between the host and device memories, and perform all processing required for the display on the graphics card.

The presented CUDA code also utilizes the CUDA/OpenGL interoperability to capture and display electromagnetic fields. The test case presented above is run with electromagnetic field animation enabled, and electric field distribution is captured and displayed on a window while the program is running. The snapshot of the display window is shown in Figure 16.7. The details for CUDA-OpenGL interoperability can be found in [98].

APPENDIX A

One-dimensional FDTD code

Listing A.1 MATLAB® code for one-dimensional FDTD

```
% This program demonstrates a one-dimensional FDTD simulation.
% The problem geometry is composed of two PEC plates extending to
% infinity in y, and z dimensions, parallel to each other with 1 meter
% separation. The space between the PEC plates is filled with air.
% A sheet of current source paralle to the PEC plates is placed
% at the center of the problem space. The current source excites fields
% in the problem space due to a z-directed current density Jz,
% which has a Gaussian waveform in time.

% Define initial constants
eps_0 = 8.854187817e-12;           % permittivity of free space
mu_0  = 4*pi*1e-7;                 % permeability of free space
c     = 1/sqrt(mu_0*eps_0);        % speed of light

% Define problem geometry and parameters
domain_size = 1;                   % 1D problem space length in meters
dx = 1e-3;                         % cell size in meters
dt = 3e-12;                        % duration of time step in seconds
number_of_time_steps = 2000;       % number of iterations
nx = round(domain_size/dx);        % number of cells in 1D problem space
source_position = 0.5;             % position of the current source Jz

% Initialize field and material arrays
Ceze      = zeros(nx+1,1);
Cezhy     = zeros(nx+1,1);
Cezj      = zeros(nx+1,1);
Ez        = zeros(nx+1,1);
Jz        = zeros(nx+1,1);
eps_r_z   = ones (nx+1,1); % free space
sigma_e_z = zeros(nx+1,1); % free space

Chyh      = zeros(nx,1);
Chyez     = zeros(nx,1);
Chym      = zeros(nx,1);
Hy        = zeros(nx,1);
My        = zeros(nx,1);
```

```
mu_r_y     = ones (nx,1); % free space
sigma_m_y = zeros(nx,1); % free space
% Calculate FDTD updating coefficients
Ceze = (2 * eps_r_z * eps_0 - dt * sigma_e_z) ...
      ./(2 * eps_r_z * eps_0 + dt * sigma_e_z);

Cezhy = (2 * dt / dx) ...
      ./(2 * eps_r_z * eps_0 + dt * sigma_e_z);

Cezj  = (-2 * dt) ...
      ./(2 * eps_r_z * eps_0 + dt * sigma_e_z);

Chyh   = (2 * mu_r_y * mu_0 - dt * sigma_m_y) ...
      ./(2 * mu_r_y * mu_0 + dt * sigma_m_y);

Chyez = (2 * dt / dx) ...
      ./(2 * mu_r_y * mu_0 + dt * sigma_m_y);

Chym   = (-2 * dt) ...
      ./(2 * mu_r_y * mu_0 + dt * sigma_m_y);

% Define the Gaussian source waveform
time          = dt*[0:number_of_time_steps-1].';
Jz_waveform  = exp(-((time-2e-10)/5e-11).^2)*1e-3/dx;
source_position_index = round(nx*source_position/domain_size)+1;

% Subroutine to initialize plotting
initialize_plotting_parameters;

% FDTD loop
for time_step = 1:number_of_time_steps

    % Update Jz for the current time step
    Jz(source_position_index) = Jz_waveform(time_step);

    % Update magnetic field
    Hy(1:nx) =    Chyh(1:nx) .* Hy(1:nx) ...
         + Chyez(1:nx) .* (Ez(2:nx+1) - Ez(1:nx))   ...
         + Chym(1:nx) .* My(1:nx);

    % Update electric field
    Ez(2:nx) = Ceze  (2:nx) .*  Ez(2:nx) ...
             + Cezhy(2:nx) .*  (Hy(2:nx) - Hy(1:nx-1))   ...
             + Cezj (2:nx)  .*  Jz(2:nx);

    Ez(1)     = 0; % Apply PEC boundary condition at x = 0 m
    Ez(nx+1) = 0; % Apply PEC boundary condition at x = 1 m

    % Subroutine to plot the current state of the fields
    plot_fields;
end
```

Listing A.2 initialize_plotting_parameters

```
% subroutine used to initialize 1D plot

Ez_positions = [0:nx]*dx;
Hy_positions = ([0:nx-1]+0.5)*dx;
v = [0 -0.1 -0.1; 0 -0.1 0.1; 0 0.1 0.1; 0 0.1 -0.1; ...
     1 -0.1 -0.1; 1 -0.1 0.1; 1 0.1 0.1; 1 0.1 -0.1];
f = [1 2 3 4; 5 6 7 8];
axis([0 1 -0.2 0.2 -0.2 0.2]);
lez = line(Ez_positions,Ez*0,Ez,'Color','b','LineWidth',1.5);
lhy = line(Hy_positions,377*Hy,Hy*0,'Color','r', ...
     'LineWidth',1.5,'linestyle','-.');
set(gca,'fontsize',12,'FontWeight','bold');
axis square;
legend('E_{z}', 'H_{y} \times 377','Location','NorthEast');
xlabel('x [m]');
ylabel('[A/m]');
zlabel('[V/m]');
grid on;
p = patch('vertices', v, 'faces', f, 'facecolor', 'g', 'facealpha',0.2);
text(0,1,1.1,'PEC','horizontalalignment','center','fontweight','bold');
text(1,1,1.1,'PEC','horizontalalignment','center','fontweight','bold');
```

Listing A.3 plot_fields

```
% subroutine used to plot 1D transient fields

delete(lez);
delete(lhy);
lez = line(Ez_positions,Ez*0,Ez,'Color','b','LineWidth',1.5);
lhy = line(Hy_positions,377*Hy,Hy*0,'Color','r', ...
     'LineWidth',1.5,'linestyle','-.');
ts = num2str(time_step);
ti = num2str(dt*time_step*1e9);
title(['time step = ' ts ', time = ' ti ' ns']);
drawnow;
```

APPENDIX B

Convolutional perfectly-matched layer regions and associated field updates for a three-dimensional domain

B.1 Updating E_x at convolutional perfectly-matched layer (CPML) regions

Initialization

Create new coefficient arrays for $C_{\psi exy}$ and $C_{\psi exz}$:

$$C_{\psi exy} = \Delta y C_{exhz} \quad C_{\psi exz} = \Delta z C_{exhy}.$$

Modify the coefficient arrays for C_{exhy} and C_{exhz} in the CPML regions:

$$C_{exhz} = (1/k_{ey})C_{exhz}, \quad \text{in the } yn \text{ and } yp \text{ regions.}$$
$$C_{exhy} = (1/k_{ez})C_{exhy}, \quad \text{in the } zn \text{ and } zp \text{ regions.}$$

Finite-difference time-domain (FDTD) time-marching loop

Update E_x in the full domain using the regular updating equation:

$$\begin{aligned} E_x^{n+1}(i,j,k) &= C_{exe}(i,j,k) \times E_x^n(i,j,k) \\ &\quad + C_{exhz}(i,j,k) \times \left(H_z^{n+\frac{1}{2}}(i,j,k) - H_z^{n+\frac{1}{2}}(i,j-1,k)\right) \\ &\quad + C_{exhy}(i,j,k) \times \left(H_y^{n+\frac{1}{2}}(i,j,k) - H_y^{n+\frac{1}{2}}(i,j,k-1)\right). \end{aligned}$$

Calculate $\psi_{exy}^{n+\frac{1}{2}}$ for the yn and yp regions:

$$\psi_{exy}^{n+\frac{1}{2}}(i,j,k) = b_{ey}\psi_{exy}^{n-\frac{1}{2}}(i,j,k) + a_{ey}\left(H_z^{n+\frac{1}{2}}(i,j,k) - H_z^{n+}(i,j-1,k)\right),$$

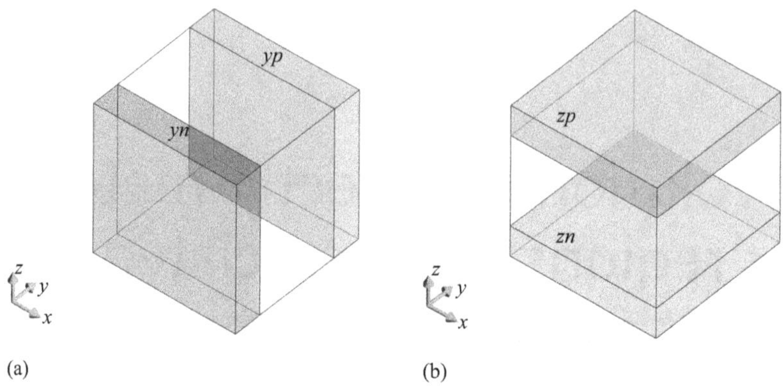

Figure B.1 CPML regions where E_x is updated: (a) E_x is updated in yn and yp using ψ_{exy} and (b) E_x is updated in zn and zp using ψ_{exz}.

where

$$a_{ey} = \frac{\sigma_{pey}}{\Delta y(\sigma_{pey}k_{ey} + \alpha_{ey}k_{ey}^2)}\left[b_{ey} - 1\right],$$

$$b_{ey} = e^{-\left(\frac{\sigma_{pey}}{k_{ey}} + \alpha_{pey}\right)\frac{\Delta t}{\varepsilon_0}}.$$

Calculate $\psi_{exz}^{n+\frac{1}{2}}$ for the zn and zp regions:

$$\psi_{exz}^{n+\frac{1}{2}}(i,j,k) = b_{ez}\psi_{exz}^{n-\frac{1}{2}}(i,j,k) + a_{ez}\left(H_y^{n+\frac{1}{2}}(i,j,k) - H_y^{n+\frac{1}{2}}(i,j,k-1)\right),$$

where

$$a_{ez} = \frac{\sigma_{pez}}{\Delta z(\sigma_{pez}k_{ez} + \alpha_{ez}k_{ez}^2)}\left[b_{ez} - 1\right],$$

$$b_{ez} = e^{-\left(\frac{\sigma_{pez}}{k_{ez}} + \alpha_{pez}\right)\frac{\Delta t}{\varepsilon_0}}.$$

Add the CPML auxiliary term to E_x in the yn and yp regions:

$$E_x^{n+1} = E_x^{n+1} + C_{\psi exy} \times \psi_{exy}^{n+\frac{1}{2}}.$$

Add the CPML auxiliary term to E_x in the zn and zp regions:

$$E_x^{n+1} = E_x^{n+1} + C_{\psi exz} \times \psi_{exz}^{n+\frac{1}{2}}.$$

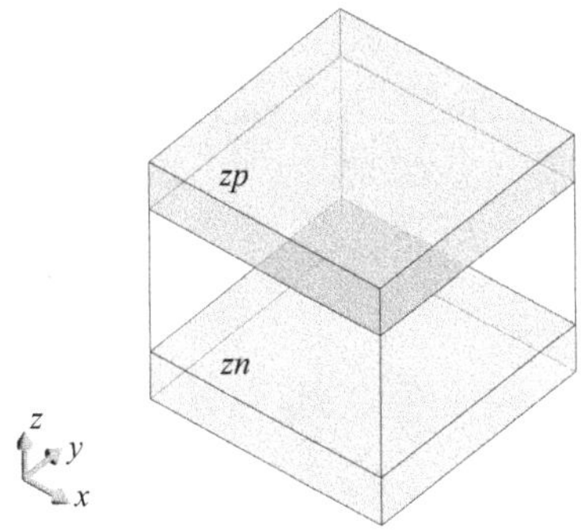

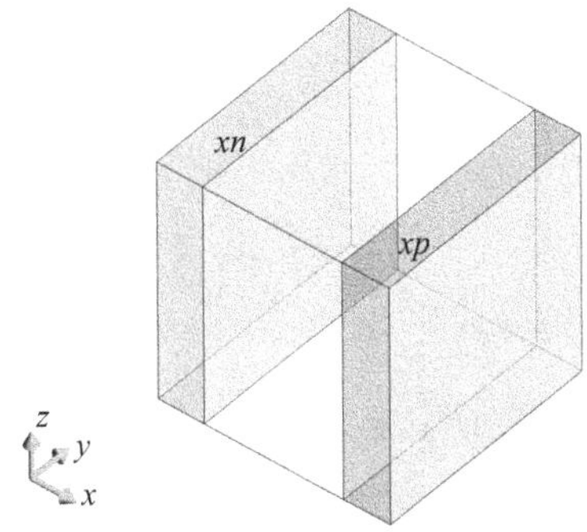

Figure B.2 CPML regions where E_y is updated: (a) E_y is updated in *zn* and *zp* using ψ_{eyz} and (b) E_y is updated in *xn* and *xp* using ψ_{eyx}.

B.2 Updating E_y at CPML regions

Initialization

Create new coefficient arrays for $C_{\psi eyz}$ and $C_{\psi eyx}$:

$$C_{\psi eyz} = \Delta z C_{eyhx} \quad C_{\psi eyx} = \Delta x C_{eyhz}.$$

Modify the coefficient arrays for C_{eyhz} and C_{eyhx} in the CPML regions:

$$C_{eyhx} = (1/k_{ez})C_{eyhx}, \quad \text{in the } zn \text{ and } zp \text{ regions.}$$
$$C_{eyhz} = (1/k_{ey})C_{eyhz}, \quad \text{in the } xn \text{ and } xp \text{ regions.}$$

FDTD time-marching loop

Update E_y in the full domain using the regular updating equation:

$$\begin{aligned} E_y^{n+1}(i,j,k) &= C_{eye}(i,j,k) \times E_y^n(i,j,k) \\ &\quad + C_{eyhx}(i,j,k) \times \left(H_x^{n+\frac{1}{2}}(i,j,k) - H_x^{n+\frac{1}{2}}(i,j,k-1)\right) \\ &\quad + C_{eyhz}(i,j,k) \times \left(H_z^{n+\frac{1}{2}}(i,j,k) - H_z^{n+\frac{1}{2}}(i-1,j,k)\right). \end{aligned}$$

Calculate $\psi_{eyz}^{n+\frac{1}{2}}$ for the *zn* and *zp* regions:

$$\psi_{eyz}^{n+\frac{1}{2}}(i,j,k) = b_{ez}\psi_{eyz}^{n-\frac{1}{2}}(i,j,k) + a_{ez}\left(H_x^{n+\frac{1}{2}}(i,j,k) - H_x^{n+\frac{1}{2}}(i,j,k-1)\right),$$

where

$$a_{ez} = \frac{\sigma_{pez}}{\Delta z(\sigma_{pez}k_{ez} + \alpha_{ez}k_{ez}^2)}[b_{ez} - 1],$$

$$b_{ez} = e^{-\left(\frac{\sigma_{pez}}{k_{ez}} + \alpha_{pez}\right)\frac{\Delta t}{\varepsilon_0}}.$$

Calculate $\psi_{eyx}^{n+\frac{1}{2}}$ for the *xn* and *xp* regions:

$$\psi_{eyx}^{n+\frac{1}{2}}(i,j,k) = b_{ex}\psi_{eyx}^{n-\frac{1}{2}}(i,j,k) + a_{ex}\left(H_z^{n+\frac{1}{2}}(i,j,k) - H_z^{n+\frac{1}{2}}(i-1,j,k)\right),$$

where

$$a_{ex} = \frac{\sigma_{pex}}{\Delta x(\sigma_{pex}k_{ex} + \alpha_{ex}k_{ex}^2)}[b_{ex} - 1],$$

$$b_{ex} = e^{-\left(\frac{\sigma_{pex}}{k_{ex}} + \alpha_{pex}\right)\frac{\Delta t}{\varepsilon_0}}.$$

Add the CPML auxiliary term to E_y in the *zn* and *zp* regions:

$$E_y^{n+1} = E_y^{n+1} + C_{\psi eyz} \times \psi_{eyz}^{n+\frac{1}{2}}.$$

Add the CPML auxiliary term to E_y in the *xn* and *xp* regions:

$$E_y^{n+1} = E_y^{n+1} + C_{\psi eyx} \times \psi_{eyx}^{n+\frac{1}{2}}.$$

B.3 Updating E_z at CPML regions

Initialization

Create new coefficient arrays for $C_{\psi ezx}$ and $C_{\psi ezy}$:

$$C_{\psi ezx} = \Delta x C_{ezhy} \quad C_{\psi ezy} = \Delta y C_{eyhx}.$$

Modify the coefficient arrays for C_{ezhx} and C_{ezhy} in the CPML regions:

$$C_{ezhy} = (1/k_{ex})C_{ezhy}, \quad \text{in the } xn \text{ and } xp \text{ regions.}$$
$$C_{ezhx} = (1/k_{ez})C_{ezhx}, \quad \text{in the } yn \text{ and } yp \text{ regions.}$$

FDTD time-marching loop

Update E_z in the full domain using the regular updating equation:

$$\begin{aligned} E_z^{n+1}(i,j,k) = {} & C_{eze}(i,j,k) \times E_z^n(i,j,k) + C_{ezhy}(i,j,k) \times \left(H_y^{n+\frac{1}{2}}(i,j,k) - H_y^{n+\frac{1}{2}}(i-1,j,k)\right) \\ & + C_{ezhx}(i,j,k) \times \left(H_x^{n+\frac{1}{2}}(i,j,k) - H_x^{n+\frac{1}{2}}(i,j-1,k)\right) \end{aligned}$$

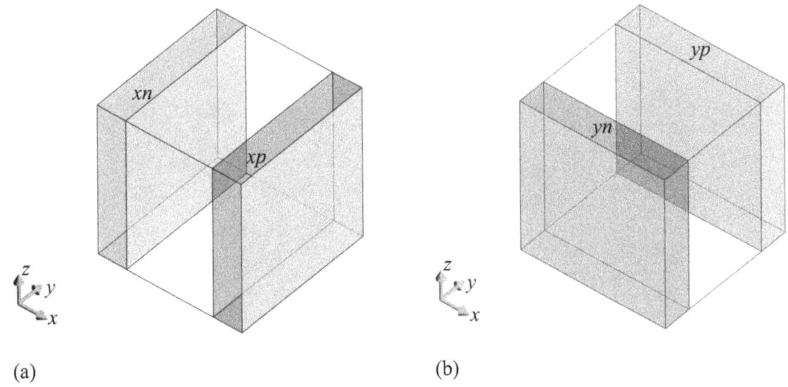

Figure B.3 CPML regions where E_z is updated: (a) E_z is updated in xn and xp using ψ_{ezx} and (b) E_z is updated in yn and yp using ψ_{ezy}.

Calculate $\psi_{ezx}^{n+\frac{1}{2}}$ for the xn and xp regions:

$$\psi_{ezx}^{n+\frac{1}{2}}(i,j,k) = b_{ex}\psi_{ezx}^{n-\frac{1}{2}}(i,j,k) + a_{ex}\left(H_y^{n+\frac{1}{2}}(i,j,k) - H_y^{n+\frac{1}{2}}(i-1,j,k)\right),$$

where

$$a_{ex} = \frac{\sigma_{pex}}{\Delta x(\sigma_{pex}k_{ex} + \alpha_{ex}k_{ex}^2)}[b_{ex} - 1],$$

$$b_{ex} = e^{-\left(\frac{\sigma_{pex}}{k_{ex}} + \alpha_{pex}\right)\frac{\Delta t}{\varepsilon_0}}.$$

Calculate $\psi_{ezy}^{n+\frac{1}{2}}$ for the yn and yp regions:

$$\psi_{ezy}^{n+\frac{1}{2}}(i,j,k) = b_{ey}\psi_{ezy}^{n-\frac{1}{2}}(i,j,k) + a_{ey}\left(H_x^{n+\frac{1}{2}}(i,j,k) - H_x^{n+\frac{1}{2}}(i,j-1,k)\right),$$

where

$$a_{ey} = \frac{\sigma_{pey}}{\Delta y(\sigma_{pey}k_{ey} + \alpha_{ey}k_{ey}^2)}\left[b_{ey} - 1\right],$$

$$b_{ey} = e^{-\left(\frac{\sigma_{pey}}{k_{ey}} + \alpha_{pey}\right)\frac{\Delta t}{\varepsilon_0}}.$$

Add the CPML auxiliary term to E_z in the xn and xp regions:

$$E_z^{n+1} = E_z^{n+1} + C_{\psi ezx} \times \psi_{ezx}^{n+\frac{1}{2}}.$$

Add the CPML auxiliary term to E_z in the yn and yp regions:

$$E_z^{n+1} = E_z^{n+1} + C_{\psi ezy} \times \psi_{ezy}^{n+\frac{1}{2}}.$$

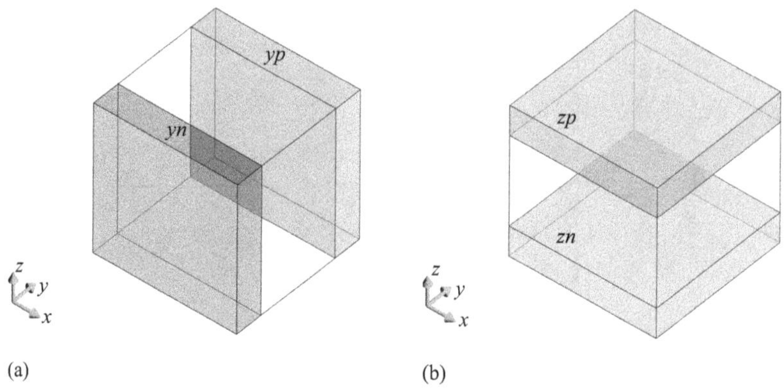

Figure B.4 CPML regions where H_x is updated: (a) H_x is updated in *yn* and *yp* using ψ_{hxy} and (b) H_x is updated in *zn* and *zp* using ψ_{hxz}.

B.4 Updating H_x at CPML regions

Initialization

Create new coefficient arrays for $C_{\psi hxy}$ and $C_{\psi hxz}$:

$$C_{\psi hxy} = \Delta y C_{hxez} \quad C_{\psi hxz} = \Delta z C_{hxey}.$$

Modify the coefficient arrays for C_{hxey} and C_{hxez} in the CPML regions:

$$C_{hxez} = (1/k_{my})C_{hxez,} \text{ in the } yn \text{ and } yp \text{ regiones.}$$

$$C_{hxey} = (1/k_{mz})C_{hxey,} \text{ in the } zn \text{ and } zp \text{ regiones.}$$

FDTD time-marching loop

Update H_x in the full domain using the regular updating equation:

$$\begin{aligned} H_x^{n+\frac{1}{2}}(i,j,k) = C_{hxh}(i,j,k) \times H_x^{n-\frac{1}{2}}(i,j,k) \\ + C_{hxez}(i,j,k) \times \left(E_z^n(i,j+1,k) - E_z^n(i,j,k)\right) \\ + C_{hxey}(i,j,k) \times \left(E_y^n(i,j,k+1) - E_y^n(i,j,k)\right). \end{aligned}$$

Calculate ψ_{hxy}^n for the *yn* and *yp* regions:

$$\psi_{hxy}^n(i,j,k) = b_{my}\psi_{hxy}^{n-1}(i,j,k) + a_{my}\left(E_z^n(i,j+1,k) - E_z^n(i,j,\ k)\right),$$

where

$$a_{my} = \frac{\sigma_{pmy}}{\Delta y(\sigma_{pmy}k_{my} + \alpha_{my}k_{my}^2)}\left[b_{my} - 1\right],$$

$$b_{my} = e^{-\left(\frac{\sigma_{pmy}}{k_{my}} + \alpha_{pmy}\right)\frac{\Delta t}{\mu_0}}.$$

Calculate ψ^n_{hxz} for the *zn* and *zp* regions:

$$\psi^n_{hxz}(i,j,k) = b_{mz}\psi^{n-1}_{hxz}(i,j,k) + a_{mz}\left(E^n_y(i,j,k+1) - E^n_y(i,j,k)\right),$$

where

$$a_{mz} = \frac{\sigma_{pmz}}{\Delta z(\sigma_{pmz}k_{mz} + \alpha_{mz}k^2_{mz})}[b_{mz} - 1],$$

$$b_{mz} = e^{-\left(\frac{\sigma_{pmz}}{k_{mz}} + \alpha_{pmz}\right)\frac{\Delta t}{\mu_0}}.$$

Add the CPML auxiliary term to H_x in the *yn* and *yp* regions:

$$H_x^{n+\frac{1}{2}} = H_x^{n+\frac{1}{2}} + C_{\psi hxy} \times \psi^n_{hxy}.$$

Add the CPML auxiliary term to H_x in the *zn* and *zp* regions:

$$H_x^{n+\frac{1}{2}} = H_x^{n+\frac{1}{2}} + C_{\psi hxz} \times \psi^n_{hxz}.$$

B.5 Updating H_y at CPML regions

Initialization

Create new coefficient arrays for $C_{\psi hyz}$ and $C_{\psi hyx}$:

$$C_{\psi hyz} = \Delta z C_{hyex} \quad C_{\psi hyx} = \Delta x C_{hyez}.$$

Modify the coefficient arrays for C_{hyez} and C_{hyex} in the CPML regions:

$$C_{hyex} = (1/k_{mz})C_{hyex}, \quad \text{in the } zn \text{ and } zp \text{ regions.}$$
$$C_{hyez} = (1/k_{mx})C_{hyez}, \quad \text{in the } xn \text{ and } xp \text{ regions.}$$

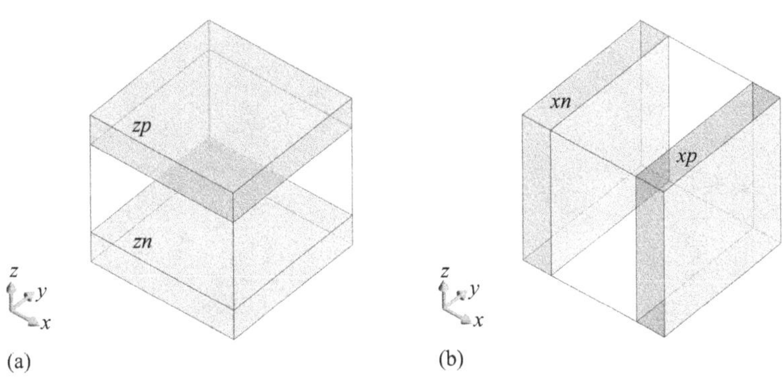

Figure B.5 CPML regions where H_y is updated: (a) H_y is updated in *zn* and *zp* using ψ_{hyz} and (b) H_y is updated in *xn* and *xp* using ψ_{hyx}.

FDTD time-marching loop

Update H_y in the full domain using the regular updating equation

$$H_y^{n+\frac{1}{2}}(i,j,k) = C_{hyh}(i,j,k) \times H_y^{n-\frac{1}{2}}(i,j,k) + C_{hyex}(i,j,k) \times \left(E_x^n(i,j,k+1) - E_x^n(i,j,k)\right) + C_{hyez}(i,j,k) \times \left(E_z^n(i+1,j,k) - E_z^n(i,j,k)\right).$$

Calculate ψ_{hyz}^n for the zn and zp regions:

$$\psi_{hyz}^n(i,j,k) = b_{mz}\psi_{hyz}^{n-1}(i,j,k) + a_{mz}\left(E_x^n(i,j,k+1) - E_x^n(i,j,k)\right),$$

where

$$a_{mz} = \frac{\sigma_{pmz}}{\Delta z(\sigma_{pmz}k_{mz} + \alpha_{mz}k_{mz}^2)}[b_{mz} - 1],$$

$$b_{mz} = e^{-\left(\frac{\sigma_{pmz}}{k_{mz}} + \alpha_{pmz}\right)\frac{\Delta t}{\mu_0}}.$$

Calculate ψ_{hyx}^n for the xn and xp regions:

$$\psi_{hyx}^n(i,j,k) = b_{mx}\psi_{hyx}^{n-1}(i,j,k) + a_{mx}\left(E_z^n(i+1,j,k) - E_z^n(i,j,k)\right),$$

where

$$a_{mx} = \frac{\sigma_{pmx}}{\Delta x(\sigma_{pmx}k_{mx} + \alpha_{mx}k_{mx}^2)}[b_{mx} - 1],$$

$$b_{mx} = e^{-\left(\frac{\sigma_{pmx}}{k_{mx}} + \alpha_{pmx}\right)\frac{\Delta t}{\mu_0}}.$$

Add the CPML auxiliary term to H_y in the zn and zp regions:

$$H_y^{n+\frac{1}{2}} = H_y^{n+\frac{1}{2}} + C_{\psi hyz} \times \psi_{hyz}^n.$$

Add the CPML auxiliary term to H_y in the xn and xp regions:

$$H_y^{n+\frac{1}{2}} = H_y^{n+\frac{1}{2}} + C_{\psi hyx} \times \psi_{hyx}^n.$$

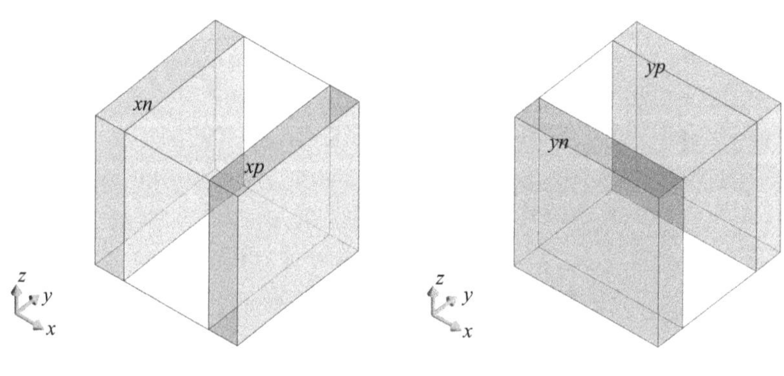

(a) (b)

Figure B.6 CPML regions where H_z is updated: (a) H_z is updated in xn and xp using ψ_{hzx} and (b) H_z is updated in yn and yp using ψ_{hzy}.

B.6 Updating H_z at CPML regions

Initialization

Create new coefficient arrays for $C_{\psi hzx}$ and $C_{\psi hzy}$:

$$C_{\psi hzy} = \Delta x C_{hzey} \quad C_{\psi hzy} = \Delta y C_{hzex}.$$

Modify the coefficient arrays for C_{hzex} and C_{hzey} in the CPML regions:

$$C_{hzey} = (1/k_{mx}) C_{hzey}, \quad \text{in the } xn \text{ and } xp \text{ regions.}$$

$$C_{hzex} = (1/k_{my}) C_{hzex}, \quad \text{in the } yn \text{ and } yp \text{ regions.}$$

FDTD time-marching loop

Update H_z in the full domain using the regular updating equation:

$$\begin{aligned} H_z^{n+\frac{1}{2}}(i,j,k) &= C_{hzh}(i,j,k) \times H_z^{n-\frac{1}{2}}(i,j,k) \\ &+ C_{hzey}(i,j,k) \times \left(E_y^n(i+1,j,k) - E_y^n(i,j,k)\right) \\ &+ C_{hzex}(i,j,k) \times \left(E_x^n(i,j+1,k) - E_x^n(i,j,k)\right). \end{aligned}$$

Calculate ψ_{hzx}^n for the xn and xp regions:

$$\psi_{hzx}^n(i,j,k) = b_{mx}\psi_{hzx}^{n-1}(i,j,k) + a_{mx}\left(E_y^n(i+1,j,k) - E_y^n(i,j,k)\right),$$

where

$$a_{mx} = \frac{\sigma_{pmx}}{\Delta x(\sigma_{pmx}k_{mx} + \alpha_{mx}k_{mx}^2)}[b_{mx} - 1],$$

$$b_{mx} = e^{-\left(\frac{\sigma_{pmx}}{k_{mx}} + \alpha_{pmx}\right)\frac{\Delta t}{\mu_0}}.$$

Calculate ψ_{hzy}^n for the yn and yp regions:

$$\psi_{hzy}^n(i,j,k) = b_{my}\psi_{hzy}^{n-1}(i,j,k) + a_{my}\left(E_y^n(i,j+1,k) - E_y^n(i,j,k)\right),$$

where

$$a_{my} = \frac{\sigma_{pmy}}{\Delta y(\sigma_{pmy}k_{my} + \alpha_{my}k_{my}^2)}[b_{my} - 1],$$

$$b_{my} = e^{-\left(\frac{\sigma_{pmy}}{k_{my}} + \alpha_{pmy}\right)\frac{\Delta t}{\mu_0}}.$$

Add the CPML auxiliary term to H_z in the xn and xp regions:

$$H_z^{n+\frac{1}{2}} = H_z^{n+\frac{1}{2}} + C_{\psi hzx} \times \psi_{hzx}^n.$$

Add the CPML auxiliary term to H_z in the yn and yp regions:

$$H_z^{n+\frac{1}{2}} = H_z^{n+\frac{1}{2}} + C_{\psi hzy} \times \psi_{hzy}^n.$$

APPENDIX C

MATLAB® code for plotting far-field patterns

Listing C.1 Code for plotting far-field patterns on a constant θ plane

```
function polar_plot_constant_theta(phi,pattern_1,pattern_2, ...
      max_val, step_size, number_of_rings,...
      line_style_1, line_style_2,constant_theta, ...
      legend_1,legend_2,scale_type)

% this function plots two polar plots in the same figure
plot_range = step_size * number_of_rings;
min_val = max_val - plot_range;

hold on;
th = 0:(pi/50):2*pi; circle_x = cos(th); circle_y = sin(th);
for mi = 1:number_of_rings
   r = (1/number_of_rings) * mi;
   plot(r*circle_x,r*circle_y,':','color','k','linewidth',1);
   text(0.04,r,[num2str(min_val+step_size*mi)],...
      'verticalalignment','bottom','color','k',...
      'fontweight','demi','fontsize',10);
end

r=[0:0.1:1];
for mi = 0:11
   th=mi*pi/6;
   plot(r*cos(th),r*sin(th),':','color','k','linewidth',1);
text(1.1*cos(th),1.1*sin(th),[num2str(30*mi)],...
      'horizontalalignment','center','color','k',...
      'fontweight','demi','fontsize',10);
end

pattern_1(find(pattern_1 < min_val)) = min_val;
pattern_1 = (pattern_1 - min_val)/plot_range;
pattern_2(find(pattern_2 < min_val)) = min_val;
pattern_2 = (pattern_2 - min_val)/plot_range;

% transform data to Cartesian coordinates
x1 = pattern_1.*cos(phi);
y1 = pattern_1.*sin(phi);

x2 = pattern_2.*cos(phi);
y2 = pattern_2.*sin(phi);
```

```
% plot data on top of grid
p = plot(x1,y1,line_style_1,x2,y2,line_style_2,'linewidth',2);
text(1.2*cos(pi/4),1.2*sin(pi/4),...
    ['\theta = ' num2str(constant_theta) '^o'],...
    'color','b','fontweight','demi');
legend(p,legend_1,legend_2,'location','southeast');
text(-1, -1.1, scale_type,'fontsize',12);
text(1.02 * 1.1,  0.13 * 1.1,'\uparrow',...
    'color','b','fontweight','demi');
text(1.08 * 1.1,  0.13 * 1.1,'\phi',...
    'fontname','arial','color','b','fontweight','demi','fontsize',12);

if constant_theta == 90
    text(1.2,  0.06,'x','fontname','arial',...
        'color','b','fontweight','demi');
    text(1.2, 0,'\rightarrow','color','b','fontweight','demi');
    text(0.06,1.23,'y','fontname','arial',...
        'color','b','fontweight','demi');
    text(0,1.23,'\uparrow','color','b','fontweight','demi');
    text(1.2*cos(pi/4),1.18*sin(pi/4)-0.12,...
        'xy plane','color','b','fontweight','demi');
end

axis([-1.2 1.2 -1.2 1.2]);
axis('equal');axis('off');
hold off;
set(gcf,'PaperPositionMode','auto');
set(gca,'fontsize',12);
```

Listing C.2 Code for plotting far-field patterns on a constant ϕ plane

```
function polar_plot_constant_phi(theta,pattern_1,pattern_2, ...
        max_val, step_size, number_of_rings,...
        line_style_1, line_style_2,constant_phi, ...
        legend_1,legend_2,scale_type)

% this function plots two polar plots in the same figure
plot_range = step_size * number_of_rings;
min_val = max_val - plot_range;

hold on;
th = 0:(pi/50):2*pi; circle_x = cos(th); circle_y = sin(th);
for mi = 1:number_of_rings
r = (1/number_of_rings) * mi;
    plot(r*circle_x,r*circle_y,':','color','k','linewidth',1);
    text(0.04,r,[num2str(min_val+step_size*mi)],...
        'verticalalignment','bottom','color','k',...
        'fontweight','demi','fontsize',10);
end

r=[-1:0.1:1];
for mi = 3:8
    th=mi*pi/6;
```

```
    plot(r*cos(th),r*sin(th),':','color','k','linewidth',1);
    text(1.1*cos(th),1.1*sin(th),[num2str(30*(mi-3))],...
        'horizontalalignment','center','color','k',...
        'fontweight','demi','fontsize',10);
    text(-1.1*cos(th),1.1*sin(th),[num2str(30*(mi-3))],...
        'horizontalalignment','center','color','k',...
        'fontweight','demi','fontsize',10);
end
    text(0,-1.1,'180',...
        'horizontalalignment','center','color','k',...
        'fontweight','demi','fontsize',10);

pattern_1(find(pattern_1 < min_val)) = min_val;
pattern_1 = (pattern_1 - min_val)/plot_range;
pattern_2(find(pattern_2 < min_val)) = min_val;
pattern_2 = (pattern_2 - min_val)/plot_range;
% transform data to Cartesian coordinates
x1 = -pattern_1.*cos(theta+pi/2);
y1 = pattern_1.*sin(theta+pi/2);

x2 = -pattern_2.*cos(theta+pi/2);
y2 = pattern_2.*sin(theta+pi/2);

% plot data on top of grid
p = plot(x1,y1,line_style_1,x2,y2,line_style_2,'linewidth',2);
text(1.2*cos(pi/4),1.2*sin(pi/4),...
    ['\phi = ' num2str(constant_phi) '^o'],...
    'color','b','fontweight','demi');
legend(p,legend_1,legend_2,'location','southeast');
text(-1, -1.1, scale_type,'fontsize',12);
text(0.2,1.02,'\rightarrow','color','b','fontweight','demi');
text(0.2, 1.08,'\theta','fontname','arial','color','b',...
    'fontweight','demi','fontsize',12);
text(-0.21,1.02,'\leftarrow','color','b','fontweight','demi');
text(-0.2, 1.08,'\theta','fontname','arial','color','b',...
    'fontweight','demi','fontsize',12);

if constant_phi == 0
    text(1.2,0.06,'x','fontname','arial','color','b','fontweight','demi');
    text(1.2,0,'\rightarrow','color','b','fontweight','demi');
    text(0.06,1.23,'z','fontname','arial','color','b','fontweight','demi');
    text(0,1.23,'\uparrow','color','b','fontweight','demi');
    text(1.2*cos(pi/4),1.18*sin(pi/4)-0.12,'xz plane',...
        'color','b','fontweight','demi');
end
if constant_phi == 90
    text(1.2,0.06,'y','fontname','arial','color','b','fontweight','demi');
    text(1.2, 0,'\rightarrow','color','b','fontweight','demi');
    text(0.06,1.23,'z','fontname','arial','color','b','fontweight','demi');
    text(0,1.23,'\uparrow','color','b','fontweight','demi');
    text(1.2*cos(pi/4),1.18*sin(pi/4)-0.12,'yz plane',...
        'color','b','fontweight','demi');
end
```

```
axis([-1.2 1.2 -1.2 1.2]);
axis('equal');axis('off');
hold off;
set(gcf,'PaperPositionMode','auto');
set(gca,'fontsize',12);
```

APPENDIX D

MATLAB® GUI for project template

The supplementary files (available on request from the publisher by emailing books@theiet.org) include a base code for Chapters 12–16 in the "codelisting\Appendix_D" folder. Also placed in this folder is a utility code with a graphical user interface (GUI), named as *create_project_template*, which helps the user create template definition files for a new project. Running the file *create_project_template.m* in MATLAB will launch the GUI shown in Figure D.1 below. Here the user can specify some properties of the project such as simulation parameters, cell sizes, boundary conditions, types of objects, sources, lumped components, and output parameters. Also, the user enters a name for the project. Clicking on the "create template file" button will generate a folder with a name which is the same as the project name that includes the problem definition files. These files are not complete yet they can be used as template files to define the problem for which the simulation is sought: the user can open these problem definition files and complete them with the details. With this code organization, the problem definition files for each project

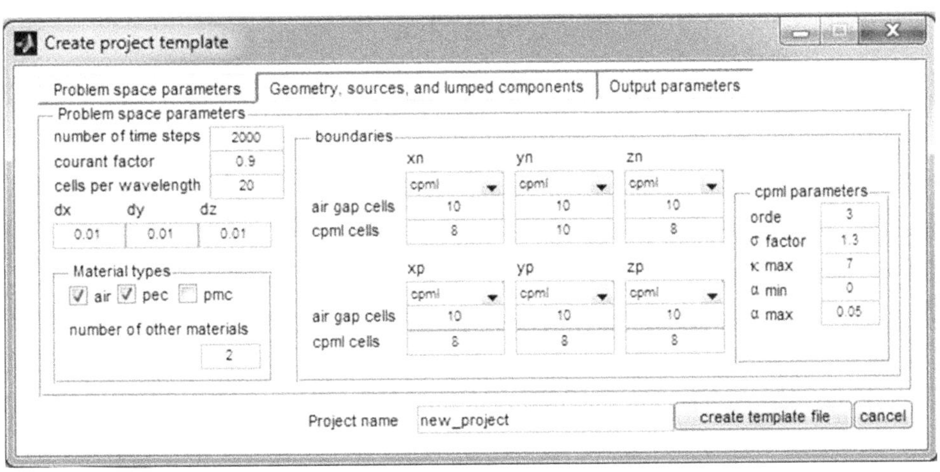

Figure D.1 The GUI to create template problem definition files for a project.

will be located in a separate folder. Meanwhile, other parts of the FDTD code are located in a folder named as "fdtd_files".

To run the simulation for a project, one needs to open the "fdtd_solve.m" file and specify the name of the project in it. Running the code "fdtd_solve.m" will add the project folder and the "fdtd_files" folder in MATLAB path list, run the problem definition files, and then run the rest of the FDTD code.

References

[1] A. Taflove and S. C. Hagness, *Computational Electrodynamics: The Finite Difference Time Domain Method*, 3rd edn. Norwood, MA: Artech House Publishers, 2005.

[2] K. S. Yee, "Numerical solution of initial boundary value problems involving Maxwell's equations in isotropic media," *IEEE Transactions on Antennas and Propagation*, vol. 14, pp. 302–307, 1966.

[3] R. Courant, K. Friedrichs, and H. Lewy, "On the partial difference equations of mathematical physics," *IBM Journal of Research and Development*, vol. 11, no. 2, pp. 215–234, 1967.

[4] J. A. Kong, *Electromagnetic Wave Theory*. Cambridge, MA: EMW Publishing, 2000.

[5] S. Dey and R. Mittra, "Conformal finite-difference time-domain technique for modeling cylindrical dielectric resonators," *IEEE Transactions on Microwave Theory and Techniques*, vol. 47, no. 9, pp. 1737–1739, 1999.

[6] R. Schechter, M. Kragalott, M. Kluskens, and W. Pala, "Splitting of material cells and averaging properties to improve accuracy of the FDTD method at interfaces," *Applied Computational Electromagnetics Society Journal*, vol. 17, no. 3, pp. 198–208, 2002.

[7] D. E. Aspnes, "Bounds on allowed values of the effective dielectric function of two-component composites at finite frequencies," *Physical Review B*, vol. 25, no. 2, pp. 1358–1361, 1982.

[8] N. Kaneda, B. Houshmand, and T. Itoh, "FDTD analysis of dielectric resonators with curved surfaces," *IEEE Transactions on Microwave Theory and Techniques*, vol. 45, no. 9, pp. 1645–1649, 1997.

[9] J.-Y. Lee and N.-H. Myung, "Locally tensor conformal FDTD method for modeling arbitrary dielectric surfaces," *Microwave and Optical Technology Letters*, vol. 23, no. 4, pp. 245–249, 1999.

[10] R. F. Harrington, *Time-Harmonic Electromagnetic Fields*. Hoboken, NJ: Wiley-IEEE Press, 2001.

[11] J. Fang and D. Xeu, "Numerical errors in the computation of impedances by the FDTD method and ways to eliminate them," *IEEE Microwave and Guided Wave Letters*, vol. 5, no. 1, pp. 6–8, 1995.

[12] R. W. Anderson, "S-parameter techniques for faster, more accurate network design," Hewlett-Packard Company, Tech. Rep., November 1996.

[13] K. Kurokawa, "Power waves and the scattering matrix," *IEEE Transactions on Microwave Theory*, vol. 13, no. 2, pp. 194–202, 1965.

[14] D. Sheen, S. Ali, M. Abouzahra, and J. Kong, "Application of the three-dimensional finite-difference time-domain method to the analysis of planar microstrip circuits," *IEEE Transactions on Microwave Theory*, vol. 38, no. 7, pp. 849–857, 1990.

[15] J.-P. Berenger, "A perfectly matched layer for the absorption of electromagnetic waves," *Journal of Computational Physics*, vol. 114, no. 2, pp. 185–200, 1994.

[16] J.-P. Berenger, "Three-dimensional perfectly matched layer for the absorption of electromagnetic waves," *Journal of Computational Physics*, vol. 127, no. 2, pp. 363–379, 1996.

[17] W. V. Andrew, C. A. Balanis, and P. A. Tirkas, "Comparison of the Berenger perfectly matched layer and the Lindman higher-order ABC's for the FDTD method," *IEEE Microwave and Guided Wave Letters*, vol. 5, no. 6, pp. 192–194, 1995.

[18] J.-P. Berenger, "Perfectly matched layer for the FDTD solution of wave-structure interaction problems," *IEEE Transactions on Antennas and Propagation*, vol. 44, no. 1, pp. 110–117, 1996.

[19] J. C. Veihl and R. Mittra, "Efficient implementation of Berenger's perfectly matched layer (PML) for finite-difference time-domain mesh truncation," *IEEE Microwave and Guided Wave Letters*, vol. 6, no. 2, pp. 94–96, 1996.

[20] S. D. Gedney, "An anisotropic perfectly matched layer-absorbing medium for the truncation of FDTD lattices," *IEEE Transactions on Antennas and Propagation*, vol. 44, no. 12, pp. 1630–1639, 1996.

[21] A. Taflove, *Computational Electrodynamics: The Finite Difference Time Domain Method.* Norwood, MA: Artech House Publishers, 1995.

[22] A. Z. Elsherbeni, "A comparative study of two-dimensional multiple scattering techniques," *Radio Science*, vol. 29, no. 4, pp. 1023–1033, 1994.

[23] J. Roden and S. Gedney, "Convolution PML (CPML): an efficient FDTD implementation of the CFS-PML for arbitrary media," *Microwave and Optical Technology Letters*, vol. 27, no. 5, pp. 334–339, 2000.

[24] F. Teixeira and W. Chew, "On causality and dynamic stability of perfectly matched layers for FDTD simulations," *IEEE Transactions on Microwave Theory and Techniques*, vol. 47, no. 6, pp. 775–785, 1999.

[25] M. Kuzuoglu and R. Mittra, "Frequency dependence of the constitutive parameters of causal perfectly matched anisotropic absorbers," *IEEE Microwave and Guided Wave Letters*, vol. 6, no. 12, pp. 447–449, 1996.

[26] J.-P. Berenger, *Perfectly Matched Layer (PML) for Computational Electromagnetics.* San Rafael, CA: Morgan & Claypool Publishers, 2007.

[27] W. C. Chew and W. H. Weedon, "A 3D perfectly matched medium from modified Maxwell's equations with stretched coordinates," *Microwave and Optical Technology Letters*, vol. 7, no. 13, pp. 590–604, 1994.

[28] S. D. Gedney and B. Zhao, "An auxiliary differential equation formulation for the complex-frequency shifted PML," *IEEE Transactions on Antennas and Propagation*, vol. 58, no. 3, pp. 838–847, 2010.

[29] W. L. Stutzman and G. A. Thiele, *Antenna Theory and Design*, 2nd edn. New York: John Wiley & Sons, 1998.

[30] R. J. Luebbers, K. S. Kunz, M. Schnizer, and F. Hunsberger, "A finite-difference time-domain near zone to far zone transformation," *IEEE Transactions on Antennas and Propagation*, vol. 39, no. 4, pp. 429–433, 1991.

[31] R. J. Luebbers, D. Ryan, and J. Beggs, "A two dimensional time domain near zone to far zone transformation," *IEEE Transactions on Antennas and Propagation*, vol. 40, no. 7, pp. 848–851, 1992.

[32] S. A. Schelkunoff, "Some equivalence theorem of electromagnetics and their application to radiation problem," *Bell System Technical Journal*, vol. 15, pp. 92–112, 1936.

[33] C. A. Balanis, *Advanced Engineering Electromagnetics*. New York: Wiley, 1989.

[34] C. A. Balanis, *Antenna Theory: Analysis and Design*, 3rd edn. Hoboken, NJ: John Wiley & Sons, 2005.

[35] H. Nakano, *Helical and Spiral Antennas—A Numerical Approach. Herts*, UK: Research Studies Press Ltd, 1987.

[36] A. Z. Elsherbeni, C. G. Christodoulou, and J. Gomez-Tagle, *Handbook of Antennas in Wireless Communications*. Boca Raton, FL: CRC Press, 2001.

[37] V. Demir, C.-W. P. Huang, and A. Z. Elsherbeni, "Novel dual-band WLAN antennas with integrated band-select filter for 802.11 a/b/g WLAN radios in portable devices," *Microwave and Optical Technology Letters*, vol. 49, no. 8, pp. 1868–1872, 2007.

[38] B. Li and K. W. Leung, "Strip-fed rectangular dielectric resonator antennas with/without a parasitic patch," *IEEE Transactions on Antennas and Propagation*, vol. 53, no. 7, pp. 2200–2207, 2005.

[39] K. Umashankar, A. Taflove, and B. Beker, "Calculation and experimental validation of induced currents on coupled wires in an arbitrary shaped cavity," *IEEE Transactions on Antennas and Propagation*, [legacy, pre-1988], vol. AP-35, no. 11, pp. 1248–1257, 1987.

[40] B. M. Kolundzija, J. S. Ognjanovic, and T. Sarkar, *WIPL-D Microwave: Circuit and 3D EM Simulation for RF & Microwave Applications: Software and Users Manual*. Norwood, MA: Artech House Publishers, 2006.

[41] R. M. Mäkinen, J. S. Juntunen, and M. A. Kivikoski, "An improved thin-wire model for FDTD," *IEEE Transactions on Microwave Theory and Techniques*, vol. 50, no. 5, pp. 1245–1255, 2002.

[42] R. M. Mäkinen and M. A. Kivikoski, "A stabilized resistive voltage source for FDTD thin-wire models," *IEEE Transactions on Antennas and Propagation*, vol. 51, no. 7, pp. 1615–1622, 2003.

[43] F. E. Terman, *Radio Engineers' Handbook*. London: McGraw-Hill, 1950.

[44] V. Demir, A. Z. Elsherbeni, D. Worasawate, and E. Arvas, "A graphical user/interface (GUI) for plane-wave scattering from a conducting, dielectric, or chiral sphere," *IEEE Antennas and Propagation Magazine*, vol. 46, no. 5, pp. 94–99, 2004.

[45] D. Worasawate, J. R. Mautz, and E. Arvas, "Electromagnetic scattering from an arbitrarily shaped three-dimensional homogeneous chiral body," *IEEE Transactions on Antennas and Propagation*, vol. 51, no. 5, pp. 1077–1084, 2003.

[46] V. Demir and A. Z. Elsherbeni, "A graphical user interface for calculation of the reflection and transmission coefficients of a layered medium," *IEEE Antennas and Propagation Magazine*, vol. 46, no. 5, pp. 94–99, February 2006.

[47] K. R. Umashankar and A. Taflove, "A novel method to analyze electromagnetic scattering of complex objects," *IEEE Transactions on Electromagnetic Compatibility*, vol. EMC-24, no. 4, pp. 397–405, 1982.

[48] R. M. Joseph, S. C. Hagness, and A. Taflove, "Direct time integration of Maxwell's equations in linear dispersive media with absorption for scattering and propagation of femtosecond electromagnetic pulses," *Optics Letters*, vol. 16, no. 18, pp. 1412–1414, 1991.

[49] O. P. Gandhi, B. Q. Gao, and J. Y. Chen, "A frequency-dependent finite difference time-domain formulation for general dispersive media," *IEEE Transactions on Microwave Theory and Techniques*, vol. 41, no. 4, pp. 658–665, 1993.

[50] T. Kashiwa and I. Fukai, "A treatment by the FDTD method of the dispersive characteristics associated with electronic polarization," *Microwave and Optical Technology Letters*, vol. 3, no. 6, pp. 203–205, 1990.

[51] M. M. Okoniewski, M. P. Mrozowski, and M. A. Stuchly, "Simple treatment of multi-term dispersion in FDTD," *IEEE Microwave and Guided Wave Letters*, vol. 7, no. 5, pp. 121–123, 1997.

[52] Y. H. Takayama and W. Klaus, "Reinterpretation of the auxiliary differential equation method for FDTD," *IEEE Microwave and Wireless Components Letters*, vol. 12, no. 3, pp. 102–104, 2002.

[53] R. J. Luebbers, F. P. Hunsberger, K. S. Kunz, R. B. Standler, and M. Schneider, "A frequency-dependent finite-difference time-domain formulation for dispersive materials," *IEEE Transactions on Electromagnetic Compatibility*, vol. 32, no. 3, pp. 222–227, August 1990.

[54] R. J. Luebbers and F. P. Hunsberger, "FDTD for N-th order dispersive media," *IEEE Transactions on Antennas and Propagation*, vol. 40, no. 11, pp. 1297–1301, 1992.

[55] D. F. Kelley and R. J. Luebbers, "Piecewise linear recursive convolution for dispersive media using FDTD," *IEEE Transactions on Antennas and Propagation*, vol. 44, no. 6, pp. 792–797, 1996.

[56] F. L. Teixeira, W. C. Chew, M. Straka, M. L. Oristaglio, and T. Wang, "Finite-difference time-domain simulation of ground penetrating radar on dispersive, inhomogeneous, and conductive soils," *IEEE Transactions on Geoscience and Remote Sensing*, vol. 36, no. 6, pp. 1928–1937, 1998.

[57] R. Siushansian and J. LoVetri, "Efficient evaluation of convolution integrals arising in FDTD formulation of electromagnetic dispersive media," *Journal of Electromagnetic Waves and Applications*, vol. 11, no. 1, pp. 101–117, 1997.

[58] D. M. Sullivan, "Frequency dependent FDTD methods using Z transforms," *IEEE Transactions on Antennas and Propagation*, vol. 40, no. 10, pp. 1223–1230, 1992.

[59] D. M. Sullivan, "Z-transform theory and the FDTD method," *IEEE Transactions on Antennas and Propagation*, vol. 44, no. 1, pp. 28–34, 1996.

[60] F. L. Teixeira, "Time-domain finite-difference and finite-element methods for Maxwell equations in complex media," *IEEE Transactions on Antennas and Propagation*, vol. 56, no. 8, pp. 2150–2166, August 2008.

[61] M. Okoniewski and E. Okoniewska, "Drude dispersion in ADE FDTD revisited," *Electronics Letters*, vol. 42, no. 9, pp. 503–504, 2006.

[62] V. Demir, A. Z. Elsherbeni, D. Worasawate, and E. Arvas, "A graphical user interface (GUI) for plane wave scattering from a conducting, dielectric or a chiral sphere," *IEEE Antennas and Propagation Magazine*, vol. 46, no. 5, pp. 94–99, 2004.

[63] D. Popovic and M. M. Okoniewski, "Effective permittivity at the interface of dispersive dielectrics in FDTD," *IEEE Microwave and Wireless Components Letters*, vol. 13, no. 7, pp. 265–267, 2003.

[64] J. Maloney and M. P. Kesler, Analysis of Periodic Structures, *Computational Electrodynamics: The Finite-Difference Time-Domain Method*, 3rd edn. Norwood, MA: Artech House, 2005.

[65] J. A. Roden, S. D. Gedney, P. Kesler, J. G. Maloney, and P. H. Harms, "Time-domain analysis of periodic structures at oblique incidence: orthogonal and nonorthogonal FDTD implementations," *IEEE Transaction on Antennas and Propagation*, vol. 46, no. 4, pp. 420–427, 1998.

[66] Y. C. A. Kao and R. G. Atkins, "A finite-difference time-domain approach for frequency selective surfaces at oblique incidence," *IEEE Antennas and Propagation Society International Symposium*, AP-S. Digest vol. 2, pp. 1432–1435, 1996.

[67] P. Harms, R. Mittra, and W. Ko, "Implementation of the periodic boundary condition in the finite-difference time-domain algorithm for FSS structures," *IEEE Transaction on Antennas and Propagation*, vol. 42, no. 9, pp. 1317–1324, 1994.

[68] F. Yang, J. Chen, R. Qiang, and A. Z. Elsherbeni, "A simple and efficient FDTD/PBC algorithm for scattering analysis of periodic structures," *Radio Science*, vol. 42, no. 4, 2007.

[69] F. Yang, J. Chen, R. Qiang, and A. Z. Elsherbeni, "FDTD analysis of periodic structures at arbitrary incidence angles: a simple and efficient implementation of the periodic boundary conditions," *IEEE Antennas and Propagation Society International Symposium, Albuquerque, NM*, pp. 2715–2718, 2006.

[70] F. Yang, A. Z. Elsherbeni, and J. Chen, "A hybrid spectral-FDTD/ARMA method for periodic structure analysis," *IEEE Antennas and Propagation Society International Symposium, Honololu, HI*, pp. 3720–3723, 2007.

[71] K. ElMahgoub, F. Yang, and A. Z. Elsherbeni, *Scattering Analysis of Periodic Structures Using Finite-Difference Time-Domain Method*, Morgan and Claypool, 2012.

[72] S. Xiao, R. Vahldieck, and H. Jin, "Full-wave analysis of guided wave structures using a novel 2-D FDTD," *IEEE Microwave Guided Wave Letters*, vol. 2, no. 5, pp. 165–167, 1992.

[73] A. Aminian and Y. Rahmat-Samii, "Spectral FDTD: a novel computational technique for the analysis of periodic structures," *IEEE Antennas and Propagation Society International Symposium, Monterey, CA*, vol. 3, pp. 3139–3142, 2004.

[74] A. Aminian, F. Yang, and Y. Rahmat-Samii, "Bandwidth determination for soft and hard ground planes by spectral FDTD: a unified approach in visible and surface wave regions," *IEEE Transaction on Antennas and Propagation*, vol. 53, no. 1, pp. 18–28, 2005.

[75] A. Aminian and Y. Rahmat-Samii, "Spectral FDTD: a novel technique for the analysis of oblique incident plane wave on periodic structures," *IEEE Transaction on Antennas and Propagation*, vol. 54, no. 6, pp. 1818–1825, 2006.

[76] Ansoft Designer v6.1.2, ANSYS Inc., 2012.

[77] P. Monk and E. Suli, "A convergence analysis of Yee's scheme on non-uniform grids," *SIAM Journal on Numerical Analysis*, vol. 31, no. 2, pp. 393–412, April 1994.

[78] J.-S. Hong and M. J. Lancaster, *Microstrip Filters for RF/Microwave Applications*, 1st edn. New York, NY: John Wiley and Sons Inc., 2001.

[79] M. J. Inman, A. Z. Elsherbeni, and C. E. Smith, "GPU programming for FDTD calculations," *The Applied Computational Electromagnetics Society (ACES) Conference*, Honolulu, HI, 2005.

[80] M. J. Inman and A. Z. Elsherbeni, "Programming video cards for computational electromagnetics applications," *IEEE Antennas and Propagation Magazine*, vol. 47, no. 6, pp. 71–78, 2005.

[81] M. J. Inman and A. Z. Elsherbeni, "Acceleration of field computations using graphical processing units," The Twelfth Biennial IEEE Conference on Electromagnetic Field Computation CEFC 2006, *Miami, FL*, April 30–May 3, 2006.

[82] M. J. Inman, A. Z. Elsherbeni, J. G. Maloney, and B. N. Baker, "Practical implementation of a CPML absorbing boundary for GPU accelerated FDTD technique," *The 23rd Annual Review of Progress in Applied Computational Electromagnetics Society, Verona, Italy*, March 19–23, 2007.

[83] M. J. Inman, A. Z. Elsherbeni, J. Maloney, and B. N. Baker, "Practical implementation of a CPML absorbing boundary for GPU accelerated FDTD technique," *Applied Computational Electromagnetics Society Journal*, vol. 23, no. 1, pp. 16–22, 2008.

[84] M. J. Inman and A. Z. Elsherbeni, "Optimization and parameter exploration using GPU based FDTD solvers," *IEEE MTT-S International Microwave Symposium Digest, Atlanta, GA*, pp. 149–152, June 2008.

[85] M. J. Inman, A. Elsherbeni, and V. Demir, "Graphics processing unit acceleration of finite difference time domain," *Ch. 12 in The Finite Difference Time Domain Method for Electromagnetics (with MATLAB Simulations)*, SciTech Publishing, 2009.

[86] N. Takada, N. Masuda, T. Tanaka, Y. Abe, and T. Ito, "A GPU implementation of the 2-D finite-difference time-domain code using high level shader language," *Applied Computational Electromagnetics Society Journal*, vol. 23, no. 4, pp. 309–316, 2008.

[87] S. E. Krakiwsky, L. E. Turner, and M. M. Okoniewski, "Graphics processor unit (GPU) acceleration of finite-difference time-domain (FDTD) algorithm," *Proceedings 2004 International Symposium on Circuits and Systems*, vol. 5, pp. V-265–V-268, May 2004.

[88] S. E. Krakiwsky, L. E. Turner, and M. M. Okoniewski, "Acceleration of finite-difference time-domain (FDTD) using graphics processor units (GPU)," *2004 IEEE MTT-S International Microwave Symposium Digest*, vol. 2, pp. 1033–1036, June 2004.

[89] S. Adams, J. Payne, and R. Boppana, "Finite difference time domain (FDTD) simulations using graphics processors," *Proceedings of the 2007 DoD High Performance Computing Modernization Program Users Group (HPCMP) Conference, Pittsburgh, PA*, vol. 1, pp. 334–338, 2007.

[90] NVIDIA CUDA ZONE, http://www.nvidia.com/object/cuda_home.html.

[91] CUDA 2.1 Quickstart Guide, http://www.nvidia.com/object/cuda_develop.html.

[92] P. Sypek, A. Dziekonski, and M. Mrozowski, "How to render FDTD computations more effective using a graphics accelerator," *IEEE Transactions on Magnetics*, vol. 45, no. 3, pp. 1324–1327, 2009.

[93] V. Demir and A. Z. Elsherbeni, "Compute unified device architecture (CUDA) based finite-difference time-domain (FDTD) implementation," *Journal of the Applied Computational Electromagnetics Society (ACES)*, vol. 25, no. 4, pp. 303–314, 2010.

[94] The OpenCL Specification, ver. 1.0, Khronos OpenCL Working Group, http://www.khronos.org/registry/cl/specs/opencl-1. 0.48.pdf, 2009.

[95] T. P. Stefanski, S. Benkler, N. Chavannes, and N. Kuster. "Parallel implementation of the finite-difference time-domain method in open computing language," in *Electromagnetics in Advanced Applications (ICEAA), 2010 International Conference* on, pp. 557–560, IEEE, 2010.

[96] CUDA 2.1 Programming Guide, http://www.nvidia.com/object/cuda_develop.html.

[97] Acceleware: www.acceleware.com.

[98] V. Demir and A. Z. Elsherbeni, "Utilization of CUDA-OpenGL interoperability to display electromagnetic fields calculated by FDTD," *Computational Electromagnetics International Workshop (CEM'11)*, Izmir, Turkey, August 10–13, 2011.

About the authors

Atef Z. Elsherbeni received an honor B.Sc. degree in Electronics and Communications, an honor B.Sc. degree in Applied Physics, and M.Eng. degree in Electrical Engineering, all from Cairo University, Cairo, Egypt, in 1976, 1979, and 1982, respectively, and a Ph.D. degree in Electrical Engineering from Manitoba University, Winnipeg, Manitoba, Canada, in 1987. He joined the faculty at the University of Mississippi in August 1987 as an Assistant Professor of Electrical Engineering. He advanced to the rank of Associate Professor in 1991 and to the rank of Professor in 1997. He was appointed as Associate Dean of Engineering for Research and Graduate Programs in 2009. He became the Dobelman Distinguished Chair and Professor of Electrical Engineering at Colorado School of Mines in August 2013. He was appointed as Adjunct Professor, at the Department of Electrical Engineering and Computer Science at Syracuse University in 2004. He spent a sabbatical term in 1996 at the Electrical Engineering Department, University of California at Los Angeles (UCLA) and was a visiting Professor at Magdeburg University during the summer of 2005 and at Tampere University of Technology in Finland during the summer of 2007. From 2009 to 2011, he was a Finland Distinguished Professor selected by the Academy of Finland and TEKES.

Dr. Elsherbeni received the 2013 Applied Computational Electromagnetics Society (ACES) Technical Achievements Award, 2012 University of Mississippi Distinguished Research and Creative Achievement Award, 2006 and 2011 School of Engineering Senior Faculty Research Award for Outstanding Performance in research, 2005 School of Engineering Faculty Service Award for Outstanding Performance in Service, 2004 ACES Valued Service Award for Outstanding Service as 2003 ACES Symposium Chair, Mississippi Academy of Science 2003 Outstanding Contribution to Science Award, 2002 IEEE Region 3 Outstanding Engineering Educator Award, 2002 School of Engineering Outstanding Engineering Faculty Member of the Year Award, 2001 ACES Exemplary Service Award for leadership and contributions as Electronic Publishing Managing Editor 1999–2001, 2001 Researcher/Scholar of the Year Award in the Department of Electrical Engineering,

University of Mississippi, and 1996 Outstanding Engineering Educator of the IEEE Memphis Section.

Dr. Elsherbeni is the co-author of the books *Antenna Analysis and Design using FEKO Electromagnetic Simulation Software (ACES Series on Computational Electromagnetics and Engineering)*, SciTech 2014, *Double-Grid Finite-Difference Frequency-Domain (DG-FDFD) Method for Scattering from Chiral Objects*, Morgan & Claypool, 2013, *Scattering Analysis of Periodic Structures Using Finite-Difference Time-Domain Method*, Morgan & Claypool, 2012, *Multiresolution Frequency Domain Technique for Electromagnetics*, Morgan & Claypool, 2012, *The Finite Difference Time Domain Method for Electromagnetics with Matlab Simulations*, SciTech, 2009, *Antenna Design and Visualization Using Matlab*, SciTech, 2006, *MATLAB Simulations for Radar Systems Design*, CRC Press, 2003, *Electromagnetic Scattering Using the Iterative Multiregion Technique*, Morgan & Claypool, 2007, *Electromagnetics and Antenna Optimization using Taguchi's Method*, Morgan & Claypool, 2007, *Scattering Analysis of Periodic Structures Using Finite-Difference Time-Domain Method*, Morgan & Claypool, 2012, *Multiresolution Frequency Domain Technique for Electromagnetics*, Morgan & Claypool, 2012, and the main author of the chapters "Handheld Antennas" and "The Finite Difference Time Domain Technique for Microstrip Antennas" in *Handbook of Antennas in Wireless Communications*, CRC Press, 2001. He was the advisor/co-advisor for 33 M.S. and 20 Ph.D. students.

Dr. Elsherbeni is a Fellow member of both IEEE and ACES. He is the Editor-in-Chief for *ACES Journal* and a past Associate Editor to the *Radio Science Journal*. He was the Chair of the Engineering and Physics Division of the Mississippi Academy of Science and was the Chair of the Educational Activity Committee for the IEEE Region 3 Section. He was the general Chair for the APS-URSI 2014 Symposium. He held the president position of ACES Society from 2013 to 2015.

Veysel Demir is an Associate Professor at the Department of Electrical Engineering at Northern Illinois University. He received his Bachelor of Science degree in Electrical Engineering from Middle East Technical University, Ankara, Turkey, in 1997. He studied at Syracuse University, New York, where he received both Master of Science and Doctor of Philosophy degrees in Electrical Engineering in 2002 and 2004, respectively. During his graduate studies, he worked as a Research Assistant for Sonnet Software, Inc., Liverpool, New York. He worked as a Visiting Research Scholar in the Department of Electrical Engineering at the University of Mississippi from 2004 to 2007. He joined Northern Illinois University in August 2007 and served as an Assistant Professor until August 2014. He has been serving as an Associate Professor since then.

Dr. Demir's main field of research is electromagnetics and microwaves. He is experienced especially in applied computational electromagnetics. He heavily participated in the development of time domain and frequency domain numerical analysis tools for new applications and contributed to research on improving the accuracy and speed of algorithms being developed. He is experienced in designing RF/microwave circuits and antennas for the related technologies, and performing experimental characterizations of these devices.

Dr. Demir is a member of IEEE, ACES, and SigmaXi and has coauthored more than 50 technical journal and conference papers. He is a coauthor of the books "Electromagnetic Scattering Using the Iterative Multiregion Technique" (Morgan and Claypool, 2007) and the first edition of "The Finite-Difference Time-Domain Method for Electromagnetics with MATLAB® Simulations" (Scitech, 2009). He served as a technical program co-chair for the 2014 IEEE International Symposium on Antennas and Propagation and USNC-URSI Radio Science Meeting and for the ACES 2015 conference.

Index

www.ingramcontent.com/pod-product-compliance
Lightning Source LLC
LaVergne TN
LVHW080645210926
848485LV00003B/327

* 9 7 8 1 6 1 3 5 3 1 7 5 4 *